IET RADAR, SONAR, NAVIGATION AND AVIONICS SERIES 20

Series Editors: Dr N. Stewart
Professor M. Griffiths

Sea Clutter: Scattering, the K Distribution and Radar Performance

Other volumes in this series:

Volume 1	**Optimised radar processors** A. Farina (Editor)
Volume 2	**Radar development to 1945** R. W. Burns (Editor)
Volume 3	**Weibull radar clutter** M. Sekine and Y. H. Mao
Volume 4	**Advanced radar techniques and systems** G. Galati (Editor)
Volume 7	**Ultrawideband radar measurements** L. Yu. Astanin and A. A. Kostylev
Volume 8	**Aviation weather surveillance systems: advanced radar and surface sensors for flight safety and air traffic management** P. Mahapatra
Volume 10	**Radar techniques using array antennas** W. D. Wirth
Volume 11	**Air and spaceborne radar systems** P. Lacomme, J. C. Marchais, J. P. Hardange and E. Normant
Volume 12	**Principles of space-time adaptive processing** R. Klemm
Volume 13	**Introduction to RF Stealth** D. Lynch
Volume 14	**Applications of space-time adaptive processing** R. Klemm (Editor)
Volume 15	**Ground penetrating radar, 2nd edition** D. J. Daniels (Editor)
Volume 16	**Target detection by marine radar** J. Briggs
Volume 17	**Strapdown inertial navigation technology** D. H. Titterton and J. L. Weston
Volume 18	**Introduction to radar target recognition** P. Tait
Volume 19	**Radar imaging and holography** A. Pasmmov and S. Zinoviev

Sea Clutter: Scattering, the K Distribution and Radar Performance

Keith D. Ward
Robert J.A. Tough
Simon Watts

The Institution of Engineering and Technology

Published by: The Institution of Engineering and Technology, London, United Kingdom

The Institution of Engineering and Technology,
Michael Faraday House,
Six Hills Way, Stevenage,
Herts., SG1 2AY, United Kingdom

www.theiet.org

British Library Cataloguing in Publication Data

Watts, Simon
Sea Clutter : scattering, the K-distribution and radar performance
1. Incoherent scatter 2. Radar - Interference -
Mathematical models 3. Boats and boating - Radar equipment
I. Title II. Tough, Robert III. Ward, Keith IV. Institution of
Electrical Engineers
623.8′933′015118

ISBN (10 digit) 0 86341 503 2
ISBN (13 digit) 978-086341-503-6

Typeset in India by Newgen Imaging Systems (P) Ltd, Chennai, India
Printed in the UK by MPG Books Ltd, Bodmin, Cornwall

Acknowledgements

Much of the work described in this book was supported by research funding from the UK MOD in many programmes undertaken over the last 25 years. This includes recent support from the EMRS DTC. Work was also supported by funding from the authors' own organisations, TW Research and Thales UK.

The authors gratefully acknowledge the support they have received and the work of many other workers in the United Kingdom and around the world who have contributed to the body of knowledge represented in this book.

Contents

List of symbols

Symbol	Meaning	Page[1]	Section
$\langle\rangle_N$	average over N trials	341	A1.2
$\langle\rangle$	expectation	20	2.4
∇	gradient operator	55	3.3
α	negative binomial bunching parameter	343	A1.3
α	threshold multiplier (relative to mean)	271	9.2
α	probability of Type 1 error (Seller's risk) in sequential testing	337	10.5
$\hat{\alpha}$	estimate of threshold multiplier, α	303	9.3
β	shape parameter of Weibull distribution	128	4.5
β	probability of Type 2 error (Buyer's risk) in sequential testing	337	10.5
$\beta(\mathrm{t})$	relaxation rate of wave growth	63	3.4
$\Gamma(.)$	Gamma function	20	2.4
$\Gamma(\nu, t)$	incomplete gamma function	238	8.4
γ	Measure of the curvature of the correlation function of a_z	118	4.4
γ	Euler's constant	271	9.2
Δf	frequency step size (frequency agility)	23	2.5
ε	dielectric constant or relative permittivity	55	3.3
ε_0	permittivity of free space	55	3.3
ζ	amplitude of the Hermitian product ξ	115	4.3
$\zeta(\mathbf{x})$	normal component of the field gradient at a surface	78	3.6
η	sea surface height	48	3.2
$\eta_\mathrm{L}\,\eta_\mathrm{s}$	sea surface heights of long and short wave components	73	3.5
θ	signal phase	106	4.2
θ_{az}	antenna azimuth beamwidth	231	8.2

Symbol	Meaning	Page[1]	Section
θ	azimuth look direction	232	8.2
θ_{sw}	direction of sea swell	237	8.3
θ_{gr}	grazing angle	92	3.8
θ_i	angle of incidence	70	3.5
θ_{loc}	local incidence angle	74	3.5
θ_w	wind direction	47	3.2
λ	discrimination ratio in sequential testing	337	10.5
λ	radar wavelength	92	3.8
Λ	likelihood ratio	170	6.3
Λ_2	matrix of derivatives of the correlation function of a random field	224	7.6
μ	relative permeability	55	3.3
μ_0	permeability of free space	55	3.3
μ	K-distribution intensity after pulse-to-pulse integration	238	8.4
$\boldsymbol{\mu}$	vector of mean values of image pixels	198	7.2
ν_{eff}	effective K-distribution shape parameter	119	4.4
ν	shape parameter of gamma and K distribution	109	4.2
$\hat{\nu}$	estimate of K-distribution shape parameter	302	9.3
ξ	Hermitian product of two complex signals	114	4.3
ρ	charge density	55	3.3
ρ	surface roughness parameter	92	3.8
ρ	range resolved length (range resolution)	231	8.2
ρ_L	correlation function of long wave component	73	3.5
ρ_j	range correlation coefficient for lag j	38	2.10
ρ_w	water density	49	3.2
σ	radar cross section (RCS) and scattered power	81	3.6
σ^0, σ^0_{HH}, σ^0_{VV}	area reflectivity (normalised clutter RCS), VV and HH	14	2.1
σ_{BW}	mean Bragg and whitecap intensity in K plus spike model	123	4.4
σ_h	root mean square wave height	48	3.2
σ_{sp}	mean spike intensity in K plus spike model	123	4.4
$\hat{\sigma}_1$	estimate of parameter σ_1	178	6.6
τ	pulse length	332	10.5
ϕ	look direction in elevation,	232	8.2
ϕ	long wave surface lateral tilt	74	3.5
ϕ	phase of individual scatterer	102	4.2
ϕ_{el}	antenna elevation beamwidth	231	8.2
ϕ_{gr}	grazing angle	231	8.2
$\varphi(t)$	wave action spectral density	63	3.4

Symbol	Meaning	Page[1]	Section
χ	discriminant used as detection test statistic, or to enhance imagery	189	6.8
ψ	scalar fields satisfying the Helmholtz equation	77	3.6
ψ	antenna depression angle	231	8.2
$\psi(z)$	digamma function	192	6.9
ω	temporal frequency	41	2.10
Ω	wave age	51	3.2
A	area of scattering surface	59	3.3
a	scale parameter of Weibull distribution	128	4.5
$\mathbf{A}$, A_I, A_Q	coherent signal with in phase and quadrature components	104	4.2
a, b	parameters of logarithmic amplifier	270	9.2
ACF	autocorrelation function	38	2.10
AGC	automatic gain control	314	10.2
a_j	cell-averager weights	293	9.2
$\mathbf{a}_n$	contribution of individual scatterer to field **E**	102	4.2
ASuW	Anti-Surface Warfare	3	1.2
ASW	Anti-Submarine Warfare	3	1.2
a_z	vertical acceleration of water surface	118	4.4
$B_\eta(q)$	wave height saturation spectrum	48	3.2
B	magnetic induction	55	3.3
B	pulse bandwidth and receiver noise bandwidth	23	2.5
b	Phillips constant	47	3.2
b	scale parameter of gamma and K distributions	109	4.3
BAM	breaking area model	93	3.8
c	sea wave phase velocity	50	3.2
$C(\mathbf{U})$ or $C(k)$	characteristic function	103	4.2
CA CFAR	cell-averaging constant false alarm rate	271	9.2
CFAR	Constant False Alarm Rate	10	1.4
CNR	clutter-to-noise ratio	249	8.5
$\mathbf{D}$	electric field displacement	55	3.3
E	amplitude	20	2.4
$\mathbf{E}$	electric field vector	27	2.7
E_I, E_Q	in phase and quadrature components of **E**	103	4.2
EM	Electromagnetic	7	1.4
$\langle E^n \rangle_{\text{Rice}}$	n^{th} moment of a Rice distribution	185	6.7
erf()	error function, defined in equation (A1.31)	106	4.2
erfc()	complementary error function	155	5.7
ERS1	Remote sensing SAR satellite	2	1.2
ESA	European Space Agency	2	1.2
E_T	threshold field amplitude	104	4.2

Symbol	Meaning	Page[1]	Section
$\mathbf{e}_z$	unit vector in vertical direction	74	3.5
${}_2F_1, {}_1F_1$	hypergeometric and confluent hypergeometric functions	122	4.4
f	Radar frequency	23	2.5
$F(x)$	cumulative distribution function or probability distribution	277	9.2
FFT	Fast Fourier transform	32	2.8
F_n	receiver noise figure	231	8.2
$G(\theta,\phi)$	one-way antenna gain	232	8.2
g	gravitational acceleration	41	2.10
G	number of gap cells either side of cell under test in CA CFAR	271	9.2
$G()$	green's function	56	3.3
$\bar{G}$	normal component of gradient of Green's function	425	A3.7.1
$g_\mathrm{HH}, g_\mathrm{VV}$	Term in Rice and composite model scattering RCS	71	3.5
g_n	Gaussian uncorrelated random variate	148	5.3
H, HH	Horizontal polarisation	18	2.3
$\mathbf{H}$	magnetic field vector	27	2.7
h	height of radar above the Earth surface	232	8.2
$H_0^{(1)}(z)$	zeroth order Hankel function	79	3.6
H_0	null hypothesis in sequential testing	337	10.5
H_1	alternative hypothesis in sequential testing	337	10.5
$h_{1/3}$	significant wave height of the sea	17	2.2
H_n	Hermite polynomial	155	5.6
$\Im(.)$	integrand in scattering formula	82	3.6
I_n	Bessel functions (modified) of the first kind and order n.	104	4.2
$i_n(z), k_n(z)$	modified spherical Bessel functions	209	7.4
$\mathbf{j}$	electric current vector	55	3.3
J_n	Bessel functions (unmodified) of the first kind and order n.	120	4.4
k	target fluctuation parameter	245	8.5
k	Boltzman's constant	231	8.2
$\mathbf{k}$	wave vector	58	3.3
k	spatial frequency or wavenumber	41	2.10
k	degree of coherence of complex Gaussian process	208	7.4
k	rank of ordered statistic	279	9.2
$\mathbf{K}$	covariance matrix of SAR image pixels	198	7.2
$\mathbf{k}_\mathrm{H}$	horizontal component of backscattered wavevector	71	3.5
k_m	wavenumber of slowest waves	49	3.2

Symbol	Meaning	Page[1]	Section
K_ν	modified Bessel function of the second kind (Hankel function of imaginary argument)	109	4.3
k_z	vertical component of backscattered wavevector	71	3.5
L	surface length in scattering calculations	80	3.5
L_n^α	associated Laguerre polynomial	158	5.7
L_μ	radar microwave loss	332	10.5
L, LL	left-hand polarisation	27	2.7
L_a	atmospheric and propagation loss	230	8.2
LGA	low grazing angle	46	3.1
L_n	Laguerre polynomial	158	5.7
L_PM	Pierson Moskowitz saturation spectrum	50	3.2
ℓ_x	lateral extent of scattering surface	89	3.7
$\hat{m}$	estimate of mean intensity	38	2.10
M	MTF value	96	3.8
M	number of averaging cells in CA CFAR	271	9.2
m	maximum value in Greater-of CFAR (GO CFAR)	277	9.2
m_0	ideal value in sequential testing	336	10.5
m_1	maximum acceptable value in sequential testing	336	10.5
MFR	Multi-Function Radar	4	1.2
m_n	n^th moment	117	4.4
MTF	modulation transfer function	96	3.8
$\mathbf{n}$	vector normal to a surface	56	3.3
n	number of scans	330	10.5
N	number of pulses integrated	275	9.2
N	number of discrete events (trials)	340	A1.2
$\bar{N}$	mean populations supported by Poisson and negative binomial processes	116	4.4
N_eff	effective number of independent samples	324	10.3
Norm(**K**)	normalisation integral of matrix **K**	354	A1.6
p	number of random variables in set modelled by Wishart distribution	203	7.3
p	probability of an event	302	9.3
$P(E_1\|E_2)$	conditional PDF	104	4.2
$P(n\|N)$	probability of throwing n heads in N trials	343	A1.3
$P(x)$	probability density function (pdf)	103	4.2
$\hat{p}$	estimate of probability of an event	302	9.3
p_c	mean clutter power	231	8.2
$P_c(x)$	clutter local power pdf	108	4.3
P_D	probability of detection	104	4.2
P_FA	probability of false alarm	105	4.2
p_n	noise power	113	4.3

Symbol	Meaning	Page[1]	Section
PPI	plan position indicator	319	10.2
p_r	power received by a radar	230	8.2
p_s	mean signal power	231	8.2
p_t	peak transmit power	230	8.2
q	wavenumber (variable)	47	3.2
$\mathbf{Q}$	upper triangular matrix from Cholesky decomposition of $\mathbf{K}$	198	7.2
q_{cut}	wavenumber separating long and short waves	73	3.5
Q_{dist}	complementary quantile function	154	5.6
R	Fresnel reflection coefficient	404	A3.3
R	range to the target	232	8.2
$\mathbf{r}, \mathbf{r}'$	points above a scattering surface	56	3.3
R, RR	Right-hand polarisation	27	2.7
RAF	Royal Air Force	3	1.2
RCS	radar cross section	14	2.1
r_e	radar Earth radius (4/3 times actual radius)	232	8.2
$R_G(t)$	correlation coefficient of Gaussian random variable	155	5.6
R_j *or* $R(j)$	range ACF, lag j	38	2.10
RN	Royal Navy	3	1.2
s	area of a breaking wave in the BAM model	118	4.4
S_η	sea surface height power spectrum	47	3.2
SAR	Synthetic Aperture Radar	2	1.2
SCR	signal-to-clutter ratio	250	8.5
SIR	signal-to-interference ratio	247	8.5
SNR	signal-to-noise ratio	272	9.2
$S(\omega)$	sea clutter Doppler power spectrum	136	4.6
$S_s S_L$	Wave spectrum of short and long waves	73	3.5
STC	sensitivity-time control	317	10.2
T_0	standard noise temperature (290°K)	231	8.2
t	detection threshold	271	9.2
T	test statistic	296	9.3
T	surface tension	49	3.2
t_a	mean time between alarms	325	10.3
t_d	mean alarm duration	325	10.3
Tr(**A**)	trace of matrix **A**	204	7.3
U	water current	60	3.4
U	log estimator of Weibull shape parameter	303	9.3
$u*$	wind friction velocity	51	3.2
U_{10}	wind speed (at 10m above the sea surface)	47	3.2
V, VV	Vertical polarisation	18	2.3
$\mathbf{v}$	a point on the mean x, y plane of a scattering surface	58	3.3

Symbol	Meaning	Page[1]	Section
var(.)	Variance	302	9.3
VH, HV, RL, LR	Cross-polarisations	27	2.7
x	Mean local clutter intensity	20	2.4
X	non-dimensional fetch	51	3.2
$\mathbf{x}$, $\mathbf{x}'$	points on a scattering surface	56	3.3
$\{x\}$	set of values of variable or parameters x	175	6.5
x_n	correlated random variate	148	5.3
Y	Threshold	238	8.4
y	Log transformed intensity	270	9.2
Y_n	Bessel function of the second kind and order n.	402	A3.2
z	clutter intensity	20	2.4

[1]The page and section denote the first occurrence of the symbol with the defined meaning.

Chapter 1
Introduction

1.1 Prologue

The largest part of the Earth's surface lies beneath the sea (see Figure 1.1); events taking place in, on and directly above the oceans have an enormous impact on our lives. Consequently maritime remote sensing and surveillance are of great importance. Since its discovery during the 1930s, radar has played a central role in these activities. Much of their military development was driven by the circumstances of the Cold War; now this era is past and a different set of imperatives holds sway. Military surveillance does, however, remain a key requirement. Great progress has been made recently on non-military applications, particularly the remote sensing of the environment, of which the sea is the most important component. These newly emerging concerns, whether they are ecological or geo-political, currently define the requirements imposed on maritime radar systems. Nonetheless, the underlying principles of the systems' operation, and the interpretation of their output, remain the same; the body of knowledge developed in the twentieth century provides us with the tools with which to address the problems facing the radar engineer of the twenty-first.

This book attempts to bring together those aspects of Maritime Radar relating to scattering from the sea surface, and their exploitation in radar systems. The presentation aims to emphasise the unity and simplicity of the underlying principles and so should facilitate their application in these changing circumstances.

1.2 Maritime radar

A wide variety of radars can be deployed to sense or interrogate a maritime environment. Space-based radars are used for Earth remote sensing, particularly for oceanography. The earliest example of this application was SEASAT [1]. The SEASAT-A satellite was launched on 26 June 1978, and continued until 9 October 1978. Its mission was to demonstrate that measurements of the ocean dynamics are feasible. It carried a synthetic aperture radar, which was intended to obtain radar imagery of ocean wave patterns in deep oceans and coastal regions, sea and fresh

Figure 1.1 A view of the Earth from space (http://images.jsc.nasa.gov/)

water ice, land surfaces and a number of other remote sensing objectives. It also carried a radar altimeter and a wind scatterometer. More recent examples of remote sensing radars have been the ESA ERS1 and ERS2 [2] satellites, which had similar missions to SEASAT and a similar range of radar instruments. The most recent example at the time of writing this book is ENVISAT [3].

Each of the radar instruments carried by these satellites exploits different aspects of scattering from the sea surface. The radar altimeter is used to make very accurate measurements of the sea height. The SEASAT-A satellite carried an altimeter having a measurement accuracy of better than 10 cm, which allowed the measurement of oceanographic features such as currents, tides and wave heights. A wind scatterometer measures the average backscatter power over large areas of the sea surface. The measurement of backscatter scattering coefficients at 3 different directions relative to the satellite track allows the surface wind direction to be calculated. Finally, a synthetic aperture radar (SAR) images the surface at much finer resolution (typically about 25 m × 25 m) over grazing angles typically in the range of 20–60°. Figure 1.2 shows an example of an ERS-1 SAR image of the sea surface, showing wave, current and weather patterns on the sea off the eastern tip of Kent in the south of England. Data such as these allow scientists to better understand air–sea interactions, which have a major effect on the World's weather patterns and overall climate. They are therefore an important component of modelling for climate change, and in particular global warming.

Figure 1.2 ERS1 radar image of North East Kent and the surrounding sea (data copyright ESA – Image processed by DERA UK)

The backscatter from the sea observed by remote sensing radars is the intended radar signal. Airborne radar may also be used for ocean imaging in this way. However, for most airborne and surface radars operating in a maritime environment, the backscatter from the sea is unwanted and is called sea clutter. A prime example of a military application that encounters problems of this kind is maritime surveillance. Typical examples include the Searchwater radar [4], in service with the UK RAF Nimrod MR2 aircraft (Figure 1.3), and the AN/APS-137 radar fitted in the US Navy P3-C aircraft. These radars have many operating modes, but in particular are used for long-range surveillance of surface ships (known as ASuW, anti-surface warfare) and detection of small surface targets, including submarine masts (ASW, anti-submarine warfare). For ASuW operation, the radar must detect, track and classify surface ships at ranges in excess of 100 nautical miles. For ASW operation, the radar must detect submarine masts just above the surface in high sea states, at ranges of many miles. In both these modes, the radar operator and the automatic detection processing must be able to distinguish between returns from wanted targets and those from the sea surface. Unlike satellite radars, the grazing angle at the sea surface for such radars is typically $<10°$ and often the area of interest extends out to the radar horizon (i.e. zero grazing angle). Under these conditions, returns from the sea can often have target-like characteristics and may be very difficult to distinguish from real targets. In order to aid discrimination between targets and clutter and, in extreme cases, prevent overload of the radar operator or signal processor, the radar detection processing must attempt to achieve an acceptable and constant false alarm rate from the sea clutter.

Air-to-air detection modes are also important for airborne radars. Obvious examples are Airborne Interceptor (AI) radars for fast jets, such as the Blue Vixen radar on the RN Sea Harriers [5], and Airborne Early Warning Radars such as the AWACS or the Searchwater 2000AEW radar fitted to the RN Mk7 Sea King helicopters (see Figure 1.4, which shows the earlier Mk 2 Sea King with the Searchwater radar).

Figure 1.3 Nimrod MR2 with Searchwater radar (© Crown Copyright/MOD, image from www.photos.mod.uk)

Figure 1.4 RN Mk2 Sea King helicopter with Searchwater AEW radar (© Crown Copyright/MOD, image from www.photos.mod.uk)

In order to detect targets that are much smaller than the sea backscatter, these radars use pulsed Doppler processing to distinguish moving targets from the sea clutter. Targets with a high radial velocity are easily distinguished from the clutter but slower targets or those on a crossing trajectory may have Doppler shifts that are not much

Figure 1.5 MESAR 2 MFR (Reproduced with the kind permission of BAESystems Integrated Systems Technology)

separated from the Doppler spectrum of the sea clutter. Once again the challenge for the radar processing is to maintain an acceptable false alarm rate from the clutter, especially in the edges of its Doppler spectrum.

Surface radars, especially ship-borne radars, must also cope with sea clutter. Ship radars are used, amongst other applications, for navigation, surface surveillance and air defence. Naval warships are increasingly being fitted with multi-function radars (MFR), such as the Sampson radar to be fitted to the RN Type 45 Destroyers. These MFRs must undertake all these modes and many others in an interleaved manner. Once again, the radar processing must distinguish between clutter returns and targets. Figure 1.5 shows the MESAR MFR which was the experimental precursor to Sampson.

1.3 The modelling of radar returns from the sea

The accurate and physically motivated modelling of radar returns from the sea surface should provide a firm foundation for the analysis of their impact on maritime radar performance. Simple models, of limited application, formed the basis of earlier work: thus the physical optics or Kirchoff approximation could be made in an EM scattering approach, while a Gaussian model might form the basis of a tractable statistical description. These models can be applied quite effectively to the analysis of sea clutter in low resolution radar systems, deployed well away from grazing incidence. It is only in the past three decades that modelling techniques able to capture the salient features of low grazing angle, high resolution radar sea clutter have been developed.

Ad hoc non-Gaussian statistical models had been proposed, chosen in the main for their analytic convenience; since the 1970s these have been supplanted by the K distribution and related models. This class of models was first introduced, in the context of radar sea clutter, by Jakeman and Pusey [6]; these authors drew on analogies between scattering at microwave and optical wavelengths to motivate their discussion. The K distribution was shown to emerge directly from a detailed analysis of sea clutter in the work of Ward [7]. This approach highlighted the usefulness of the compound representation of the clutter process, which in the hands of Ward, Watts and others, made possible the systematic analysis of effects of thermal noise and the spatial and temporal correlation properties displayed by the clutter and their impact on maritime surveillance radar [8]. Oliver developed similar ideas in the context of synthetic aperture radar which have contributed significantly to interpretation of two dimensional radar imagery [9]. These developments in the application of the K distribution in the microwave regime have been complemented by work at optical wavelengths [10]; a further interchange of ideas between these two areas occurred when the generalised K model, developed by Jakeman and Tough for weak forward scattering in the visible regime, proved invaluable in the analysis of polarimetric and interferometric SAR performance [11,12]. This brief review of the development of the K distribution would be incomplete if it did not mention the careful statistical analysis that has ensured that the estimation of the parameters defining the K distribution has been carried out accurately and meaningfully; Oliver and co-workers contributed significantly in this area.

The electromagnetic scattering by the sea surface at close to grazing incidence presents many problems. Perhaps the simplest model that makes direct contact with both EM theory and a reasonable statistical description of the sea surface is Rice's perturbation theory analysis of scattering by a slightly rough surface, from which a picture of resonant or Bragg scattering from small scale structure emerges [13]. The composite model [14] incorporated the imperfect electrical conductivity of sea water and allowed for the modulation of local grazing angle by the large scale structure of the sea surface. Subsequent attempts to include shadowing and multipath effects, which become important at low grazing angles, have been in the main essentially uncontrolled, or ad hoc. More recently, however, the problems inherent in the direct numerical calculation of scattering by a surface wave profile have been addressed; the past decade or so has seen the development of methods able to cope with low grazing angle geometries and realistic conductivities [15]. These can now be used in the careful analysis of grazing angle and polarisation dependence of the fluctuations and dynamics of scattering in this regime; detailed numerical results of this type have led to significant insights into the observed properties of sea clutter.

1.4 Outline of the book

Many types of radar, with both civil and military applications, will encounter backscatter from the sea, either as wanted signal or as clutter. In all cases the radar designer must understand the characteristics of the backscatter from the sea in order to either relate the signal to the ocean characteristics, or to achieve the performance required by the radar users. These characteristics are found to vary very widely, according to the radar parameters, viewing geometry and weather conditions. This book describes the very many different aspects that must be studied in order to design and develop new radars and to predict and understand the performance of operational systems. As a result it is addressed to a wide and varied audience, not all of whom may wish, at least initially, to read it in its entirety. The investigation of the mechanisms that give rise to radar sea clutter and the ways in which its impact on performance is modelled and calculated will be of great use to radar engineers, be they graduate students, academics or professional research and development workers. The discussion of the physical principles that underpin this work may be of wider interest: our presentation of low grazing angle scattering from rough surfaces can be adapted to other wavelengths and might well, for example, be of relevance to workers in optics and sonar. The use of non-Gaussian statistical models has become increasingly important in many areas; in several of these, the K-distributed process has proved to have wide applicability. Our discussion of the K-distribution is motivated by its relevance to radar clutter; nonetheless we feel that it contains insights that should be of more general interest and so will complement the recent publication by Jakeman and Ridley [16]. The numerical simulation of non-Gaussian noise processes also has widespread application; our discussion of this topic may again attract the interest of those outside the immediate radar community.

Some prior knowledge is expected of a reader who intends to work through the whole book. A reasonable acquaintance with the fundamentals of radar operation, a knowledge of EM theory and probability and sufficient 'mathematical maturity' to look an unfamiliar equation in the eye without flinching, are all required. The Appendices provide much of this background and should be useful to the beginner, while still containing something of interest to the expert. The specialist reader, whose interest is confined to the contents of individual chapters, can happily 'jump straight in', as each chapter is reasonably self contained and it is hoped, achieves a logical and consistent presentation of its subject matter. The authors also hope that some readers will, once they have worked through a topic of immediate interest, wish to venture further afield in the book.

The scene is set by a qualitative description of the phenomena observed when various maritime radars are deployed; this catalogues many of the problems encountered in practice and highlights those that are incompletely or only recently understood. The mechanisms that give rise to these effects are elucidated in a review of the physics that governs the scattering of microwaves by the sea surface and other objects; where a detailed deterministic description becomes intractable or uninformative, probabilistic models are developed. The statistical theory underlying clutter characterisation and target detection is also presented in sufficient detail for it to be practically useful.

These developments are illustrated throughout by detailed and realistic applications of the emerging principles to experimental data and radar performance. It is important that our discussion makes direct and frequent contact with the realities of radar operation in order that it is both motivated and validated appropriately. Once these essential preliminaries have been covered they are applied to radar design, optimisation and calculation of radar system performance; the methods we present here are chosen for their efficiency, versatility and accuracy, rather than a summary of everything described in the literature. Finally we describe application of the knowledge to the specification and measurement of radar performance, which are parts of the process of radar system procurement.

The layout of the book is shown in Figure 1.6 and summaries of the topics covered in each chapter are given below.

Chapter 2 describes the experimental evidence of sea clutter characteristics that has been collected over many years of observation, mainly based on data collected from high-resolution airborne and ship-borne radars operating in G/H-band (5.3 GHz) and I-band (9 GHz). The features used to characterise sea clutter, such as reflectivity, amplitude statistics, temporal statistics (Doppler spectra) and spatial features, are defined and illustrated. In addition, evidence is presented on the dependencies of these features on radar parameters (such as spatial resolution, frequency agility, polarisation and so on), and on the environment (sea state, sea swell, wind speed, viewing geometry and so on). All these observed characteristics provide the foundations from which the subsequent chapters of the book are developed.

The need to understand the empirical observations of sea clutter has led to an extensive study of the theory of electromagnetic scattering from the sea surface. Much progress has been made on this subject, as described in **Chapter 3**. In particular, it is increasingly possible to match electromagnetic modelling results with observed data, although difficulties remain at low grazing angles where the effects of multiple scattering become important. The mathematical background to the theory of electromagnetic scattering from corrugated surfaces, the basis for the development of much of the theory for scattering from the sea surface, is described in **Appendix 3**.

Statistical models for target and clutter radar returns form the basis of the performance calculations presented in this book. In **Chapter 4** we develop these models, starting with a simple, physically motivated analysis of the Rayleigh clutter encountered in low-resolution radar systems. The effects of target returns and increasingly higher radar resolution can be accommodated within this framework, which leads us to the compound representation of the K distribution. Other widely used non-Gaussian clutter models, such as the Weibull and lognormal distributions, are also considered briefly. The effects of discrete spikes are analysed and the relation of this description of clutter to the continuous K model is discussed. Finally some properties of coherent clutter are considered, again within the compound model. Background material for this chapter is contained in **Appendices 1 and 2**.

The realistic simulation of clutter plays a vital role in the development of new detection algorithms and the testing of systems with controlled and well-understood data. **Chapter 5** describes how this can be done. Much as in Chapter 4, Gaussian

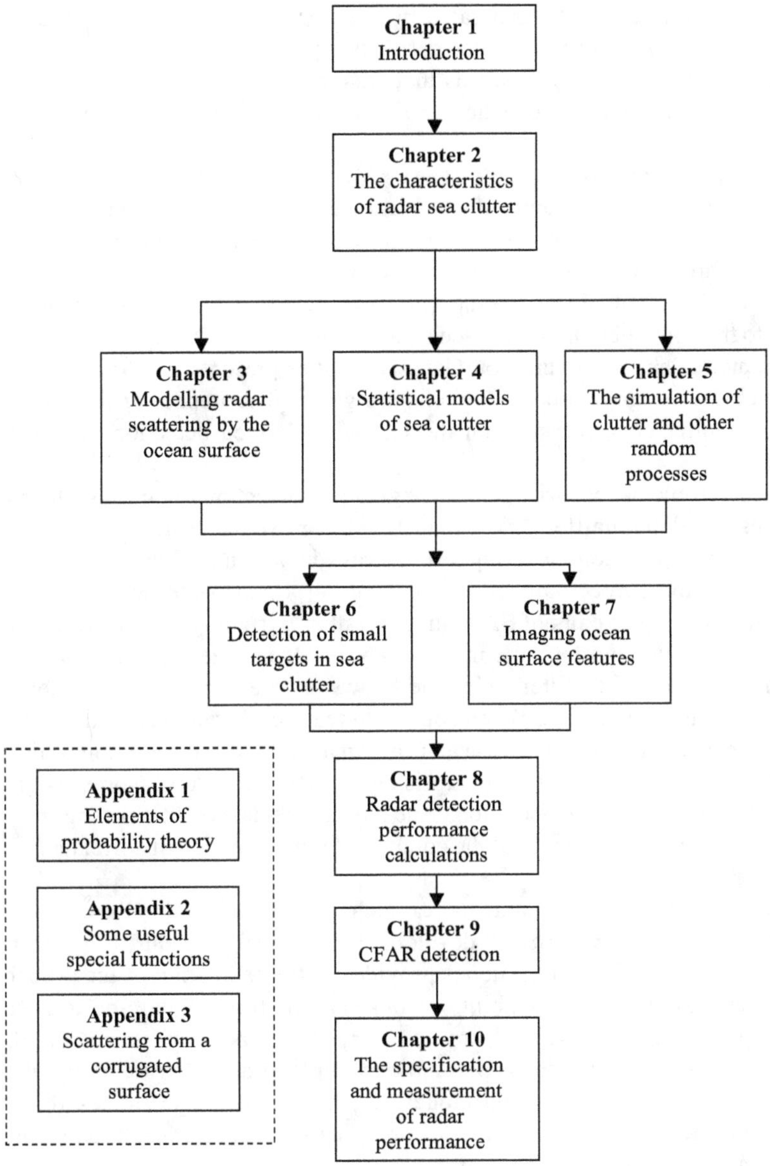

Figure 1.6 Summary of chapters

processes are considered first; their single point statistics are preserved under linear operations while their correlation properties can be controlled through simple recurrence relations and by Fourier synthesis. In the compound model, clutter is represented as a Gaussian process of constant power that is, in turn, multiplied by a non-Gaussian process. The latter can be generated by the non-linear transformation of an input Gaussian process, whose controlled correlation properties are related unambiguously to those of its output. The analysis that enables us to do this is described in some detail, and illustrated by realistic simulations of one- and two-dimensional clutter processes.

Chapter 6 addresses several issues relevant to the detection of small targets in clutter. The pertinent elements of statistical theory underlying the problems of detection and estimation are reviewed with particular attention being paid to likelihood based methods. As well as being optimal, in a well-defined sense, these are quite tractable when applied to Gaussian processes. Consequently they provide a useful guide to the construction of localised small target detection strategies. The estimation of parameters characterising non-Gaussian clutter presents less tractable problems, to which useful approximate solutions are given. The impact of this prior knowledge of clutter characteristics on the Gaussian derived detection methods is also discussed.

Some problems that arise in the imaging and detection of ocean surface features are considered in **Chapter 7**. The detection theory concepts developed in Chapter 6 find further application here, suggesting ways in which anomalous features can be enhanced by local processing. Particular attention is paid to interferometric and polarimetric processing of pairs of SAR images; rather intriguingly, K distribution related concepts play an important role in this analysis. Once a feature in an image has been enhanced, the matched filter technique allows us to exploit our prior knowledge of its structure in its detection. When applied to features that are barely discernible, this technique can easily produce false alarms. An analysis of the variation of false alarm rate with threshold is given that is surprisingly tractable, and gives some insight into its insensitivity to prevailing operational and environmental conditions. The chapter concludes with a brief discussion of the compound representation of correlated structures in clutter.

Using the background mathematical techniques developed in the earlier chapters and the Appendices, **Chapter 8** develops the practical procedures for radar performance prediction. These techniques provide for the estimation of the probability of target detection and the probability of false alarms from clutter and thermal noise, taking into account their dependency on radar parameters, signal processing strategies and environmental conditions. Using the example of an airborne, maritime surveillance radar, the techniques and models are used to evaluate 'trade-offs' between resolution, frequency agility, scan rate and polarisation. The results show the extent to which optimal radar design is affected by clutter statistics.

Chapter 9 describes and analyses some of the techniques used to achieve constant false alarm rate (CFAR) detection. The radar must continuously adapt its detection thresholds to both the local clutter power and its amplitude statistics. The performance of cell-averaging CFAR detectors is analysed for both thermal noise and sea clutter, including the effects of spatial correlation of the sea clutter. Different techniques for adapting to changing amplitude statistics are discussed, particularly in terms of the K distribution model. Some techniques using other characteristics of clutter or the local environment, such as site-specific techniques and the exploitation of the transient coherence of clutter, are briefly reviewed. Finally, some guidance on the development of practical detectors is provided.

Chapter 10 looks at some of the issues facing manufacturers and procurers of radar. For them, the ultimate aim of all the work presented in this book must be to support the design of radar systems that meet the requirements of their users. The very complex nature of radar performance in sea clutter, combined with the advanced adaptive radar signal processing techniques that may be employed, mean that the specification and measurement of radar performance for procurement processes is very difficult. This chapter reviews the specification of the detection performance of complex adaptive radars, techniques for predicting detection performance, the measurement of detection performance and the accuracy of trials measurements. Although users will ideally want to quantify detection performance in practical trials, an important conclusion from this chapter is that in a complex sea clutter environment this will often be either impossible or too time consuming to be affordable. In this event, the use of modelling and simulation, using some of the techniques presented elsewhere in this book, may be the only practical solution.

References

1 L.J. Cantafio, '*Space-based radar handbook*', Artech House, Boston, 1989.

2 S.J. Austin, D.J.Q. Carter and J.D. McCaughey, 'ERS-1 spacecraft – active microwave instrumentation: a major achievement', *GEC J. Res.*, **9**(1), 2–12, 1991.

3 http://envisat.esa.int/

4 S. Watts and A.J. Gordon, 'The Searchwater family of radars', IEE International Conference, Radar 97, 14–16 October 1997, IEE Publication No. 449, pp. 334–338.

5 A.M. Kinghorn, P.E. Holbourn, M.J. Smith and J.F. Roulston, 'The Blue Vixen Radar for Sea Harrier FRS.2', *GEC Review*, **9**(3), p. 131–145, 1994.

6 E. Jakeman and P.N. Pusey, 'A model for non-Rayleigh sea echo', *IEEE Trans. Antennas Propag.*, **AP-24**, 808, 1976.

7 K.D. Ward, 'Compound representation of high resolution sea clutter', *Electron. Lett.*, **17**(16), 561–563, 1981.

8 K.D. Ward, C.J. Baker and S. Watts, 'Maritime surveillance radar part 1: Radar scattering from the ocean surface', *Proc. IEE*, **137** (part F, No. 2), 51–62, 1990; S. Watts, C.J. Baker and K.D. Ward, 'Maritime surveillance radar part 2: Detection performance prediction in sea clutter', *Proc. IEE*, **137** (part F, No. 2), 63–72, 1990.

9 C.J. Oliver and S. Quegan, '*Understanding synthetic aperture radar images*', Artech House, Boston, 1998.

10 E. Jakeman and R.J.A. Tough, 'Non-Gaussian models for the statistics of scattered waves', *Adv. Phys.*, **37**, 471, 1988.

11 E. Jakeman and R.J.A. Tough, 'The generalised K distribution: a statistical model for weak scattering', *J. Opt. Soc. Am.*, **A4**, 1764, 1987.

12 R.J.A. Tough, D. Blacknell and S. Quegan, 'A statistical description of polarimetric and interferometric synthetic aperture data', *Proc. Roy. Soc. Series A*, **449**, 567, 1995.
13 S.O. Rice, 'Reflection of electromagnetic waves from slightly rough surfaces', *Commun. Pure Appl. Math.*, **4**, 351–478, 1951.
14 G.R. Valenzuela, 'Theories for the interaction of electromagnetic waves and oceanic waves – a review', *Boundary-Layer Meteorol.*, **13**, 61–85, 1978.
15 *IEEE Trans. Antennas propagation* (special issue on LGA scattering by the sea surface), **AP-46**(1), 1998.
16 E. Jakeman and K. Ridley, '*Modelling fluctuations in scattered waves*', Taylor & Francis, London, 2006.

Chapter 2
The characteristics of radar sea clutter

2.1 Overview

In this chapter, sea clutter is described in terms of its observed characteristics. These observations will provide the foundations upon which the themes of subsequent chapters are developed.

Radars operating in a maritime environment inevitably encounter backscattered radar signals from the sea surface, usually referred to as sea clutter. For some applications, such as remote sensing systems, the reception of this backscattered signal may be the main purpose of the radar. Space-borne synthetic aperture radars, with spatial resolutions of a few meters, are used for oceanographic studies, gathering data on waves and currents, sea ice and so on. Scatterometers measure average backscatter over hundreds of square kilometres to measure wind speed and direction over the sea surface.

For most other applications, the backscatter from the sea is unwanted and may interfere with the radar's operation. For example, almost all airborne radar operation over the sea encounters sea clutter, whether the radar is used for long range surveillance of shipping, detection of other aircraft or detecting very small targets such as submarine periscopes. Similarly, ship-borne radars, whether used for navigation or surveillance of the air and sea, must contend with unwanted sea clutter.

The characteristics of sea clutter over this range of applications may vary extremely widely. The radar system designer needs to understand these characteristics in order to develop suitable signal processing strategies and to predict performance under different conditions. An important step in achieving these goals is the development of accurate statistical models of the clutter return. These models must be able to incorporate the spatial and temporal properties of the backscatter, under a wide range of environmental conditions and for different radar waveforms and viewing geometries.

Figure 2.1 The sea surface before the onset of significant wave breaking (courtesy of Charlotte Pagendam)

A number of features are used to characterise the clutter returns. These are

- the area reflectivity, σ^0
- the amplitude distribution of the clutter amplitudes or power
- the spectrum of the clutter returns
- the spatial variation of the clutter return
- the polarisation scattering matrix
- discrete clutter spikes

The average radar cross-section (RCS) of the returns per unit area is defined by the area reflectivity σ^0. For a surface area A illuminated by the radar resolution cell (e.g. for a pulsed radar, the area defined by the compressed pulse length and the antenna azimuth beamwidth), the RCS of the clutter is given by σ^0 A. The constantly changing and complex nature of the sea surface means that the instantaneous RCS of the returns fluctuates widely around the mean value determined by σ^0. The statistics of these fluctuations are an important characteristic of sea clutter. The single point amplitude statistics are described using families of probability density functions (pdf), with a specific pdf being appropriate for a given set of observations. The manner in which these amplitude fluctuations vary with time is characterised by the spectrum of the returns.

It is apparent from casual observation of the sea surface (as shown in Figures 2.1 and 2.2) that it is not a random rough surface but contains significant structure. When the wind blows it generates small ripples, which grow and transfer their energy to longer waves. At some point the waves become large enough to break (as shown, in Figure 2.2) and this redistributes the wave energy further. When the wind has been blowing for some time, an equilibrium is established between the input of energy and its dissipation. There is then a wide spectrum of waves propagating on the sea, which can be added to by swell travelling into the area from remote rough weather. All of these waves and the breaking events are reflected in the spatial variation of clutter returns and the nature of such variation needs to be characterised.

As will be described, each of these characteristics is dependent in complex ways on the prevailing environment, radar waveforms and viewing geometry. One aspect of the radar waveform that requires particular attention is its polarisation and the dependence

Figure 2.2 Wave breaking on the sea surface (courtesy of Charlotte Pagendam)

on polarisation of the back-scattered radar signals; this can be represented by the polarisation scattering matrix. While sea clutter can generally be described in terms of the statistical behaviour of the returns from multiple distributed scatterers, individual discrete scatterers or isolated clutter spikes are sometimes observed. These spikes can also be characterised and incorporated into the standard distributed clutter models.

2.2 The sea surface

As might be expected, observations of radar sea clutter are usually associated with particular characteristics of the sea surface and the environment, such as sea waves, sea swell or wind speed. Before discussing these observations in more detail, some of the basic terms used to describe the sea surface are presented here. Further detail can be found in classic texts such as References 1–3. The following definitions are given in Reference 3:

Wind wave: a wave resulting from the action of the wind on a water surface. While the wind is acting upon it, it is a sea; thereafter, it is a swell.

Gravity wave: a wave whose velocity of propagation is controlled primarily by gravity. Water waves of a length greater than 5 cm are considered gravity waves.

Capillary wave (also called ripple or capillary ripple): a wave whose velocity of propagation is controlled primarily by the surface tension of the liquid in which the wave is travelling. Water waves of a length of less than 2.5 cm are considered capillary waves.

Fetch: (1) (also called generating area) an area of the sea surface over which seas are generated by a wind having a constant direction and speed; (2) the length of the fetch area, measured in the direction of the wind in which the seas are generated.

Duration: the length of time the wind blows in essentially the same direction over the fetch.

Fully developed sea (also called a fully arisen sea): the maximum height to which ocean waves can be generated by a given wind force blowing over sufficient fetch, regardless of duration, as a result of all possible wave components in the spectrum being present with their maximum amount of spectral energy.

Sea state: the numerical or written description of ocean-surface roughness. Ocean sea state may be defined more precisely as the average height of the highest one-third of the waves (the significant wave height) observed in a wave train (see Tables 2.1 and 2.2).

Clutter reflectivity is often characterised in terms of sea state and the description of sea state is a common source of confusion in the specification of radar performance. Often the Beaufort wind scale is used, but this wind speed only gives a reliable estimate of wave height if the duration (length of time the wind has been blowing) and the fetch (the range extent of the sea over which the wind has been blowing) are known. A standard estimate of wave height is the Douglas Sea State (this is the scale used by

Table 2.1 Douglas sea state

Douglas sea state	Description	Wave height $h_{1/3}$ (ft)	Wind speed (kn)	Fetch (nmi)	Duration (h)
1	Smooth	0–1	0–6		
2	Slight	1–3	6–12	50	5
3	Moderate	3–5	12–15	120	20
4	Rough	5–8	15–20	150	23
5	Very rough	8–12	20–25	200	25
6	High	12–20	25–30	300	27
7	Very high	20–40	30–50	500	30
8	Precipitous	>40	>50	700	35

Table 2.2 World Meteorological Organisation sea state

Sea state	Wave height $h_{1/3}$ (ft)	Description
0	0	Calm, glassy
1	0–1/3	Calm, rippled
2	1/3–2	Smooth, wavelets
3	2–4	Slight
4	4–8	Moderate
5	8–13	Rough
6	13–20	Very rough
7	20–30	High
8	30–45	Very high
9	>45	Phenomenal

Nathanson [1] in his tables of reflectivity), which is shown in Table 2.1. The wave height, $h_{1/3}$, is the 'significant wave height', defined as the average peak-to-trough height of the highest one third of the waves. In Table 2.1 the stated wind speed produces the stated wave height with the fetch and duration shown.

Other descriptions of sea state are also used, such as the World Meteorological Organisation sea state; this is listed in Table 2.2. When modelling or predicting radar performance it is very important to agree what scale of measurement is being used.

Assessing the environment in an actual radar trial is notoriously difficult and the sea state estimated by observers often shows a wide variation. Wave-rider buoys are used to estimate local wave height but these only give a rough guide to the likely clutter characteristics, which also depend on such things as 'wind friction velocity', 'wave age' and other factors discussed in Chapter 3.

2.3 Sea clutter reflectivity

Radar backscatter from the sea is derived from a complex interaction between incident electromagnetic waves and the sea surface. There are many theoretical models for backscatter, based on different descriptions of the rough surface and approximations to the scattering mechanism. Attempts have been made to represent the sea surface in terms of many small segments, called facets, with orientations modulated by the sea waves and swell. Scattering from local wind-derived ripples may be approximated by Bragg (or resonant) scattering. The tilting of the ripples by longer waves on the sea changes the scattered power, and this is incorporated in the so-called Composite model. These models, which are discussed further in Chapter 3, tend to give satisfactory results at high and medium grazing angles. At low grazing angles the scattering mechanisms are much more complex and the observed values of reflectivity deviate from the models. Here scattering is complicated by factors such as shadowing from up-range surface features, diffraction over edges and interference between scattered signals travelling over different paths. Some progress has recently been made in our understanding of scattering at low grazing angles; this is described in Chapter 3 and Appendix 3. Generally speaking however, the practical development of models for sea clutter still relies mainly on empirical measurements and observations.

Figure 2.3 illustrates the form of the variation in sea clutter reflectivity with grazing angle that is observed in practice for microwave radars. This example is typical of backscatter for a wind speed of around 15 kn (sea state 3 in a fully developed sea) for an I-band (9 GHz) radar. At near vertical incidence, backscatter is quasi-specular. In this region, the backscatter varies inversely with surface roughness, with maximum backscatter at vertical incidence from a perfectly flat surface. At medium grazing angles, in this example below about 50° from grazing, the reflectivity shows a lower dependence on grazing angle; this is often called the plateau region. Below a critical angle (typically around 10° from grazing, but dependent on the sea roughness) it is found that the reflectivity reduces much more rapidly with smaller grazing angles; this region is shown in Figure 2.3 as the interference region.

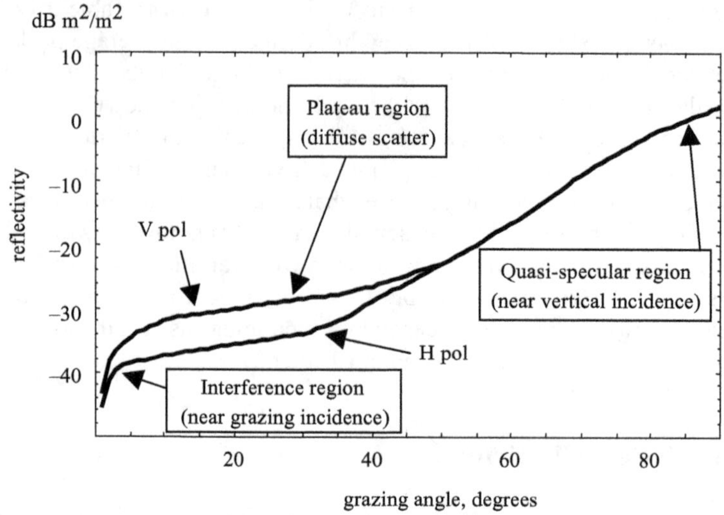

Figure 2.3 Typical variation with grazing angle and polarisation of sea clutter reflectivity at I-band (for a wind speed of about 15 kn)

Also shown in Figure 2.3 is the dependence of the reflectivity on radar polarisation. In the plateau and interference regions the backscatter from horizontal (H) polarisation is significantly less than that for vertical (V) polarisation. As will be seen in Section 2.7, many other clutter characteristics are dependent on radar polarisation and often quite different scattering mechanisms are present for different polarisations.

In addition to the variation with grazing angle and polarisation discussed above, it is found that the average reflectivity of sea clutter is dependent on many other factors. Empirical models often present reflectivity as a function of sea state. However, as discussed earlier, this is not always a reliable indicator as it is dominated by the wave height of long waves and sea swell. Local wind causes small-scale surface roughening, which responds quickly to changes in wind speed and changes the backscatter. A strong swell with no local wind gives a low reflectivity, whilst a local strong wind gives a high backscatter from a comparatively flat sea. Further deviation from a simple sea state trend may be caused by propagation effects such as ducting, which can affect the illumination grazing angle.

Figure 2.4 is an illustration of the variability of reflectivity that can occur within a simple sea state description. It shows an example of an image from the ERS1 satellite [4]. This was a synthetic aperture radar operating in G/H-band (5.3 GHz), achieving a spatial resolution of about 30 m × 30 m. The image is of the ocean surface, over which a very large variation of reflectivity is apparent. This variation is due to many causes, including local wind speed variations, rain and cloud, currents and internal waves. The ocean currents are influenced by the topology of the sea bottom and cause a roughening on the sea surface, and hence a variation in radar reflectivity, that appears to replicate the sea bottom features below. (This effect is discussed and

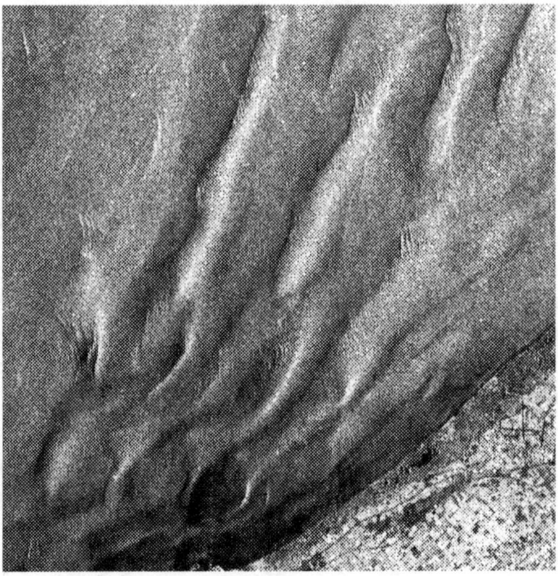

Figure 2.4 ERS1 image of the sea surface (data copyright ESA – Image processed by DERA UK)

modelled in Chapter 3.) The lower right hand corner of the image shows returns from land, actually part of Belgium.

Specific empirical models for the reflectivity of sea clutter are discussed in Chapter 8. It is important to realise that the spread in values of σ^0 observed between the various models for a given set of conditions is not dissimilar to that which might be encountered in practice, especially given the difficulty in characterising the actual sea conditions. The value of these models lies in their use for radar design. They allow the radar systems engineer to determine the range of values to be expected. The models of σ^0 discussed in Chapter 8 provide good estimates of the range of values of σ^0 likely to be encountered in different conditions and the expected variation with radar parameters and viewing geometry.

Some models are defined in terms of the median reflectivity, rather than the mean. For a uniform scattering surface, characterised by a Rayleigh amplitude distribution, the mean reflectivity is about 1.6 dB greater that the median. This is negligible in practice, in view of the uncertainties of measuring the average reflectivity. However, as will be discussed later the ratio of mean to median reflectivity in very spiky clutter can be much larger and it is then important to establish which value is presented by the model.

2.4 Amplitude statistics

Many radars use only the envelope of the received signal in their signal processing. Since they do not use the signal phase, these systems need not be coherent from pulse to pulse. Non-coherent statistics may be described in terms of either the envelope (i.e. following a linear detector) or the power (i.e. following a square law detector) of the return. For coherent processing, where the phase of the returns is retained, the statistics are described in terms of the temporal variation of the complex amplitude and its associated Fourier Transform, or spectrum; this is discussed in Section 2.9.

As discussed earlier, the area reflectivity σ^0 represents the mean clutter power. The instantaneous power received from a single radar resolution cell varies about this mean value. This variation is characterised by the pdf of the returns. There are two main contributions to the fluctuations. First, the variation of local surface shape, grazing angle, ripple density, and other factors associated with the passage of long waves and swell, cause the backscatter from local small areas to vary widely about the mean reflectivity. Second, within a radar resolution patch, scattering occurs from many small structures (often called scatterers), which move relative to each other and create interference in the scattered signal (often referred to as speckle).

Speckle is often described as resulting from a uniform field of many random scatterers, which exhibits Gaussian scattering statistics (as discussed in Chapter 4). The pdf of the amplitude, E, of these returns, as observed by a linear detection, is given by the Rayleigh distribution:

$$P(E) = \frac{2E}{x} \exp\left(\frac{-E^2}{x}\right); \quad 0 \leq E \leq \infty \tag{2.1}$$

where the average value of E, $\langle E \rangle = \sqrt{\pi x}/2$, the mean square, $\langle E^2 \rangle = x$, and the higher moments, $\langle E^n \rangle = x^{n/2}\Gamma(1 + n/2)$.

The PDF of the power of the returns (i.e. as observed by a square-law detector) is given by

$$P(z) = \frac{1}{x} \exp\left(\frac{-z}{x}\right); \quad 0 \leq z \leq \infty \tag{2.2}$$

where $z = E^2$, $\langle z \rangle = x$ and $\langle z^n \rangle = n!x^n$.

For radars that have a low spatial resolution, where the resolution cell dimensions are much greater than the sea swell wavelength, and for grazing angles greater than about 10°, clutter is usually modelled simply as speckle and the amplitude is Rayleigh distributed. The clutter returns have fairly short temporal decorrelation times, typically in the range 5–10 ms, and are fairly well decorrelated from pulse to pulse by a radar employing frequency agility (provided that the radar frequency is changed from pulse to pulse by at least the transmitted pulse bandwidth – see Section 2.5). There is no inherent correlation of returns in range, beyond that imposed by the transmitted pulse length.

As the radar resolution is increased, and for smaller grazing angles, the clutter amplitude distribution is observed to develop a longer 'tail' and the returns are often described as becoming spiky. In other words, there is a higher probability of large amplitude values (relative to its mean) being observed, compared to that expected for a Rayleigh distribution. This is a result of the first effect described above, which in essence is the radar resolving the 'structure' of the sea surface.

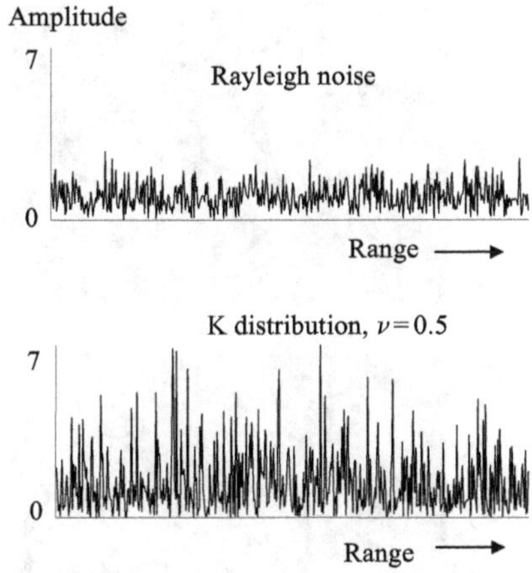

Figure 2.5 Gaussian clutter contrasted with spiky clutter

The causes of this spikiness are quite complex and a number of different mechanisms contribute to the effects that are observed; descriptions of data are given in Section 2.6 and the mechanisms giving rise to these features are discussed in more detail in Chapter 3. Figure 2.5 illustrates the essential difference between the amplitude variation of Gaussian clutter and spiky clutter. The data in each plot have the same mean level of 1 unit but the different ranges of variation about the mean are quite apparent. In this case the spiky clutter has been modelled as having a K distribution pdf, which is discussed in detail in Chapter 4.

In addition to the deviations of high-resolution clutter amplitude from the Rayleigh distribution, the temporal and spatial correlation properties also change from those of speckle. In particular, frequency agility no longer decorrelates clutter and the longest correlation times stretch to seconds rather than milliseconds. Also, there is range correlation beyond the radar resolution patch that is not evident in speckle. In the literature, much work on understanding the non-Gaussian nature of sea clutter concentrates on the amplitude distribution of the envelope (or intensity) of the returns and on the average temporal power spectrum. These features are sufficient to describe Gaussian processes but in general provide an incomplete description of non-Gaussian processes. Therefore they may result in modelling that omits features of importance concerning the effects of the non-Gaussian nature. This is the case for sea clutter, where correlation properties not evident in the average power spectrum (nor in the complex autocorrelation function, which is equivalent) severely affect radar performance and signal processing optimisation.

The evidence for these complex properties of sea clutter is presented in the next section, using experimental observations of sea clutter from airborne and cliff top radars. Nearly all the results presented in this chapter were collected from radars operating at I-band, with a range of pulse bandwidths and polarisations. These observations form the basis of the compound K distribution mathematical model for sea clutter, which is described in detail in Chapter 4.

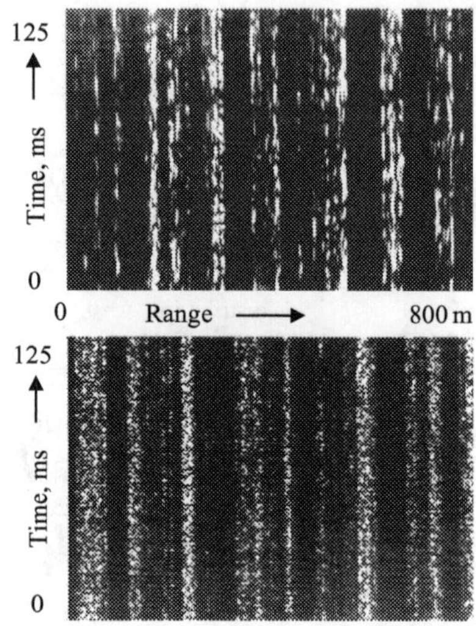

Figure 2.6 Range-time intensity plots of sea clutter; upper plot shows returns from a fixed frequency radar and lower plot shows returns with pulse-to-pulse frequency agility

2.4.1 The compound nature of sea clutter amplitude statistics

Measurements of high resolution, low grazing angle sea clutter returns have identified two components of the amplitude fluctuations [5,6]. The first component is a spatially varying mean level that results from a bunching of scatterers associated with the long sea waves and swell structure. This component has a long correlation time and is unaffected by frequency agility. A second 'speckle' component occurs due to the multiple scatterer nature of the clutter in any range cell. This decorrelates through relative motion of the scatterers or through the use of frequency agility.

These properties can be appreciated with the aid of Figures 2.6 and 2.7, which show radar sea echo data, collected from a cliff top radar at I-band (8–10 GHz) incorporating frequency agility, with a 1.2° beamwidth and 28 ns pulse length. Figure 2.6 shows range-time intensity plots of pulse-to-pulse clutter from a range window of 800 m at a range of 5 km and grazing angle of 1.5°. The upper plot of Figure 2.6 is for fixed frequency and shows that at any range the return fluctuates with a characteristic time of approximately 10 ms as the scatterers within the patch move with the internal motion of the sea and change their phase relationships. This varying speckle pattern is decorrelated from pulse-to-pulse by frequency agility as shown in the lower plot of Figure 2.6. Both of the figures show that the local mean level varies with range due to the bunching of scatterers. This is unaffected by frequency agility. The total time (1/8 s) of Figure 2.6 is not sufficient for the bunching to change at any given range. In Figure 2.7, averaging has been used to remove the speckle component, and the plot therefore shows the 'bunching' term over a longer time period of about 120 s. Over the first 60 s, the radar was transmitting and receiving with vertical

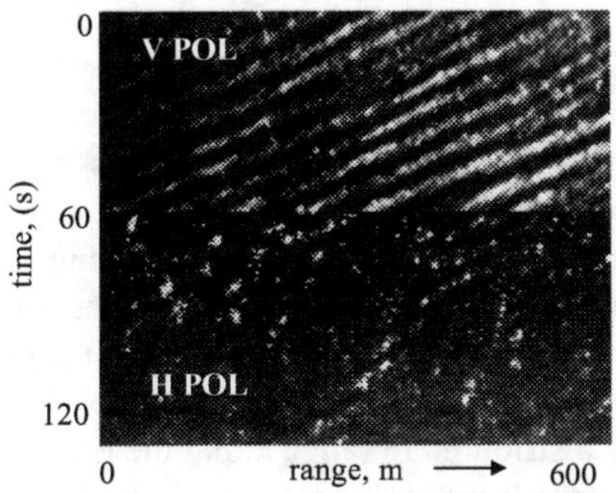

Figure 2.7 *Range-time intensity plot of sea clutter averaged over 250 successive pulses to remove the speckle component, revealing the underlying mean level. After 60 s, the radar was switched from vertical to horizontal polarisation*

polarisation. The plot demonstrates the wavelike nature of this component. It can be deduced that the dominant long waves were moving toward the radar at a speed of about 10 m/s. After 60 s, the radar polarisation was switched to horizontal. Some vestiges of the long wave pattern can still be detected, but now the clutter appears much more patchy, with isolated clutter spikes having a lifetime of about 1 s or so. In both polarisations the spikes and modulations shown in Figure 2.7 appear to be associated with long waves and associated breaking events. The mechanisms for producing spiky backscatter may be different for the two polarisations (as discussed in Chapter 3). The horizontally polarised spikes illustrated in Figure 2.7 are seen in this data to have a much higher amplitude relative to the mean level, compared to those for vertical polarisation. Also the overall mean level for horizontal polarisation is lower than for vertical polarisation. Both these trends are regularly observed in sea clutter, as discussed in Chapter 8.

The compound K distribution model allows the separate characterisation of these two components of the clutter envelope fluctuations, as shown in Chapter 4. The returns from a given patch of sea (a radar resolution cell) have an amplitude distribution corresponding to that of the speckle component of the clutter with a mean given by the local value of the spatially varying mean level component of the clutter.

2.5 Frequency agility and sea clutter

As discussed in Section 2.4, Gaussian clutter can be characterised as comprising multiple scatterers distributed in range over the clutter patch. If the radar frequency changes, the vector sum of the scatterer returns also changes. For a patch of Gaussian clutter illuminated by a radar pulse of bandwidth B, returns on two different frequencies are decorrelated if the change in frequency, Δf, causes the change of phase from the leading edge of the clutter patch to the trailing edge, to be at least 2π. For a radar frequency f, the phase change over the two-way path across the clutter patch is $2\pi f/B$ and so a frequency step $\Delta f = B$ causes successive pulse returns to be decorrelated.

Consequently, in uniform clutter with Rayleigh amplitude statistics, pulse-to-pulse frequency agility with frequency steps of at least the pulse bandwidth should ensure that returns are decorrelated from pulse to pulse. This ability to decorrelate the returns is very important in the design of radar processing for detecting small targets in sea clutter. This is discussed in detail in Chapters 8 and 9.

As discussed in Section 2.4.1, the speckle component of the compound representation of sea clutter is observed to be decorrelated by frequency agility, implying many scatterers evenly distributed in range along the clutter resolution cell. Equally, it can be understood that the large scale spatial variation of the underlying mean level, associated with the changing area reflectivity induced by long waves the sea swell, cannot be decorrelated by frequency agility. Again, the exploitation of this effect in the design of radar detection algorithms is described in Chapters 8 and 9.

To be more precise, the presence of the underlying spatial correlation means that the scatterers cannot be *completely* uniformly distributed along the clutter cell. For this reason, pulse-to-pulse frequency steps equal to the pulse bandwidth cannot *fully* decorrelate the clutter speckle component. This is described further in Reference 7. However, in practice this effect is often small and it is usually quite accurate to assume pulse-to-pulse decorrelation with frequency agility steps of the pulse bandwidth. This is especially true for radar with an antenna scanning in azimuth across the sea surface. Here, additional decorrelation is introduced due to the changing composition of the scatterers in the clutter cell [8]. This is discussed further in Chapter 8.

2.6 Observations of amplitude distributions

As mentioned previously, the non-Gaussian amplitude statistics of sea clutter vary considerably, dependent on the prevailing sea and weather conditions, the viewing geometry and the radar parameters. Some insight into the causes of these variations is given in Chapter 3, whilst Chapter 4 shows how empirical models have been developed based on practical measurements.

An initial appreciation of the range of these variations may be obtained from observation of range-time intensity plots of the underlying modulation of clutter, as shown in Figure 2.8(a)–(f), using the same type of I-band radar as for Figures 2.6 and 2.7. These indicate how the clutter structure is dependent on radar polarisation, sea state and grazing angle. Figure 2.8(a) is taken looking at 30° to the dominant wave direction, with a sea condition of medium roughness (sea state 3, significant wave height 1.2 m). The grazing angle is 1°. If this is compared with Figure 2.8(b), where the sea condition is rougher (sea state 4, significant wave height 2 m) and the radar is looking upwind, we find that in the latter the wave pattern is more pronounced and the polarisation change has less effect on the clutter. The calm sea plot, Figure 2.8(c), shows both vertical and horizontal polarisation with large spikes of a few seconds duration giving the dominant return. Figure 2.8(a–c) all show a grazing angle in the region of 1°. If this is reduced to 0.1°, as shown in Figure 2.8(d) for a rough sea (sea state 5, significant wave

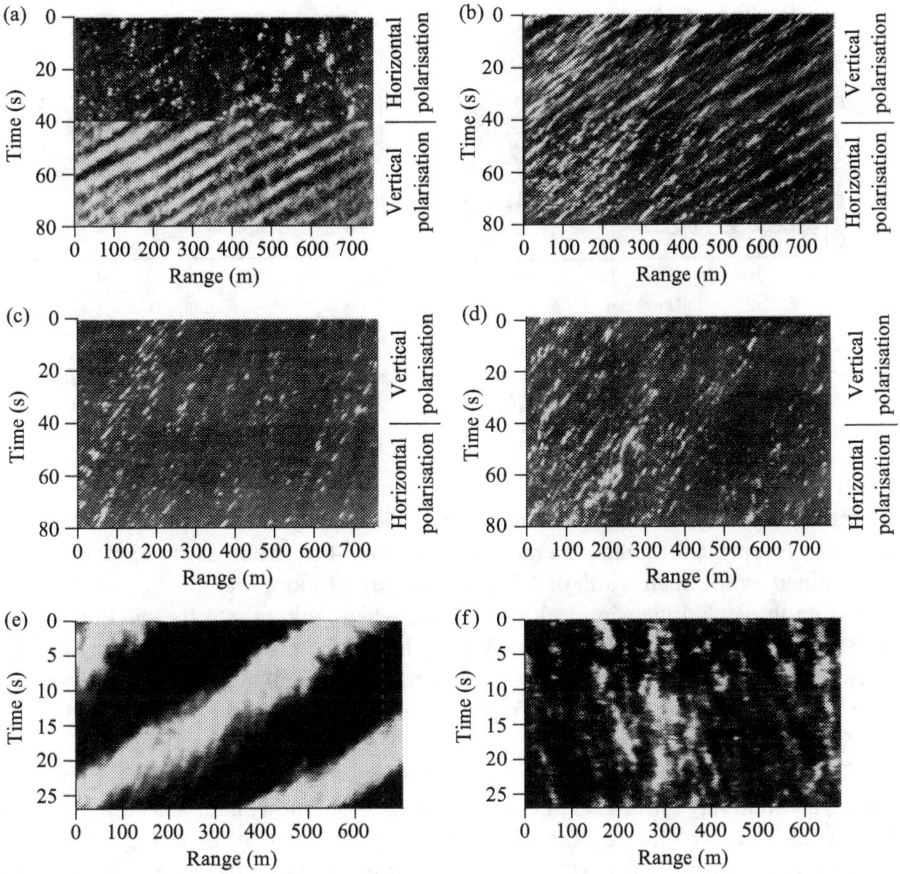

Figure 2.8 *Range-time intensity plots of averaged clutter showing the dependence of the structure on radar and environmental parameters. (a) Sea state 3, 30° to upwind, 1° grazing, (b) Sea state 4, upwind, 1° grazing, (c) Sea state 1, upwind, 1° grazing, (d) Sea state 5, upwind, 0.1° grazing, (e) Sea state 3, up swell, 1° grazing and (f) Sea state 3, across swell, 1° grazing*

height 3 m), the well-defined wave structure of Figure 2.8(b) becomes concentrated into spikes of much higher amplitude (relative to the overall mean). However, there is still a fairly regular structure. Occasionally there is very long-wavelength swell present in the sea wave spectrum. It may not be easily identified visually, but the grazing incidence radar backscatter is dominated by its effects. Figure 2.8(e) shows an averaged plot at 1° incidence looking into the swell. The sea roughness is medium (sea state 3) but the swell produces a very clear pattern. As can be seen, the sea wavelength is much longer than any of the other plots. An interesting effect is the dependence of the structure on viewing aspect. Looking down the swell propagation direction produces a plot similar to Figure 2.8(e), but looking across the swell

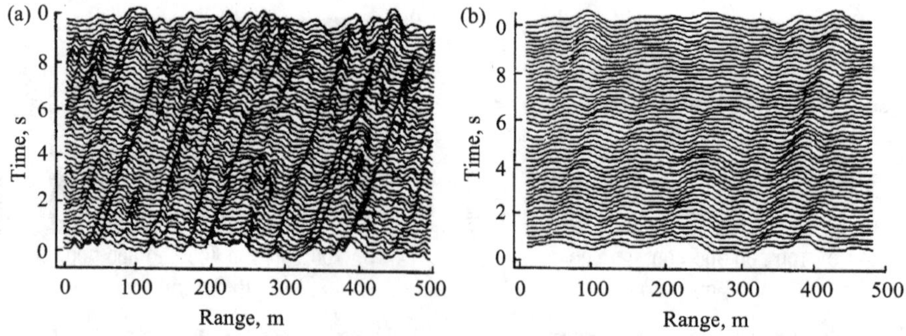

Figure 2.9 Range-time plots of averaged sea clutter showing the effect of different synthesised pulse widths. (a) pulse width = 28 ns and (b) pulse width = 200 ns

direction breaks up the wave pattern (Figure 2.8(f)). This can be explained in terms of the radar footprint. Although the range resolution is 4 m, the cross-range resolution is determined by the beamwidth of 1.2°. At a range of 16 km, used for Figures 2.8(e,f), the azimuth patch dimension is 320 m. Thus, when looking into the swell, the radar resolves structure greater than 4.2 m. Across the swell, only structure greater than 320 m is resolved, and then not in range but in time as the swell moves through the beam. The 'across swell' picture is therefore not expected to show the wavelike pattern of the 'into swell' pictures.

The effect of range resolution can be considered in a similar manner. Figure 2.9 shows qualitatively the effects of changing range resolution on the sea clutter returns. Figure 2.9(a) was collected from a radar with a pulse length of 28 ns (equivalent to a range resolution of 4.2 m) and the long wave features in the clutter are clearly resolved. In Figure 2.9(b) the resolution has been reduced to 200 ns (a range resolution of 30 m) and it is clear that the high-resolution spatial features have been averaged out.

These varying characteristics can all be described within the compound formulation for sea clutter. Empirical results relating the parameters of the compound amplitude distribution to the prevailing conditions are described in Chapter 8, where the following results emerge:

- Sea clutter is more spiky at lower grazing angles and higher spatial resolution.
- It is more spiky when there is a sea swell present, with the spikiest conditions observed when looking up or down swell.
- Horizontally polarised returns are generally more spiky than those for vertical polarisation.
- No significant statistical trend has been established for variations of spikiness with sea state, wind speed or aspect angle relative to wind direction.

Further observations on the variation of distribution shape with radar and environmental parameters are reported by Hair *et al.* [9]. This work shows that the compound formulation is a good model over radar frequencies from 1 to 16 GHz and for range resolutions from 0.375 to 15 m.

2.7 Polarisation characteristics

We have seen that sea clutter characteristics depend on the polarisation of the radar signal. Polarisation is defined by the direction of the electromagnetic fields and if these are deterministic, the wave is defined as being polarised. When the direction of the electric and magnetic field vectors, **E** and **H** respectively, reside in fixed planes as the wave progresses it is said to be linearly polarised, the plane of polarisation being defined by the **E** vector. In general, however, the directions of **E** and **H** vary with time and the wave is said to be elliptically polarised or, as a special case of this, circularly polarised. In circular polarisation the directions of **E** and **H** rotate, whilst the magnitudes remain constant. (Circular polarisation may be synthesised by adding vertical and horizontal linear polarisations with a phase difference of 90°.[1]) On scattering the polarisation of the wave may change. The polarisation scattering matrix, **S**, describes the amplitude and relative phase of returns from different combinations of polarisations on transmit and receive.

In order to investigate the effects of polarisation further, it is of interest to examine polarisation agile recordings; these provide more insight into the polarisation characteristics evident in Figures 2.8(a–f). The data are from a clifftop radar with the following basic parameters:

- I band (9.75 GHz)
- 100 MHz pulse bandwidth, giving about 2 m range resolution
- pulse to pulse switching of transmitter polarisation between V and H at 1 kHz
- two receiver channels for recording simultaneous V and H

Thus, every 2 pulses, recordings are made of VV, VH, HV and VV scattering. A flat plate reflector is used for calibration. Figure 2.10 shows range-time intensity plots of the four polarisation combinations from a 12 s record of data recorded using V and H polarisations. The similarity of the VH and HV is evident, as is expected from reciprocity. Also the characteristic spikiness of HH compared to VV can be seen. The dominant spikes on the HH record persist for about one or two seconds and have associated depolarised signals in VH and HV. In contrast, the shorter spikes in HH (only lasting a small fraction of a second) do not tend to have cross-polar returns.

As discussed earlier, circular polarisation can be synthesised by the linear transformation of linear polarisations. This requires careful balancing of the gain and phase responses of the transmitter and receiver channels. This may be achieved by measuring the returns from a flat plate and a depolarising target, such as a tilted dihedral. Figures 2.11–2.13 compare VV and HH data, recorded with the same radar as above, with RR and RL circularly polarised data, obtained by transforming the coherent

[1] Left-hand circular polarisation is defined as $E_{\mathrm{V}} = iE_{\mathrm{H}}$; Right-hand circular is $E_{\mathrm{V}} = -iE_{\mathrm{H}}$.

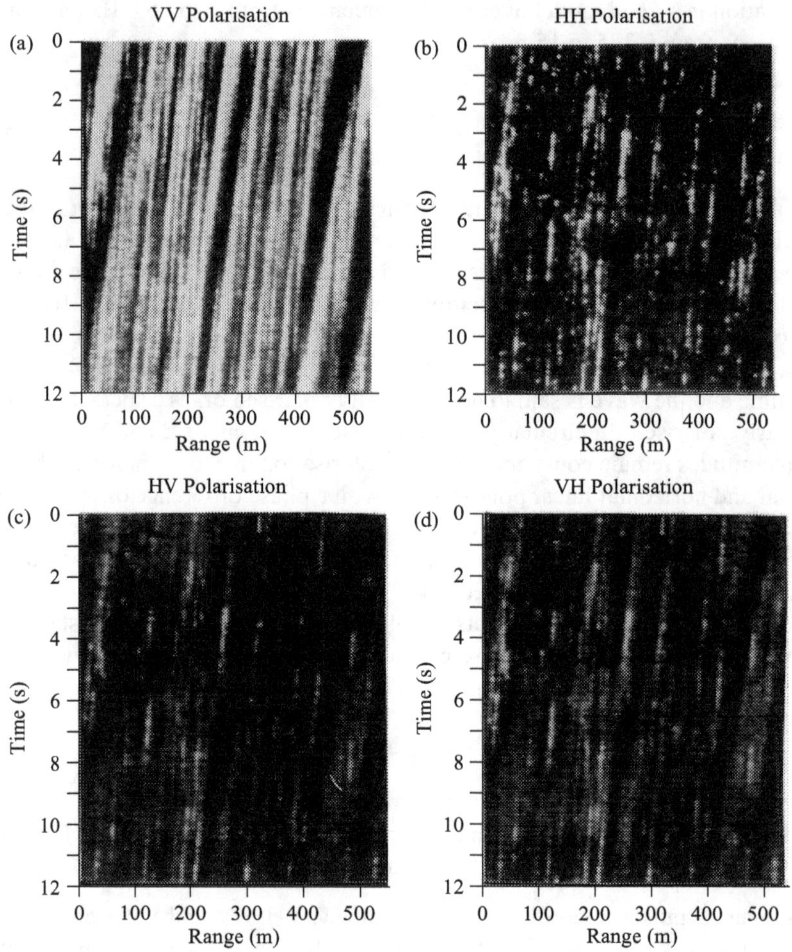

Figure 2.10 Range-time intensity plots for HH/VV/HV/VH polarisations

scattering matrix of the linearly polarised data. These figures are range-time intensity plots of three sea conditions, recorded using the same radar as above:

Figures 2.11, wind speed 0–5 kn
Figures 2.12, wind speed 6–10 kn
Figures 2.13, wind speed 18–20 kn

All of the images show 7 s of data from 256 range cells (384 m of range). The low wind speed data are very spiky. As the wind speed increases the contributions from spatially distributed scattering increase, tending to mask the discrete-like spikes. The dominant sea wavelength increases with the wind speed, and at high wind speed the wave pattern is very clear, with different wavelengths evident, travelling at different speeds due to

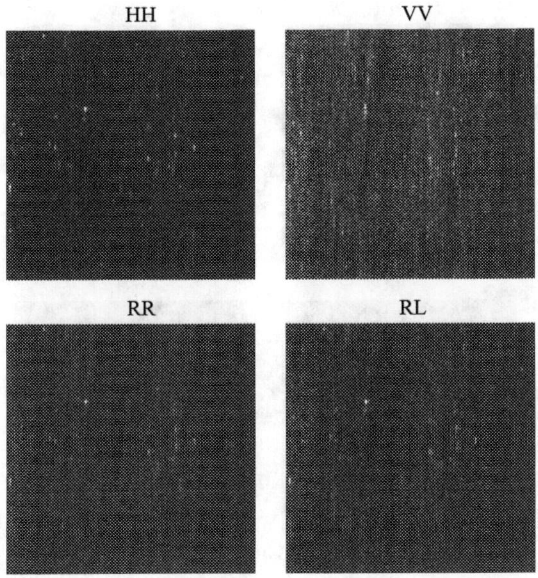

Figure 2.11 *Range-time intensity plots for VV and HH, compared with RR and RL for the same data, with a wind speed of 0–5 kn (the time axis is vertical and displays 7 s, starting from the top; the range axis is horizontal, and displays 256 range cells, i.e. 384 m)*

their dispersion relation. The co- and cross-circular polarisation images look similar to each other, containing the features present in both the VV and HH images.

As we will see later in Section 2.9, the shape of the Doppler spectrum of sea clutter is also dependent on the polarisation used. Horizontally polarised returns tend to have a higher absolute Doppler frequency than vertically polarised returns. Taken together, all of these observations of polarisation effects suggest that there are a number of different mechanisms at work.

2.8 Clutter spikes and modulations

The data illustrated in Figures 2.6–2.13 show examples of smooth modulations and what may broadly be classified as clutter 'spikes'. Together, the modulation and spikes contribute to the non-Gaussian nature of the clutter, a feature that is generally referred to as 'spikiness'. Thus we must remember that wave-like modulation, for example as shown in Figure 2.8(e), contributes to the clutter 'spikiness' despite not actually being made up of spikes.[2] The spikes themselves seem to divide into two classes that are most easily differentiated by their time evolution. Short spikes, which we will refer to as 'bursts', last for a small fraction of a second (200 ms is typical) and do not

[2] This may seem perverse, but we cannot change 50 years of common usage in the literature.

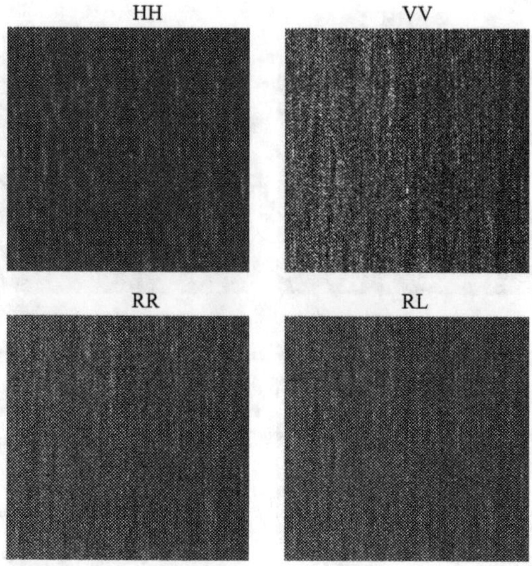

Figure 2.12 *Range-time intensity plots for RR and RL, compared with RR and RL for the same data, with a wind speed of 6–10 kn (the time axis is vertical and displays 7 s, starting from the top; The range axis is horizontal and displays 256 range cells, i.e. 384 m)*

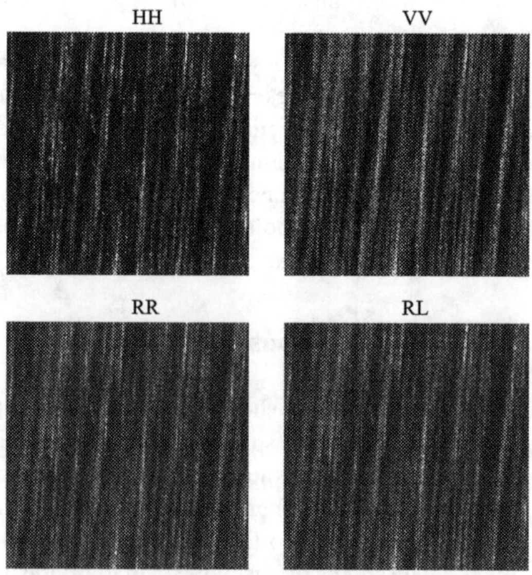

Figure 2.13 *Range-time intensity plots for RR and RL, compared with RR and RL for the same data, with a wind speed of 18–20 kn (the time axis is vertical and displays 7 s, starting from the top; The range axis is horizontal, and displays 256 range cells, i.e. 384 m)*

fluctuate much within that time. Longer spikes, which we will refer to as 'whitecaps', last for a second or so and have a noise-like fluctuation that decorrelates in a few milliseconds (at a fixed radar frequency). Thus we have three distinct features and may consider potential mechanisms for their origin [10].

The smooth modulations seem to arise from Bragg scattering, which is associated with resonant capillary waves. This gives rise to many scattering regions within the radar resolution cell, which account for the speckle (decorrelated by frequency agility) and local Gaussian statistics. The backscatter is modulated by longer waves and sea swell; these alter the local slope of the sea surface, modulate the short wave amplitudes (via long wave advection currents) and cause shadowing and multipath illumination. This type of scattering has a much higher radar cross-section for VV polarisation than for HH polarisation. The Doppler spectrum (see Chapter 4, Figure 4.12) is quite broad, approximately symmetrical and centred close to zero frequency.

Burst scattering appears to arise from the crests of waves, just before they spill. This gives rise to a specular reflection of short duration (typically about 200 ms), with a strong cross-section with HH polarisation, but much smaller cross-section with VV polarisation. The polarisation ratio seems to be caused by constructive multipath interference at HH (between the direct illumination and forward scattering from the sea surface in front of the wave). At VV the forward scattering is suppressed by the Brewster effect, and is subject to a different phase change from HH. Bursts have a Doppler shift consistent with the velocity of the wave top and a narrow Doppler spectrum (see Chapter 4, Figure 4.14). They are not decorrelated by frequency agility due to their short range-extent.

Finally, whitecap scattering seems to arise from the very rough surfaces of waves as they break. (The term whitecap is used to describe the visual occurrence of a whitecap on the sea coincident with these events). These spikes are more noise-like than specular burst spikes and have a broader Doppler spectrum (Chapter 4, Figure 4.16). Nevertheless, some are restricted in range compared with the radar range resolution and so may not decorrelate from pulse to pulse with frequency agility. They have similar radar cross-sections for VV and HH polarisations due to the diffuse scattering from very rough surfaces.

Figure 2.14 illustrates these three features from recordings of real data, using the same radar as for Figure 2.10, which employed pulse-to-pulse frequency agility. The plots show two examples of returns from single range cells; the time histories of simultaneous returns from VV and HH are compared. In Figure 2.14(a), the overall behaviour of the two polarisations is similar. However, the detailed structure is different, indicating that different scatterers are contributing. This is an example of whitecap scattering. An example of another area of the record, which does not contain dominant HH spikes is shown in Figure 2.14(b). Here there are occasional large amplitudes in the HH record. These 'bursts' are highly polarisation sensitive and appear to be discrete in nature because of the lack of fluctuation in their time history. The effect of the bursts on the average radar cross-section and the distribution appears to be small compared to the whitecaps of Figure 2.14(a). These plots also show the modulated Bragg scattering. We see rapid fluctuations in both VV and HH polarisations, but VV returns have a much higher amplitude than the HH returns. (Comparing

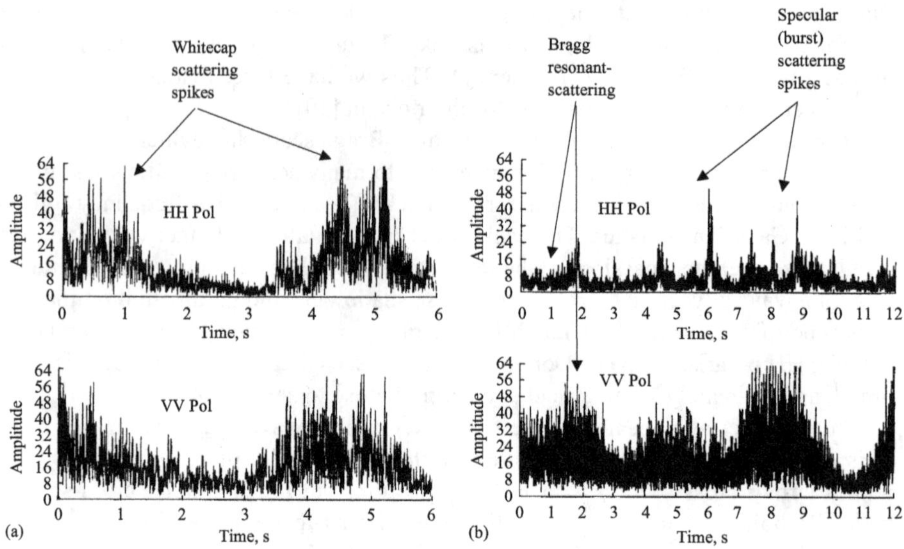

Figure 2.14 Pairs of simultaneous pulse-by-pulse time histories of a range clutter cell with VV and HH polarisations (using pulse-to-pulse frequency agility; sea state 3)

the two VV traces one may conclude that they both show the same type of scattering. However, comparison of the Doppler spectra reveals the presence, in the left-hand plot, of whitecap scattering).

A more detailed discussion of the scattering mechanisms that contribute to clutter spikes can be found in Chapter 3 and their incorporation into clutter models is described in Chapter 4. There is also work in the literature relating to the observations of sea spikes in data. Early investigations of clutter spikes by Olin and Ewell *et al.* may be found in References 11 and 12. Werle [13] used 2D FFTs of range-time intensity plots from a staring radar (ω–k plots – see Section 2.10.2) to investigate the presence of wave-group processes contributing to dominant radar scattering wavelengths. Posner [14–16] has investigated the spatial and temporal characteristics of sea clutter spikes for very low grazing angle clutter, with both VV and HH polarisation.

2.9 Coherent properties of radar sea clutter

Coherent radars employing Doppler processing can distinguish targets from clutter if the target's radial velocity is sufficiently high. However, some targets of interest will have Doppler shifts that are not significantly different from the Doppler spectrum of the clutter. In these cases, it is important for the radar designer to have a detailed understanding of the characteristics of the Doppler spectrum for all the different conditions likely to be encountered. In this section we consider the properties of coherent high-range-resolution radar sea clutter with reference to those properties

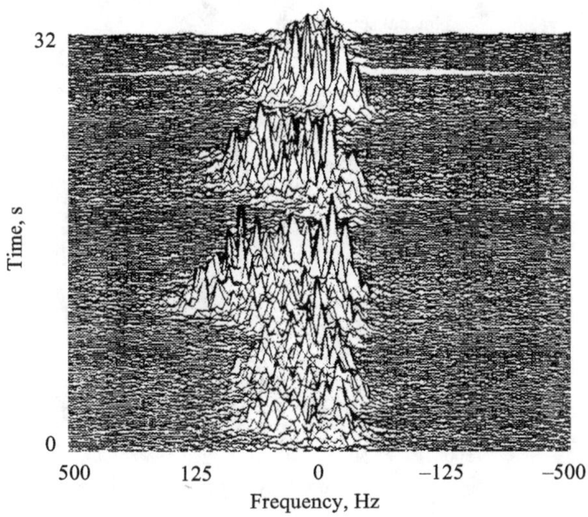

Figure 2.15 Time history of sea clutter Doppler spectra, from a single range cell

observed non-coherently, as described in the earlier sections of this chapter. The mathematical modelling of sea clutter Doppler spectra is discussed in Chapter 4.

An example of a typical Doppler spectrum for sea clutter is shown in Figure 2.15, which is a time history of the coherent spectrum from a single range cell, obtained from an I-band radar with a pulse length of 28 ns and an azimuth beamwidth of 1.2°. Each spectrum is generated from a 128-point Fourier transform. The polarisation is vertical, the range is approximately 2.5 km and the sea state was observed to be 2. The modulation of intensity is evident as changes in the integrated power of the spectrum as a function of time in Figure 2.15. However, an important point to note is that the normalised form of the spectrum is not constant, but has a changing shape and offset. Between approximately 11 and 12 s into the time history the spectra exhibit a particularly high Doppler frequency content, which is thought to be a result of local wind gusting effects. This highlights the complexity of the relationship between the intensity modulation and the form of the spectrum, the former being dominated by the swell structure in the sea surface and the latter being additionally affected by the local gusting of the wind and, as discussed above, the detailed scattering mechanism. However, despite this complexity it should be noted that the compound modulated Gaussian process is still applicable in the spectral domain and has as direct an effect on the performance of coherent radars as it does on non-coherent radars.

The non-Gaussian nature of the sea clutter spectra is illustrated in Figure 2.16. Here the normalised second intensity moments, $\langle z^n \rangle / \langle z \rangle^n$, (see Chapter 4) of the Doppler component intensities (corresponding to those of Figure 2.15) are plotted against frequency and compared to the mean spectrum amplitude. As described in Chapter 4, we would expect the normalised second intensity moment to have a value of 2 in Gaussian clutter or noise (see equation (4.18)). Values higher than 2 are indicative of clutter with non-Gaussian statistics, with the most spiky clutter having the largest

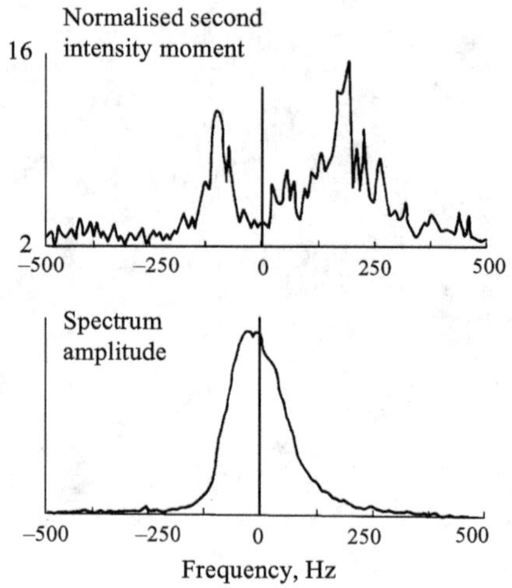

Figure 2.16 Comparison of sea clutter spectrum, averaged over many time periods, with the normalised second intensity moment for each frequency cell as derived from the sample of time periods, for vertical polarisation

values. As expected, at the extremes of the frequency span the power is dominated by radar noise because of the relatively small amount of backscatter at these frequencies (or radial velocities). Where there is significant power in the radar return the second moment is larger. This signifies a non-Gaussian distribution and is due to a modulation caused by changes in the average intensity, spectral shape and offset. Further, the value of the second intensity moment changes across the clutter spectrum, indicating that a more sophisticated model than that of filtered Gaussian noise modulated by a gamma distribution is required to describe these observations. Figure 2.16 also shows the time-averaged spectrum to be somewhat asymmetrical, with a bias towards positive Doppler frequencies. This bias is associated with the direction of the prevailing wind and the largest values of normalised second intensity moments occur predominantly on the side of the spectrum where the biasing is most pronounced. However, the value of the normalised second intensity moments also takes large values in the opposite wings of the spectrum. This complicates the detection problem still further as the clutter Doppler cells with the least power, where the detection of targets might be expected to occur, have the 'spikiest' signals.

Figure 2.17 shows a plot of the same type as Figure 2.16 with the exception that the measurement was made approximately 30 s earlier and that the transmit and receive polarisations were horizontal; otherwise the experimental conditions were identical. Clearly the plots are very similar in terms of the shape, bandwidth and distribution of the normalised second intensity moment although the values of the latter are generally higher for horizontal polarisation. In addition, the mean Doppler offset of the averaged spectrum is significantly different for the two polarisations. The dependency of the Doppler offset of the averaged spectrum, for both horizontal

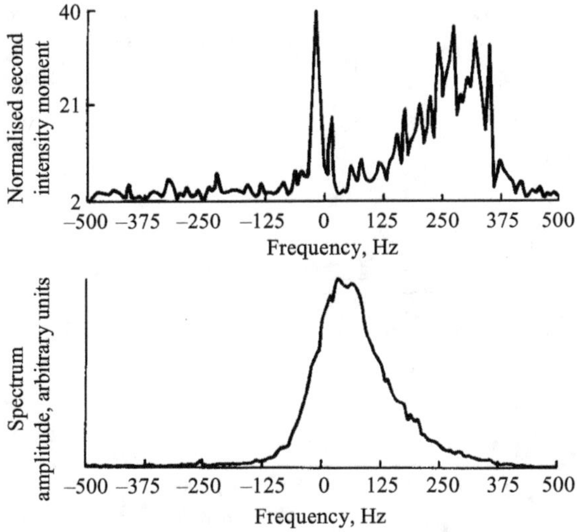

Figure 2.17 *Comparison of sea clutter spectrum, averaged over many time periods, with the normalised second intensity moment for each frequency cell as derived from the sample of time periods, for horizontal polarisation*

and vertical polarisations, on the direction of look of the radar with respect to the sea surface is examined in more detail in Figure 2.18. Here it is shown that data for both polarisations show a cosinusoidal dependence on the direction of the wind, with a zero Doppler offset when looking across-wind. In all cases, except that of cross-wind, the Doppler offset is larger for horizontal polarisation than vertical, suggesting that a different set of scatterers are contributing to the received signals. The characteristics described above are in agreement with observations reported elsewhere [17–19]. Further examples of sea clutter Doppler spectra are shown in Figure 2.19 [20]. These were collected on the same occasion, for slightly different look directions relative to the wind. The sea was very rough with 6 m waves and a 32 kn wind. Figure 2.19(a) shows the effect of wind-driven spray on top of the spatially varying spectral shape. Figure 2.19(b) shows a more slowly varying spectrum. In both cases, large variations in spectrum shape and intensity are observed over the time period of the observation.

The autocorrelation functions for a number of Doppler frequency components from the data in Figure 2.15 are shown in Figure 2.20. Initially in each case there is a fast drop-off, which is followed by a slower periodic decay. The initial fast decay can be associated with the speckle component and the slower periodic decay may be associated with the modulation. Similar effects on performance to those observed with non-coherent detection are therefore expected for high resolution coherent processing. It is thus clear that the statistical and correlation properties of range-Doppler cells must be taken into account in the design of coherent radar processors operating in a sea clutter environment.

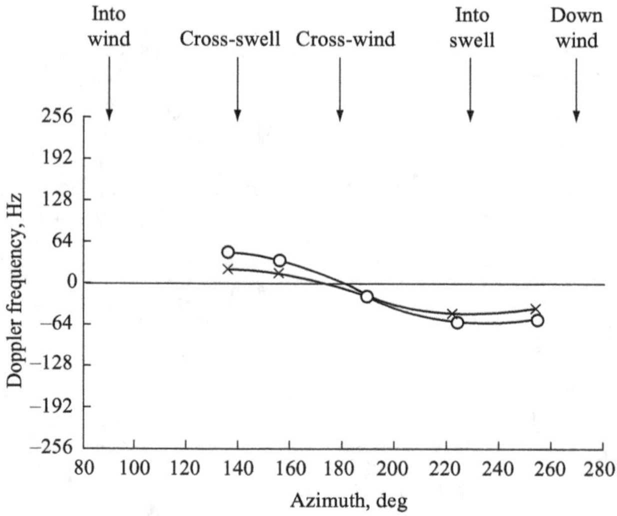

Figure 2.18 *Plot showing the trend of mean Doppler frequency against azimuth angle* o*: horizontal polarisation;* ×*: vertical polarisation*

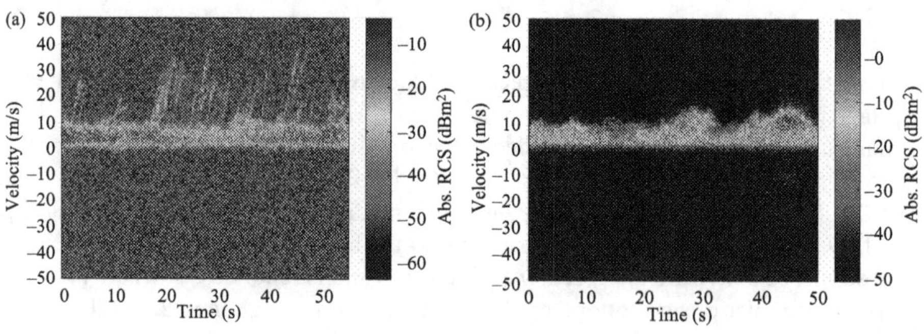

Figure 2.19 *Clutter spectra, showing temporal variation in intensity and the effects of wind-driven spray; radar frequency 3 GHz, VV polarisation, wind speed 32 kn; wave height 6 m (Figures courtesy of QinetiQ). (a) Radar looking at 45° to wind direction and (b) Radar looking upwind*

2.10 Spatial characteristics

It will be seen in Chapters 8 and 9 that the spatial variation of clutter characteristics can have a significant effect on the performance of radar detection processing. For a Gaussian process the full statistical properties are characterised by the first- and second-order statistics. The many-point statistics may then be derived from these two. For a non-Gaussian process this is not always, and is in fact rarely, the case. It is therefore inappropriate to base clutter models on just the amplitude distribution and autocorrelation function or spectrum. The compound formulation for clutter

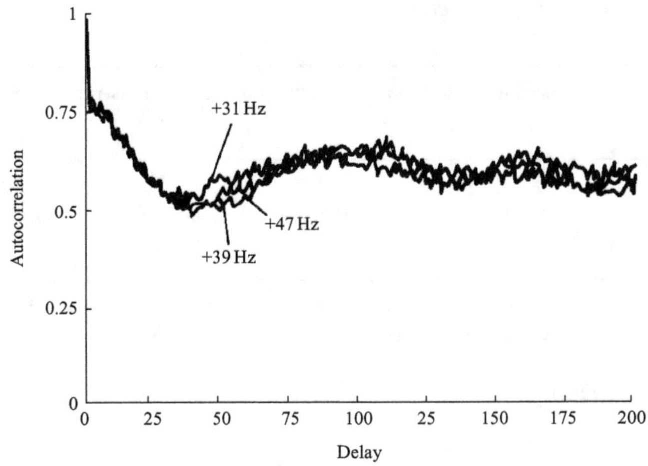

Figure 2.20 Temporal autocorrelation functions of different Doppler frequency components, taken from individual range cells, plotted against delay measured in units of spectral sampling time (128 ms)

amplitude statistics identifies the process as the product of two random variables. This cannot be deduced from the autocorrelation function or spectrum. The model also shows that the spatial and long-term temporal correlation characteristics are solely dependent on the modulating underlying intensity component. Considering this component in detail it is shown here that approximating this component from its first- and second-order statistics alone provides a reasonable and useful fit to more complex models and real data. This, it must be emphasised, is not the same as approximating the overall clutter return from the two statistics.

2.10.1 Range ACF

The range resolution of real aperture radars is often several orders of magnitude better than the cross-range resolution. The spatial variations of sea clutter are therefore most readily observed as variations with range. The results from three typical recordings of real clutter are presented in Reference 21. The basic radar parameters are the same as those described in Section 2.4.1. The specific parameters relating to the data to be discussed here are given in Table 2.3.

The data consist of a series of range profiles of the envelope of the clutter return taken from successive pulses of a frequency-agile radar. Each profile consists of approximately 250 range samples, with a sampling interval of 2.6 m in range. The radar range resolution is 4.2 m. For the purposes of the present analysis, each point in range is averaged over 100 successive pulses to remove the speckle component of the clutter and yield profiles of the mean amplitude. These data are then squared to provide profiles of the clutter mean intensity x_i . Some typical plots of these range profiles of the clutter reflectivity are shown in Figure 2.21.

Table 2.3 Data file parameters

File number	Polarisation	Direction	Grazing angle, deg	Across-range resolution, m	Sea state	ν
1	H	D	0.9	720	4/5	1.0
2	V	U	0.7	750	5	5.0
3	V	U	0.3	300	3	0.5

H = horizontal polarisation; V = vertical polarisation; D = radar looking down-swell; and U = radar looking up-swell; ν = K distribution shape parameter (see Chapter 4)

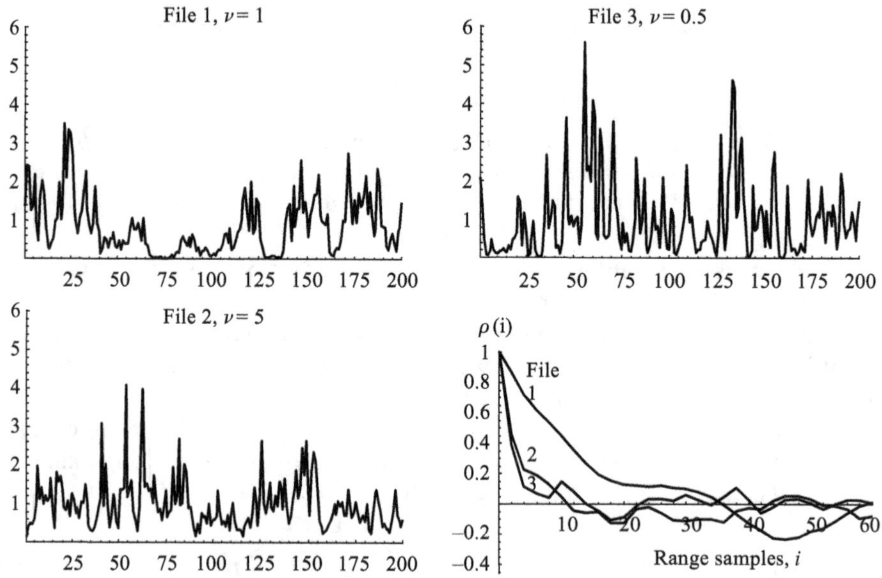

Figure 2.21 Recorded data exhibiting different spatial correlations

The range ACF of x, R_j, is estimated for a number of profiles derived from each file using

$$R_j = \sum_{i=1}^{M/2} (x_i - \hat{m})(x_{i+j} - \hat{m}) \tag{2.3}$$

where $\hat{m} = \frac{1}{M} \sum_{i=1}^{M} x_i$ and M is the number of range samples in each profile.

The range correlation coefficient ρ_j is then given by

$$\rho_j = \frac{R_j}{R_0} \tag{2.4}$$

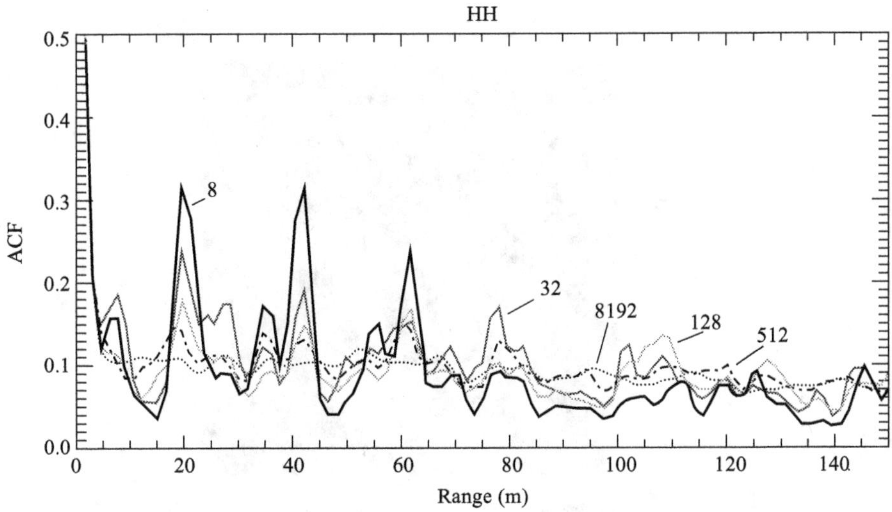

Figure 2.22 *RCS range ACF of sea state 1 data, averaged over 8 to 8192 pulses and showing evidence of transient coherence: black line – 8, dark grey – 32, light grey – 128, dashed dotted – 512, dotted – 8192*

The final values of ρ_j used for each file are obtained by averaging equation (2.4) over groups of 100 successive profiles of x. The first 40 terms of the averaged range correlation coefficient for the three data files are shown in Figure 2.21. It can be seen that the form of ρ_j is different in each case. File 1 contains clutter with a very long periodic fluctuation due to a heavy sea swell, while at the other extreme File 3 represents very spiky clutter with little spatial correlation.

The results in Figure 2.21 are obtained by averaging over relatively short intervals of time (about 0.1 s) and range (about 750 m). Other work [22] has looked at the longer-term spatial and temporal coherence of clutter. Figure 2.22 shows the range ACF of HH polarised clutter data for different degrees of averaging, ranging from 8 pulses through to 8192 pulses; the PRF is 1 kHz. It can be seen that over short time periods the data exhibit periodic behaviour, in a similar manner to the data in Figure 2.21. However, when the data are averaged over longer time periods (e.g. over up to 8 s in Figure 2.22) it can be seen that the periodic behaviour is no longer apparent. This is typical of many data records. It is suggested in Reference 22 that this is due to the existence of transient coherent structures in the clutter. These are evident from visual inspection of data and from the formation of autocorrelation functions determined from records whose duration is comparable with the lifetime of these features. Averaging over significantly longer times yields ACFs that do not exhibit periodic behaviour.

In Chapter 3, the properties of sea clutter are examined in terms of the electromagnetic scattering from hydrodynamic models of the sea surface. Long-term range ACFs that do not exhibit periodic behaviour are reasonably well described by models based on power spectra derived from equilibrated oceanographic models [22].

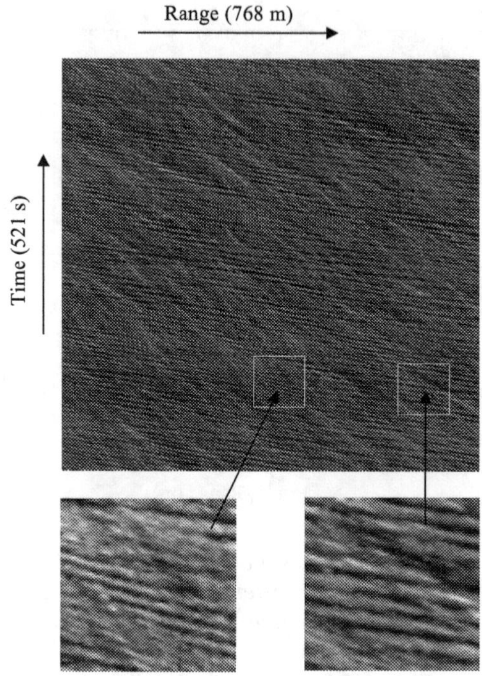

Figure 2.23 Range-time intensity presentation of a sea clutter record (sea state 4; 512 s, 512 1.5 m range cells), with enlargements of two subsets of the data

2.10.2 Power spectrum analysis of range-time intensity plots

Considerable insight into the time evolution of sea clutter can be obtained from range-time intensity plots, similar to those presented in Sections 2.6 and 2.7, that display the variation of intensity in a set of range gates over rather longer periods of time. In these one sees explicitly both the spatial variation in the clutter returns along one dimension, and their development over a time scale in which significant changes in the local correlation structure can be seen. As a consequence of the bulk motion of the sea these plots caricature the two-dimensional (2D) spatial variation of the clutter. Figure 2.23 shows a typical data set of this type; the clutter intensity variation in 512, 1.5 m range cells (horizontal axis) over a period of 512 s (vertical axis) as measured in sea state 4 by a stationary radar.

The seemingly non-stationary character of spatial frequency of the clutter within the range cells are evident in Figure 2.23; two rather different frequencies are highlighted in the enlarged snapshots of subsets of the data. The power spectrum and correlation function of a stochastic process are related through the Wiener Khintchine theorem. Figure 2.24 shows the amplitude of the 2D Fourier transform of the data presented in Figure 2.23.

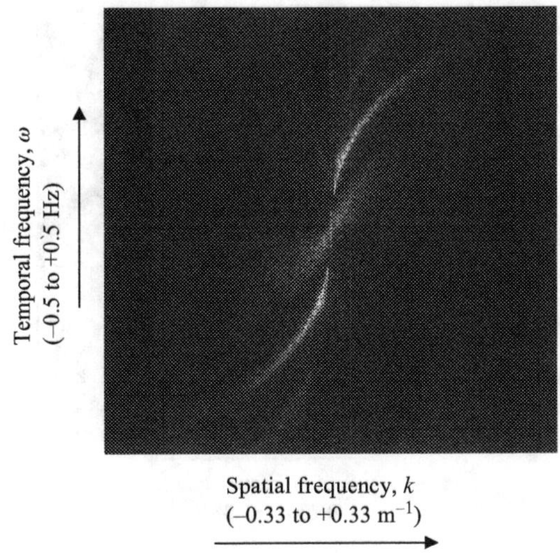

Figure 2.24 *Fourier transformation of the data shown in Figure 2.23. Temporal frequencies over the range -0.5 to 0.5 Hz are plotted vertically, the horizontal axis displays spatial frequencies from -0.33 to 0.33 m^{-1}*

The gross features of this plot can be understood in relatively simple terms. The sea surface structure is itself a 2D random field, as is its Fourier transform. The low across-range resolution of the radar effectively filters out all Fourier components other than those in the direction of the line of sight; thus we may identify the spatial frequencies displayed in Figure 2.24 with the selected wavenumbers k. The spatial and temporal frequencies of a wave motion are related through its characteristic dispersion relationship; in the case of gravity waves this takes the simple form

$$\omega = \pm\sqrt{gk} \tag{2.5}$$

and is evident in the approximately parabolic shape of the ω–k plot shown in Figure 2.24. One branch of (2.5) is much more prominent than the other; this is a consequence of the directionality of the wave motion (either towards or from the radar). If the sea surface dynamics were linear the ω–k plot would merely map out the relationship (2.5); the broadening of these lines, and the presence of a discernible feature in the region of the origin, are consequences of non-linear interactions in both the wave dynamics and the imaging process itself. Thus, even at this crude level, a simple Fourier transform is able to capture signatures of the non-linearity that may be responsible for the apparently non-stationary behaviour of the clutter. To examine these signatures in more detail we present the data shown in Figure 2.24 on a logarithmic intensity scale, in Figure 2.25 .

Rather more features of the ω–k plot are discernible in Figure 2.25, the principal among these are labelled (1–5). Feature (1) is the 'parabolic' gravity wave dispersion line observed in Figure 2.24; this is associated with advancing waves, moving towards the radar. The extent to which this dispersion 'line' is broadened is also more evident in

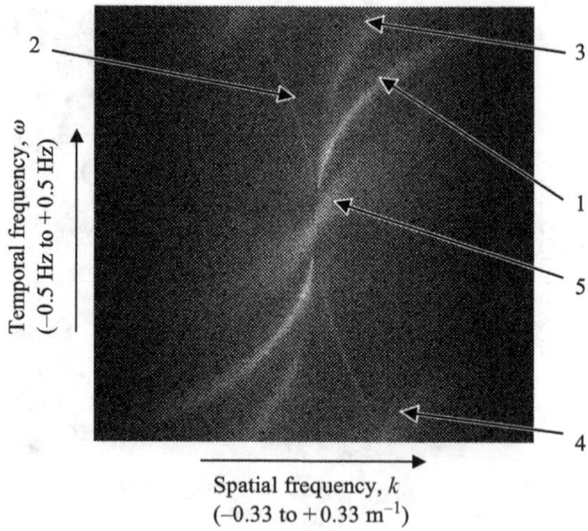

Figure 2.25 Data shown in Figure 2.9, plotted on a log intensity scale. Salient features are numbered 1–5 (see text)

this plot. The companion dispersion line, associated with waves travelling away from the radar, is identified with feature (2); we note an asymmetry in both the intensities and shapes of the advancing and receding wave lines. The latter can be understood fairly simply in terms of a 'Doppler shift' in the temporal frequency as a result of tidal currents [23] and wind drifts producing an overall sea surface velocity v along the line of sight of the radar. As a consequence the dispersion relation is modified and now takes the form

$$\omega = kv \pm \sqrt{kg} \tag{2.6}$$

Thus we see two distinct branches, which can be identified with the advancing and receding waves. In fact it is possible to obtain a satisfactory fit to the ω–k lines in Figure 2.25 using equation (2.6), and deduce that a sea surface velocity of 0.8 ms^{-1} is manifest in this data set.

The remaining features in Figure 2.25 can be identified with signatures of non-linearity. The line (3) is the second harmonic of feature (1); the 'wrap around' behaviour of the discrete Fourier transform induces the alias of this line seen in feature (4). The generation of a second harmonic through a non-linear interaction can be understood in terms of the simple trigonometric identity

$$\cos^2(\omega t) = \frac{1 + \cos(2\omega t)}{2} \tag{2.7}$$

Similarly, the coupling of components with different frequencies yields sum and difference terms

$$\cos(\omega_1 t)\cos(\omega_2 t) = \frac{\cos((\omega_1 + \omega_2)t) + \cos((\omega_1 - \omega_2)t)}{2} \tag{2.8}$$

These latter are evident in feature (5) – the so-called inter-modulation product 'line'.

The features shown in Figure 2.25 can be simulated to reproduce the non-stationary behaviour of the sea clutter. It will be seen in Chapter 9 that it may be possible to exploit this to predict the radar returns from the clutter, for the purposes of setting a target detection threshold.

References

1 F.E. Nathanson, *'Radar design principles'*, McGraw Hill, New York, 1969.
2 M.W. Long, *'Radar reflectivity of land and sea'*, Artech House, Boston, 1983.
3 M.I. Skolnik, *'Radar handbook'*, McGraw Hill, New York, 1970.
4 S.J. Austin, D.J.Q. Carter and J.D. McCaughey, 'ERS-1 spacecraft – active microwave instrumentation: a major achievement', *GEC J. Res.*, 9(1), 2–12, 1991.
5 K.D. Ward, 'Compound representation of high resolution sea clutter', *Electron. Lett.*, **17**(16), 561–563, 1981.
6 K.D. Ward, C.J. Baker and S. Watts, 'Maritime surveillance radar part 1: Radar scattering from the ocean surface', *Proc. IEE*, **137**(part F, No. 2), 51–62, 1990.
7 K.D. Ward, R.J.A. Tough and P.W. Shepherd, 'Modelling sea clutter: correlation, resolution and non-Gaussian statistics', IEE Conference Publication 449, Radar 97, pp. 95–99.
8 S. Watts and G. Knight, 'Performance prediction for scanning radars with correlated returns from K-distributed sea clutter', Radar 2004 International Conference on Radar Systems, Toulouse 19–21 October 2004.
9 T. Hair, T. Lee and C.J. Baker, 'Statistical properties of multifrequency high-range-resolution sea reflections', *Proc. IEE*, **138**(part F, No. 2), 75–79, 1991.
10 T. Lamont-Smith, K.D. Ward and D. Walker, 'A comparison of EM scattering results and radar sea clutter', IEE Radar 2002 International Conference, October 2002, IEE Publication 490, pp. 439–443.
11 I.D. Olin, 'Characterisation of spiky sea clutter for target detection', Proceedings of the 1984 IEEE National Radar Conference, 1984, pp. 27–31.
12 G.W. Ewell, M.T. Tuley and W.F. Horne,' Temporal and spatial behaviour of high resolution sea clutter spikes', Proceedings of the 1984 IEEE National Radar Conference, 1984, pp. 100–104.
13 B.O. Werle, 'Sea backscatter, spikes and wave group observations at low grazing angles', IEEE 1995 International Radar Conference, Washington DC, May 1995, pp. 187–195.
14 F.L. Posner, 'Distribution function for temporal measures of sea spike characterization', International Conference Radar 2004, Toulouse France, October 2004, paper 3B-PHENO-3.
15 F.L. Posner, 'Spiky sea clutter at high range resolutions and very low grazing angles', *IEEE Trans. AES*, **38**(1), 58–73, 2002.
16 F.L. Posner and K. Gerlach, 'Sea spike demographics at high range resolutions and very low grazing angles', Proceedings of the 2003 IEEE Radar Conference, Huntsville, AL, May 2003, pp. 38–45.

17 C.J. Baker, 'K-distributed radar sea clutter', *Proc. IEE*, **138**(part F, No. 2), 89–92, 1991.
18 D.G. Money *et al.*, 'Sea clutter power spectral lineshape measurements', Radar 97, 140–16 October 1997, IEE Publication No. 449, pp. 85–89.
19 R.J. Miller and W.N. Dawber, 'Analysis of spectrum variability in sea clutter', Radar 2002, 15–17 October 2002, IEE Conference Publication No. 490, pp. 444–448.
20 W.N. Dawber and N.M. Harwood, 'Comparison of Doppler clutter cancellation techniques for naval multi-function radars', Radar 2002, 15–17 October 2002, IEE Publication No. 490, pp. 424–428.
21 S. Watts and K.D. Ward, 'Spatial correlation in K-distributed Sea Clutter', *Proc. IEE* Part F, **134**(6), 526–532, 1987.
22 R.J.A. Tough, K.D. Ward and P.W. Shepherd, 'The modelling and exploitation of spatial correlation in spiky sea clutter', EMRS DTC 2nd Annual Conference, 16–17 June 2005, paper A1.
23 T. Lamont Smith, 'Radar Imaging of Strain Rate', IEE Int Radar Conference RADAR97, IEE publication 449, pp. 80–84, 1997.

Chapter 3

Modelling radar scattering by the ocean surface

3.1 Introduction

In the previous chapter we considered the phenomenology of sea clutter and its impact on the operation of microwave radar systems. To exploit this knowledge and improve radar performance, we need to understand the underlying physical mechanisms responsible for these clutter properties, so that they can be modelled realistically. At first sight this might seem, in principle at least, to be a simple matter particularly if use is made of a computer. The fundamental physics of the generation and transmission of microwaves, their interaction with the ocean surface and scattering to the radar receiver is well understood. The ocean and atmosphere have been subject to intensive study for at least a century; the underlying laws of fluid motion have been known for much longer. Can clutter modelling be any more than a matter of assembling these constituent parts and turning the handle?

In practice this optimism turns out to be ill-founded. While the mathematical formulation of the underlying physics of clutter generation has been understood for many years, that period has not seen the development of generally effective techniques for the solution of the equations emerging from this framework. For example, the computation of scattering from extended objects at close to grazing incidence presents problems that have only just begun to be addressed effectively in the last decade, and yet calculations of this kind must lie at the heart of any *a priori* modelling of sea clutter. Attempts to develop approximate analytical models of scattering by other than slightly rough planar surfaces again encounter serious difficulties in the low grazing angle regime. The direct modelling of the sea surface itself with structure manifest on many length scales, ranging from the tiniest capillary ripples through crashing breakers, to large scale fully developed swell, has also proved to be extremely intractable. Furthermore, even if these problems were tractable, their detailed solution would provide us with an abundance of information, from which it would be virtually impossible to extract useful insights into radar performance.

If however we are more modest in our expectations, significant and useful progress can be made with this direct modelling approach. For example, a quantitative understanding of the variation of the mean clutter power with environmental and operational conditions can be achieved, while detailed modelling of specific features of both the sea surface and EM scattering provides real insight into the origin of sea spike clutter features. In this chapter we will describe this less ambitious program, highlighting its underlying physical and methodological principles, its successes and its limitations.

There are many books and papers on electromagnetism, and there is not room here to present all of the theory and derivations underlying rough surface scattering results. However, in order to help the reader to be able to understand, extend and apply the techniques, we provide sufficient detail to allow the numerical results to be reproduced. This is achieved by excluding most of the detail relating to the vector derivations of scattering from two-dimensional (2D) sea surfaces, whilst providing a more complete scalar analysis of scattering from one-dimensional (1D), corrugated surfaces (much of this material is in Appendix 3). The approach is justified for two reasons: first the main physical effects are present in both the 1D and 2D cases; and second the scalar derivations demonstrate all of the main analysis techniques without obscuring the development with vector notation. Once the reader has mastered the analysis in this chapter and Appendix 3, papers in the scattering literature should be readily accessible.

Following this introduction there is a Section 3.2, which describes features of the sea surface useful for modelling radar scattering. The remaining material in the chapter is then divided into radar grazing angle[1] regimes; high (90° to about 60°), medium (about 60–15°), and low (about 15–0°). Section 3.3 introduces electromagnetic scattering theory and shows how the simple approximation of 'physical optics' scattering from a perfectly conducting (PC) rough surface may be used to model radar scattering at high grazing angles. This is illustrated in Section 3.4, where the model is applied to satellite radar imaging of sea bottom topography. The modulations on the radar imagery appear to match the shape of the sea bottom, rather like the image in Figure 2.4; this is not due to the radar 'seeing' through the sea, but to tidal current variation with sea depth. In Section 3.5 EM scattering theory is extended to the composite model for scattering from an imperfectly conducting dielectric rough surface. In contrast to Physical Optics, the EM polarisation sensitivity of the composite model is able to explain the difference between vertical (V) and horizontal (H) polarisations at medium grazing angles. However, at low grazing angles there are many characteristics, as described in Chapter 2, which are not consistent with the composite model. Section 3.6 therefore considers extending low grazing angle (LGA) scattering theory beyond the composite model. The principal LGA clutter phenomena of multipath interference and scattering from breaking waves (and their

[1] The grazing angle is the angle between the mean sea surface plane and the radar illumination wave vector, measured in the plane of incidence. Sometimes an alternative term, the 'angle of incidence', is used. This is the angle between the average sea surface normal vector and the radar wave vector. The angle of incidence is thus the grazing angle minus 90°.

manifestation in radar 'sea spikes') may be understood from scattering off corrugated (1D) surfaces. Thus it is here that we show how for corrugated surfaces vector EM scattering simplifies to scalar scattering. We introduce numerical methods for calculating the backscatter from rough surfaces at LGA and compare the accuracy of various techniques for idealised surfaces. In Section 3.7 the radar cross-sections (RCS) of spikes and the effect of multipath are calculated for realistic surfaces. These are put together in Section 3.8 with a model for the occurrence of breaking waves to derive trends for the average RCS of sea clutter versus grazing angle and sea state. It is shown that the results are well matched with a wide range of experimental LGA data. Finally, in Section 3.9, we return to the remote sensing problem of Section 3.4 and illustrate how the imaging of tidal flow over bottom topography is affected by LGA scattering.

3.2 The sea surface

When modelling the sea surface one faces conflicting requirements of realism and tractability. The hydrodynamic description of the sea surface and its interaction with the wind is a problem of such complexity that its complete solution is essentially impossible; the detail required to both define the problem (through appropriate boundary conditions) and specify its solution would be overwhelming. To make progress towards a physically useful model, simplifying assumptions must be made.

The most common approximation is to linearise the hydrodynamic equations for surface waves and to obtain the solutions of small amplitude, sinusoidal waves. As the system is approximated to be linear the overall sea structure may be modelled as the superposition of many such waves generated by the interaction of the wind and sea over an area of much greater dimensions than the correlation lengths typical of the sea surface. Thus we would expect Gaussian statistics, as a consequence of the central limit theorem. Experimental measurements of the sea surface slope and height, for example, approximate well to a Gaussian form.

The key property that defines a Gaussian process is the average power spectrum. This expresses how the wave energy is distributed (on average) across spatial and temporal frequencies. There are many experimentally derived models for the average power spectrum for the sea. A widely used formula based on the Pierson-Moskowitz [1] spectrum is:

$$S_\eta(\mathbf{q},\omega) = \frac{b}{4\pi} q^{-4} \exp\left(-0.6\left(\frac{g}{qU_{10}^2}\right)^2\right) \cos^{2n}\left(\frac{\theta_w}{2}\right) \frac{\left(2n\right)!!}{\left(2n-1\right)!!} \delta(\omega - \sqrt{gq}) \tag{3.1}$$

where

$\mathbf{q}$ is the spatial frequency vector (and q is its magnitude),
ω is the temporal frequency,

b is the Phillips equilibrium parameter (taken here to be 0.0081),
g is the gravitational acceleration (9.8 ms^{-2}),
U_{10} is the wind speed (ms^{-1}) measured at 10 m above the sea surface,
θ_{w} is the angle between $\mathbf{q}$ and the wind direction,
n is an angular spreading parameter (we use $n = 4$), $(2n)!! = 2n(2n-2)\times(2n-4)\ldots 2$ and $(2n-1)!! = (2n-1)(2n-3)\ldots 1$.

This spectrum is only really suitable for waves down to about 5 cm wavelength, because it only describes gravity waves and not the shorter capillary waves.[2] The δ function imposes the gravity wave dispersion relation (i.e. relates the wavelength to the wave speed and frequency). It can be removed by integrating over ω to give

$$S_\eta(\mathbf{q}) = \frac{b}{4\pi} q^{-4} \exp\left(-0.6\left(\frac{g}{qU_{10}^2}\right)^2\right)\cos^{2n}\left(\frac{\theta_{\mathrm{w}}}{2}\right)\frac{(2n)!!}{(2n-1)!!} \tag{3.2}$$

along with the additional constraint (the dispersion relation)

$$\omega(q) = \sqrt{gq} \tag{3.3}$$

The mean square height deviation of the sea from its mean level is given by the two-dimensional integral of the average power spectrum over q. Thus,

$$\sigma_h^2 = \langle \eta^2 \rangle = \int d^2q\, S_\eta(\mathbf{q}) \tag{3.4}$$

Integrating equation (3.2) over the angular dependence, θ, gives the so-called omnidirectional elevation spectrum,

$$S_\eta(q) = \frac{b}{2} q^{-3} \exp\left(-0.6\left(\frac{g}{qU_{10}^2}\right)^2\right) \tag{3.5}$$

This is shown in Figure 3.1 for wind speeds from 3 to 21 ms^{-1}, along with the so-called saturation spectrum, which is defined as

$$B_\eta(q) = S_\eta(q)\, q^3 \tag{3.6}$$

The Pierson-Moskowitz saturation spectrum is effectively constant except for a low q (long wavelength) cut-off, which limits the longest wave phase velocity to about the wind speed. The saturation spectrum being constant implies that waves of all wavelengths are the same shape.

An example simulation of the sea surface at sea state 3 (wind speed 7 ms^{-1}) is shown in Figure 3.2. The images are at a resolution of 0.5 m and show a spatial extent of 256 m by 256 m. The slope image (on the right) looks the most like an optical image of the sea because optical reflection depends on slope. The root mean

[2] Gravity waves are so named because gravity provides the restoring force that makes the waves oscillate and propagate. For capillary waves the restoring force is surface tension.

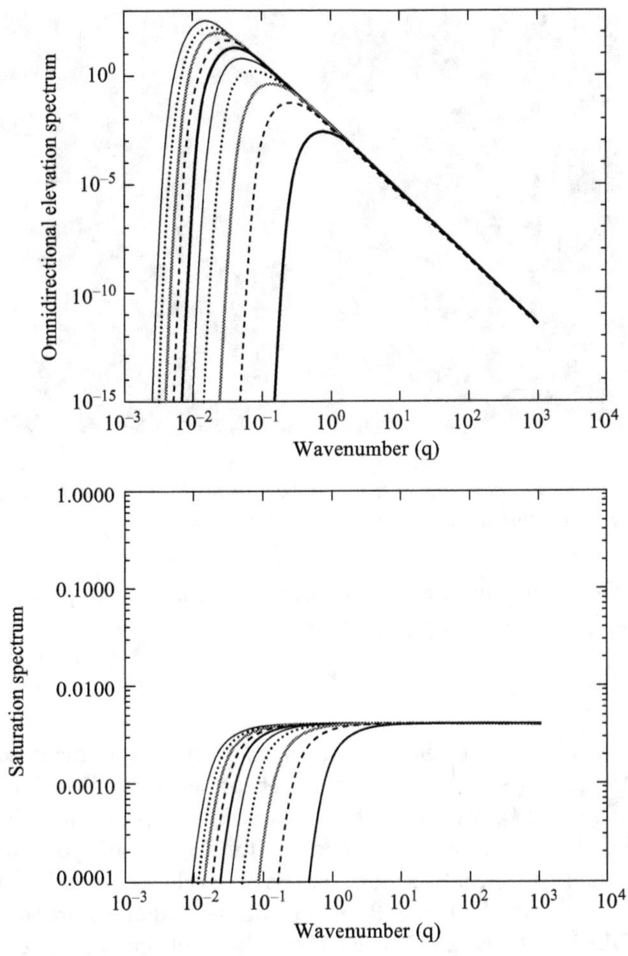

Figure 3.1 *Omnidirectional and saturation elevation spectra derived from equations (3.5) and (3.6) for wind speeds 3 ms^{-1} (right-most curve) to 21 ms^{-1} in steps of 2 ms^{-1}*

square (rms) height is 0.3 m, which corresponds to the value give by Skolnik [2] for sea state 3.

In order to introduce capillary waves, the dispersion relation is altered from equation (3.3) to

$$\omega^2(q) = gq\left(1 + \left(\frac{q}{k_\mathrm{m}}\right)^2\right) \tag{3.7}$$

$$k_m = \sqrt{\frac{\rho_\mathrm{w} g}{T}} = 370 \text{ rad m}^{-1}$$

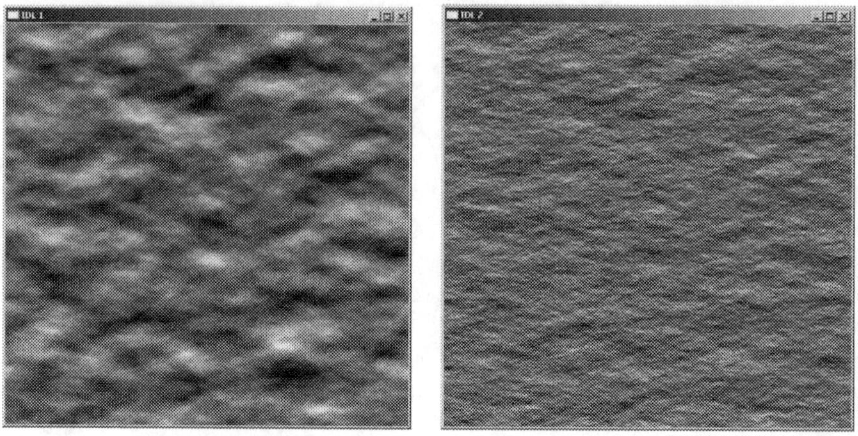

Figure 3.2 A sea surface simulation for sea state 3. Window 1 (on the left) is the height and Window 2 (on the right) is the slope

where ρ_{w} is the water density and T is the surface tension. The phase velocity of the waves is

$$c(q) = \frac{\omega(q)}{q} \tag{3.8}$$

and k_{m} is the wavenumber of the slowest waves. Recent experimental evidence from Jähne and Riemer [3] shows that, in contrast to the Pierson-Moskowitz spectrum, wavenumbers around k_{m} are modulated by the wind speed. As we will show in later sections of this chapter this feature is important for the modelling of low grazing angle microwave scattering. Recently a model has been developed (by Elfouhaily and others [4]), which incorporates the Jähne–Riemer results and other features discovered since the Pierson-Moskowitz work [1]. The Elfouhaily spectrum is given by

$$S_{\eta}(q) = L_{\mathrm{PM}}(q)\, q^{-3}(B_l(q) + B_h(q)) \tag{3.9}$$

where $L_{\mathrm{PM}}(q)$ is the Pierson-Moskowitz saturation spectrum

$$L_{\mathrm{PM}}(q) = \exp\left\{-\frac{5}{4}\left(\frac{k_{\mathrm{p}}}{q}\right)^2\right\} \tag{3.10}$$

The peak wavenumber is

$$k_{\mathrm{p}} = \frac{g}{c_p^2} \tag{3.11}$$

with corresponding phase speed

$$c_{\mathrm{p}} = \frac{U_{10}}{\Omega} \tag{3.12}$$

Ω is the so-called dimensionless 'wave-age' and is related to the development of the sea. Thus, a value of 0.84 (the fourth root of a half) corresponds to a fully developed sea, a value of one is a mature sea, and values up to five are for young seas. The model includes a relationship between the wave age and fetch, thus allowing seas to be represented in sheltered and littoral environments. Here,

$$\Omega = 0.84 \tanh\left\{-\left(\frac{X}{X_0}\right)^{0.4}\right\}^{-0.75} \tag{3.13}$$

where X is the non-dimensional fetch (i.e. the distance, x, over which the wind has been blowing, measured in units of 2π times the number of wavelengths of waves travelling at the wind speed),

$$X = \frac{gx}{U_{10}^2} \tag{3.14}$$

and X_0 is a constant,

$$X_0 = 2.2 \times 10^4 \tag{3.15}$$

The other terms in equation (3.9) are the low and high wavenumber saturation spectra given by

$$B_l(q) = \frac{\alpha_p}{2}\frac{c_p}{c}F_p(q) \quad \text{and} \quad B_h(q) = \frac{\alpha_m}{2}\frac{c_m}{c}F_m(q) \tag{3.16}$$

Here the minimum wave phase velocity is

$$c_m = \sqrt{\frac{2g}{k_m}} \tag{3.17}$$

The scaling factors are

$$\alpha_p = 0.006\sqrt{\Omega} \tag{3.18}$$

and

$$\begin{aligned} \alpha_m &= 0.01[1 + \log_e(u^*/c_m)] \qquad u^* > c_m \\ &= 0.01[1 + 3\log_e(u^*/c_m)] \quad u^* < c_m \end{aligned} \tag{3.19}$$

u^* in equation (3.19) is the wind friction velocity and relates to the ability of the wind to generate short waves. It is affected by atmospheric stability and is increased when there is an inversion layer near the sea surface. In our work we tend to use the Inoue model [5] for friction velocity (which applies to neutral stability), where

$$U_{10} = \frac{u_*}{40}\log_e\left(\frac{1000}{z(u_*)}\right) \tag{3.20}$$

and

$$z(u_*) = 0.00684u_*^{-1} + 42.8u_*^2 - 0.0443 \tag{3.21}$$

We reduce u^* by a factor 0.8 before substitution into (3.19) in order to match the results presented in Reference 4. The long wave 'side effect' function in equation (3.16) is

$$F_{\rm p}(q) = J_{\rm p}(q) \exp\left\{-\frac{\Omega}{\sqrt{10}}\left(\sqrt{\frac{q}{k_{\rm p}}} - 1\right)\right\} \tag{3.22}$$

Here, $J_{\rm p}$ is the peak enhancement factor introduced by Hasselmann *et al.* [6],

$$J_{\rm p}(q) = \gamma^{\Gamma(q)} \tag{3.23}$$

$$\begin{aligned} \gamma &= 1.7 \quad 0.84 < \Omega < 1 \\ &= 1.7 + 6 \log_e \Omega \quad 1 < \Omega < 5 \end{aligned} \tag{3.24}$$

$$\Gamma(q) = \exp\left(-\frac{\left(\sqrt{q/k_p} - 1\right)^2}{2\sigma^2}\right) \quad \text{and} \quad \sigma = 0.08(1 + 4\Omega^{-3}) \tag{3.25}$$

The short wave 'side effect' function in equation (3.16) is

$$F_m(q) = \exp\left\{-\frac{1}{4}\left[\frac{q}{k_{\rm m}} - 1\right]^2\right\} \tag{3.26}$$

Equations (3.9) to (3.26) make up the Elfouhaily spectrum [4], and as such it is much more complicated than the basic Pierson-Moskowitz model. Most of the constituent parts are the result of experimental data or theoretical considerations, and the others are necessary to maintain overall consistency. The incorporation of wave-age and wind friction velocity allows the model to be adapted to different fetch and atmospheric conditions. Putting everything together, for a mature sea at the same wind speeds as Figure 3.1, results in the omnidirectional elevation and saturation spectra as plotted in Figure 3.3. The main difference from Figure 3.1 (the PM spectrum) is modulation of the short waves by the wind speed.

As discussed at the beginning of this section, many of the salient features of the sea's behaviour cannot be extracted from a linear, Gaussian theory. In particular the breaking of waves that appears to be associated with spike-like returns in low grazing angle radar sea clutter can only be described by a non-linear theory. The detailed modelling of breaking waves is really beyond the scope of this book, but we need to use some results to help us with understanding sea clutter at low grazing angles.

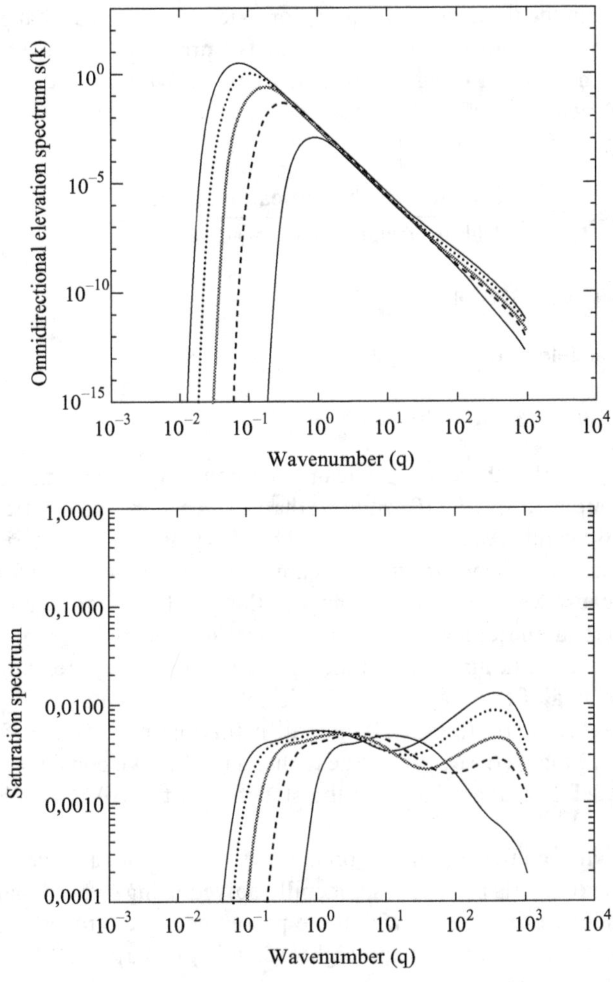

Figure 3.3 *Omnidirectional and saturation elevation spectra derived from the Elfouhaily [4] model for wind speeds 3 ms*$^{-1}$ *(right-most curve) to 21 ms*$^{-1}$ *in steps of 2 ms*$^{-1}$

In the linear modelling we characterise the features of the motion (velocities, accelerations, slopes) of the sea surface statistically as Gaussian random variables with correlation properties derived from established power spectrum models, we now seek to relate these properties to the process of wave breaking. Several criteria for the occurrence of wave breaking have been cited in the literature. These include [7] that: the fluid velocity at the wave crest exceeds the phase velocity of the wave; the vertical acceleration at the crest exceeds some threshold; the surface slope exceeds some threshold; and that the pressure at the free surface exceeds that allowed by Bernoulli's equation (Kelvin-Helmholtz instability). To a certain extent these may be quantified within the framework of an appropriate model such as that provided by the Stokes surface wave; this has been identified as a precursor to wave breaking events [8] and forms the basis of Longuet-Higgins' semi-analytical model of breaking wave profiles [9]. A detailed analysis of the geometry of the breaking Stokes wave provides the following conditions for breaking:

$$\text{Wave steepness: } \frac{\text{Peak to trough (vertical) distance}}{\text{Peak to peak (horizontal) distance}} > 0.142$$

$$\text{Fluid velocity-phase speed ratio: } > 0.47$$

$$\text{Vertical acceleration: } > -g/2$$

$$\text{Surface slope: } > \tan(\pi/6) = 0.58 \tag{3.27}$$

It has been argued that there is significant commonality between these threshold criteria. For example Xu *et al.* [10] suggest that surface slope and vertical acceleration provide equally valid breaking criteria; Srokosz [11] claims that criteria based on vertical acceleration and the wave steepness are also equivalent. Partly for these reasons and partly because we are able to characterise the quantity involved straightforwardly as a Gaussian random field, we adopt this vertical acceleration threshold as a criterion for surface wave breaking as a first step towards modelling breaking waves. Thus we model the area of the sea surface that is subject to wave breaking as that which exceeds the acceleration threshold and call it (in common with Reference 12) the Breaking Area Model (BAM). In the next chapter, when we consider statistical models, various useful results relating to the statistics of the BAM will be derived and discussed.

Having identified the quantity of breaking waves, we need to model the form that the breaking takes. This problem is generally solved using hydrodynamic numerical computer models; examples of which are presented in References 13 and 14. It is found that, as a wave increases in height, it tends to sharpen at the top and flatten out at the bottom. The sharp wave crest then steepens rapidly and curls over, with a jet of water crashing onto the front wave slope. In smaller breakers the 'curling' process is much diminished and the breaking manifests itself mainly as an increase of roughness on the front face; these are often called 'micro-breakers'. Calculating the shape of the breaking wave up to the point that the top turns is readily achieved with the numerical models; an example is given in Figure 3.4. After the jet has hit the front face the process becomes much more confused and is best modelled as an area of very rough water.

3.3 EM scattering from the sea at high grazing angles

Given this specification of the sea surface, we must now capture its interaction with the incident microwave radiation. The behaviour of this electromagnetic field is governed

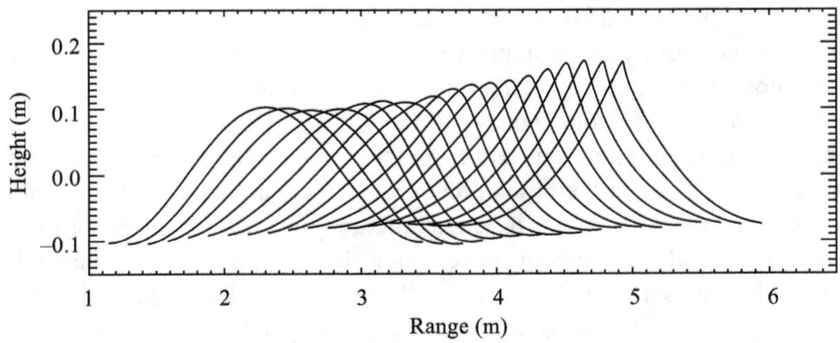

Figure 3.4 *The shape of a wave as it starts to break, as derived by the numerical hydrodynamics code LONGTANK [14]*

by Maxwell's equations [15]

$$\begin{aligned}\nabla \cdot \mathbf{D} &= \rho \\ \nabla \cdot \mathbf{B} &= 0 \\ \nabla \wedge \mathbf{E} &= -\frac{\partial \mathbf{B}}{\partial t} \\ \nabla \wedge \mathbf{H} &= \mathbf{j} + \frac{\partial \mathbf{D}}{\partial t}\end{aligned} \tag{3.28}$$

Here **E** and **H** are the electric and magnetic field intensities, **D** is the electric displacement and **B** is the magnetic induction; ρ and **j** are charge and current densities satisfying the conservation relation

$$\nabla \cdot \mathbf{j} + \frac{\partial \rho}{\partial t} = 0 \tag{3.29}$$

The field intensities are related to the associated displacement and induction through constitutive relations

$$\begin{aligned}\mathbf{D} &= \varepsilon_0 \varepsilon\, \mathbf{E} \\ \mathbf{B} &= \mu_0 \mu\, \mathbf{H}\end{aligned} \tag{3.30}$$

If harmonic time dependence $\exp(-i\omega t)$ is assumed in (3.28) the equations become

$$\begin{aligned}\nabla \cdot \mathbf{D} &= \rho \\ \nabla \cdot \mathbf{B} &= 0 \\ \nabla \wedge \mathbf{E} &= i\omega \mathbf{B} \\ \nabla \wedge \mathbf{H} &= \mathbf{j} - i\omega\, \mathbf{D}\end{aligned} \tag{3.31}$$

We are interested in the free space above the sea (where $\mu = \varepsilon = 1$) and the sea itself, which is an imperfect electrical conductor. We can describe the finite electrical conductivity of sea water by setting $\mu = 1$ and taking ε to have a large imaginary part and a real part greater than unity. This allows us to suppress the **j** in equation (3.31) as it is accounted for by the imaginary part of ε. At the sea surface the fields satisfy boundary conditions that require that the tangential components of **E** and **B** and the normal components of **E** and **D** are continuous.

Calculation of the scattered field from the sea surface for a particular radar illumination may be done in the time domain by using equation (3.28) and a numerical method such as FDTD (finite difference time domain). FDTD steps the field values forwards in time, at all points in space, using the differential equations (3.28). It is well suited to a pulsed waveform, where there are no fields at time zero; otherwise there is the problem of how to calculate the starting fields. FDTD is however only a numerical technique and is difficult to apply to random surfaces, where we are interested in the statistics of the scattered field.

It is much more common to express the problem in the frequency domain and to express (3.31) as integral equations, where the integrals are just over the sea surface. As we shall see, this allows more analysis to be undertaken and simple approximate models to be developed. Ultimately it is often necessary to resort to a numerical solution, but the result may be used to refine the approximate analytical models. The integral equations are derived from (3.31) using the vector Green's theorem. The process is described in detail in Stratton [15] and leads to the so-called Stratton-Chu equations[3]

$$\mathbf{E}(\mathbf{r}) = \mathbf{E}_{\text{in}}(\mathbf{r}) + \int_{S'} dS'[i\omega\mathbf{n}(\mathbf{x}') \wedge \mathbf{B}(\mathbf{x}')G(\mathbf{r},\mathbf{x}') + (\mathbf{n}(\mathbf{x}') \wedge \mathbf{E}(\mathbf{x}')) \wedge \nabla' G(\mathbf{r},\mathbf{x}') + \mathbf{n}(\mathbf{x}').\mathbf{E}(\mathbf{x}')\nabla' G(\mathbf{r},\mathbf{x}')] \tag{3.32}$$

The corresponding result for the magnetic field is

$$\mathbf{B}(\mathbf{r}) = \mathbf{B}_{\text{in}}(\mathbf{r}) - \int_{S'} dS'[i\omega\mu_0\varepsilon\varepsilon_0\mathbf{n}(\mathbf{x}') \wedge \mathbf{E}(\mathbf{x}')G(\mathbf{r},\mathbf{x}') - (\mathbf{n}(\mathbf{x}') \wedge \mathbf{B}(\mathbf{x}')) \wedge \nabla' G(\mathbf{r},\mathbf{x}') - \mathbf{n}(\mathbf{x}').\mathbf{B}(\mathbf{x}')\nabla' G(\mathbf{r},\mathbf{x}')] \tag{3.33}$$

Here the point **r** is in free space above the sea surface, S', and $\mathbf{x}'$ is on that surface; the normal **n** points up from S'. The Stratton-Chu equations are usually expressed in terms of an integral over a surface S' enclosing a source-free volume. In equations (3.32) and (3.33) we have performed the integral over a large hemisphere above the sea surface and found its value to be the incoming field ($\mathbf{E}_{\text{in}}$ or $\mathbf{H}_{\text{in}}$), which is then added to the remaining integral over the sea surface. The function $G(\mathbf{r}, \mathbf{r}')$ is the Green's function, which is given by

$$G(\mathbf{r},\mathbf{x}') = -\frac{\exp(ik|\mathbf{r}-\mathbf{x}'|)}{4\pi|\mathbf{r}-\mathbf{x}'|} \tag{3.34}$$

[3] Although we do not give here the full derivation of the 3D vector Stratton-Chu equations, the reader is referred to Appendix 3, where the 2D scalar equations are derived for use later in this chapter. The basic principles are the same for 2D and 3D; and the 2D derivation demonstrates those principles most clearly because they are not obscured by the vector notation.

and

$$k = \omega\sqrt{\mu_0\varepsilon_0} \tag{3.35}$$

In order to obtain radar scattering results using (3.32) and (3.33), the point **r** has to be moved onto the surface S'. This results in two coupled integral equations for the fields on the top of the surface. Two similar equations are formed on the underside of the surface by completing the enclosed surface with a hemisphere under the surface. Along with the boundary conditions, these underside equations are used to relate **E** and **B** at the surface, so that the topside equations may be solved for the actual field values. The scattered fields at the radar are then evaluated using equations (3.32) and (3.33) with **r** at the radar antenna.

Very little progress has been made over the years using this general approach in three dimensions with realistic sea surfaces. Therefore it is necessary to make some simplifying assumptions. The first is to note that at high grazing angles the imperfect conductivity of the sea surface does not have a very large effect and we can obtain a useful answer by assuming that it is a perfect conductor.

Inside a perfect conductor the fields are zero. At the surface there are large currents but no electric field and so all of the terms in equations (3.32) and (3.33) go to zero at the surface except for $\mathbf{n} \wedge \mathbf{B}$. Thus the equations decouple and (3.33) becomes

$$\mathbf{B}(\mathbf{r}) = \mathbf{B}(\mathbf{r})_{\text{in}} - \int_{S'} dS'(\mathbf{n}(\mathbf{x}') \wedge \mathbf{B}(\mathbf{x}')) \wedge \nabla' G(\mathbf{r}, \mathbf{x}') \tag{3.36}$$

Bringing the point **r** down to the point **x** an the surface S' leads to

$$\mathbf{n}(\mathbf{x}) \wedge \mathbf{B}(\mathbf{x}) = 2\mathbf{n}(\mathbf{x}) \wedge \mathbf{B}(\mathbf{x})_{\text{in}} - 2\mathbf{n}(\mathbf{x}) \wedge \int_{S'} dS'(\mathbf{n}(\mathbf{x}') \wedge \mathbf{B}(\mathbf{x}')) \wedge \nabla' G(\mathbf{r}, \mathbf{x}') \tag{3.37}$$

The factors of 2 in equation (3.37) are the result of **x** being on the surface (loosely speaking, half on one side and half on the other); the derivation of the Stratton-Chu equations [15] and the scalar equivalent (here in Appendix 3) provide a more formal justification. The solution of (3.37) is difficult for large, realistic sea surfaces. Therefore, at high grazing angles, a further approximation is that the integral term may be ignored. This is equivalent to assuming that the field at any point on the surface is that which would be present were the surface to be an infinite plane at a tangent to the surface at that point. Hence it is often called the 'tangent plane approximation'; other terms used are 'physical optics' and the Kirchoff approximation. The result is

$$\mathbf{n}(\mathbf{x}) \wedge \mathbf{B}(\mathbf{x}) = 2\mathbf{n}(\mathbf{x}) \wedge \mathbf{B}(\mathbf{x})_{\text{in}} \tag{3.38}$$

where **x** is on the sea surface. If we now place **r** at the radar, equation (3.38) may be substituted back into (3.36). We can ignore $\mathbf{B}_{\text{in}}(\mathbf{r})$ as this is just the radar transmission, and obtain the scattered field

$$\mathbf{B}(\mathbf{r})_{\text{scat}} = -2 \int_{S'} dS' \, \mathbf{n}(\mathbf{x}') \wedge \mathbf{B}(\mathbf{x}')_{\text{in}} \wedge \nabla' G(\mathbf{r}, \mathbf{x}') \tag{3.39}$$

This may be simplified for the calculation of radar cross section because we will require

$$\text{Lim}[r^2 |\mathbf{B}_{\text{scat}}(\mathbf{r})|^2]; \quad r \to \infty$$

where we have assumed the origin to be in the centre of the scattering surface. Hence we can simplify $\nabla' G$ by keeping only terms that are of order $1/r$ as follows:

$$\begin{aligned} \nabla' \frac{\exp(ik|\mathbf{r} - \mathbf{x}'|)}{4\pi |\mathbf{r} - \mathbf{x}'|} &= \frac{\mathbf{x}' - \mathbf{r}}{4\pi |\mathbf{r} - \mathbf{x}'|^3} \exp(ik|\mathbf{r} - \mathbf{x}'|)\,(ik|\mathbf{r} - \mathbf{x}'| - 1) \\ &\sim i\mathbf{k} \frac{\exp(ikr)}{4\pi r} \exp(-i\mathbf{k}.\mathbf{x}') \quad \text{where } \mathbf{k} = k\frac{\mathbf{r}}{r} \end{aligned} \tag{3.40}$$

As in Section 3.2, the sea surface S' is defined by its height $\eta(\text{v})$, where $\mathbf{v}$ is the component in the x–y plane of the position vector $\mathbf{x}$, i.e. we can write $\mathbf{x} = \{\mathbf{v}, \eta(\mathbf{v})\}$. The normal to this surface is then given by

$$\mathbf{n} = \frac{1}{\sqrt{1 + (\nabla\eta)^2}} \{-\nabla\eta, \mathbf{e}_z\} \tag{3.41}$$

which allows us to write

$$\mathbf{n}\, dS = dx\, dy(-\nabla\eta + \mathbf{e}_z) \tag{3.42}$$

For an incoming plane wave from the radar,

$$\mathbf{B}_{\text{in}}(\mathbf{r}) = \exp(-i\mathbf{k} \cdot \mathbf{r})\mathbf{B}_0 \tag{3.43}$$

where $\mathbf{k}$ is the backscattered wavevector to be consistent with (3.40), and $\mathbf{B}_0 \cdot \mathbf{k} = 0$. Substituting (3.40) to (3.43) into (3.39) and multiplying out the triple vector product gives

$$\mathbf{B}_{\text{scat}}(\mathbf{r}) = -\frac{i \exp(ikr)}{2\pi r} \mathbf{B}_0 \int dxdy[-\nabla\eta.\mathbf{k} + k_z] \exp(-i2\mathbf{k}.\mathbf{x}) \tag{3.44}$$

where $k_z = \mathbf{k} \cdot \mathbf{e}_z$. This can be reduced yet further by integration by parts; the final Kirchoff formula is

$$\mathbf{B}_{\text{scat}}(\mathbf{r}) = -\frac{i \exp(ikr)}{2\pi r} \frac{k^2}{k_z} \mathbf{B}_0 \int dxdy \exp(-i2\mathbf{k}_{\text{H}}.\mathbf{v}) \exp(-i2k_z\eta(\mathbf{v})) \tag{3.45}$$

where $\mathbf{k}_{\text{H}}$ is horizontal component of $\mathbf{k}$. The scattered field vector $\mathbf{B}_{\text{scat}}$ is in the direction as the incoming field $\mathbf{B}_0$, i.e. the polarisations are the same. Thus there is no cross-polar return and the RCS is independent of polarisation. The RCS, σ, of an object is defined as the cross-sectional area of an isotropic scatterer that reflects the same amount of power back to the radar as the real object. Thus the power (in appropriate units) scattered by the isotropic scatterer is σB_0^2, which leads to a power density (i.e. power per unit area) at the radar of $\sigma B_0^2/(4\pi r^2)$. Equating this to the power density from the rough surface of $|\mathbf{B}_{\text{scat}}(\mathbf{r}')|^2$, results in

$$\sigma = \text{Lim}\left[4\pi r \frac{|\mathbf{B}_{\text{scat}}(\mathbf{r})|^2}{B_0^2}\right]; \quad r \to \infty \tag{3.46}$$

It is usual in the case of sea clutter to define an average RCS, normalised by the illuminated sea surface area. Thus,

$$\sigma^0 = \frac{\langle\sigma\rangle}{A} \tag{3.47}$$

where A is the area over which the integral in (3.45) is taken. To evaluate σ^0 we form

$$\langle|\mathbf{B}_{\text{scat}}(\mathbf{r})|^2\rangle = \frac{1}{4\pi^2 r^2}\frac{k^4}{k_z^2}B_0^2 \iint d^2x_1 d^2x_2 \exp(-2i\mathbf{k}_{\text{H}}.(\mathbf{v}_1 - \mathbf{v}_2))$$
$$\times \langle\exp(-2ik_z(\eta(\mathbf{v}_1) - \eta(\mathbf{v}_2)))\rangle \tag{3.48}$$

The expectation value is taken as

$$\langle\exp(-2ik_z(\eta(\mathbf{v}_1) - \eta(\mathbf{v}_2)))\rangle = \exp(-2k_z^2\langle(\eta(\mathbf{v}_1) - \eta(\mathbf{v}_2))^2\rangle)$$
$$= \exp(-4k_z^2\langle\eta^2\rangle(1 - \rho(\mathbf{v}_1 - \mathbf{v}_2))) \tag{3.49}$$

Here we have exploited the Gaussian statistics[4] of the surface height fluctuations and have identified their mean square value and normalised autocorrelation function ρ. This latter depends only on relative co-ordinates if, as we have assumed here, the height fluctuations have spatially stationary statistics. If, in much the same spirit, we now adopt sum and difference co-ordinates in the fourfold integral we see that the area, A, results from the integral over the sum co-ordinate and can then be cancelled out in (3.47). So, all told, we end up with

$$\sigma^0 = \frac{k^4}{k_z^2}\frac{1}{\pi}\int d^2x \exp(-2i\mathbf{k}_{\text{H}}.\mathbf{v}) \exp(-4k_z^2\langle\eta^2\rangle(1 - \rho(\mathbf{v}))) \tag{3.50}$$

If the surface roughness is very small, then the second exponential in the integrand can be expanded in powers of the mean square surface height and, identifying the Fourier representation of their power spectrum (as is defined in Section 3.2):

$$\sigma^0 = 8\pi k^4[S_\eta(-2\mathbf{k}_{\text{H}}) + S_\eta(2\mathbf{k}_{\text{H}})] \tag{3.51}$$

where

$$\langle\eta^2\rangle\rho(\mathbf{v}) = \langle\eta(0)\eta(\mathbf{v})\rangle = \int d^2k \cos(\mathbf{k}.\mathbf{v})S_\eta(\mathbf{k}) \tag{3.52}$$

This (equation (3.51)) is an example of resonant, or Bragg, scattering.

[4] Details of these manipulations of Gaussian statistics may be found in Appendix 1, equation (A1.74).

3.4 Imaging ocean currents at high grazing angles

Equation (3.50) for Kirchoff scattering from a perfect conductor is quite effective for modelling sea clutter at high grazing angles. The use of either equation (3.1) or (3.9) with appropriate angular dependence produces results that are consistent with experiment for angles from close to normal (90°) out to about 60° grazing angle. Perhaps the most interesting application of this scattering model, however, is to predict the change of RCS due to local changes in the wave spectrum, S_η, which are in turn caused by ocean currents. This is the mechanism by which it appears that a space based imaging radar, such as ERS1, is able to 'see' though the ocean to the sea-bed. What is actually happening is that the tidal currents vary as they flow over submarine topology; these change the surface wave spectrum, and this is observed by the radar. Understanding this process requires some knowledge of hydrodynamics, which is not the main subject of this book. Therefore only a brief description is given here. The reader is referred to the references for more detail.

In order to model the effect of ocean currents on the surface waves we use the theory of '*action balance*'. According to Hughes [16], application of the principles of conservation of waves and conservation of energy leads to equations in Reference 17 that may be used to derive the principle of conservation of wave action. This principle is that, in the absence of energy sources and sinks (such as input from wind, molecular and turbulent dissipation, or non-linear transfer among spectral components), the wave action spectral density is conserved along ray paths defined by the following coupled equations:

$$\frac{d\mathbf{x}}{dt} = \nabla_{\mathbf{q}}\omega_0(\mathbf{q}) + \mathbf{U}$$
$$\frac{d\mathbf{q}}{dt} = -\mathbf{q}\cdot\nabla\mathbf{U} \tag{3.53}$$

where $\mathbf{x}$ is the ray position, $\mathbf{q}$ is the wavenumber and $\mathbf{U}$ is the current. The first equation in (3.53) corresponds to the wave energy travelling at the group velocity plus the water current. The second equation accounts for the change of wavelength as the current changes (the Doppler effect).

The wave action spectrum, $\phi(\mathbf{q})$, is derived from the wave height spectrum, $S_\eta(\mathbf{q})$, using the relation

$$\phi(\mathbf{q}) = \frac{S_\eta(\mathbf{q})\omega(\mathbf{q})}{q} \tag{3.54}$$

where $\omega(q)$ is the dispersion relation defined in equation (3.7). We use the wave spectrum defined in equation (3.2).

To model the South Falls sandbank in the English Channel, off the Kent Coast, we approximate the flow over the sandbank as a sech^2 function; where the water current is given by

$$U(\mathbf{x}) = U_0 + A\,\text{sech}^2(Kx) \tag{3.55}$$

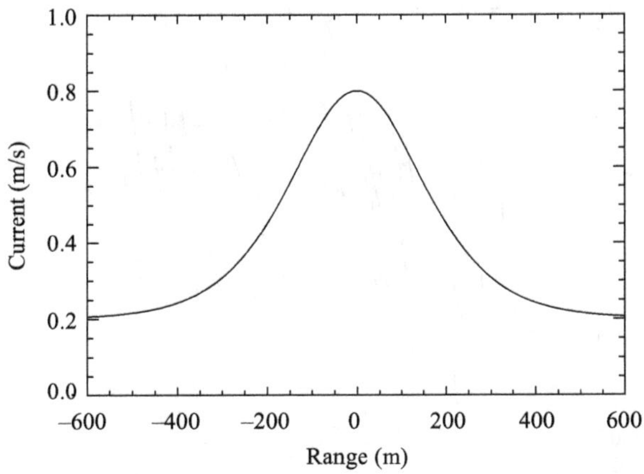

Figure 3.5 *Current flow over sandbank*

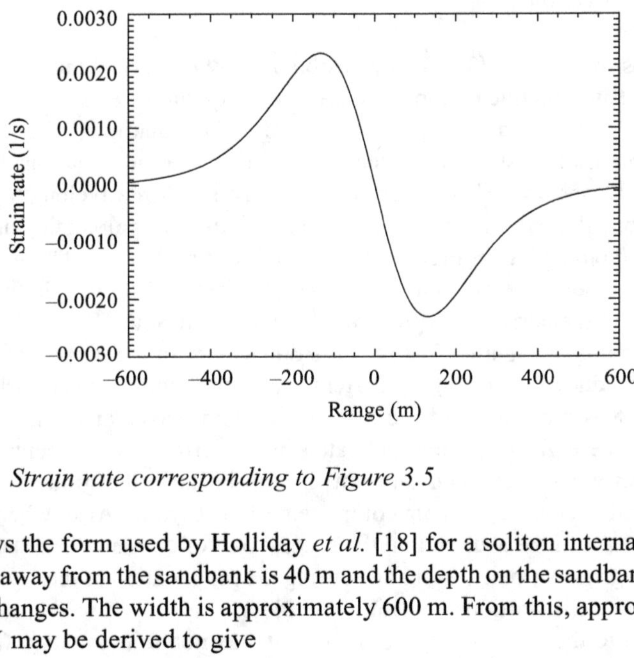

Figure 3.6 *Strain rate corresponding to Figure 3.5*

(This follows the form used by Holliday *et al.* [18] for a soliton internal wave). The water depth away from the sandbank is 40 m and the depth on the sandbank is 10 ± 2 m as the tide changes. The width is approximately 600 m. From this, appropriate values for A and K may be derived to give

$$U\left(\mathbf{x}\right) = U_0\left(1 + 3\,\mathrm{sech}^2\left(0.005x\right)\right) \tag{3.56}$$

Figures 3.5 and 3.6 show the current and strain rate[5] for an ambient tidal current component across the sandbank of 0.2 ms^{-1}, which is a typical value for that area.

[5] The *strain rate* is the change of current per unit of distance; it is therefore measured in ms^{-1} per metre, or, more succinctly, s^{-1}. Hence the use of the term 'rate'.

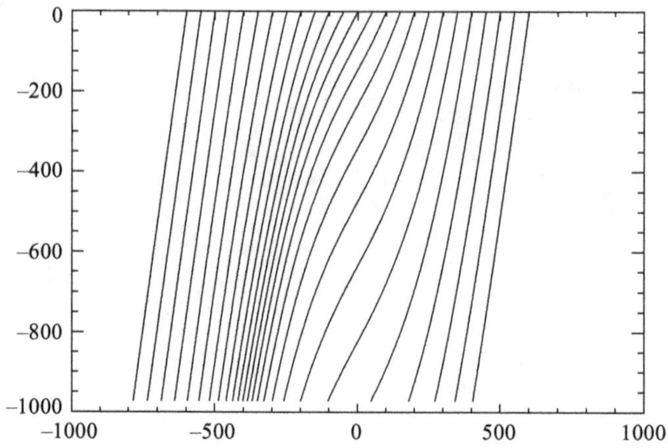

Figure 3.7 *600 s ray paths for 1m waves arriving at $y = 0$. (q vector in the y direction)*

Solutions of equation (3.53) may be obtained by numerical integration, using for example the Runge-Kutte method. We are generally interested in the history of a ray in order to calculate the action spectrum (as defined in equation (3.54)) at a particular point. To obtain the history it is necessary to integrate the equations backwards in time. Figure 3.7 shows the histories of rays for waves of 1 m wavelength (q pointing in the y direction, along the sandbank) reaching various points along the line $y = 0$. The sandbank is modelled as in Figures 3.5 and 3.6, but in this case the ambient current is 0.3 ms^{-1} in the x direction (across the sandbank) and 1.0 ms^{-1} in the y direction. This angling of the current is typical of the situation at South Falls.

The ray paths in Figure 3.7 are for the 600 s prior to the waves reaching $y = 0$. The distortion due to the changing current across the sandbank is clearly visible.

Figure 3.8 is a montage of the histories of 1 m waves arriving at $y = 0$ with $\mathbf{q}$ vectors in the eight directions indicated by the arrows in the centre. The traces to the bottom are shorter than those at the top because the group velocity of the waves is fighting against the main component of the current. Also, when the $\mathbf{q}$ vector has a component in the x direction, the second part of equation (3.53) causes $\mathbf{q}$, and therefore the group velocity, to change during the propagation of the rays. This causes the refraction and distortions evident in the plots.

To return to the 'action-balance' theory, it was noted in equation (3.53) that, in the absence of energy sinks and sources, the action spectral density is conserved. The assumption of lack of energy sinks and sources is, of course, not generally valid. The wind causes waves to grow, there are non-linear interactions between wave-numbers in the spectrum, and there is energy dissipation through turbulence and wave breaking. Existing models for these processes are fairly qualitative. The basic concept is that there is an equilibrium spectrum that exists in the absence of varying currents. The equilibrium spectrum balances the various energy fluxes, and is therefore a function of wind conditions. As described above, when a varying current is present, the intrinsic

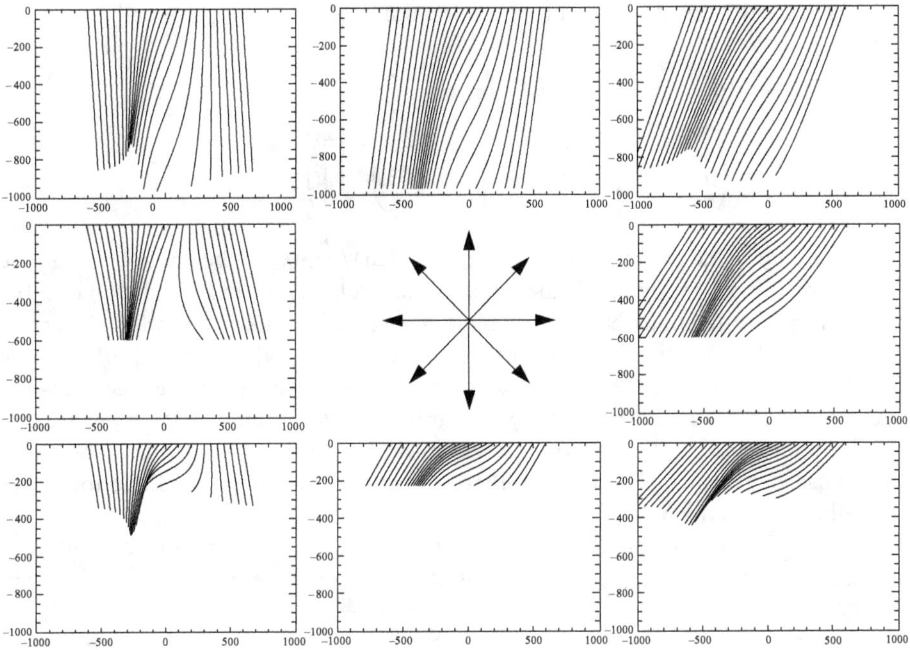

Figure 3.8 Ray histories of waves arriving at y = 0 with q vectors in eight different directions

action spectral density is conserved along the ray paths. However, since **q** changes through (3.53), the action spectrum moves away from equilibrium. This causes the energy sources and sinks to be out of balance, and the action spectrum is forced back towards equilibrium.

There are two differential equations commonly in use to describe the time evolution of the action spectral density under the process outlined above. The first, introduced by Alpers and Hennings [19], is a linear relaxation equation returning the action spectral density ϕ to its equilibrium value ϕ_0.

$$\frac{d\varphi}{dt} = \beta(t)\,(\varphi_0(t) - \varphi(t)) \tag{3.57}$$

The second equation is a non-linear relaxation introduced by Hughes [16]:

$$\frac{d\varphi}{dt} = \beta(t)(\varphi(t) - (\varphi(t)^2/\varphi_0(t))) \tag{3.58}$$

The time dependence in (3.57) and (3.58) of both ϕ and ϕ_0 include the implicit variation of **q** and **x**, which are due to the wave travelling along the ray paths defined by (3.53). The relaxation rate, β, in (3.57) and (3.58) is generally approximated by the wave growth constant. A simple empirical model for this is ([16] and [20]),

$$\beta(\mathbf{q}) = \omega(\mathbf{q})\frac{u_*|\cos(\theta_\mathrm{w})|}{c_\mathrm{p}(\mathbf{q})}\left(0.01 + \frac{0.016\, u_*|\cos(\theta_\mathrm{w})|}{c_\mathrm{p}(\mathbf{q})}\right) \times \left(1 - \exp\left[-8.9\left(\frac{u_*}{c_\mathrm{p}(\mathbf{q})} - 0.03\right)^{1/2}\right]\right) \tag{3.59}$$

where the c_p is the phase velocity ($c_\mathrm{p}(\mathbf{q}) = \omega(\mathbf{q})/q$), θ_w is the angle between the wind direction and $\mathbf{q}$, and u_* is the wind friction velocity given by equation (3.20).

Minor modifications have been made here to (3.59) from the original version derived by Hughes [16]. In his paper the first $\cos(\theta_\mathrm{w})$ is without the absolute value brackets. It therefore renders β to be negative for a wave travelling against the wind. This is reasonable in the context of wave growth constant, but not for application to a relaxation equation, since it would result in exponential growths. A lower bound of 0.002 has also been applied to β in order to keep the relaxation time finite for waves travelling across the wind.

Although the two relaxation equations (3.57) and (3.58) seem to be quite different, and are considered as such in the literature, they are both cases of the more general form

$$\frac{d}{dt}F(\varphi(t)) = -\beta(t)(F(\varphi(t)) - F(\varphi_0(t))) \tag{3.60}$$

where, for (3.57), $F(\phi) = \phi$, whilst for (3.58), $F(\phi) = 1/\phi$. There are many other possible functional forms of F that may be used. Potentially these may allow more physics to be introduced to the modelling of the relaxation process. Equation (3.60) can be solved using the integrating factor method to give

$$F(\varphi(t)) = F(\varphi_0(t)) + \int_{t_0}^{t} dt'\, \frac{\partial F(\varphi_0(t'))}{\partial \varphi_0}(\mathbf{q}\cdot\nabla\mathbf{U})\nabla_\mathbf{q}\varphi_0(\mathbf{q}(t'))\exp\left[-\int_{t'}^{t} dt''\beta(t'')\right] \tag{3.61}$$

This is the required solution. Results are obtained by integrating (3.61) along paths defined by the ray equations (3.53). The resulting wave spectrum is used to calculate radar scattering using the Kirchoff model, equation (3.50). Details of a numerical scheme to speed up the scattering calculations are given in Reference 20; a 'brute force' numerical integration method, however, works reasonably well. In the literature approximations are often made in the evaluation of the wave spectrum. The formulae used may be derived simply from equation (3.61).

The first approximation is to assume that the wavevector, $\mathbf{q}$, does not vary over the t' integral in (3.61), and that $\mathbf{x}$ only moves with the group velocity and ambient current. This is essentially the approximation used by Holliday *et al.* [18], and is appropriate for small variations in current. It removes the need to track the rays using (3.53) and therefore speeds up the computation. As $\mathbf{q}$ is constant, various terms can be moved outside the integral to give

$$F(\varphi(t)) = F(\varphi_0(t)) + \frac{\partial F(\varphi_0(t))}{\partial \varphi_0} \nabla_{\mathbf{q}} \varphi_0(\mathbf{q}) \int_{t_0}^{t} dt' (\mathbf{q} \cdot \nabla \mathbf{U}(\mathbf{x}'(t')))$$

$$\times \exp\left[- \int_{t'}^{t} dt'' \beta(t'') \right] \tag{3.62}$$

where $\mathbf{x}'(t') = \mathbf{x}(t) - (\mathbf{c}_g + \mathbf{U}_0)(t - t')$ and $\mathbf{c}_g = \nabla_{\mathbf{q}} \omega_0(\mathbf{q})$. This approximation is referred to here as fix_q.

A second, more severe, approximation may be made by assuming that both $\mathbf{q}$ and $\mathbf{x}$ are constant within the t' integral in (3.61). This provides a result equivalent to the formula used by Alpers and Hennings in Reference 19. In our terms it is

$$F(\varphi) = F(\varphi_0) + \frac{\partial F(\varphi_0)}{\partial \varphi_0} \nabla_{\mathbf{q}} \varphi_0(\mathbf{q}) \cdot (\mathbf{q} \cdot \nabla \mathbf{U}(\mathbf{x})) \beta^{-1} \tag{3.63}$$

Equation (3.63) is referred to here as fix_xq.

From the above we have six ways of calculating the surface wave spectrum in the current field over the sandbank. We can either use the linear (L) differential equation (3.57) or the non-linear (NL) differential equation (3.58). In both cases we have three options: to integrate the solution along the rays (3.61); to assume that $\mathbf{q}$ is constant (3.62); or to assume both $\mathbf{q}$ and $\mathbf{x}$ are constant (3.63). The six methods are to be called

L-rays, L-fix_q, L-fix_xq, NL-rays, NL-fix_q, NL-fix_xq.

Results are presented below to compare the various approximations. These are followed by a specific comparison with ERS-1 satellite radar data.

Figure 3.9 shows the effect of the changing water current on waves of 1 m wavelength travelling across the sandbank. All of the approximations show that where the current is increasing the waves are attenuated, and where it is decreasing the waves are amplified. The complete ray calculation shows that, when the current is perturbed, the linear differential equation L produces a higher spectral density than the non-linear equation NL. This is to be expected because NL has a greater 'restoring force' than L for positive excursions and a smaller 'restoring force' than L for negative excursions. When either of the two approximations fix_q or fix_xq are used, the effect is the opposite and the wave spectrum for NL is increased above the value for L. At this wavenumber the L-fix_q and L-fix_xq approximations are not far from the full calculation results, but the NL approximations fail rather badly.

The same conclusions cannot be applied to the RCS modulations shown in Figure 3.10 for the same conditions. Here the only approximation which is close to the full calculation is L-fix_q. Even in this case there are fairly large errors. The failure of the approximations may be put down to a combination of the fairly large strain rates and the large spatial extent of the sandbank structure. Where the approximations have been applied in previous publications, either the strain rate or the spatial

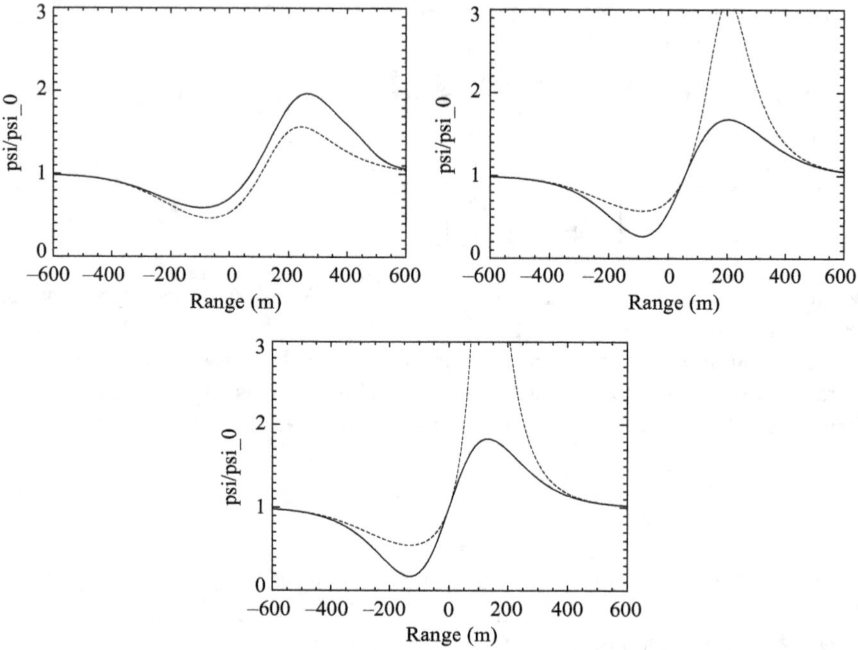

Figure 3.9 *Plots showing the variation of the wave spectral density across the sandbank. (The wavelength is 1 m, with q in the x direction. The three graphs show different approximations: the top left is the full ray calculation; the top right is fix_q; and the bottom is fix_xq. On each graph the solid line uses L and the dotted line uses NL)*

extent has been smaller and therefore the approximations may have been accurate. However this has not been confirmed by a comparison of the type presented here.

Figure 3.11 shows overlaid plots of four values of background current (which is proportional to the strain rate, as may be derived from equation (3.55)) for the L-rays and NL-rays. In both cases there is a gradual increase of RCS modulation as the strain rate increases.

All of the above results correspond to the wind blowing in the same direction as the variation of current. For other angles there should be more RCS modulation due to the longer relaxation times (smaller β) for waves travelling at an angle to the wind (see equation (3.59)). Figure 3.12 shows results for the wind at three angles ($\theta_W = 0°$, $\theta_W = 45°$ and $\theta_W = 90°$). The RCS modulation increases as the angle increases. The graph on the right (for NL-rays, which is the favoured model from the previous figures) shows a reasonable change between 0° and 45°. However, at 90° the modulation is rather large and may indicate problems with the model for β in the across wind direction.

In order to compare the modelling with experiment, data from ERS-1 have been analysed. 46 images from passes over the South Falls sandbank were collected over

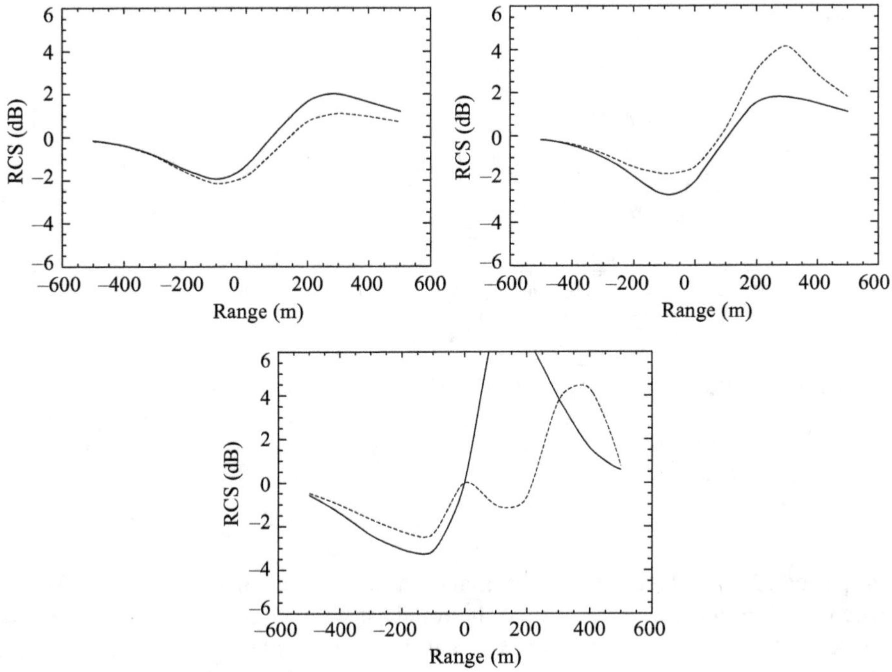

Figure 3.10 *Plots showing RCS profiles across the sandbank for an ambient transverse current of 0.2 ms^{-1}. ($\theta_w = 0$, $U_{10} = 5\ ms^{-1}$, and the radar frequency is 5 GHz. The three graphs show different approximations: the top left is the full ray calculation; the top right is fix_q; and the bottom is fix_xq. On each graph the solid line uses L and the dotted line uses the NL)*

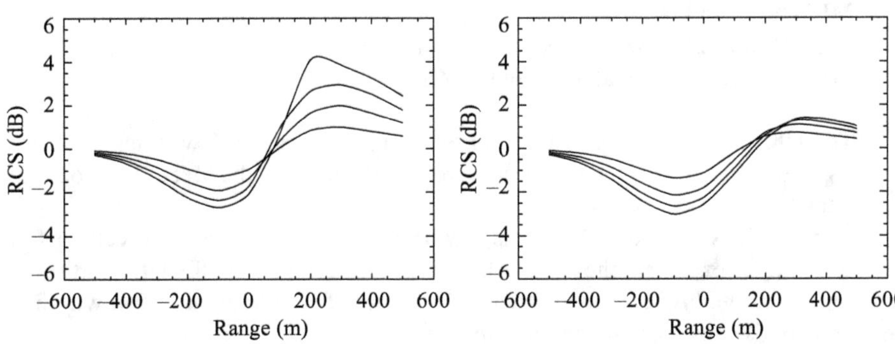

Figure 3.11 *RCS profiles from ray calculations for four values of ambient current (0.1, 0.2, 0.3, 0.4 ms^{-1}). The plot on the left uses L-rays and the plot on the right uses NL-rays. ($\theta_w = 0$, $U_{10} = 5\ ms^{-1}$ and the radar frequency is 5 GHz)*

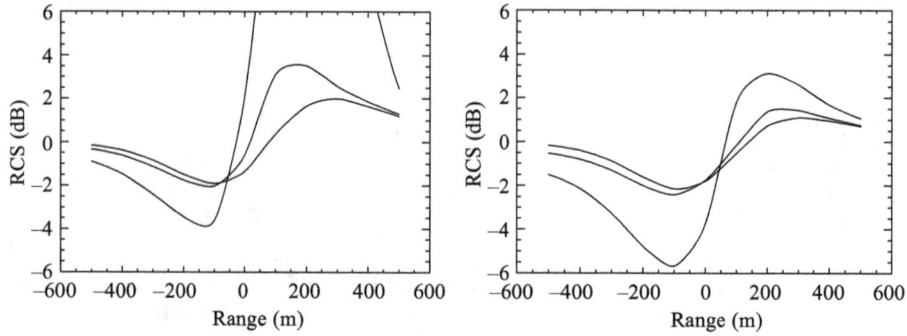

Figure 3.12 *RCS profiles from ray calculations for three wind directions ($\theta_W = 0°$, $\theta_W = 45°$ and $\theta_W = 90°$). The plot on the left uses L and the plot on the right uses NL. ($U_0 = 0.2\ ms^{-1}$, $U_{10} = 5\ ms^{-1}$ and the radar frequency is 5 GHz)*

the period 1994–1996. In 95% of the images the sandbank appears quite clear. The 5% where it is not visible correspond to the roughest weather conditions. A survey was carried out in June 1994 to measure the water conditions at the sandbank. The results show that the tidal flows across the sandbank are fairly predictable. The transverse current varies in a periodic manner between ± 40 cm s^{-1}. The slack water times are half an hour before high water and 6 h after high water. The depth of the bank is 10 m at MLWS (mean low water Springs) at its shallowest point. This increases to a level floor of 40 m at approximately 600 m to either side of the shallowest ridge. The shape is fairly uniform along the 7.5 km chosen for analysis. The tidal levels relative to MLWS are

- MHWS (mean high water Springs) 5m
- MLWN (mean low water Neaps) 1m
- MHWN (mean high water Neaps) 4m

The UK Met Office produced reports detailing the sea and weather conditions corresponding to the 46 images. These were calculated using the Met Office computer 'hindcast' model.

Figure 3.13 shows an ERS-1 image with the transverse current travelling from left to right. As expected, the RCS is lower than the ambient on the left hand side of the sandbank and higher than ambient on the right. The RCS modulation within the marked box is shown on the graph below the image.

Figure 3.14 contains results from action balance and radar scattering model for the conditions and geometry corresponding to Figure 3.13. The full ray calculation from the non-linear differential equation (NL-rays) provides the best match with the data, therefore reinforcing the conclusions drawn from the previous figures. The ambient transverse current value that is necessary for the prediction to match the data is just

Orbit	Date	Time	Illumination direction
21267 (21)	09-08-95	10:46:56	103°

Wind	Wave height	High water	Low water	Image time
NE 10–15kt	1.0–1.5m	09:44	17:05	HW+1:03

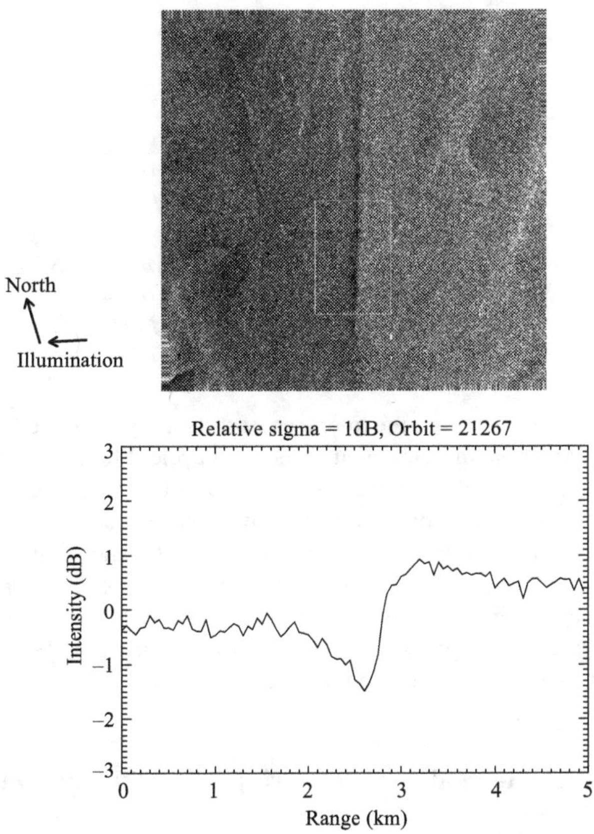

Figure 3.13 *Sandbank image and RCS modulation for orbit 21267 (the box on the image shows the portion of data analysed to produce the measure of RCS modulation, ERS data copyright ESA – Image processed by DERA UK)*

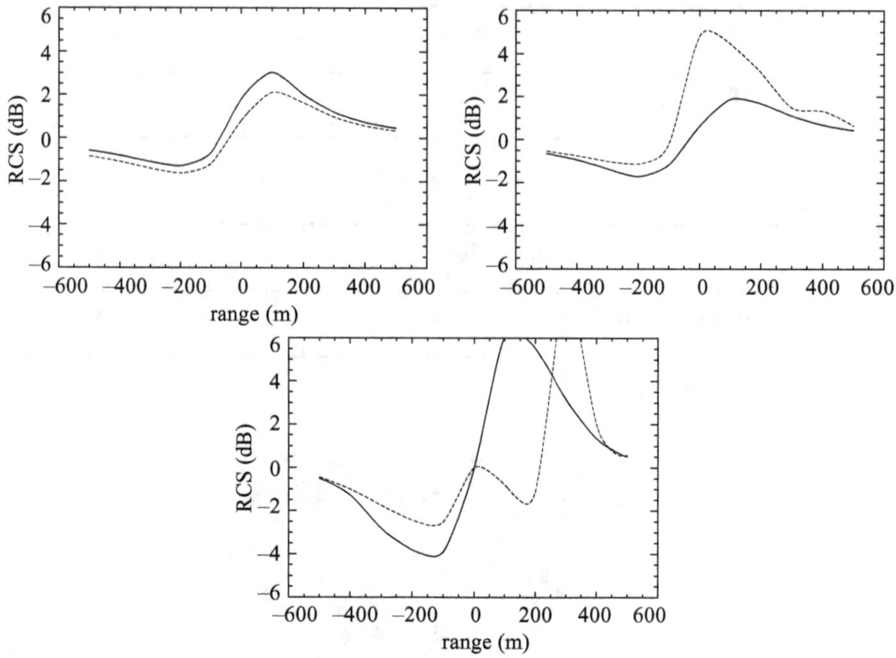

Figure 3.14 Plots showing RCS profiles predicted for orbit 21267 of ERS-1, using an ambient transverse current of 0.1 ms^{-1}, $\theta_w = 241°$, $U_{10} = 7.5$ ms^{-1}, illumination direction = 183°. The plots are in the same sequence as for Figure 3.10

under 0.1 ms^{-1}. This is within the tolerance of the survey data and tidal flow charts, given the position of the measurement in the tidal cycle. Most of the ERS-1 data can be matched to the model in this way. The most difficult data to fit correspond to either a wind direction along the sandbank or to a low wind speed.

Figure 3.15 is an example of a low wind speed (3 knots) image and shows a modulation of 15 dB above a very low ambient backscatter level. This very large RCS modulation is not predicted by the action balance and Kirchoff scattering modelling described above, and is an interesting, as yet, unsolved problem. Imagery of this type occurs about 15–20% of the time in this location.

3.5 The composite model for scattering at medium grazing angles

When the grazing angle is reduced below about 60°, sea clutter begins to show a dependence on polarisation. The assumptions behind the perfect conductor Kirchoff model no longer apply. In particular, the imperfect conductivity of sea water has a significant effect and the 'tangent plane' approximation for the fields on the surface is very inaccurate. Without these simplifications, it is difficult to apply the general

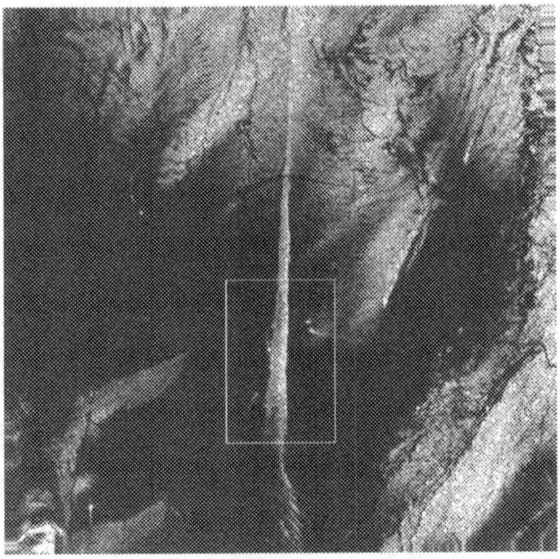

Figure 3.15 *Sandbank image with a low wind speed; 3 knots (the box on the image shows the portion of data analysed to produce the measure of RCS modulation, ERS data copyright ESA – Image processed by DERA UK)*

solution (i.e. equations (3.32) and (3.33)) to realistic sea surfaces. Progress can be made, however, for the case where the roughness is very small. It was shown in equation (3.51) that the small perturbation limit for the perfect conductor Kirchoff model depends only upon the 'resonant' wavenumber value of the surface wave spectrum. Thus

$$\sigma^0 = 8\pi k^4 [S_\eta(-2\mathbf{k}_\mathrm{H}) + S_\eta(2\mathbf{k}_\mathrm{H})] \tag{3.51}$$

where $\mathbf{k}_\mathrm{H}$ is the horizontal component of the backscattered wave-vector $\mathbf{k}$, and $2\mathbf{k}_\mathrm{H}$ is the resonant wavenumber. A similar result may be obtained for an imperfect conductor without the Kirchoff approximation,

$$\begin{aligned} \sigma^0_\mathrm{HH} &= 8\pi k^4 \cos^4\theta_i [S_\eta(-2\mathbf{k}_\mathrm{H}) + S_\eta(2\mathbf{k}_\mathrm{H})] |g_\mathrm{HH}|^2 \\ \sigma^0_\mathrm{VV} &= 8\pi k^4 \cos^4\theta_i [S_\eta(-2\mathbf{k}_\mathrm{H}) + S_\eta(2\mathbf{k}_\mathrm{H})] |g_\mathrm{VV}|^2 \end{aligned} \tag{3.64}$$

The factors g_HH and g_VV are given by

$$g_\mathrm{HH} = \frac{\varepsilon - 1}{\left(\cos\theta_i + \sqrt{\varepsilon - 1 + \cos^2\theta_i}\right)^2}, \quad g_\mathrm{VV} = \frac{(\varepsilon - 1)(\varepsilon(1 + \sin^2\theta_i) - \sin^2\theta_i)}{\left(\varepsilon\cos\theta_i + \sqrt{\varepsilon - 1 + \cos^2\theta_i}\right)^2} \tag{3.65}$$

and θ_i is the incidence angle (i.e. 90° minus the grazing angle). This SPM (small perturbation method) result, which is exact in the limits of small height and slope, was originally obtained by Rice [21] and Peake [22], using a wave expansion technique. Wright [23] obtained the same formulae using a reciprocal field method (the Lorentz Reciprocity Principle, LRP). Holliday [24] showed that the perfect conductor limit of (3.64), i.e.

$$\sigma^0_{\mathrm{HH}} = 8\pi k^4(1-\sin^2\theta_i)^2[S_\eta(-2\mathbf{k}_\mathrm{H}) + S_\eta(2\mathbf{k}_\mathrm{H})]$$

$$\sigma^0_{\mathrm{HH}} = 8\pi k^4(1+\sin^2\theta_i)^2[S_\eta(-2\mathbf{k}_\mathrm{H}) + S_\eta(2\mathbf{k}_\mathrm{H})] \tag{3.66}$$

can be obtained by one iteration of the surface current integral equation (equation (3.37)), which produces a term to correct the Kirchoff approximation.[6] (Unfortunately the analytical evaluation of the iteration only works in the SPM limit although, as we will find in the next section, iteration is a useful numerical technique.)

In order to apply the SPM results to realistic sea surfaces, Wright [23] and Valenzuela [25] proposed the so-called 'composite' model, where ripples, which are small enough to satisfy SPM, ride on larger smooth waves. The tilting by the large waves changes the resonant wave vector and the local grazing angle, and thus results in a modulation of equation (3.64). The composite model was originally motivated by purely physical arguments, but some justification may be obtained from the Kirchoff approximation. Thompson [26] showed that the Kirchoff RCS integral (i.e. equation (3.50)) can be factored into the Fourier transform of the product of a low **q** (long wave) function and a high **q** (short wave) function. The Fourier transform (FT) of the short wave part is approximated to the Kirchoff SPM result (equation (3.51)), and the FT of the long wave part is approximated to the distribution of wave slopes. Thus, the overall result is the convolution of SPM with the slope distribution, which picks out the resonant wave vector

$$\mathbf{k}_{\mathrm{res}} = 2(\mathbf{k}_\mathrm{H} + k_z\nabla\eta_\mathrm{L}) \tag{3.67}$$

(where $\nabla\eta_\mathrm{L}$ is the slope of the long waves) at each value of slope. This is therefore a form of 'Kirchoff composite model'. No analogous derivation of the Wright-Valenzuela composite model is available, but it is possible to extend the LRP derivation of SPM for a flat plane, to the perturbation of the fields on a smooth curved surface. The Wright-Valenzuela composite model can then be identified as the result obtained when the tangent plane approximation is applied to the smooth long wave surface. This is a reasonable approximation for medium grazing angles, but is poor at low grazing angles where multipath propagation and shadowing are important when deriving the fields on the smooth surface. We will not derive the composite model using these ideas at this point, but wait until we consider 2D scalar scattering in the next section and in Appendix 3. Here we just present the model in its 3D form.

[6] The iteration is achieved by substituting the Kirchoff current from (3.38) into the integral in (3.37), and then evaluating it in the SPM limit.

The initial step is to separate the sea surface wave spectrum into long and short waves.

$$\begin{aligned} S_s(\mathbf{q}) &= S_\eta(\mathbf{q}) \quad q \geq q_{\text{cut}} \\ &= 0 \qquad\quad q < q_{\text{cut}} \\ S_L(\mathbf{q}) &= S_\eta(\mathbf{q}) \quad q < q_{\text{cut}} \\ &= 0 \qquad\quad q \geq q_{\text{cut}} \end{aligned} \tag{3.68}$$

The cut-off frequency q_{cut} is somewhat arbitrary, but should be a few times the resonant wavenumber. The long wave spectrum, $S_L(\mathbf{q})$, corresponds to the wave height $\eta_L(\mathbf{x})$ used in equation (3.67). The local normal on this smooth surface is given by

$$\begin{aligned} \mathbf{n}(\mathbf{x}) &= \mathbf{n}(x, y, \eta(x, y)) = \frac{\mathbf{e}_z - \nabla\eta_L(x, y)}{\sqrt{1 + |\nabla\eta_L(x, y)|^2}} \\ \nabla\eta_L(x, y) &= \mathbf{e}_x\frac{\partial\eta_L(x, y)}{\partial x} + \mathbf{e}_y\frac{\partial\eta_L(x, y)}{\partial y} = \mathbf{e}_x\eta_{L,x}(\mathbf{x}) + \mathbf{e}_y\eta_{L,y}(\mathbf{x}) \end{aligned} \tag{3.69}$$

As the surface height is taken to be a Gaussian random variable (see Section 3.2), its derivatives are therefore (correlated) Gaussian random variables as well. (An account of Gaussian random variables is contained in Appendix 1.) The correlation properties of these derivatives can be related to those of the surface height correlation function, assuming η_L to be spatially homogeneous i.e.

$$\langle\eta_L(\mathbf{x}_1)\eta_L(\mathbf{x}_2)\rangle = \rho_L(\mathbf{x}_1, \mathbf{x}_2) = \rho_L(\mathbf{x}_1 - \mathbf{x}_2) \tag{3.70}$$

From this we find that

$$\begin{aligned} \langle\eta_{L,x}^2\rangle &= -\left.\frac{\partial^2\rho_L(x, y)}{\partial x^2}\right|_{x,y=0} \\ \langle\eta_{L,y}^2\rangle &= -\left.\frac{\partial^2\rho_L(x, y)}{\partial y^2}\right|_{x,y=0} \\ \langle\eta_{L,x}\eta_{L,y}\rangle &= -\left.\frac{\partial^2\rho_L(x, y)}{\partial x\partial y}\right|_{x,y=0} \end{aligned} \tag{3.71}$$

The correlation function is related to the spectrum by

$$\rho_L(\mathbf{x}) = \int d^2q\, \cos(\mathbf{q}\cdot\mathbf{x})S_L(\mathbf{q}) \tag{3.72}$$

in terms of which we have

$$m_{xx} = \langle \eta_{L,x}^2 \rangle = \int d^2q q_x^2 S_L(\mathbf{q})$$

$$m_{yy} = \langle \eta_{L,y}^2 \rangle = \int d^2q q_y^2 S_L(\mathbf{q})$$

$$m_{xy} = \langle \eta_{L,x} \eta_{L,y} \rangle = \int d^2q q_x q_y S_L(\mathbf{q}) \tag{3.73}$$

So, once we have identified the power spectrum S_L the elements of the covariance matrix of the surface height derivatives can be calculated by numerical quadrature. These can then be used to construct the joint pdf of the derivatives as

$$P(\nabla\eta_L) = \frac{\exp(-(\eta_{L,x}^2 m_{yy} + \eta_{L,y}^2 m_{xx} - 2\eta_{L,x}\eta_{L,y} m_{xy})/2(m_{xx}m_{yy} - m_{xy}^2))}{2\pi\sqrt{m_{xx}m_{yy} - m_{xy}^2}} \tag{3.74}$$

Equation (3.69) expresses the normal $\mathbf{n}$ in terms of the long wave slope $\nabla\eta_L$. The local incidence angle θ_{loc} as a function of $\mathbf{n}$ (and therefore of $\nabla\eta_L$) is

$$\cos\theta_{loc} = -\frac{\mathbf{k}\cdot\mathbf{n}(\nabla\eta)}{k} \tag{3.75}$$

The local 'vertical' and 'horizontal' polarisations on the slope $\nabla\eta_L$ are defined with respect to the vectors **k** and **n**. A local Hpol field is perpendicular to both **k** and **n**, while a local Vpol field is co-planar with **k** and **n** and is perpendicular to **k**. The radar measures polarisation states relative to a basis defined in terms of **k** and $\mathbf{e}_z$; this is generated by an appropriate rotation, ϕ, about an axis coincident with the direction of the wave-vector **k**.

$$\cos\phi = \frac{\mathbf{n}\cdot\mathbf{e}_z - \mathbf{i}\cdot\mathbf{e}_z\mathbf{i}\cdot\mathbf{n}}{\sqrt{(1-(\mathbf{n}\cdot\mathbf{i})^2)(1-(\mathbf{e}_z\cdot\mathbf{i})^2)}} \qquad \mathbf{i} = -\frac{\mathbf{k}}{k} \tag{3.76}$$

The normalised backscatter RCS from the ripples on the slope $\nabla\eta_L$ is calculated by replacing θ_{inc} and $2\mathbf{k}_H$ in equations (3.64) and (3.65) with θ_{loc} and $\mathbf{k}_{res}$ (from (3.67))[7] respectively, and by rotation through ϕ. Thus,

$$\begin{aligned}
\sigma_{HH}^0(\nabla\eta_L) &= 8\pi k^4 \cos^4\theta_{loc}[S_s(-\mathbf{k}_{res}) + S_s(\mathbf{k}_{res})]|g_{HH}\cos^2\phi + g_{VV}\sin^2\phi|^2 \\
\sigma_{VV}^0(\nabla\eta_L) &= 8\pi k^4 \cos^4\theta_{loc}[S_s(-\mathbf{k}_{res}) + S_s(\mathbf{k}_{res})]|g_{HH}\sin^2\phi + g_{VV}\cos^2\phi|^2 \\
\sigma_{HV}^0(\nabla\eta_L) &= 8\pi k^4 \cos^4\theta_{loc}[S_s(-\mathbf{k}_{res}) + S_s(\mathbf{k}_{res})]|(g_{HH} - g_{VV})\sin\phi\cos\phi|^2
\end{aligned} \tag{3.77}$$

[7] Wright [24] and Valenzuela [25] have different formulae for the resonant wave-vector. This is because they fail to account for the fact that the sea surface wave spectrum is specified in terms of wave-vectors in the x–y plane, not a locally tilted plane.

where,

$$g_{HH} = \frac{\varepsilon - 1}{\left(\cos\theta_{loc} + \sqrt{\varepsilon - 1 + \cos^2\theta_{loc}}\right)^2}$$

$$g_{VV} = \frac{(\varepsilon - 1)(\varepsilon(1 + \sin^2\theta_{loc}) - \sin^2\theta_{loc})}{\left(\varepsilon\cos\theta_{loc} + \sqrt{\varepsilon - 1 + \cos^2\theta_{loc}}\right)^2} \tag{3.78}$$

(We note that, while the underlying perturbation theory results do not induce cross-polar scattering, the variation in $\nabla\eta_L$ results, in a significant HV RCS). The RCS appropriate to scattering by the two-scale surface is obtained by averaging this result over the distribution of $\nabla\eta_L$. Thus, the final composite model result is

$$\sigma^0_{HH} = \int d^2(\nabla\eta_L)P(\nabla\eta_L)\sigma^0_{HH}(\nabla\eta_L)$$

$$\sigma^0_{VV} = \int d^2(\nabla\eta_L)P(\nabla\eta_L)\sigma^0_{VV}(\nabla\eta_L)$$

$$\sigma^0_{VH} = \int d^2(\nabla\eta_L)P(\nabla\eta_L)\sigma^0_{VH}(\nabla\eta_L) \tag{3.79}$$

The composite model has found wide application to the modelling of sea clutter at moderate grazing angles and has provided much insight into radar remote sensing of the ocean environment. However, as the grazing angle is reduced, discrepancies between experimental data and the composite model become evident. These are addressed in the next section.

3.6 Scattering at low grazing angles: beyond the composite model

The characteristics of sea clutter at low grazing angles have caused problems to radar designers since the discovery of radar. The description of clutter has been essentially empirical and so it has not been possible to use the data effectively for remote sensing. The scattering models described thus far in this chapter are ineffective at evaluating the average normalised RCS, σ^0, as shown in Figures 3.16 and 3.17. Here a comparison is made of an empirical model, the Georgia Tech (GIT) model [27] (which is described in detail in Chapter 8), and the composite model as described above. The sea surface wave spectrum is the Pierson-Moskowitz formula equation (3.1).

There are three main differences between the experimental data (GIT model, Figure 3.16) and the composite scattering model (Figure 3.17). The composite model:

- has much too large a difference between σ^0_{VV} and σ^0_{HH},
- approaches a limiting value at very low grazing angles, rather than decaying as $((\pi/2) - \theta_i)^4$, and
- has too little variation with sea state.

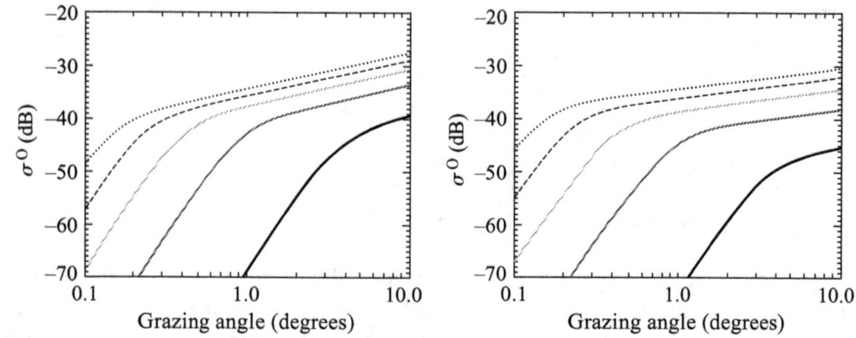

Figure 3.16 *VV (left) and HH (right) clutter normalised RCS as a function of grazing angle from the GIT model [27] for sea states 1 to 5 (black line: 1, dark grey line: 2, light grey line: 3, dashed line: 4, dotted line: 5)*

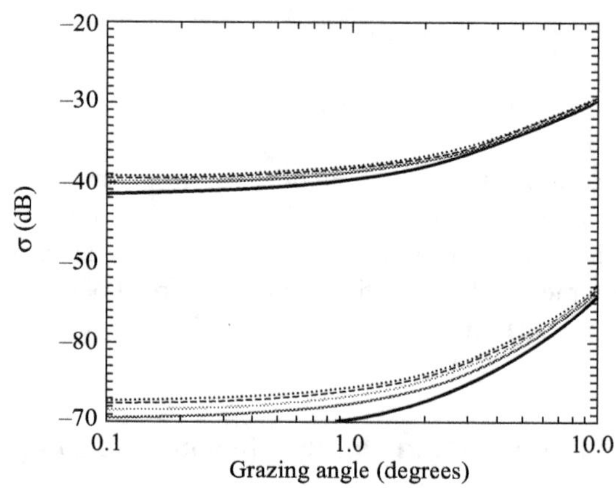

Figure 3.17 *Clutter normalised RCS as a function of grazing angle using the composite model for sea states 1 to 5 (black line: 1, dark grey line: 2, light grey line: 3, dashed line: 4, dotted line: 5). The top set of curves are VV and the bottom set are HH polarisation*

These can be addressed by using a better wave spectrum for the capillary waves, introducing breaking waves and improving the scattering model.

As we have already noted in Section 3.3, the full answer to the EM scattering problem requires the solution of the coupled Stratton-Chu integral equations for the sea surface fields on both sides of the interface. Even if the surface is idealised to a perfect conductor, the single resulting integral equation is very difficult to solve due to the extended nature of the surface. As we move close to grazing incidence the size of surface illuminated, by a given radar beamwidth, gets larger. Many alternative approximations to the composite model have been proposed, but it is not possible to put any bounds on their accuracy unless we have a means of calculating the correct answer.

One possible technique is to use numerical methods to solve the integral equations. This is very difficult for a large surface, but recently techniques have been developed to perform the task. Unfortunately, however, the methods are currently limited by computing power to 1D, corrugated surfaces. At first sight this may seem to be a severe limitation, but on reflection it is not found to be so. The principal limitations of the composite model are the lack of multipath propagation, shadowing and the reflections from breaking waves coming towards the radar. All these features can be reproduced with corrugated surfaces, and so it seems reasonable to analyse them numerically and to compare the results with approximate models. When the models are sufficiently accurate for 1D surfaces, we may then extend to 2D for comparison with data. As an intermediate check, a wave-tank is a good environment to generate real scattering data from corrugated surfaces for comparison with numerical results and models.

Thus we need to develop a EM scattering theory for corrugated surfaces. This reduces to a scalar problem and so is much more straightforward than the general vector approach. We therefore take the opportunity to provide more detail, mostly in Appendix 3, on some of the material glossed over in the earlier presentation. The corrugated scattering surface varies only along one direction, y, and so we model the wave height as

$$z = \eta(y) \tag{3.80}$$

For horizontally polarised incident radiation, the electric field vector points in the x direction (along the grooves) and we can write

$$\mathbf{E}(x, y, z, t) = \mathbf{e}_x \psi_{\mathrm{H}}(y, z) \exp(-i\omega t) \tag{3.81}$$

We expect the tangential **E** and **B** fields to be continuous at the sea surface. These constraints allow us to place boundary conditions on $\psi_{\mathrm{H}}(y, z) = E_x(y, z)$. The continuity in the tangential electric field is directly interpreted as continuity in ψ_{H}. The associated magnetic field is deduced from Maxwell's equations as

$$\mathbf{B} = \frac{1}{i\omega}(\nabla \wedge \mathbf{E}) = \frac{1}{i\omega}\left(\mathbf{e}_y \frac{\partial \psi_{\mathrm{H}}}{\partial z} - \mathbf{e}_z \frac{\partial \psi_{\mathrm{H}}}{\partial y}\right) \tag{3.82}$$

The normal to the interface is denoted by $\mathbf{n}$; the tangential component of the magnetic field can be identified as

$$\mathbf{n} \wedge (\mathbf{n} \wedge \mathbf{B}) = \frac{i}{\omega} \mathbf{n} \wedge \mathbf{e}_x \mathbf{n} \cdot \nabla \psi_{\mathrm{H}} \tag{3.83}$$

Thus continuity in the tangential component of the magnetic field is manifest in continuity in the normal derivative of ψ_{H}. These are our boundary conditions for the horizontally polarised case.

In the case of vertically polarised radiation, the **B** field points along the grooves in the corrugated surface. We can therefore write

$$\mathbf{B}(x, y, z, t) = \mathbf{e}_x \psi_V(y, z) \exp(-i\omega t) \tag{3.84}$$

Continuity in the tangential component of **B** is manifest as continuity in $\psi_V(y, z)$. The electric field is given by

$$\mathbf{E} = \frac{i}{\omega\mu_0\varepsilon_0\varepsilon}(\nabla \wedge \mathbf{B}) = \frac{i}{\omega\mu_0\varepsilon_0\varepsilon}\left(\mathbf{e}_y \frac{\partial\psi_V}{\partial z} - \mathbf{e}_z \frac{\partial\psi_V}{\partial y}\right) \tag{3.85}$$

Arguing as in the horizontally polarised case we now see that continuity in the tangential electric field imposes the following boundary condition on ψ_V:

$$\frac{1}{\varepsilon_1}\mathbf{n} \cdot \nabla\psi_V^{(1)} = \frac{1}{\varepsilon_2}\mathbf{n} \cdot \nabla\psi_V^{(2)} \tag{3.86}$$

Thus there is a discontinuity in the normal derivative of ψ_V at the interface. A summary of the vector to scalar conversions and the boundary conditions is given in the tables below.

	V	H
ψ	$\mathbf{B}_x$	$\mathbf{E}_x$
$\mathbf{n} \cdot \nabla\psi$	$i\omega\mu_0\varepsilon\varepsilon_0(\mathbf{n} \wedge \mathbf{E})_x$	$-i\omega(\mathbf{n} \wedge \mathbf{B})_x$
α	ε	1

	ψ	$\mathbf{n} \cdot \nabla\psi$
Boundary condition	$\psi^{(1)} = \psi^{(2)}$	$\frac{1}{\alpha_1}\mathbf{n} \cdot \nabla\psi_V^{(1)} = \frac{1}{\alpha_2}\mathbf{n} \cdot \nabla\psi_V^{(2)}$

The scalar equivalent of the Stratton-Chu equations (3.32) and (3.32) may be derived either by substitution of the above conversions, or by the direct use of Green's theorem as shown in Appendix 3. The result is the Helmholz integral. If we define the normal derivative as $\zeta(\mathbf{x}') = \mathbf{n}(\mathbf{x}') \cdot \nabla'\psi(\mathbf{x}')$, the integral may be written as

$$\psi(\mathbf{x}) = \psi_{in}(\mathbf{x}) + \int (G_0(\mathbf{x}, \mathbf{x}')\zeta(\mathbf{x}') - \psi(\mathbf{x}')\mathbf{n}(\mathbf{x}') \cdot \nabla' G_0(\mathbf{x}, \mathbf{x}'))\, dS' \tag{3.87}$$

S' is the 2D surface of the sea and G_0 is given by equation (3.34). This may be simplified by integrating along x, the grooves of the corrugated surface. The fields are constant with respect to x, and so the integrand remains unchanged except for

G_0, which integrates up to become

$$G_0^{(1,2)}(\mathbf{x}-\mathbf{x}') = -\frac{i}{4}H_0^{(1)}(k_{1,2}|\mathbf{x}-\mathbf{x}'|) \tag{3.88}$$

where $H_0^{(1)}$ is a Hankel function of zeroth order and the first kind and where 1 and 2 refer to free space and sea water. Thus,

$$k_1 = \omega\sqrt{\mu_0\varepsilon_0} \quad \text{and} \quad k_2 = k_1\sqrt{\varepsilon} \tag{3.89}$$

S' is now a one dimensional integral along the surface. As shown in Appendix 3, taking the point $\mathbf{x}$ down onto the sea surface results in equations above and below the surface as follows

$$\psi(\mathbf{x}) = 2\psi_{\text{in}}(\mathbf{x}) + 2\int (G_0^{(1)}(\mathbf{x},\mathbf{x}')\zeta(\mathbf{x}') - \psi(\mathbf{x}')\mathbf{n}(\mathbf{x}')\cdot\nabla' G_0^{(1)}(\mathbf{x},\mathbf{x}'))\,dS' \tag{3.90}$$

and

$$\psi(\mathbf{x}) = -2\int (\alpha\, G_0^{(2)}(\mathbf{x},\mathbf{x}')\zeta(\mathbf{x}') - \psi(\mathbf{x}')\mathbf{n}(\mathbf{x}')\cdot\nabla' G_0^{(2)}(\mathbf{x},\mathbf{x}'))\,dS' \tag{3.91}$$

where α is as defined in the table above. In order to solve these coupled equations it is convenient to exploit the rapid decay of the Green's function in the sea, and to use the impedance boundary condition

$$\psi(\mathbf{x}) \approx \frac{i\alpha\zeta(\mathbf{x})}{k_2} \tag{3.92}$$

in order to relate the field to its normal derivative on the surface. This leads to a single integral equation for the unknown field.

Another way to decouple the surface field equations is to assume that the sea is a perfect conductor. In this case ψ_{H} and $\mathbf{n}\cdot\nabla\psi_{\text{V}}$ are zero on the surface, and we can write

$$\psi_{\text{V}}(\mathbf{x}) = 2\psi_{\text{Vin}}(\mathbf{x}) - 2\int (\psi_{\text{V}}(\mathbf{x}')\mathbf{n}(\mathbf{x}')\cdot\nabla' G_0^{(1)}(\mathbf{x},\mathbf{x}'))\,dS'$$

$$\zeta_{\text{H}}(\mathbf{x}) = 2\zeta_{\text{Hin}}(\mathbf{x}) + 2\int (\zeta_{\text{H}}(\mathbf{x}')\mathbf{n}(\mathbf{x})\cdot\nabla G_0^{(1)}(\mathbf{x},\mathbf{x}'))\,dS' \tag{3.93}$$

for V and H respectively.

The relative simplicity of these scalar equations for 1D surfaces means that it is practical to consider numerical solution methods. The low grazing angle geometry introduces significant problems due to the necessarily finite size of computation surface that can be accommodated in the computer; the unavoidable truncation of the surface induces a spurious edge scattering, which is particularly pronounced in the low grazing angle, back-scatter configuration. One way to avoid these effects is to apply a smooth weighting to the incoming and outgoing beams. This, however, suppresses multiple scattering unless the beam pattern is very wide. In an alternative scheme, Holliday *et al.* [28] have suggested that a finite portion of sea surface be joined smoothly on either side to adjunct planes extending to infinity. The fields established on these planes can be approximated by the Fresnel results, appropriate to an infinite planar interface. The adjunct plane contributions to the surface field on the sea surface portion can then be calculated; (3.90) is thus reduced to an integral equation over a finite support. This can be discretised and solved by a variant of the Gauss Seidel method of iterative matrix inversion, developed in this context by Holliday and his co-workers [28] and called Forward-back. Appendix 3 provides explicit details of these numerical techniques for perfectly conducting and for complex dielectric surfaces, along with some potential improvements.

Once the equations for the surface field and its normal gradient have been solved, the scattered field can be calculated relatively straightforwardly by substitution back into equation (3.78), with $\mathbf{r}$ at the radar antenna. $\psi_{\text{in}}(\mathbf{r})$ can be omitted, since it is the transmitted wave. The RCS for a corrugated surface has a similar definition to the 3D RCS given in equation (3.46), except that an isotropic scatterer spreads the energy over a line of length $2\pi r$ rather than an area of $4\pi r^2$. Hence the corrugated RCS, which is measured in length rather than area, is given by

$$\sigma = \text{Lim}\left[2\pi r \frac{|\psi_{\text{scat}}(\mathbf{r})|^2}{\psi_0^2}\right]; \quad r \to \infty \tag{3.94}$$

Using the asymptotic form of the Green's function (3.88) and its derivative leads, as shown in Appendix 3, to

$$\sigma = \frac{1}{4k}\left|\int \exp(i\mathbf{k}\cdot\mathbf{x})(\zeta(\mathbf{x}) - \psi(\mathbf{x})\mathbf{n}(\mathbf{x})\cdot\mathbf{k})\, dS\right|^2 \tag{3.95}$$

The normalised average RCS is

$$\sigma^0 = \frac{\langle\sigma\rangle}{L} \tag{3.96}$$

where L is the length of the patch of corrugated sea.

Having a means to calculate numerically the scattering from a rough sea surface allows us to compare the accuracy of various approximate models. The scattering surface is carefully controlled to demonstrate the low grazing angle effects of multi-path and shadowing. We start with a large perfectly conducting Gaussian as shown in Figure 3.18, which is smooth at the radar wavelength, and then superimpose a small Gaussian at different positions. We illuminate the surface at 10° grazing angle with a radar wavelength of 1 cm. The large Gaussian (hump) gives negligible backscatter because it is smooth and there are no specular points. The small hump is about quarter of a radar wavelength long, and therefore produces measurable backscatter.

The height η, as a function of the distance along the profile, y, is given by

$$\begin{aligned} z = \eta(y) &= \eta_{\ell g}(y) + \eta_{sg}(y) \\ &= a_\ell \exp\left(-\frac{(y-0.7)^2}{2c_\ell^2}\right) + a_s \exp\left(-\frac{(y-p)^2}{2c_s^2}\right) \end{aligned} \tag{3.97}$$

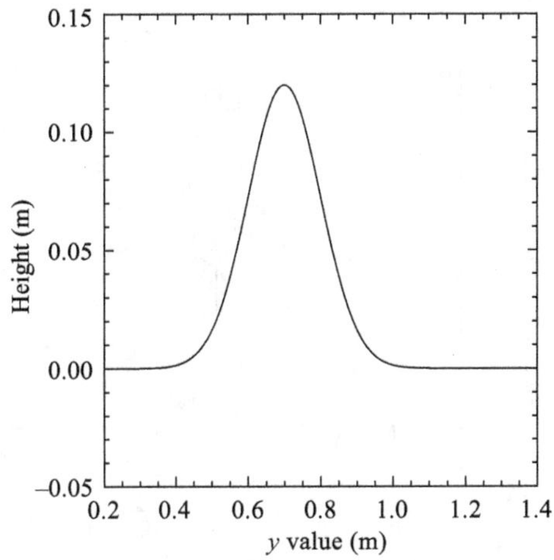

Figure 3.18 The smooth surface height profile

where $\eta_{\ell g}$ is the large Gaussian profile with height $a_\ell = 0.12$ and sd $c_\ell = 0.1$, and η_{sg} is the small Gaussian profile with height $a_s = 5 \times 10^{-4}$ and sd $c_s = 10^{-3}$.

As illuminating radiation has a wavelength of 0.01, the large hump has $ka = 75$ and the small hump has $ka = 0.314$. The large hump is centred at 0.7 and the position p of the small hump is varied from 0.3 to 1.3. The radar illuminates from the right (i.e. from large positive y).

As described in Appendix 3, the numerical code Forward Backward (FB) calculates the RCS from

$$\sigma = \frac{1}{4k}\left|\int_{-L/2}^{L/2} [(q(y) + q_{\mathrm{ap}}(y))\exp(-2ik_z\eta(y)) - 2k_z]\exp(-2ik_{\mathrm{H}}y)\,dy\right|^2 \tag{3.98}$$

where $q(y)$ is the contribution of the surface field at y, which is calculated by FB iteration of the surface field integral equation (SFIE), and $q_{\mathrm{ap}}(y)$ is the contribution to the scattered field of the field on the adjunct planes (beyond $\pm L/2$) induced by the field at y. Figure 3.19 shows results from FB of the variation of RCS of the double Gaussian as a function of the position of the small hump. The top trace is Vpol and the lower trace is Hpol.

With the small hump on the plane in front of the large one (i.e. at 1.2 m), the Vpol and Hpol RCS values correspond to the Rice-Peake flat plane perturbation result of equation (3.64), and thus Hpol is much greater than Vpol (over 30 dB). As the small hump is moved up the front slope of the large one (1.1–0.7 m) interference fringes are evident from the sum of the direct illumination and the forward specular scatter from the plane in front of the large hump. The Vpol and Hpol fringes are out of phase because the specular reflection coefficient is 1 for Vpol and –1 for Hpol. At various

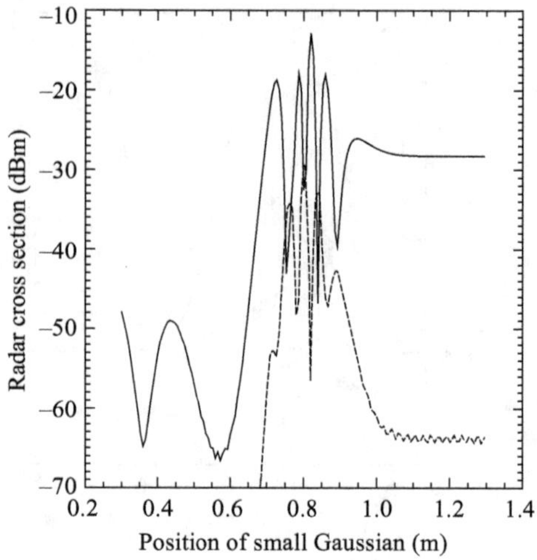

Figure 3.19 FB RCS results for V pol (solid line) and H pol (dashed line). The scattering surface is the double Gaussian defined in equation (3.97)

positions on the slope the RCS of Hpol exceeds that of Vpol. As the small hump moves past the peak of the large hump (which is at 0.7 m) it goes into shadow and the RCS drops significantly.

The method used at high grazing angles in Section 3.2 is the Kirchoff or Physical Optics approximation. The RCS for a 1D surface is given (following Appendix 3) by

$$\sigma = \frac{1}{4k}\left|\frac{2k}{k_z}\int_{-L/2}^{L/2}[\exp(-2ik_z\eta(y)) - 1]\exp(-2ik_{\mathrm{H}}y)\,dy\right|^2 = \frac{1}{4k}|\Im_{\mathrm{PO}}|^2 \tag{3.99}$$

The small height limit of the Physical Optics approximation gives the Bragg result

$$\sigma_{\mathrm{Bragg}} = \frac{1}{4k}\left|4k^2\int_{-L/2}^{L/2}\eta(y)\exp(-2ik_{\mathrm{H}}y)\,dy\right|^2 = \frac{1}{4k}|\Im_{\mathrm{Bragg}}|^2 \tag{3.100}$$

The Rice perturbation approximation is

$$\sigma_{\mathrm{Rice}} = \frac{1}{4k}\left|\Im_{\mathrm{Bragg}}\left(1 \pm \frac{k_H^2}{k^2}\right)\right|^2 \tag{3.101}$$

where '+' corresponds to V pol and '−' to H pol. The composite model is calculated for this surface from the Rice RCS evaluated at the local grazing angle corresponding

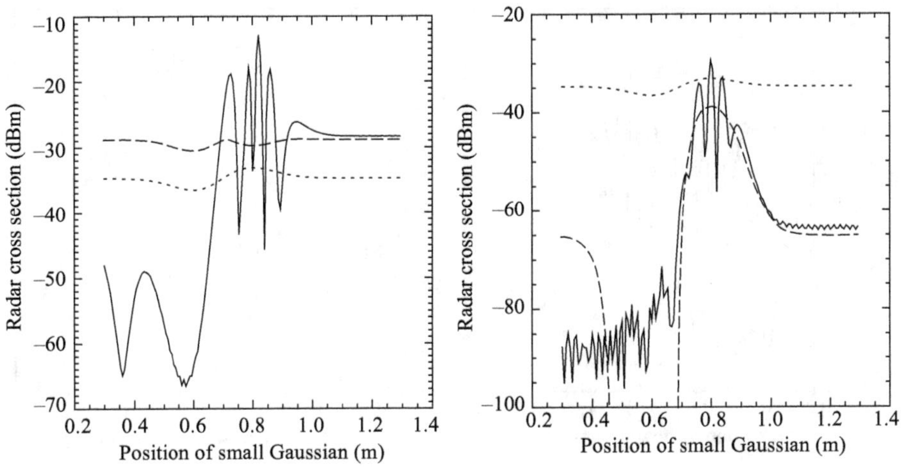

Figure 3.20 *RCS versus small Gaussian position for FB calculations (solid lines), PO calculations (dotted lines) and composite model (dashed lines). The graph on the left is Vertical polarisation and the graph on the right is Horizontal polarisation*

to the small hump position on the large hump. Figure 3.20 shows the FB numerical results compared with Kirchoff (Physical Optics) and the composite model. It is clear that neither of the approximate models, which are both accurate in other grazing angle regimes (as shown in the previous sections), is able to reproduce the low grazing angle multipath and shadowing.

As explained in Appendix 3, the Half Space model [29] provides a fairly straightforward way of introducing some features of multipath through the use of a half space Green's function. The RCS from this technique for a 1D surface may be written as

$$\sigma_{\mathrm{HS}} = \frac{1}{4k}\left|\Im_{\mathrm{PO}} \pm \Im_{\mathrm{Bragg}}\frac{k_{\mathrm{H}}^{2}}{k^{2}}\right|^{2} \tag{3.102}$$

where $\pm$ is again for the two polarisations. The Half Space (HS) results are shown with the FB and Rice results in Figure 3.21. Where the roughness is in front of the smooth profile, both FB and HS show large polarisation ratios approximately consistent with Rice. On the illuminated face of the smooth profile, HS correctly follows the positions of the peaks and troughs but does not match the amplitudes very well. On the shadow side neither HS nor Rice predict the low RCS.

The LRP method uses an estimate for the surface field on a smooth surface ($q_{\ell\mathrm{g}}$; here derived by a Forward stepping of the SFIE along $\eta_{\ell\mathrm{g}}$ and expressed relative to the phase of the incoming wave) and accounts for the roughness η_{sg} through perturbation. The complete formulation involves the surface derivative of the current, a path length integral along the smooth surface, and a requirement for the roughness to be normal to the smooth surface. Here we use an approximation that is of a similar form to

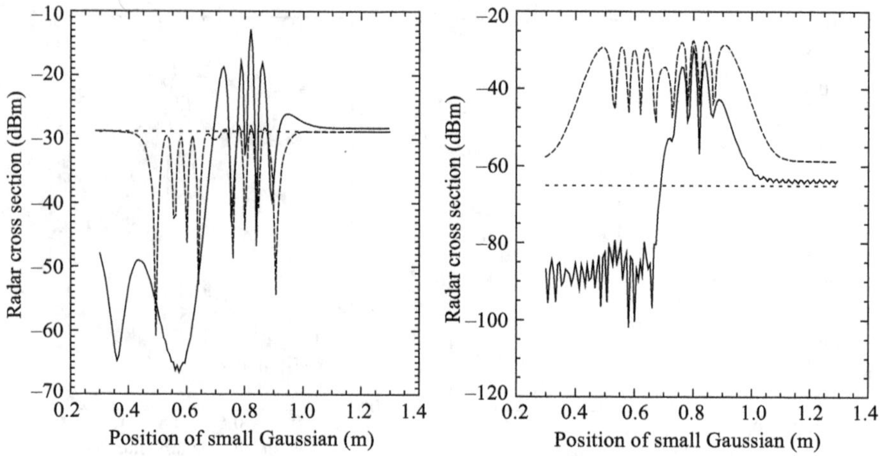

Figure 3.21 RCS versus small Gaussian position for FB (solid lines), Half Space (dashed lines) and Rice (dotted lines). The graph on the left is Vertical polarisation and the graph on the right is Horizontal polarisation

equations (3.99) to (3.102),

$$\sigma_{\text{LRP(H)}} = \frac{1}{4k}\left|\int_{-L/2}^{L/2} q_{\ell g\text{H}}^2(y)\eta_{sg}(y)\exp(-2ik_{\text{H}}y)\,dy\right|^2 \tag{3.103}$$

$$\sigma_{\text{LRP(V)}} = \frac{1}{4k}\left|\int_{-L/2}^{L/2} q_{\ell g\text{V}}^2(y)\left[1+\frac{k_{\text{H}}^2}{k^2}\right]\eta_{sg}(y)\exp(-2ik_{\text{H}}y)\right|^2 dy \tag{3.104}$$

Figure 3.22 is a comparison of FB and LRP, which match very well except in the shadow region of HH (where FB is at its accuracy limit). Although the LRP method clearly does better than the other approximations, it uses a numerical solution for the fields on the smooth surface. It therefore is limited with respect to further direct analysis, such as the derivation of average RCS for a surface defined by its wave spectrum. However, it demonstrates clearly where the composite model is deficient at low grazing angles, in terms of the lack of multipath and shadowing. As was noted in the previous section, the composite model may be derived from the LRP method by assuming that the smooth surface fields satisfy the tangent plane approximation, and therefore only depend upon the local slope. The results here for the two Gaussians clearly show that this approximation is poor at low grazing angle.

The results presented in Figures 3.18–3.21 are for a perfectly conducting surface. The analysis may be carried forward to an imperfect conductor, which is more representative of sea water. The Hpol results are very similar for both materials. The Vpol results differ because of the absorption that occurs on forward reflection near to the

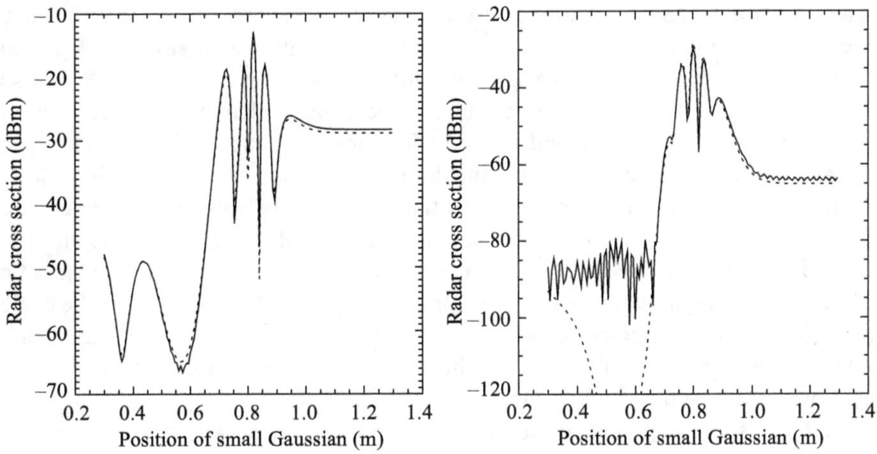

Figure 3.22 RCS versus small Gaussian position for FB (solid lines) and LRP (dotted lines). The graph on the left is Vertical polarisation and the graph on the right is Horizontal polarisation

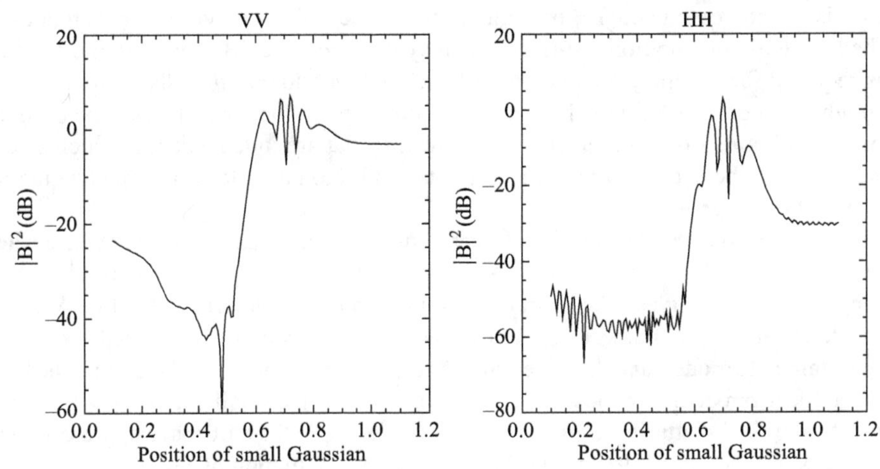

Figure 3.23 Dielectric conductor RCS versus small Gaussian position for FB ($|B|^2$ can be converted to RCS by the subtraction of 35 dB)

Brewster angle. This reduces the multipath interference. Figure 3.23 shows the FB results for sea water and demonstrates this Brewster angle effect.

The problems of multipath and shadowing at low grazing angle have been subjects of interest for many years. A 'critical angle' phenomenon in sea clutter was first observed by Katzin [30] in 1957. Usually sea clutter power falls off as R^{-3}, where R is the range; this is due to the combination of the radar equation R^{-4} and the linear increase in clutter patch size with range. Katzin noticed that when the grazing angle is smaller than a 'critical angle' the clutter power law changes to R^{-7}. This he described as being due to multipath interference; his explanation was viewed with some scepticism at the time and is not widely accepted even today. An alternative explanation from Wetzel [31] is that the dominant mechanism is shadowing and that multipath is unimportant. A key differentiating feature between these two explanations is the variation of the critical angle with sea state. For the multipath effect, the path length differences between the direct and indirect paths tend to be less than $\lambda/2$ at grazing angles below the critical angle and greater than $\lambda/2$ above the critical angle. Thus, the rougher the sea, the lower the multipath critical angle. Rougher seas, however, produce more shadowing. Therefore if there were a critical angle due to shadowing, it would be higher, the rougher the sea.

In the absence of good experimental data covering the necessary range of grazing angles and sea states (and not corrupted by anomalous propagation effects), it has been difficult to examine the critical angle behaviour in a controlled way. However, the LRP method described above allows us to investigate the phenomenon numerically. In order to do this we simulate many realisations of the long-wave sea surface spectrum, calculate the surface fields using Forward-Back (FB), and then calculate the backscatter from the short wave spectrum using the LRP method. The advantage of this method over using FB on simulations of the full sea wave spectrum relates to the length of computational surface that may be used. The FB sampling interval that is required for accuracy on a smooth surface is much longer than that required on a rough surface. Therefore the LRP method allows much longer surfaces to be used. In order to obtain normalised average RCS results that are independent of patch size it was found to be necessary to use a computation region of at least 150 m at a grazing angle of 0.1 degrees.

Figure 3.24 shows the results of averaging many realisations of an ocean surface using equation (3.2) for the wave spectrum and a wind speed of 2 ms^{-1} over a range of grazing angles. The short black solid line is the numerical result for VV and the dotted line is the numerical result for HH. Also shown are: calculated results for the composite model; and the Rice limit. The Rice limit is achieved by setting the long wave spectrum in the composite model to zero. This is the small grazing angle limit of most analytical scattering approximations; for example the half space approximation of Shaw and Dougan [29] and the small slope approximation of Voronovich [32].

The plot shows that, at a wind speed of 2 ms^{-1}, the normalised RCS deviates from the composite model at a grazing angle of about 10° and then follows a fourth power law at lower grazing angles; this is consistent with the R^{-7} behaviour observed by Katzin [30]. For VV this behaviour follows the Rice result, but for HH it is offset by approximately 10 dB.

The next step is to extend the grazing angle down to 0.1° and increase the wind speed (and hence the ocean wave height). Unfortunately the numerical calculations using our imperfect conductor FB code become unstable if either of these extensions are attempted. However, the results for a perfect conductor are generally

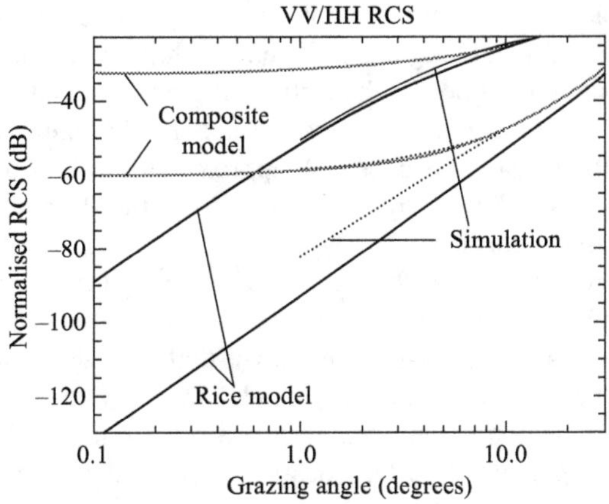

Figure 3.24 Average normalised RCS results obtained using the LRP method for an imperfectly conducting surface

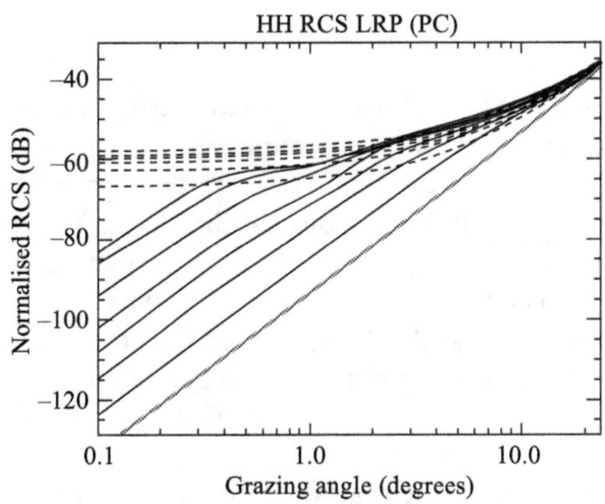

Figure 3.25 Average normalised RCS results obtained using the LRP method for a perfectly conducting surface

indistinguishable from an imperfect conductor for HH polarisation and so we are able to extend the grazing angle and wave height using the more robust perfect conductor code.

Figure 3.25 shows results for HH from 0.1° to 30° grazing angle for wind speeds of 2–8 ms^{-1} (in steps of 1 ms^{-1}). The dotted lines correspond to the composite model, the black lines to the numerical results and the grey line is the Rice result. For both the dashed and the black lines the higher RCS's correspond to the higher wind speeds. These results clearly show that the critical angle decreases with increased sea-state. This means that the interference model of Katzin [30] appears to be favoured over the 'Threshold Shadow' model of Wetzel [31]. There might be some evidence of shadowing; as the grazing angle is reduced, the higher sea-state RCS values drop below the composite model before the critical angle is reached. The effect, however is relatively small. The results also show that the small grazing angle behaviours of the Half Space [29] and Small Slopes [32] approximations appear to be many orders of magnitude (up to 50 dB) in error, as shown from the Rice result, for medium to high sea-state Gaussian surfaces.

Clearly these results need to be extended to VV polarisation by improving the imperfect conductor calculations. At the time of writing this is ongoing work.

To summarise this section, we have shown that at low grazing angles it is necessary to extend beyond the composite model in order to model scattering from ocean-like surfaces, which have roughness at a wide range of length scales. However, the modelling presented so far does not explain the spikes and bursts shown in the data in Chapter 2. For these effects it is necessary to look beyond the linear Gaussian representation of the sea and to consider the non-linear effects of breaking waves.

3.7 Scattering from breaking waves

In Section 3.2 we discussed how a breaking wave criterion can be applied to a Gaussian model of the sea surface. For this we assume that regions of the Gaussian random field where the downward vertical acceleration of the surface is greater than $g/2$ (g is the acceleration due to gravity) are subject to breaking. This breaking area model [12] provides a simple means to calculate the proportion of the sea surface that is breaking, and since it is derived directly from the wave spectrum, it can be related to sea state, wind speed and fetch. More detailed features of the breaking area, such as the properties of the local and global maxima and the supports of above threshold excursions, may be derived from the underlying Gaussian model. These are described in Reference 12.

In the mean time we will concentrate on the contribution the breaking waves make to the clutter RCS. As shown in Figure 3.4, as a wave approaches breaking it sharpens at the top and flattens at the bottom. The sharp crest then steepens rapidly and curls over, before the water tumbles down the front wave slope.

Using the numerically derived waveshapes of Figure 3.4, scattering calculations have been performed by Forward-Back at a grazing angle of 6° [33]. The results are shown in Figure 3.26. The first wave profile is at $T = 0$ s and the wave breaks at $T = 1.36$ s. The wave profiles are shown in Figure 3.4 in timesteps of 0.2 s. Figure 3.26 shows that initially the wave has a very small RCS, but as it approaches breaking the RCS increases dramatically. During this rise σ_{VV} is initially greater than σ_{HH}, but towards the top Hpol overtakes Vpol, and at the point of breaking is significantly higher. The mechanism for this is believed to be that the wave steepens until there is a specular point near the top of the wave. This specular point is illuminated by a complex multipath pattern, due to the concave shape of the wave below the peak. This, and the height of the wave, causes σ_{HH} to be significantly greater

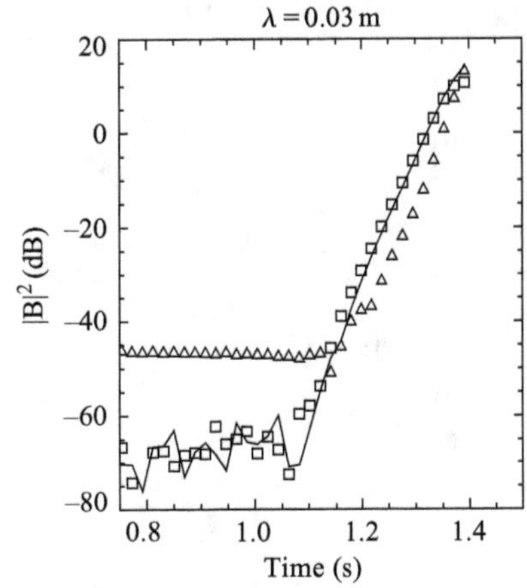

Figure 3.26 *Uncalibrated RCS of breaking wave sequence shown in Figure 3.4 (Triangles HH, Squares VV) backscatter calculated using imperfect conductivity FB. The black line corresponds to a model described in [33] (figure courtesy of QinetiQ).*

than σ_{VV}. This type of scattering may explain the 'bursts' in the real data described in Chapter 2.

The scattering from the wave can be understood further by varying the radar frequency. The Hpol frequency response of a 2.3 m wavelength breaking-wave (i.e. a wave at the peak of Figure 3.26) has been calculated using FB and is plotted (using triangular symbols) in Figure 3.27. Data have also been collected using a 4–8 GHz FMCW radar at the Ocean Engineering Lab, UC Santa Barbara by Fuchs, Lamont-Smith and Tulin [34,35]. Results from that experiment are reproduced in Figure 3.27 as a grey line. The numerical calculations and the measurements match very well, and the cyclical variation of RCS with frequency is suggestive of the proposed multipath mechanism.

The numerical results in Figure 3.27 have been calculated using a 1D corrugated surface. They are converted from the corrugated result σ to the radar cross-sectional area, σ_{3D}, in m^2 by using an estimate of the lateral coherent extent of the scattering surface ℓ_x according to the formula

$$\sigma_{3D} \cong \frac{k\ell_x^2}{\pi}\sigma \tag{3.105}$$

In this case, ℓ_x is assumed to be 0.375 m, which is a reasonable estimate given the lateral coherence of the waves in the tank and the radar beamwidth.

The x-axis of Figure 3.27 and 3.28 are in terms of the ratio of water wavelength to radar wavelength. This is because waves 'scale' above a wavelength of a few centimetres (where the effects of surface tension are negligible) and maintain approximately the same shape. Thus, to first order (neglecting the change of dielectric constant with frequency), the shape of the RCS curve should only depend on the wavelength ratio.

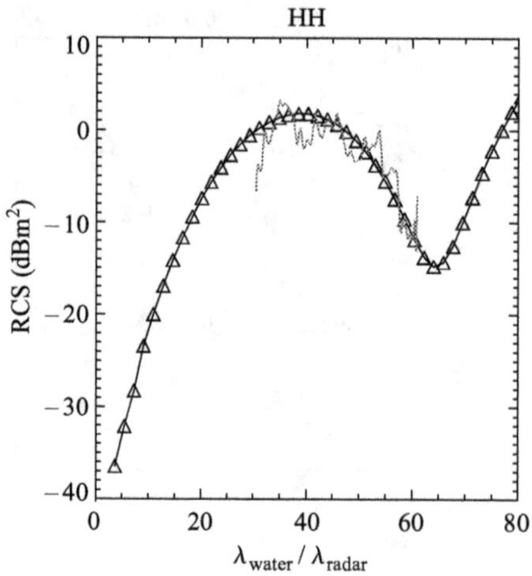

Figure 3.27 Frequency response of a 2.3 m breaking wave (figure courtesy of QinetiQ)

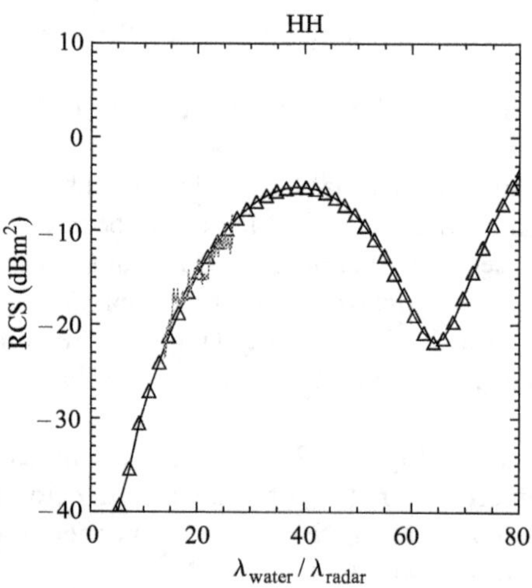

Figure 3.28 Frequency response of a 1.0 m breaking wave, presented in the same format as Figure 3.27 (figure courtesy of QinetiQ)

Fuchs and Tulin [34] also made measurements at other water wavelengths and Figure 3.28 shows the frequency response from a 1.0 m wavelength breaking wave. The value ℓ_x is now assumed to be 0.33 m. There is excellent agreement again between the experimental and numerical results, now on a different part of the curve. This is especially significant given the sensitivity of the frequency response to the height of

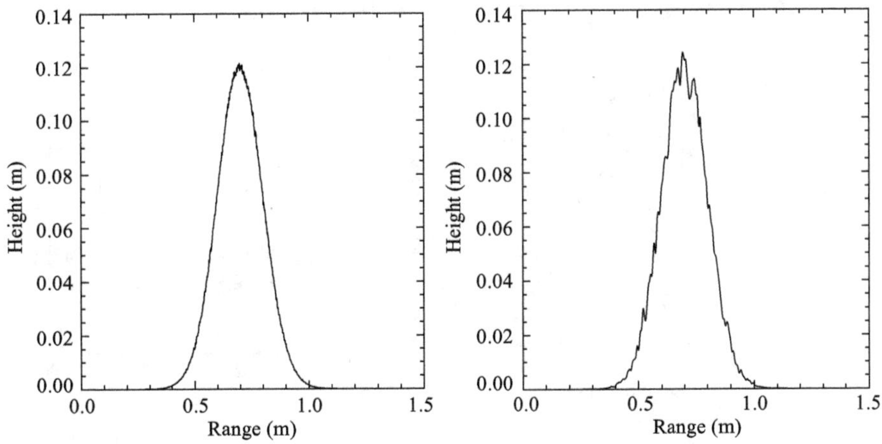

Figure 3.29 *Gaussian wave profiles superimposed with Gaussian noise with rms heights of 0.001 m (left plot) and 0.005 m (right plot) at the crest*

the surface. The height of the waves was measured using a wave wire array, and so it is known that the peak wave height is very close to that of the numerical wave.

When a wave has broken and the water is tumbling down the front face, the wave often appears as a 'whitecap'. The radar returns from these whitecaps are much longer than the short bursts caused by steepening waves. In Chapter 2 we saw that the bursts last for about 100–200 ms, whilst the whitecaps typically last for a second or so. The whitecaps also tend to be noiselike in temporal behaviour, and have roughly equal powers in both polarisations. In order to model the whitecaps we assume that after the wave has broken the very steep gradients on the profile disappear very quickly, and the shape assumes a Gaussian-like profile, with roughness on the front face. Simulations of Gaussian profiles (with the same dimensions as Figure 3.18) superimposed with random Gaussian noise are used in perfect conductor FB scattering calculations. The Gaussian noise is defined by its rms height, and has a Gaussian spectrum and rms slope of one. A Hanning weighting is applied to the noise to concentrate it at the crest of the wave. Figure 3.29 shows two realisations with different Gaussian rms heights.

The results from the scattering calculations are shown in Figure 3.30. Ten simulations are performed at each rms height, and the RCS values from FB are plotted as '+' symbols. Five values of rms height are used, from 0.001 to 0.005 m. The lines on the plots correspond to the Rice SPM result for the surfaces.

The most striking feature of the results in Figure 3.30 is that as the roughness increases the RCS tends to drop. This is a consequence of keeping the rms slope of the noise constant. Thus, as the roughness increases the spectrum becomes narrower and the resonant wave component reduces. For our purposes, we can see that it is quite easy to get a polarisation ratio of one at quite a high RCS value. Thus this simple simulation has the potential of modelling the whitecap following breaking.

The scattering calculations presented in this section and the previous one, suggest that the observed feature of low grazing angle sea clutter can be reproduced from

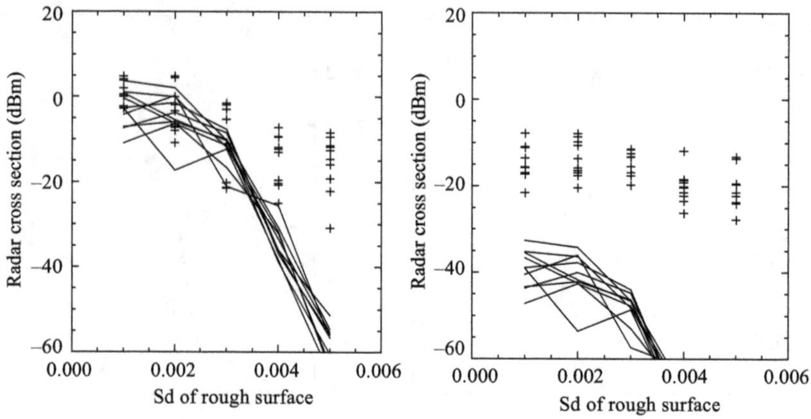

Figure 3.30 *RCS as a function of rms roughness for Gaussian noise added to a smooth Gaussian wave. The graph on the left is VV and the graph on the right is HH. Forward-Backward results are shown as + symbols and the Rice SPM approximation as lines*

electromagnetic modelling. The next section considers the extent to which these features can be put together to model the average clutter RCS versus polarisation, grazing angle and sea state.

3.8 Average backscatter from the ocean at low gazing angles

As we described at the beginning of Section 3.6, the composite model is unable to reproduce the observed sea clutter average backscatter trends at low grazing angles. The GIT empirical model [27], as presented in Figure 3.16, differs from the composite model in that it has less polarisation dependence, more variation with sea state, and a different limiting behaviour at small grazing angle.

The lack of variation with sea-state may be overcome by improving the model of the sea surface, and incorporating the change of wave spectrum in the gravity-capillary wavenumber regime, which was discovered and modelled by Jähne and Riemer [3]. This is discussed in Section 3.2, where the Elfouhaily spectrum [4] is proposed as a means to incorporate these effects. With this modified wave spectrum the variation of σ^0 with sea-state is more realistic as shown in Figure 3.31.

As shown in Section 3.6 further improvements to the composite model may be made through the introduction of shadowing and multipath interference. Shadowing reduces the amount of sea surface that is illuminated. At 10° grazing angle shadowing is negligible; below 10° we introduce an attenuation of 4 dB per decade. Following Katzin's explanation for the critical angle effect [30], multipath interference may be modelled on average using the roughness parameter

$$\rho = \frac{4\pi\sigma_h \sin\theta_{gr}}{\lambda} \tag{3.106}$$

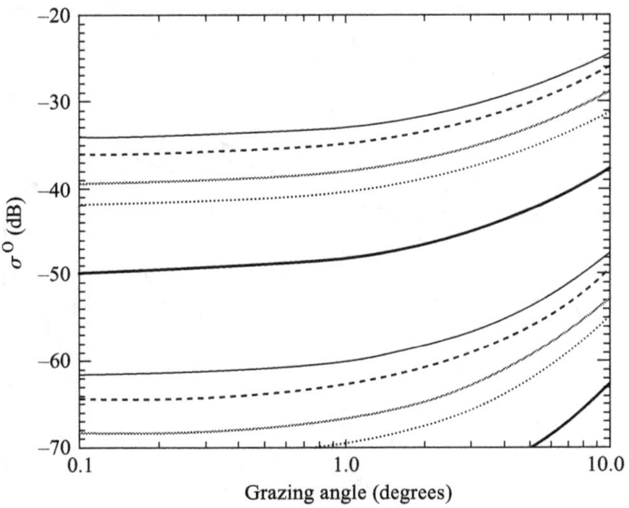

Figure 3.31 *Composite model σ^0 for Vpol (upper lines) and Hpol (lower lines) using the Elfouhaily wave spectrum [4] for sea states 1–5, (black, 1: dotted, 2: light grey, 3: dashed, 4: dark grey: 5)*

where σ_h is the rms wave height, λ is the EM wave length and θ_{gr} is the grazing angle. The normalised RCS, σ^0, is then modified approximately by a factor

$$F = \frac{\rho^4}{1 + \rho^4} \tag{3.107}$$

The effect of the multipath factor F is to introduce a critical angle where $\rho = 1$. Above this angle σ^0 is largely unaffected, below it falls off by α^{-4}. With these additional factors the composite model is as shown in Figure 3.32.

Results in the previous section of scattering calculations show that the RCS of a breaking wave whitecap is largely independent of polarisation, and is somewhat less (−3 dB) than the surface area covered by the whitecap. We can use the BAM [12] discussed earlier to evaluate the area covered by breaking waves; this is derived from the sea wave spectrum. Thus we can evaluate the mean σ^0 of breaking waves and apply the shadowing and multipath factors discussed above. When the breaking wave results are added to the composite model, Figures 3.33 and 3.34 are obtained. These graphs are remarkably similar to the GIT results plotted in Figure 3.16.

Furthermore, it is evident from our model that V pol is dominated by composite model scattering, whilst H pol is dominated by breaking wave events. The intermittent nature of the latter provides a credible explanation for the more spiky nature of Hpol compared to Vpol.

The calculations for σ^0 described above do not include a contribution for the specular scattering that is responsible for the 'bursts' observed predominantly in Horizontal polarisation. This is because they do not appear to provide a significant

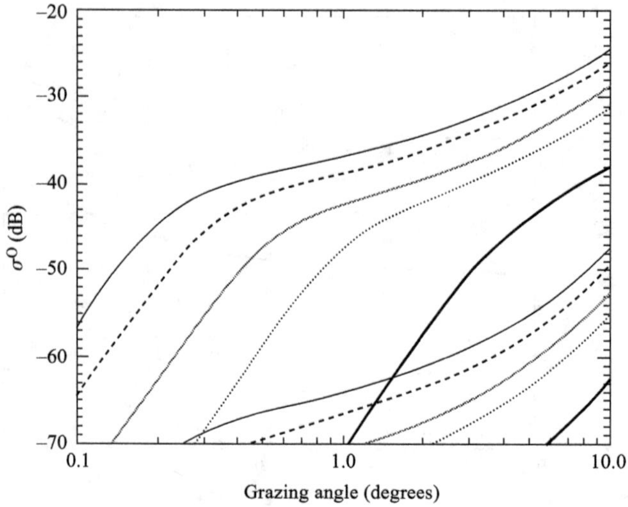

Figure 3.32 *Composite model σ^0 for Vpol (upper lines) and Hpol (lower lines) using the Elfouhaily wave spectrum [4] and including shadowing and multipath for sea states 1–5, (black, 1: dotted, 2: light grey, 3: dashed, 4: dark grey: 5)*

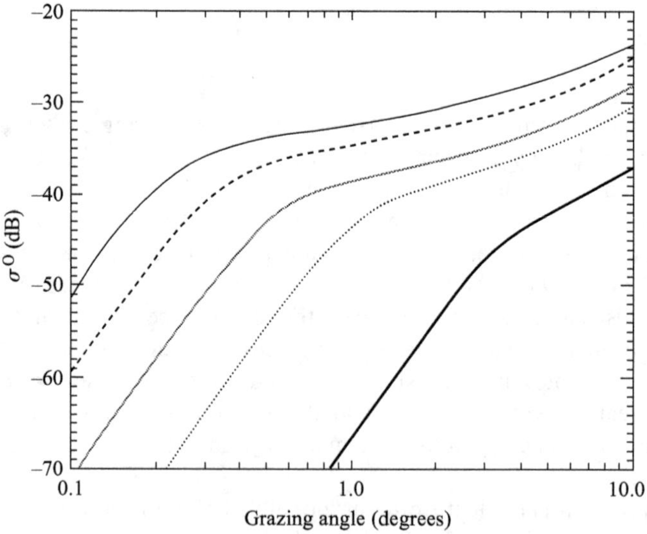

Figure 3.33 *The sum of scattering from the composite model and breaking waves, giving overall σ^0 for Vertical polarisation for sea states 1–5, (black, 1: dotted, 2: light grey, 3: dashed, 4: dark grey: 5)*

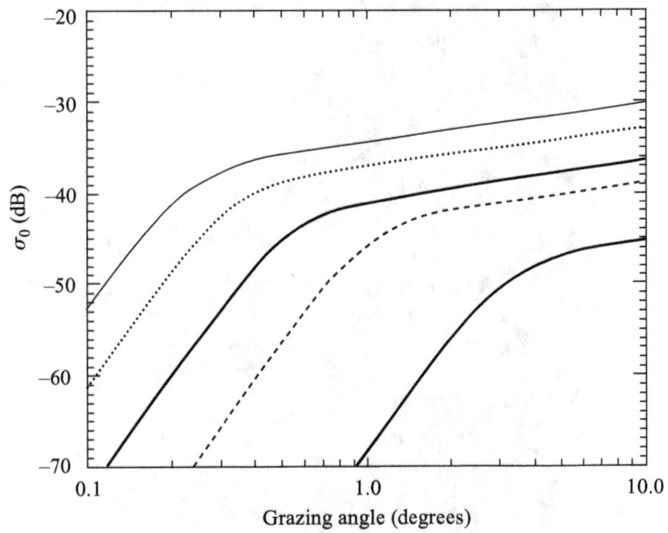

Figure 3.34 *The sum of scattering from the composite model and breaking waves, giving overall σ^0 for Horizontal polarisation for sea states 1–5, (black, 1: dotted, 2: light grey, 3: dashed, 4: dark grey: 5)*

contribution to the overall RCS. They do, however, have a significant effect on the tail of the probability distribution and cause false alarms in target detection systems.

3.9 Imaging ocean currents at low grazing angles

Having developed some understanding of the scattering process at low grazing angles, it is worthwhile returning to the problem of imaging ocean currents. In Section 3.4 we described the process whereby an ocean current field changes the radar backscatter characteristics of the sea surface and produces a distinct signature on radar images at high grazing angles. Recently, Reference 36 reported a set of experiments where it was shown that these signatures increase in magnitude as the grazing angle decreases to the low grazing angle regime (i.e. less than 10°).

The paper describes 12 airborne radar data collections: nine using the Enhanced Surveillance Radar (ESR), which is a synthetic aperture radar (SAR); and three using Searchwater, which is a scanning radar, on the Nimdod MR2. Both radar systems operated at I band. In common with the analysis presented in Section 3.4, the experiments investigated the effects of tidal flows over the South Falls sandbank. A radar image of the area, as seen from the ERS2 satellite, is shown in Figure 3.35. Example results from the low grazing angle radar systems are shown in Figures 3.36 and 3.37.

The change of radar cross section caused by the change of current as the tide flows over the sandbank is characterised in Reference 36 in terms the '*modulation transfer*

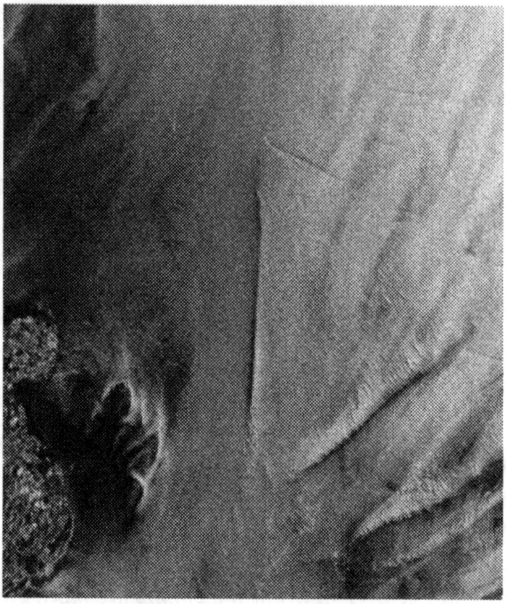

Figure 3.35 ERS2 C-band radar satellite image acquired on 26 September 2000. The South Falls sandbank is the approximately vertical line in the centre of the image (ERS data copyright ESA – Image processed by DERA UK)

function (MTF)', M. This is given by

$$\frac{\Delta\sigma}{\sigma} = -M\frac{dU}{dx} \tag{3.108}$$

where σ is the ambient clutter RCS, $\Delta\sigma$ is the change in RCS, and dU/dx is the current gradient (or strain rate) that causes the RCS change; the detailed process is described in Section 3.4. Equation (3.108) assumes a linear relationship between the modulation and the strain rate. Although this is not obtained in all conditions, it is a reasonable approximation on most occasions, and provides a straightforward method to compare results. Thus, if we evaluate results of the type presented in Section 3.4 in the manner of equation (3.108), we find the following: at high grazing angle of 50–70°, and at moderate wind speeds of 5 ms^{-1} (10 knots), typical MTF values are between 50 and 150 s. In contrast, Reference 36 reports values in the order of 500 s at grazing angles below 6° with HH polarisation. VV polarisation is reported as giving much smaller values.

These results are quite straightforward to interpret in terms of the scattering and RCS models described in earlier sections of this chapter. As the tidal current flows over the shallow sandbank, its speed increases. On the approach side of the sandbank the speed is increasing (positive strain), the wave amplitude is suppressed, and thus the RCS decreases below ambient level. On the other side of the sandbank, the water current is decreasing (negative strain), the wave amplitude is enhanced and the RCS

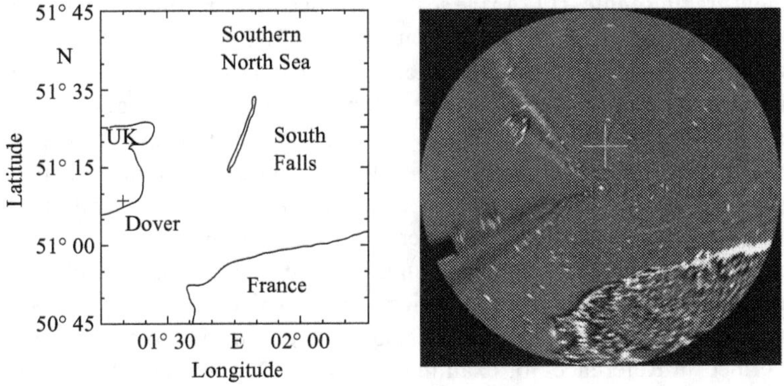

Figure 3.36 *Single scan PPI of the North Sea, with the cursor showing the position of interest centred on the South Falls sandbank. The polarisation is HH and the grazing angle at the sandbank is about 2.5° (figures courtesy of QinetiQ)*

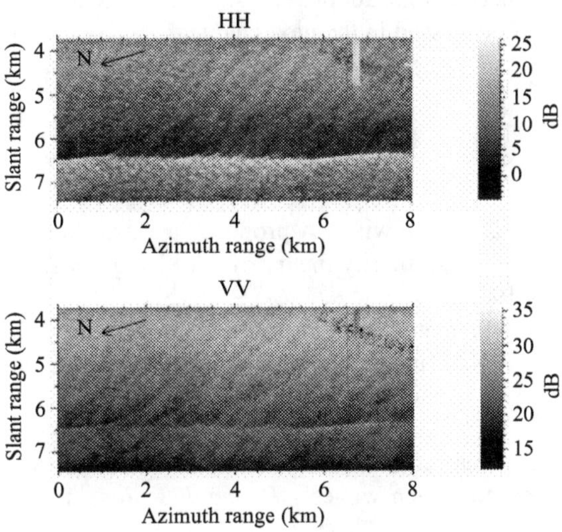

Figure 3.37 *HH and VV SAR images of the sandbank at 6° grazing angle. The wind was 2 ms^{-1} from 120° (figures courtesy of QinetiQ)*

increases. At high grazing angles, radar scattering is essentially from resonant waves riding on longer waves (i.e. the composite model). The increase in RCS with wave height is a smooth process and results in a low MTF. The RCS from very rough patches caused by breaking waves is, on an average at these angles, much less than the resonant scattering.

At low grazing angles (LGA) the scattered power from resonant waves is much less than at higher angles. LGA resonant scattering for HH polarisation is also much less than for VV. The scattered power from very rough patches caused by breaking waves is not reduced by grazing angle to the same extent, and so for HH this dominates the LGA scattered power as modelled in the previous section. The increase of breaking events caused by increase in wave amplitude is an abrupt process, and may be modelled (as we have earlier) in terms of threshold crossings of vertical acceleration. Thus, for HH in particular, the increase of scattered power with increase of wave amplitude is also an abrupt process. This results in a large MTF, and this is as observed in the experiments. The Doppler characteristics of radar spikes caused by breaking waves (results are shown here in Chapter 2) also provide a means to test this hypothesis. As described in Reference 36, examination of the Doppler signatures confirms the model.

We have, therefore, a physical model that is consistent with a wide range of experimental evidence. Much work remains to be done on the detailed characterisation of low-grazing-angle radar scattering from the sea, but we believe that we understand, in outline, the main processes involved. However, translating this directly to radar performance and the assessment of signal processing schemes is beyond current capability. Therefore it is necessary to adopt a statistical and essentially empirical approach to modelling. This is addressed in the next chapter.

References

1 W.J. Pierson and L. Moskowitz, 'A proposed spectral form for fully developed wind sea based on the similarity theory of SA Kitaigorodskii', *J. Geophys. Res.*, **69**, 5181–5190, 1964.

2 M.I. Skolnik, *'Radar Handbook'*, second edition, McGraw-Hill, 1990.

3 B. Jähne and K.S. Riemer, 'Two-dimensional wave number spectra of small-scale water surface waves', *J. Geophys. Res.*, **95**, 11531–11546, 1990.

4 T. Elfouhaily, B. Chapron and K. Katsaros, 'A unified directional spectrum for long and short wind-driven waves', *J. Geophys. Res.*, **102**(C7), 15781–15796, 1997.

5 T. Inoue, '*On the growth of the spectrum of a wind generated sea according to a modified Miles-Phillips mechanism and its application to wave forecasting*', Report TR-67-5, Geophysics Science Lab., New York University, New York, 1967.

6 K. Hasselmann, 'Measurements of wind-wave growth and swell during the Joint North Sea Wave Project (JONSWAP)', *Deutch. Hydrogr. Z.*, **12**, 95, 1973.

7 B. LeMehaute, '*An introduction to hydrodynamics and water waves*', Springer-Verlag, New York, 1976.

8 R.K. Price, 'The breaking of water waves', *J. Geophys. Res.*, **76**, 1576–1581, 1971.

9 M.S. Longuet-Higgins, 'On the overturning of gravity waves', *Proc. Roy. Soc.*, **A376**, 377–400, 1981.

10 D. Xu, P.A. Hwang and J. Wu, 'Breaking of wind generated waves', *J. Phys. Oceanogr.*, **16**, 2172–2178, 1986.

11 M.A. Srokosz, 'On the probability of wave breaking in deep water', *J. Phys. Oceanogr.*, **16**, 382–385, 1986.

12 M. Clements and U. Yurtsever, 'The Breaking Area model for LGA radar sea clutter', Dynamics Technology Inc, DTW-9234-97009, 1997.

13 J.W. Dold, 'An efficient surface-integral algorithm applied to unsteady gravity waves', *J. Comp. Phys.*, **103**, 90–115, 1992.

14 P. Wang, Y. Yao and M.P. Tulin, 'An efficient numerical tank for non-linear water waves, based on the multi-subdomain approach with BEM', *Int. J. Num. Math. Fluids*, **20**, 1315–1336, 1995.

15 J.A. Stratton, '*Electromagnetic theory*', Section 8.14, McGraw Hill, New York, 1945.

16 B. Hughes, 'The effect of internal waves on surface wind waves: 2. Theoretical Analysis', *JGR*, **83**(C1), 455–465, 1978.

17 O.M. Phillips, '*The dynamics of the upper ocean*', Cambridge University Press, New York, 1966.

18 D. Holliday, G. St-Cyr and N.E. Woods, 'Comparison of a new radar ocean imaging model with SARSEX internal wave image data', *Int. J. Remote Sensing*, **8**(9), 1423–1430, 1987.

19 W. Alpers and I. Hennings, 'A theory of the imaging mechanism of underwater bottom topography by real and synthetic aperture radar', *J. Geophys. Res.*, **89**(C6), 10529–10546, 1984.

20 D. Holliday, G. St-Cyr and N.E. Woods, 'A radar ocean imaging model for small to moderate incidence angles', *Int. J. Remote Sensing*, **7**(12), 1809–1834, 1986.

21 S.O. Rice, 'Reflections of electromagnetic waves from slightly rough surfaces', *Commun. Pure Appl. Math.*, **4**, 351–378, 1951.

22 W.H. Peake, 'Theory of radar return from terrain', *IRE Nat. Conv. Rec.*, **7**, 27–41, 1959.

23 J.W. Wright, 'A new model for sea clutter', *IEEE Trans. AP*, **16**, 217–223, 1968.

24 D. Holliday, 'Resolution of a controversy surrounding the Kirchhoff approach and the small perturbation method in rough surface scattering theory', *IEEE Trans. Antennas Propagat.*, **AP-35** , 120–122, 1987.

25 G.R. Valenzuela, 'Theories for the interaction of electromagnetic waves and oceanic waves – a review', *Boundary-Layer Meteorol.*, **13**, 61–85, 1978.

26 D.R. Thompson, 'Calculation of radar backscatter modulation from internal waves', *J. Geophys. Res.*, **93**(C10), 12371–12380, 1988.

27 M.M. Horst, F.B. Dyer and M.T. Tuley, 'Radar Sea Clutter model', International IEEE AP/S URSI symposium, 1978.

28 D. Holliday, L.L. DeRaad, Jr. and G.J. St.-Cyr, 'Forward-backward: a new method for computing low grazing angle scattering', *IEEE Trans. Antennas Propagat.*, **44**, 722–729, 1996.

29 W.T. Shaw and A.J. Dougan, 'Green's function refinement as an approach to radar back-scatter: general theory and applications to LGA scattering', *IEEE Trans. Antenna Propagat.*, **46**, 57, 1998.
30 M. Katzin, 'On the mechanisms of radar sea clutter', *Proc. IRE*, **45**, 44–54, 1957.
31 L.B. Wetzel, 'Electromagnetic scattering from the sea at low grazing angles', Vol. 2, in *Surface Waves and Fluxes*, G.L. Geernaert and W.J. Plant (Eds.), Kluwer Academic Publishers, pp. 109–171, 1990.
32 A. Voronovich, 'Small slope approximation for electromagnetic wave scattering at a rough interface of two dielectric half-spaces', *Waves Random Media*, **4**, 337, 1994.
33 T. Lamont-Smith, 'An empirical model for EM scattering from steeping wave profiles derived from numerical computation', *IEEE Trans. Geosci. Remote Sens*, **41**(6), 1447–1454, 2003.
34 J. Fuchs and M.P. Tulin, 'Laboratory measurements of the joint dependence of radar frequency and water wavelength on sea spike returns', IGARSS '99 Conference Proceedings, Seattle, 1999.
35 J. Fuchs, T. Lamont-Smith, and M.P. Tulin, 'Laboratory Measurements of LGA' Doppler Spectral components and their physical saivces unveiled' *Proc IGARSS 98*, Seattle 1998, 2276–2278.
36 T. Lamont-Smith, A.M. Jackson, P.W. Shepherd and R.D. Hill, 'Low grazing angle radar imaging experiments over the South Falls sandbank', *Int. J. Remote Sensing*, **26**(5), 937–966, 2005.

Chapter 4
Statistical models of sea clutter

4.1 Introduction

In this chapter we develop the models of sea clutter that will provide a basis for our subsequent discussion and calculation of radar performance. As we saw in the previous chapter, a direct approach to the modelling of sea clutter provides many useful insights. Nonetheless the interactions between the ocean, atmosphere and microwave radiation are far too complicated to be described usefully in strictly deterministic terms. Consequently the models we develop in this chapter are unashamedly statistical in character.

The development of these statistical models is driven by the observations of real sea clutter described in Chapter 2, where (Section 2.4) it was noted that low-resolution, spatially uniform, clutter has Gaussian amplitude statistics. By introducing a simple random walk model for the clutter process we can construct this familiar Gaussian (Rayleigh) model for low-resolution sea clutter from first principles and characterise its interaction with target returns. This analysis, in turn, can be modified to take account of the effects of increased radar resolution; in this way we are led to the K distribution and related models. An advantage of this approach is that it demonstrates clearly how performance analysis germane to low resolution Rayleigh clutter can be extended quite straightforwardly to the K-distributed case. These few simple principles underpin much of the radar performance analysis presented in the rest of this book.

We then discuss how K distribution models can be modified to take account of discrete spike-like features in the clutter, and their relationship to Weibull, log normal, Class A, BAM and other models described in the literature. Finally we extend the compound K distribution approach to the modelling of coherent clutter; while this discussion is not as complete as that of incoherent clutter, it does provide a framework for a tractable and physically motivated analysis of an increasingly important subject.

4.2 Gaussian clutter models

We begin with a simple description of low-resolution radar sea clutter that will, nonetheless, underpin a large part of our subsequent analysis of high resolution clutter. Most of the discussion will focus on the analytic characterisation of statistical properties of the clutter; for this an account of the necessary background in probability theory can be found in Appendix A1. Nonetheless, much of what we do is motivated by experimental data, and makes direct contact with the phenomenology presented in Chapter 2.

When the radar illuminates a collection of scatterers, every one makes a contribution to the returned field; the total field is simply the superposition of these. Each of these individual scattered fields is represented as a combination of two components (much as a harmonic signal with a given frequency can be constructed as the sum of a sine and a cosine). In practice these are recognised as its in-phase and quadrature components. Consequently each field can be represented as a two-dimensional vector or, equivalently, as a complex number in the Argand plane. These can then be summed up to give a resultant vector that is identified with the total scattered field. In the limit of a large number of scatterers, this simple picture leads us to a model for Gaussian or thermal noise. Thus we take the scattered field $\mathbf{E}$ to be composed of contributions from discrete scatterers:

$$\mathbf{E} = \sum_{n=1}^{N} \mathbf{a}_n \tag{4.1}$$

and to be visualised as the resultant of a random walk with N steps. The individual scatterer contributions can be expressed in terms of scattering cross section σ and phase ϕ

$$\begin{aligned} a_{n,I} &= \sqrt{\sigma_n}\cos(\phi(t)), \quad a_{n,Q} = \sqrt{\sigma_n}\sin(\phi(t)) \\ \mathbf{a}_n &\equiv \sqrt{\sigma_n}\exp(i\phi(t)) \end{aligned} \tag{4.2}$$

Taking this simple, but physically motivated, model as our starting point we can develop the statistics of $\mathbf{E}$ in a way that highlights both the model's potential utility and its possible shortcomings. In general, for an arbitrary N, it would be impossible to give a useful description of these statistics, without a detailed knowledge of the statistics of the individual scatterer contributions, $\mathbf{a}_n$. In the case where N is large however, progress can be made, initially at the price of restricting the model to clutter found in low-resolution radar systems.

The probability density function of the scattered field $\mathbf{E}$, or of its amplitude E, provides us with a simple statistical model, from which useful features of radar performance can be calculated. In Appendix A1 we introduce the characteristic function (i.e. the Fourier transform of the pdf) as a useful tool in statistical modelling. We note that the pdf of the sum of two or more independent random variables is provided by the convolution of their individual pdfs; consequently the characteristic function of this resulting pdf is the product of those of its constituent, convolved, pdfs. So if the scattered field $\mathbf{E}$ has the form (4.1), we can construct the characteristic function of its distribution as

$$C(\mathbf{U}) = \langle \exp(i\mathbf{U}\cdot\mathbf{E})\rangle = \left\langle \exp\left(i\mathbf{U}\cdot\sum_{n=1}^{N}\mathbf{a}_n\right)\right\rangle = \left\langle \prod_{n=1}^{N}\exp(i\mathbf{U}\cdot\mathbf{a}_n)\right\rangle$$
$$= \langle \exp(i\mathbf{U}\cdot\mathbf{a})\rangle^N \tag{4.3}$$

where we have assumed the **a** to be independent and to have identical statistics. These simplifying assumptions can be relaxed if required; the outcome of the analysis remains the same. To ensure that the power in the scattered field remains finite as the number of scatterers becomes large, we scale a with the square root of N, $a \rightarrow a/\sqrt{N}$. Having done this we expand the characteristic function of the distribution of **a** as

$$\left\langle \exp\left(i\frac{\mathbf{U}\cdot\mathbf{a}}{\sqrt{N}}\right)\right\rangle \approx 1 + i\frac{\langle\mathbf{U}\cdot\mathbf{a}\rangle}{\sqrt{N}} - \frac{\langle(\mathbf{U}\cdot\mathbf{a})^2\rangle}{2N}; \tag{4.4}$$

Higher order terms give rise to contributions that vanish in the large N limit. We expect the **a** to be distributed isotropically (i.e. with random phase) and $\langle\mathbf{U}\cdot\mathbf{a}\rangle$ to vanish as a result; as N becomes ever larger the characteristic function (4.3) tends to a simple limiting form

$$\langle\exp(i\mathbf{U}\cdot\mathbf{E})\rangle = \lim_{N\rightarrow\infty}\left(1 - \frac{U^2\langle a^2\rangle}{4N}\right)^N = \exp(-U^2\langle a^2\rangle/4) \tag{4.5}$$

This is the familiar Gaussian characteristic function of the normal distribution; the forgoing argument can be thought of as a simplified sketch of the derivation of the central limit theorem [1]. We can now recover the required pdf of **E** by Fourier inversion. This gives us the Gaussian or Rayleigh model for the clutter

$$P(E_\mathrm{I}, E_\mathrm{Q}) = \frac{1}{\pi x}\exp(-(E_\mathrm{I}^2 + E_\mathrm{Q}^2)/x); \quad -\infty \leq E_\mathrm{I}, E_\mathrm{Q} \leq \infty \tag{4.6}$$

The corresponding pdfs of the envelope and intensity of the signal are

$$P(E) = \frac{2E}{x}\exp(-E^2/x); \quad E = \sqrt{E_\mathrm{I}^2 + E_\mathrm{Q}^2}; \quad 0 \leq E \leq \infty$$
$$P(z) = \frac{1}{z}\exp(-z/x); \quad z = E^2; \quad 0 \leq z \leq \infty \tag{4.7}$$

Here x is the mean intensity of the clutter return, and is the single parameter that specifies the Rayleigh clutter model. This parameter can be linked directly to the received clutter power in the radar equation for performance prediction, as will be described in Chapter 8.

As we have derived this model by considering the limit in which there is a large number of contributions to the scattered field, we expect it to be a valid representation of the clutter encountered in a low-resolution radar system, whose footprint is sufficiently large to contain many independent scatterers. For such a system, the probability that the envelope exceeds some threshold E_{T} is given by

$$\mathrm{Prob}(E > E_{\mathrm{T}}) = \int_{E_{\mathrm{T}}}^{\infty} dE\, P(E) = \exp(-E_{\mathrm{T}}^2/x) \tag{4.8}$$

This calculation provides us with a simple estimate of a false alarm rate.

We now briefly consider the case of a coherent signal (i.e. a steady target return) embedded in this Gaussian clutter. Once we have done this we will be able to calculate probabilities of detection, much as we have found that of false alarm in (4.8). In this case we can represent the received field by the sum of clutter contributions and a vector $\mathbf{A}$ modelling the steady signal. Thus we express $\mathbf{E}$ as

$$\mathbf{E} = \mathbf{A} + \sum_{n=1}^{N} \mathbf{a}_n \tag{4.9}$$

and evaluate its characteristic function

$$\langle \exp(i\mathbf{U}\cdot\mathbf{E})\rangle = \exp(i\mathbf{U}\cdot\mathbf{A})\langle \exp(i\mathbf{U}\cdot\mathbf{a})\rangle^N \tag{4.10}$$

Performing the same scaling and limiting processes as before we find that

$$\langle \exp(i\mathbf{U}\cdot\mathbf{E})\rangle = \exp(i\mathbf{U}\cdot\mathbf{A})\exp(-U^2\langle a^2\rangle/4) \tag{4.11}$$

The pdf of the field components in this signal plus clutter model is obtained by Fourier inversion as

$$P(E_{\mathrm{I}}, E_{\mathrm{Q}}) = \frac{1}{\pi\langle a^2\rangle}\exp\left(-\frac{1}{\langle a^2\rangle}((E_{\mathrm{I}} - A_{\mathrm{I}})^2 + (E_{\mathrm{Q}} - A_{\mathrm{Q}})^2)\right) \tag{4.12}$$

while that of the envelope takes the form

$$P(E|A) = \frac{2E}{x}\exp\left(-\frac{E^2 + A^2}{x}\right) I_0\left(\frac{2EA}{x}\right)$$

$$x \equiv \langle a^2\rangle \tag{4.13}$$

The corresponding pdf of the intensity z is

$$P(z|A) = \frac{1}{x}\exp\left(-\frac{z + A^2}{x}\right) I_0\left(\frac{2A\sqrt{z}}{x}\right) \tag{4.14}$$

Equation (4.13) is the so-called Rice distribution [2]. I_0 is a modified Bessel function of the first kind and is discussed in Appendix A2. When used for radar performance calculations, the target is often characterised in terms of the signal-to-clutter ratio, $\mathrm{SCR} = A^2/x$. Given the threshold E_{T} we can now calculate the probability of detecting the signal as

$$P_{\mathrm{D}}(E_{\mathrm{T}}) = \frac{2}{x}\int_{E_{\mathrm{T}}}^{\infty} E\exp\left(-\frac{E^2 + A^2}{x}\right) I_0\left(\frac{2EA}{x}\right) dE \tag{4.15}$$

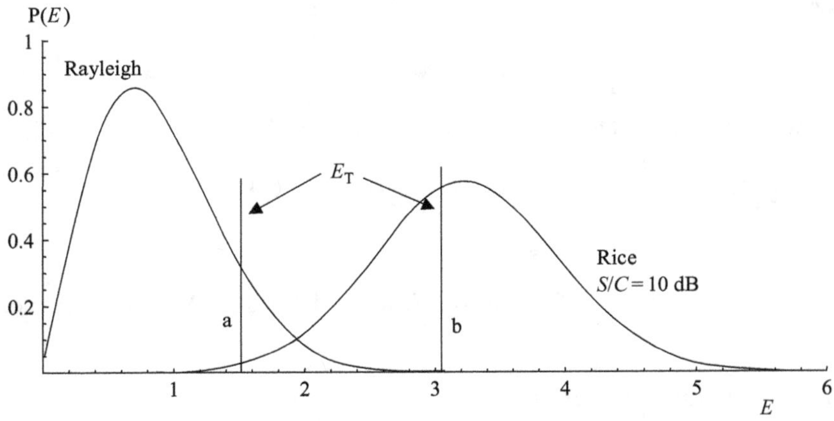

Figure 4.1 *Illustration of probability of false alarm and probability of detection for Rayleigh clutter and a coherent target signal embedded in the clutter with S/C = 10 dB*
(a) $P_{FA}(E_T) = 10^{-1}$, $P_D(E_T) = 0.95$
(b) $P_{FA}(E_T) = 10^{-4}$, $P_D(E_T) = 0.60$

Taken together the clutter and signal plus clutter pdfs (4.6) and (4.13) provide us with the basis of a simple performance calculation, comparing the probabilities of false alarm and detection from a given signal to noise ratio and threshold value. It should be noted that these models for Gaussian clutter are the same as those for thermal noise in the radar receiver.

Some implications of these results can be seen in Figure 4.1, which plots a Rayleigh pdf and the equivalent Rice distribution for a signal-to-clutter (or signal-to-noise) ratio of 10 dB. Two different values of threshold, E_T, are shown. The values of probabilities of detection and false alarm are given by the areas under the curves of the respective distributions for values of $E > E_T$. The lower value (labelled **a**) gives a high probability of detecting the target, with $P_D(E_T) > 0.95$. However the probability of a false alarm, $P_{FA}(E_T)$, in clutter alone is about 0.1, which would be completely unacceptable. Setting a higher value of E_T in this example (labelled **b**) gives a more acceptable value of $P_{FA}(E_T)$ of about 10^{-4} (a typical operating value in a practical radar). However, the probability of detection has fallen to about 0.6. As will be seen in Chapter 6, the Neyman-Pearson strategy for radar detection sets an acceptable value of probability of false alarm and tries to maximise the probability of detection.

So far we have paid very little attention to the distribution of the phase of the returned signal; the **a** were assumed to have a uniform distribution of phases and, as a consequence, the clutter model (4.6) delivers a uniform distribution for the phase of the resultant field **E**. The introduction of the steady signal **A** establishes a non-uniformity in the phase of the combined clutter and signal, measured relative to that of the coherent signal; the joint pdf of the envelope and relative phase takes the form

$$P(E,\theta|A,x) = \frac{2E}{x}\exp(-A^2/x)\exp(-(E^2 - 2EA\cos\theta)/x) \tag{4.16}$$

from which we obtain the phase pdf by integration on E

$$P(\theta|A,x) = \frac{1}{2\pi}\exp(-A^2/x) + \frac{A\cos\theta}{2\sqrt{\pi x}}(1 + \text{erf}(A\cos\theta/\sqrt{x}))$$
$$\times\exp(-a^2\sin^2\theta/x) \tag{4.17}$$

The erf(·) occurring in (4.17) is the error function defined in (A1.31) in Appendix 1.

4.3 Non-Gaussian clutter

The complex Gaussian process discussed in the previous section is a reasonable model for sea clutter in a low-resolution maritime radar system. In a high-resolution radar system, however, a given range cell may contain only a few independent scattering structures. In this case, the arguments put forward in support of the Gaussian model may break down as the radar effectively resolves some of the larger scale structure of the sea surface. The clutter observed is now subject to much greater fluctuations (sea spikes) than are seen in Gaussian clutter. Thus the improvement in signal to clutter ratio and potential detection achieved by increasing the resolution of the system can be potentially offset by the clutter becoming effectively target-like and regularly exceeding thresholds based on a Gaussian model of the clutter. Realistic non-Gaussian clutter models are needed if the performance lost to the sea spikes is to be understood and to some extent recovered. To test a non-Gaussian clutter model against available data we focus on quantities, the normalised moments of intensity data, which highlight this 'spiky' non-Gaussian behaviour. In the Gaussian limit we have

$$\frac{\langle z^n\rangle}{\langle z\rangle^n} = n!; \tag{4.18}$$

values of normalised intensity moments greater than these are characteristic of the spiky behaviour often shown by high resolution sea clutter.

The structure of the sea surface is very complex and is characterised by many length scales, ranging from 1 cm or less, typical of foams and ripples, to tens of metres, typical of swell structures. Similarly, the time-scales characteristic of sea surface motions range from less than a second to many seconds, if not minutes. Consequently there will be many effectively independent small-scale structures within a single high-resolution range cell that give rise to a speckle-like clutter, described locally by (4.7). This decorrelates over a short period of time in which the small scale structures move through a distance of the order of half a radar wavelength. However these small-scale structures are modulated by the more slowly changing large-scale structures; this results in a changing 'local power' of the Rayleigh-like returns from the small-scale structure. This decomposition of the non-Gaussian clutter into a rapidly decorrelating, locally Rayleigh process, whose power

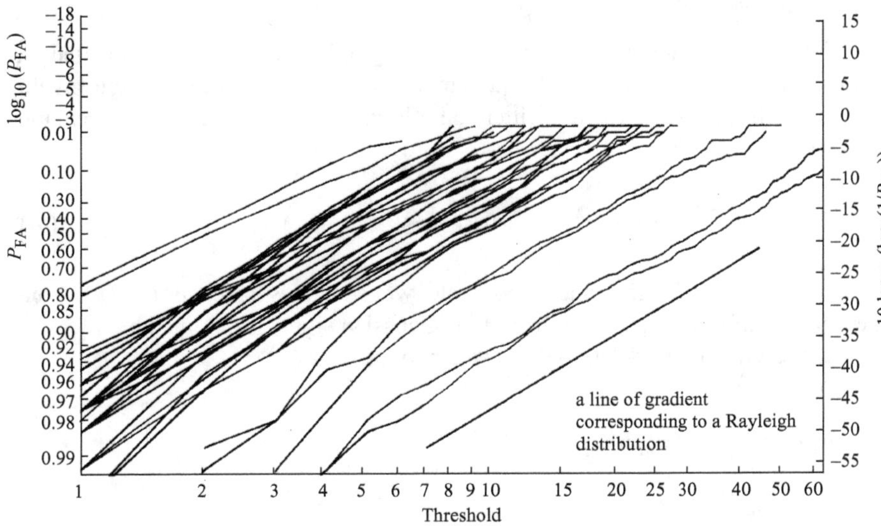

Figure 4.2 Pulse to pulse distributions of individual range cells taken over 250 ms plotted on Weibull paper; vertical polarisation and pulse-to-pulse frequency agility

is modulated by a much more slowly varying (but nonetheless random) process, forms the basis of the K distribution and other compound models for clutter [3]. We should stress that much of the value of the compound model derives from its representation of the non-Rayleigh amplitude as a product of two separate and independent components which, because of their differing spatial and temporal correlation properties, impact on different stages in the signal-processing and our evaluation of its performance.

Compelling evidence for the validity of this locally Gaussian speckle model is provided by results acquired during extensive campaigns of clutter measurement under a wide range of conditions. Some typical examples of such measurements are given in Chapter 2. Figure 4.2 shows the cumulative distribution of 250 ms time sequences taken from individual range cells of recorded I-band clutter data, plotted as a function of threshold on 'Weibull' paper. (This displays the log of the log of the reciprocal of the cumulative distribution against the log of the threshold.) On referring to (4.8) we see that, when presented in this format, a Rayleigh distribution corresponds to a straight line of gradient 2, whose intercept gives an estimate of the mean intensity or local power. Figure 4.2 shows that the time sequences each have Rayleigh distributed amplitudes; the horizontal displacements of the plots vary with the local power. The plots to the right of the figure correspond to clutter returns with high local means, which are often referred to as sea spikes. Taken together, these observations reassure us that sea clutter can be represented realistically as a locally Gaussian speckle, modulated by a varying underlying mean level or local power.

4.3.1 Compound models of non-Gaussian clutter

The physical arguments of the previous paragraphs can be accommodated in a relatively transparent and tractable mathematical framework. The pdf of the envelope of a Rayleigh process can be re-written to highlight its dependence on the local mean power x:

$$P(E|x) = \frac{2E}{x}\exp(-E^2/x); \quad 0 \le E \le \infty \tag{4.19}$$

This local power is itself a random variable, whose values have a pdf $P_c(x)$. Consequently the pdf of the envelope of the non-Gaussian signal is obtained by integrating the local Rayleigh pdf (4.19), with power x, over $P_c(x)$

$$\begin{aligned} P(E) &= \int_0^\infty dx\, P(E|x)P_c(x) \\ &= 2E\int_0^\infty \frac{dx}{x}\exp(-E^2/x)P_c(x) \end{aligned} \tag{4.20}$$

This gives us a formal realisation of the compound form of the pdf of non-Gaussian clutter, and provides us with a powerful tool with which to model and analyse its impact on radar performance.

We have already identified normalised intensity moments as a useful diagnostic for non-Gaussian behaviour. These can be calculated using the compound model (4.20) as follows

$$\begin{aligned} \langle z^n\rangle &= \int_0^\infty dx\, P_c(x)\left\{2\int_0^\infty \frac{dE}{x}E^{2n+1}\exp(-E^2/x)\right\} \\ &= n!\int_0^\infty dx\, P_c(x)x^n \\ &= n!\langle x^n\rangle_c \end{aligned} \tag{4.21}$$

so that

$$\frac{\langle z^n\rangle}{\langle z\rangle^n} = n!\frac{\langle x^n\rangle_c}{\langle x\rangle_c^n} \tag{4.22}$$

If this result is compared with (4.18), we see that the Gaussian result $n!$ is modified by multiplication by the nth normalised moment of the distribution of the local power x. The simple picture underlying the compound model, of a locally Gaussian process with a modulated power, is manifest in these formulae. In much the same way, the probability that the signal envelope exceeds a threshold E_{T} is given by

$$\mathrm{Prob}(E > E_{\mathrm{T}}) = \int_0^\infty dx \exp(E_{\mathrm{T}}^2/x) P_c(x) \tag{4.23}$$

where we see that the familiar Gaussian result (4.8) has been averaged over the distribution of local power.

These results show how straightforward the generalisation of standard Gaussian-derived results is within the compound representation of non-Gaussian clutter. One simply takes the required Gaussian based result for a given local speckle power, then averages this over the appropriate distribution of local power. This remark applies equally well to modelling target plus clutter returns and evaluating probabilities of detection. However, before any real progress can be made, we must identify the local power pdf $P_c(x)$.

4.3.2 The gamma distribution of local power and the K distribution

It was noted in Chapter 2 that, in uniform clutter with Rayleigh amplitude statistics, pulse-to-pulse frequency agility with frequency steps of at least the pulse bandwidth should ensure that returns are decorrelated from pulse to pulse. Thus we are able to obtain many independent samples of the local power in a time in which it does not change appreciably. These can be averaged to give an estimate of the local power x. By analysing a sufficiently large quantity of data we can build up a large set of independent measurements of x. These can be used to identify a good model for $P_c(x)$. It has been found that the gamma distribution provides the best fit to most of the available data, i.e.

$$P_c(x) = \frac{b^\nu}{\Gamma(\nu)} x^{\nu-1} \exp(-bx); \quad 0 \le x \le \infty \tag{4.24}$$

This distribution is characterised by a scale parameter, b, and a shape parameter, ν. which depend on sea conditions and the radar parameters (see Chapter 2, Section 2.6). When the gamma distribution of local power (4.24) is substituted into the compound form (4.20) we find that in this case, the pdf of the clutter envelope is given by

$$\begin{aligned} P(E) &= \frac{2Eb^\nu}{\Gamma(\nu)} \int_0^\infty dx x^{\nu-2} \exp(-bx - E^2/x) \\ &= \frac{4b^{(\nu+1)/2} E^\nu}{\Gamma(\nu)} K_{\nu-1}(2E\sqrt{b}) \end{aligned} \tag{4.25}$$

The corresponding pdf of the intensity $z = E^2$ is

$$P(z) = \frac{2b^{(\nu+1)/2} z^{(\nu-1)/2}}{\Gamma(\nu)} K_{\nu-1}(2\sqrt{bz}) \tag{4.26}$$

We see that the integral in (4.25) has been evaluated in terms of the modified Bessel or K function which gives its name to the model. Fortunately, no special knowledge of the properties of these Bessel functions is required to evaluate quantities of interest, such as probabilities of false alarm or intensity moments. In each case we merely take the Gaussian result and integrate it over the gamma distribution of x. (The few, possibly unfamiliar, mathematical results required to do this are reviewed in Appendix A2.) In this way we find that

$$\begin{aligned}\text{Prob}(E > E_\text{T}) &= \frac{b^\nu}{\Gamma(\nu)} \int_0^\infty dx \exp(-E_\text{T}^2/x) x^{\nu-1} \exp(-bx) \\ &= \frac{2b^{\nu/2}}{\Gamma(\nu)} E_\text{T}^\nu K_\nu(2E_\text{T}\sqrt{b}) \end{aligned} \tag{4.27}$$

much as in (4.23), and

$$\frac{\langle z^n \rangle}{\langle z \rangle^n} = n! \frac{\Gamma(n+\nu)}{\Gamma(\nu)\nu^n} = n! \prod_{k=1}^{n-1} (1 + k/\nu) \tag{4.28}$$

The moments of the envelope, E, are given by

$$\langle E^n \rangle = b^{-n/2} \frac{\Gamma(1+n/2)\Gamma(\nu+n/2)}{\Gamma(\nu)} \tag{4.28a}$$

The fit of real data to the Gamma distribution family is demonstrated in Figure 4.3(a,b) where the normalised intensity moments of the local power of many records of vertically and horizontally polarised clutter are compared to those of a gamma distribution. The parameter ν is derived from the second normalised intensity moment and it can be seen that the third and fourth moments are consistent over a wide range of values, indicating a good match to the model. Also shown on the plots are the expected values for the lognormal distribution. This distribution (see Section 4.6), which has been proposed in the past as a model for sea clutter, clearly gives a very poor fit to these data.

Many authors have investigated the fit of sea clutter data to the compound formulation, although often this work has been limited to assessing the global fit to the overall amplitude statistics. The need for this type of model has long been recognised [4]–[6]; the K distribution model, as it is described here, was developed by Ward [3,7]. A recent example of work reporting the fit to the compound formulation itself is found in Reference 8.

Some plots of the K distribution pdf (4.25) are shown in Figure 4.4, for various values of shape parameter, ν, and with the scale parameter set to give unity mean value of the envelope, E. Typically, the value of ν in sea clutter falls in the range $0.1 \le \nu \le \infty$. When $\nu = \infty$ it reduces to the Rayleigh distribution, while small values of ν, say $\nu < 1$, correspond to spiky clutter. We also note (see A2.41) that when the shape parameter ν takes values between 0 and 1/2, the envelope pdf exhibits

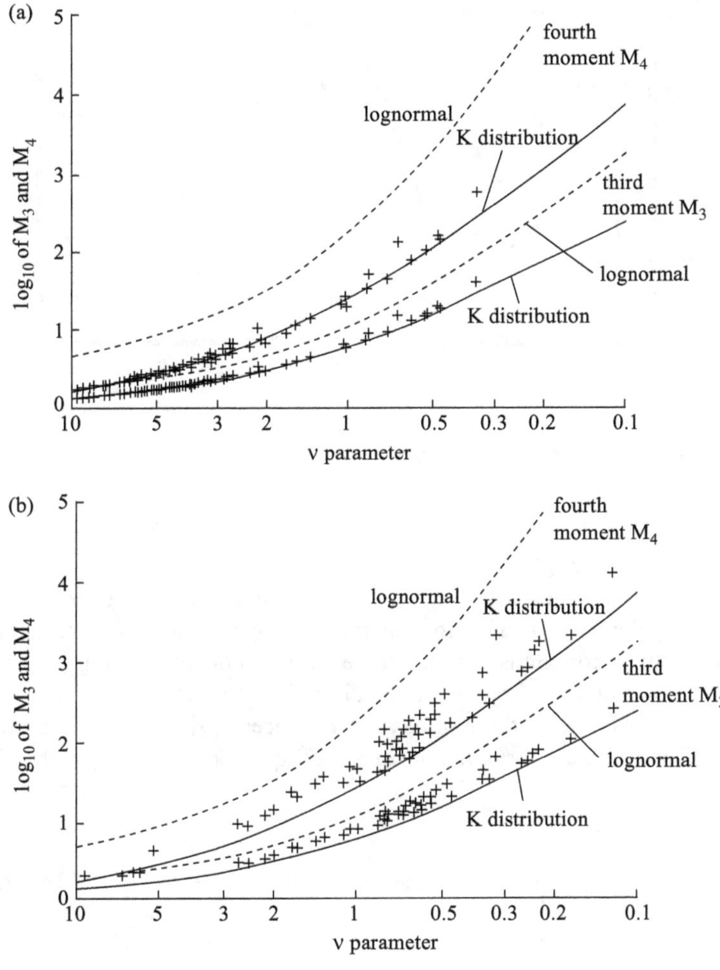

Figure 4.3 *The plotted curves show moments of modulation corresponding to the overall pulse-to-pulse distributions as labelled; moments corresponding to the Gamma distribution of modulation are labelled as the K distribution. Experimental results are shown as +. The parameter, ν, is $1/(M_2 - 1)$ and $M_n = \langle z^n \rangle / \langle z \rangle^n$ are the nth intensity moments of the data. (a) vertical polarisation, (b) horizontal polarisation*

an integrable singular behaviour at the origin; a corresponding singularity in the intensity pdf is found when $0 < \nu < 1$. This preferential weighting of small values of E reflects the apparent quiescence of spiky clutter (relative to its mean) that is occasionally interrupted by a large excursion.

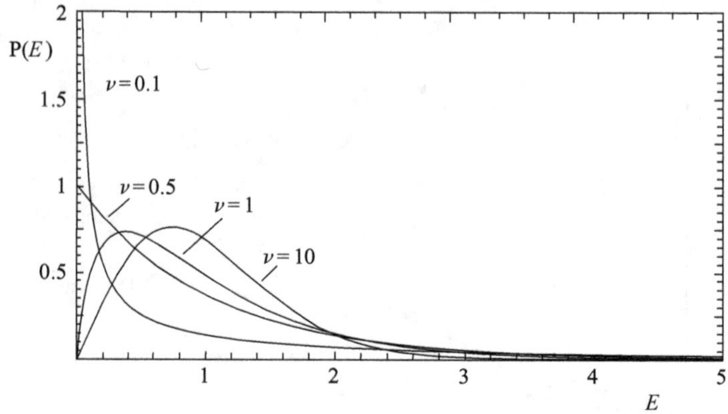

Figure 4.4 The K distribution for various values of ν

4.3.3 A coherent signal in K-distributed clutter

In Section 4.2 we considered the case of a coherent signal embedded in Rayleigh clutter, deriving the Rice distribution for the envelope of the resultant signal. The compound model of non-Gaussian clutter we have just developed suggests a simple extension of this model to a coherent signal in K-distributed clutter. The resultant is sometimes referred to as the homodyned K process [9]. To obtain the required envelope pdf we merely form the integral of equation (4.14) over the Gamma distribution of x

$$P(E|A,\nu,b) = \frac{2Eb^{\nu}}{\Gamma(\nu)} \int_0^{\infty} x^{\nu-2} \exp(-bx) \exp(-(A^2 + E^2)/x) I_0(2EA/x) dx \tag{4.29}$$

This pdf cannot in general be expressed in terms of tabulated functions. Nonetheless it is readily evaluated numerically and forms the basis of performance calculations for non-fading targets (commonly called Swerling type 0 targets, as discussed in detail in Chapter 8). Another variant of the K process, introduced by Jakeman and Tough [10] as a model for weak forward scattering, again consists of a coherent signal embedded in K-distributed clutter. In this case, however, the amplitude of the coherent signal is also modulated by the gamma variate x. Thus the pdf of the envelope of this, the generalised K process, can be constructed from (4.14) as

$$P(E|A,\nu,b)_{GK} = \frac{2Eb^{\nu}}{\Gamma(\nu)} I_0(2EA) \int_0^{\infty} x^{\nu-2} \exp(-(b + A^2)x) \exp(-E^2/x) dx \tag{4.30}$$

In this case the integral can be evaluated in closed form as

$$P(E|A,\nu,b)_{GK} = \frac{4E^{\nu}b^{\nu}}{\Gamma(\nu)(b+A^2)^{(\nu-1)/2}} I_0(2EA)K_{\nu-1}\left(2E\sqrt{b+A^2}\right) \tag{4.31}$$

4.3.4 K-distributed clutter with added thermal noise

As a final example of the application of the compound model we consider a signal consisting of thermal noise and K-distributed clutter. If the thermal noise power is P_{n} then the pdf of the envelope of the combined signal, for a given local clutter power x, is

$$P(E|p_{\mathrm{n}},x) = \frac{2E}{p_{\mathrm{n}}+x}\exp(-E^2/(p_{\mathrm{n}}+x)) \tag{4.32}$$

To obtain the pdf appropriate to K-distributed clutter we integrate this result over a gamma distribution of x

$$P(E|p_{\mathrm{n}},b,\nu) = \frac{2Eb^{\nu}}{\Gamma(\nu)}\int_0^{\infty}\frac{x^{\nu-1}\exp(-bx)}{x+p_{\mathrm{n}}}\exp(-E^2/(x+p_{\mathrm{n}}))dx \tag{4.33}$$

Once again this integral cannot be expressed in a simple closed form but is readily evaluated numerically. This process, which was first discussed in the literature by Watts [11], can be used to model both a combination of system noise and K clutter and a Swerling 2 target in K clutter, and is developed further in Chapter 8. The first three intensity moments $\langle z^n\rangle = \langle E^{2n}\rangle$ are given by:

$$\langle z\rangle = \frac{\nu}{b} + p_{\mathrm{n}}$$

$$\langle z^2\rangle = \frac{2\nu(\nu+1)}{b^2} + \frac{4p_{\mathrm{n}}\nu}{b} + 2p_{\mathrm{n}}^2$$

$$\langle z^3\rangle = \frac{6\nu(\nu+1)(\nu+2)}{b^3} + \frac{18p_{\mathrm{n}}\nu(\nu+1)}{b^2} + \frac{18p_{\mathrm{n}}^2\nu}{b} + 6p_{\mathrm{n}}^3$$

Higher order moments may be derived as shown in Section A1.12 in Appendix 1. It is interesting to note that Lamont-Smith [12] has investigated the amplitude statistics of high resolution, low-grazing angle clutter and the applicability of the compound model. He found that the data were best represented by the compound form (4.33), where the power of the additional Rayleigh distributed component was significantly greater than the measured thermal noise level. In the tail of the distribution, which is of interest for detection processing and false alarms (see Chapter 8), this feature has little effect and can be ignored. It may be more important in understanding the statistics of images of the sea, as might be found with synthetic aperture radar.

Figure 4.5 shows examples of the probability of false alarm $P_{\mathrm{FA}}(E_{\mathrm{T}})$ plotted as a function of E_{T} for $\nu = 1$ and different values of clutter-to-noise ratio, CNR. Also shown for comparison is $\nu = 4$ for CNR $= 30$ dB and noise alone. The mean level of clutter-plus-noise is normalised to unity in each case.

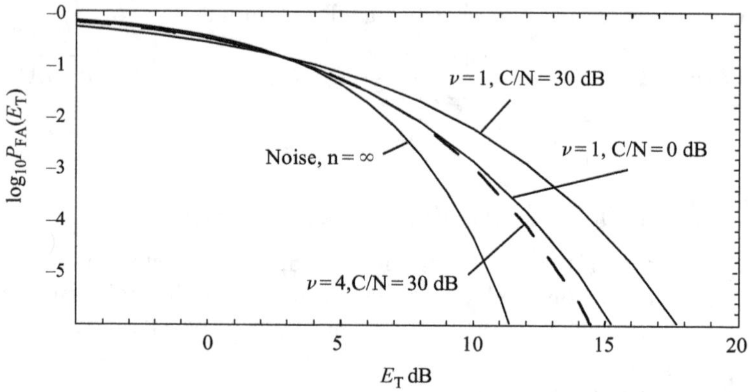

Figure 4.5 Cumulative K distributions for different values of clutter-to-noise ratio

4.3.5 Phases of homodyned and generalised K processes

In Section 4.2, we discussed the distribution of the phase of a coherent signal embedded in Rayleigh clutter; the corresponding results for the phases of the homodyned and generalised K processes (see Section 4.3.3) can be obtained by averaging an appropriately formed (4.17) over a gamma distribution of local mean power. Thus we have

$$P_{\mathrm{GK}}(\theta|A,b,\nu) = \frac{b^{\nu}}{\Gamma(\nu)}\int_0^{\infty} P(\theta|Ax,x)\exp(-bx)x^{\nu-1}dx \tag{4.34}$$

in the generalised K case while, for the homodyned K process,

$$P_{\mathrm{HK}}(\theta|A,b,\nu) = \frac{b^{\nu}}{\Gamma(\nu)}\int_0^{\infty} P(\theta|A,x)\exp(-bx)x^{\nu-1}dx \tag{4.35}$$

Significant progress can be made in the evaluation of these pdfs in terms of standard functions. The results, however, are rather involved and will not be reproduced here; the interested reader is referred to Reference 13 for details of their derivation.

4.3.6 Applications to interferometric and polarimetric processing

To conclude our discussion of the generalised K and related processes we note that this model is of considerable importance in another context. The joint pdf of the amplitudes and phases of a complex Gaussian process, measured at times separated by the interval τ, is given in Appendix 1, equation (A1.132). The hermitian product of the two signals is a complex number of the form

$$\xi = (E_I(t+\tau) + iE_Q(t+\tau))(E_I(t) - iE_Q(t)) \tag{4.36}$$

Quantities of this kind play a central role in interferometric and polarimetric processing of SAR imagery. We note that (A1.132) depends only on the phase difference $\theta(t+\tau)-\theta(t)$; the redundant phase variable can be integrated out. If we now change variables from $\{E(t), E(t+\tau)\}$ to ζ, E, where

$$\zeta = \mathrm{mod}(\xi) = E(t)E(t+\tau), \quad E = E(t) \tag{4.37}$$

we see that the pdf of the amplitude of the Hermitian product has a form

$$P(\zeta) = \frac{1}{\psi^2(1-k^2)}\zeta I_0\left(\frac{\zeta k}{\psi(1-k^2)}\right)\int_0^\infty \frac{dE}{E} \times \exp\left(-\frac{1}{2\psi(1-k^2)}(E^2+\zeta^2/E^2)\right) \tag{4.38}$$

If we identify E^2 with x, the equivalence of this pdf with the generalised K result (4.31) with $\nu = 1$ is evident; the amplitude of the coherent sum of N independent Hermitian products also has a pdf of this form, but with $\nu = N$. The phase distribution (4.34), which can be expressed in a simple closed form when ν takes integer values, is also of use in the description of multi-channel SAR processing; we will discuss this important topic further in Chapter 7.

4.4 Spikes in clutter

4.4.1 The Class A and breaking area models

So far we have considered a continuous distribution of local power, modelled by the gamma distribution, which modulates noise-like speckle. As discussed in Chapter 2 (Section 2.8), the mechanisms that give rise to this seem to be a combination of Bragg scattering from resonant capillary waves (which is then modulated by the long waves and sea swell) and whitecap scattering from the localised roughness from breaking waves. Other scattering events can occur which give rise to discrete spike-like features in clutter returns, and cannot be modelled in this way. In particular, scattering from the crests of incipiently breaking waves, referred to as bursts, results in discrete clutter spikes that have no speckle fluctuations; thus they are not included in the compound K distribution model. Also, when illuminated by a frequency agile radar, the speckle from some whitecaps does not decorrelate as the frequency is stepped (due to the scattering being localised in range); these are not modelled (for this type of radar) by the compound model.

As we will see in this section, discrete spikes may be added to the K distribution using the Class A model. We will also see that a related model, the BAM (Breaking Area Model), may be used to describe the occurrence of whitecaps as used in Chapter 3 (Section 3.8) to model the normalised clutter RCS (σ^0).

In Appendix 1 we introduce the binomial and Poisson distributions of numbers of discrete events: the $P(N)$ arising in these cases can be interpreted usefully as the probabilities of there being N spike-like returns in a sample of clutter. The Poisson distribution describes fluctuations in a population of independent, un-bunched objects. These are characterised by the normalised variance (cf. Appendix 1 equations (A1.13) and (A1.14))

$$\frac{\langle N^2\rangle}{\langle N\rangle^2} - 1 = \frac{1}{\bar{N}} \tag{4.39}$$

which we see vanishes in the limit of a large mean population.

Perhaps the simplest model incorporating discrete spikes is the so-called Class A model introduced by Middleton [14]. This assumes a Poisson population of spikes, each contributing a Rayleigh distributed field to the clutter. Each of these has a pdf of the form (4.7) with a mean intensity that, for the moment, we set to unity. So, when m of these are combined coherently, the pdf of their resultant is

$$P(E) = \frac{2E}{m} \exp\left(-\frac{E^2}{m}\right) \tag{4.40}$$

The inclusion of an external source of thermal noise (cf. equation (4.32)) leads to the modified form

$$P(E) = \frac{2E}{m + p_{\mathrm{n}}} \exp\left(-\frac{E^2}{m + p_{\mathrm{n}}}\right) \tag{4.41}$$

The corresponding distribution of intensity is

$$P(z) = \frac{1}{m + p_{\mathrm{n}}} \exp\left(-\frac{z}{m + p_{\mathrm{n}}}\right) \tag{4.42}$$

The mean value of this intensity, taken over the distribution of populations, is given by

$$\langle z\rangle = p_{\mathrm{n}} + \bar{N} \tag{4.43}$$

while the ratio of the powers of the noise and the spike contributions is

$$\Gamma = \frac{p_{\mathrm{n}}}{\bar{N}} \tag{4.44}$$

We now construct the pdf of the intensity of the process as

$$\begin{aligned} P(z) &= \exp(-\bar{N}) \sum_{m=0}^{\infty} \frac{\bar{N}^m}{m!} \frac{1}{I_m} \exp\left(-\frac{z}{I_m}\right) \\ I_m &= \frac{\langle z\rangle}{1+\Gamma}\left(\frac{m}{\bar{N}} + \Gamma\right) \end{aligned} \tag{4.45}$$

An indication of the flexibility of the compound formulation of clutter models we are developing can be found in the ease with which this Class A model can be modified to incorporate 'spikes' directly into the K model. We identify the p_{n} in (4.42) with the gamma distributed local power; the required pdf is then obtained through the familiar 'average over the gamma distribution':

$$P(z) = \frac{\exp(-\bar{N})b^{\nu}}{\Gamma(\nu)} \sum_{m=0}^{\infty} \frac{\bar{N}^m}{m!} \int_0^{\infty} dx \frac{x^{\nu-1}\exp(-bx)\exp(-z/(m+x))}{m+x} \tag{4.46}$$

This 'Class A plus K' (also known as the KA) model has been discussed by Middleton [15] and Ward and Tough [16]

The non-Gaussian character captured by the Class A model arises from the finite value of $\bar{N}$. It is interesting to compare this model (with Γ set to zero, i.e. no noise) with the K distribution (4.26). The pdf now takes the form

$$\begin{aligned} P(z) &= \exp(-\bar{N}) \sum_{m=0}^{\infty} \frac{\bar{N}^m}{m!} \frac{\bar{N}}{m\langle z\rangle} \exp(-\bar{N}z/m\langle z\rangle) \\ &= 2\exp(-\bar{N})\delta(z) + \exp(-\bar{N}) \sum_{m=1}^{\infty} \frac{\bar{N}^m}{m!} \frac{\bar{N}}{m\langle z\rangle} \exp(-\bar{N}z/m\langle z\rangle) \end{aligned} \tag{4.47}$$

The first term in this series is indeterminate and is interpreted as a delta function; the rest are well-defined. This delta function contribution becomes more significant as $\bar{N}$ becomes small, a behaviour that is reminiscent of the emergence of an integrable singularity at the origin in the K distribution intensity pdf when the shape parameter ν takes values less than unity.

The intensity moments of the distribution (4.47) can be evaluated as

$$\frac{\langle z^n\rangle}{\langle z\rangle^n} = n!\exp(-\bar{N}) \sum_{m=0}^{\infty} \frac{\bar{N}^m}{m!} \left(\frac{m}{\bar{N}}\right)^m = n! f_n(\bar{N}) \tag{4.48}$$

where

$$f_n(\bar{N}) = \exp(-\bar{N}) \frac{1}{\bar{N}^n} \left(\bar{N}\frac{\partial}{\partial \bar{N}}\right)^n \exp(\bar{N}) \tag{4.49}$$

From this we see that

$$\begin{aligned} &f_1(\bar{N}) = 1, \quad f_2(\bar{N}) = 1 + \frac{1}{\bar{N}}, \quad f_3(\bar{N}) = 1 + \frac{3}{\bar{N}} + \frac{1}{\bar{N}^2}, \\ &f_4(\bar{N}) = 1 + \frac{6}{\bar{N}} + \frac{7}{\bar{N}^2} + \frac{1}{\bar{N}^3} \end{aligned} \tag{4.50}$$

The corresponding quantities for the K model

$$m_n(\nu) = \frac{\langle z^n\rangle}{\langle z\rangle^n n!} = \frac{1}{\nu^n} \frac{\Gamma(\nu+n)}{\Gamma(\nu)} \tag{4.51}$$

are

$$m_1(\nu) = 1, \quad m_2(\nu) = 1 + \frac{1}{\nu}, \quad m_3(\nu) = 1 + \frac{3}{\nu} + \frac{2}{\nu^2},$$

$$m_4(\nu) = 1 + \frac{6}{\nu} + \frac{11}{\nu^2} + \frac{6}{\nu^3} \tag{4.52}$$

We see that the two models have the same Gaussian (large $\bar{N}, \nu$) limit and that, if we identify ν with $\bar{N}$, the first order corrections to this limit are the same for both. In the extreme non-Gaussian limit (small $\bar{N}, \nu$), rather different behaviours are found; for the Class A model

$$f_n(\bar{N}) \sim \frac{1}{\bar{N}^{n-1}}, \quad \bar{N} \to 0 \tag{4.53}$$

while the K model yields

$$m_n(\nu) \sim \frac{(n-1)!}{\nu^{n-1}}, \quad \nu \to 0 \tag{4.54}$$

A Poisson distribution of scatterers also underlies another non-Gaussian model that relates to sea clutter, the so-called BAM introduced by Clements and Yurtsever [17]. As discussed in Chapter 3, this model identifies contributions to sea clutter as arising from breaking wave events. The sea surface is modelled as a Gaussian random field, with spatial and temporal correlation functions determined by a power spectrum and dispersion relation. A simple criterion for wave breaking to occur (that minus twice the vertical acceleration a_z of the sea surface exceeds that due to gravity) allows the whitecaps to be identified with above threshold excursions of a Gaussian random field. In the limit where this threshold is large a statistical characterisation of these events becomes tractable; a detailed review of the progress made in this area can be found in Reference 18. In particular the number of such events within an area A has a Poisson distribution, whose mean $\bar{N}$ incorporates the threshold $g/2$ on the vertical acceleration a_z of the sea surface.

$$\bar{N} = \frac{A}{\sqrt{2\pi}} \gamma u \exp(-u^2/2) \qquad u = \frac{g}{2\sqrt{\langle a_z^2 \rangle}} \tag{4.55}$$

where γ is a measure of the curvature of the spatial correlation function of a_z. The area s of an individual excursion has the exponential distribution

$$P(s|1) = \gamma u^2 \exp(-\gamma u^2 s) \tag{4.56}$$

If there are m breaking area events in A, then the pdf of their sum, added incoherently, is

$$P(s|m) = \frac{(\gamma u^2)^m}{(m-1)!} s^{m-1} \exp(-\gamma u^2 s) \tag{4.57}$$

We see that the parameters characterising these distributions are given in terms of the power spectrum of the surface height fluctuations and so can be related directly to prevailing environmental conditions. Much as in the development of the Class A model this pdf, derived for a fixed number of breaking areas, is averaged over the Poisson distribution (Appendix 1 equation A1.12). The pdf of the total broken area within A is then given by (cf. equation (4.47))

$$P(s) = \exp(-\bar{N}) \sum_{m=0}^{\infty} P(s|m) \frac{\bar{N}^m}{m!}$$

$$= \exp(-\bar{N}) \left(2\delta(s) + \exp(-\gamma u^2 s) \sqrt{\frac{\gamma u^2 \bar{N}}{s}} I_1 \left(2\sqrt{\gamma u^2 \bar{N} s} \right) \right) \tag{4.58}$$

where we have identified the modified Bessel function of the first kind (A2.56). We stress that the principal difference between the Class A and BAM models is that, in the former the fields scattered by the Poisson population are added coherently, while in the latter case, their powers are added incoherently.[1] To make more direct contact with the K distribution analysis we identify a local RCS (analogous to that in (4.19)) as

$$x = \frac{s}{A}, \tag{4.59}$$

whose mean and variance take the forms

$$\langle x \rangle = \frac{\bar{N}}{A\gamma u^2}, \qquad \langle x^2 \rangle - \langle x \rangle^2 = \frac{2\bar{N}}{(A\gamma u^2)^2} \tag{4.60}$$

From these we can identify an effective shape parameter as

$$\nu_{\text{eff}} = \frac{\bar{N}}{2} \tag{4.61}$$

Introducing these results we write the pdf of x as

$$P(x) = 2\exp(-2\nu_{\text{eff}}) \left(\delta(x) + \frac{\exp(-2\nu_{\text{eff}} x/\langle x \rangle)}{\sqrt{x\langle x \rangle}} I_1 \left(4\nu_{\text{eff}} \sqrt{\frac{x}{\langle x \rangle}} \right) \right) \tag{4.62}$$

The moments of this distribution are

$$m_n = \frac{\langle x^n \rangle}{\langle x \rangle^n} = \frac{n!}{(2\nu_{\text{eff}})^{n-1}} {}_1F_1(1-n, 2, -2\nu_{\text{eff}}) \tag{4.63}$$

the first few of which take the explicit forms

$$m_1 = 1; \quad m_2 = 1 + \frac{1}{\nu_{\text{eff}}}; \quad m_3 = 1 + \frac{3}{\nu_{\text{eff}}} + \frac{3}{2\nu_{\text{eff}}^2};$$

$$m_4 = 1 + \frac{6}{\nu_{\text{eff}}} + \frac{9}{\nu_{\text{eff}}^2} + \frac{3}{\nu_{\text{eff}}^3} \tag{4.64}$$

[1] This is consistent with our use of the models. The Class A model is used to add bursts, which have no speckle and therefore add coherently. The BAM is used for the mean power (i.e. after speckle averaging) of the whitecaps; thus non-coherent addition is appropriate.

that can be compared with those obtained earlier for the gamma and Class A models.

4.4.2 Clutter spike models and K phenomenology

In both the BAM and Class A model deviations from Gaussian behaviour result from the finite size of the mean population $\bar{N}$. In the limit of a large mean population, familiar Gaussian statistics are recovered. It is interesting to see how this result emerges from the characteristic function analysis of Section 4.2. Assuming the population of scatterers to be Poisson distributed we see that the characteristic function of the scattered radiation field takes the form

$$C(U) = \exp(-\bar{N}) \sum_{n=0}^{\infty} \frac{\bar{N}^n \langle J_0(Ua)\rangle^n}{n!} = \exp(-\bar{N}(1 - \langle J_0(Ua)\rangle)) \tag{4.65}$$

To maintain a finite energy in the large mean population limit we scale the magnitude a as follows

$$a \to \frac{a}{\sqrt{\bar{N}}} \tag{4.66}$$

and again obtain the Gaussian characteristic function, equation (4.5), even in this fluctuating population case. Might it be possible, however, to impose some statistics other than Poisson on the population, that in turn impart a non-Gaussian statistical character to the scattered radiation that persists, even in the limit of a large mean population number? Jakeman [9] showed that it is possible to do so, in such a way that the K distribution emerges in this limit. Jakeman's analysis also provides further insight into the compound form of the K distribution described in Section 4.3. We see from the results (A1.16) that the negative binomial population has a normalised variance that does not (in contrast to the result (4.39) for the Poisson distribution) vanish in the limit of a large mean population; this suggests that we consider this model in a little more detail. Following Jakeman, we evaluate the characteristic function of the radiation scattered from such a population as

$$\begin{aligned} C(U) &= \frac{1}{(1+\bar{N}/\nu)^{\nu}} \sum_{n=0}^{\infty} \frac{(\nu)_n}{n!} \left(\frac{\langle J_0(Ua)\rangle \bar{N}/\nu}{1+\bar{N}/\nu} \right)^n \\ &= \frac{1}{\left(1 + (\bar{N}/\nu)(1 - \langle J_0(Ua)\rangle)\right)^{\nu}} \end{aligned} \tag{4.67}$$

If we once again introduce the scaling (4.66), and let the mean population increase without limit we find that

$$C(U) = \frac{1}{(1 + (U^2\langle a^2\rangle)/(4\nu))^{\nu}} \tag{4.68}$$

This characteristic function no longer takes the form (4.5) associated with Gaussian statistics; to retrieve the corresponding pdf of the scattered field envelope we invert the Fourier transform implicit in the construction of (4.3) to give

$$P(E) = E \int_0^\infty U J_0(UE) \left(1 + \frac{U^2\langle a^2\rangle}{4\nu}\right)^{-\nu} dU \tag{4.69}$$

It is possible to evaluate this integral directly [19]; to make more direct contact with the compound form of the K distribution encountered earlier we introduce the integral representation

$$\left(1 + \frac{U^2\langle a^2\rangle}{4\nu}\right)^{-\nu} = \frac{1}{\Gamma(\nu)}\left(\frac{\nu}{\langle a^2\rangle}\right)^{\nu} \int_0^\infty x^{\nu-1} \exp(-\nu x/\langle a^2\rangle) \exp(-U^2 x/4) dx \tag{4.70}$$

into (4.69). Fourier inversion with respect to U now yields a result identical to (4.25), with the scale parameter

$$b = \frac{\nu}{\langle a^2\rangle} \tag{4.71}$$

The homodyned K and generalised K models can also be derived by an analysis much like this one; details can be found in Reference 10.

One feature of particular note that emerges from this characteristic function analysis is its highlighting of the infinite divisibility of the K distribution. It is well known that the normal distribution is stable, that is, that a sum of two of more Gaussian random variables is itself a Gaussian random variable. The stability of the Gaussian distribution is therefore reflected in the observation that the product of two Gaussian characteristic functions retains the characteristic Gaussian functional form. If now we coherently sum two signals, whose fields are each K distributed, with characteristic functions of the form (4.68), we see that the characteristic function of the pdf of the resultant is

$$C(U) = \left(1 + \frac{U^2\langle a^2\rangle}{4\nu}\right)^{-2\nu} \tag{4.72}$$

Comparison of this result with (4.68) reveals that the resultant is itself K distributed; the shape and scale parameters now take the values 2ν and $\nu/\langle a^2\rangle$ respectively. In much the same way we see that the coherent sum of N K-distributed fields is again K distributed; while the scale parameter remains unchanged, the shape parameter is now $N\nu$. Each of these K distribution pdfs is a different function; as N becomes large we see that the characteristic function tends to a Gaussian limit (cf. (4.5)). Nonetheless they remain in the same 'family' of functions; loosely speaking, this behaviour is identified with the property of infinite divisibility which first focussed the attention of Jakeman and Pusey [20] on the K distribution. This observation also makes contact with our earlier compound representation of the clutter process; reference to (A1.53) in Appendix 1 or alternatively, a relatively simple exercise in the manipulation of Laplace transforms will convince the reader that the gamma distribution is also infinitely divisible.

The non-Gaussian character implicit in the BAM and Class A models is due to the finite size of the mean scatterer number $\bar{N}$; that of the K distribution model persists as $\bar{N}$ increases without bound, and can be ascribed to the bunching in the scatterer population that is still manifest in this limit. By introducing the negative binomial distribution in place of the Poisson in the class A model and BAM we obtain generalised models that incorporate effects of both finite scatterer number and bunching. For the generalised Class A model we have an intensity pdf of the form

$$P(z) = \frac{1}{(1+\bar{N}/\nu)^{\nu}}\left[2\delta(z) + \sum_{m=1}^{\infty} \exp(-z/I_m)\frac{1}{I_m}\frac{\Gamma(\nu+m)}{m!\Gamma(\nu)}\left(\frac{\bar{N}/\nu}{1+\bar{N}/\nu}\right)^{m}\right] \tag{4.73}$$

while the RCS pdf of the generalised BAM is given by

$$P(s) = \frac{1}{(1+\bar{N}/\nu)^{\nu}} \times \left[2\delta(s) + \frac{\bar{N}}{1+\bar{N}/\alpha}u^2\gamma\exp(-u^2\gamma s)\,{}_1F_1\left(\alpha+1, 2, \frac{\bar{N}/\alpha}{1+\bar{N}/\alpha}u^2\gamma s\right)\right] \tag{4.74}$$

We can evaluate the lower order moments of these distributions and find that, in the former case we have

$$\frac{\langle z^2\rangle}{2\langle z\rangle^2} = 1 + \frac{1}{\nu} + \frac{1}{\bar{N}} \tag{4.75}$$

while in the latter case

$$\langle s\rangle = \frac{\bar{N}}{u^2\gamma}, \quad \frac{\langle s^2\rangle}{\langle s\rangle^2} = 1 + \frac{2}{\bar{N}} + \frac{1}{\nu} \tag{4.76}$$

Contributions from both finite number and bunching effects can be seen in each. We note that, as ν becomes large, the negative binomial distribution tends to the Poisson limit and the earlier results (4.47) and (4.58) are retrieved. In the continuous limit, where $\bar{N}$ becomes large, (4.73) tends to the K distribution form (4.26), while the generalised BAM reduces to the gamma pdf. The former, generalised Class A result follows directly from the analysis leading to (A1.124) in Appendix 1; the characteristic function provides us with a convenient route to the latter. Thus we note that

$$C(k) = \langle\exp(-sk)\rangle = \frac{1}{(1+((k\bar{N}/\nu)/(k+u^2\gamma)))^{\nu}} \tag{4.77}$$

In the large $\bar{N}$ limit this reduces to

$$C(k) = \frac{1}{(1+k\langle s\rangle/\nu)^{\nu}} \tag{4.78}$$

from which the gamma distribution can be recovered straightforwardly by Laplace inversion. It is interesting to note the intimate connection between the BAM, the Class

A model and the K distribution model revealed by this simple picture of scattering by a fluctuating population.

We now illustrate these remarks with a discussion of some data that exhibit spike-like features, and assess the extent to which the models we have just developed are applicable to them.

4.4.3 An analysis of spiky clutter data

The presence of clutter spikes that do not decorrelate with pulse-to-pulse frequency agility is particularly evident following pulse-to-pulse non-coherent integration of radar clutter data. This can best be examined by using the KA model for clutter-plus-spikes-plus-noise. The development of the KA model (see Section 4.4) was prompted by the observation on some trials of large numbers of discrete spikes, which did not conform to the standard K distribution.

The pdf of clutter intensity, z, can be represented by equation (4.46). Here, the clutter comprises the speckle component, modulated by the local power, x, as in the standard K distribution. Now in addition, there are added spikes, which are assumed to have Rayleigh statistics (i.e. an Exponential distribution of intensity). However, only the speckle component of the clutter is assumed to decorrelate with frequency agility, whilst the local power and spike intensities are assumed to be constant over short time periods (e.g. a radar dwell) and unaffected by frequency agility. The speckle component is assumed to be due to Bragg scattering or range-extensive whitecaps. The mean power of this Bragg/Whitecap component is designated σ_{BW} and in terms of equations (4.24) and (4.46) $\sigma_{\text{BW}} = \nu/b$. In (4.46) the mean spike intensity was arbitrarily set to unity; for generality it is represented here as σ_{sp}. Finally, (4.46) has been modified to include thermal noise of power p_{n}. Of course, the noise component decorrelates completely from pulse-to-pulse, with or without frequency agility.

For a single pulse return we obtain

$$P(z|x) = \sum_{m=0}^{\infty} \frac{1}{x + p_{\text{n}} + m\sigma_{\text{sp}}} \exp\left(\frac{-z}{x + p_{\text{n}} + m\sigma_{\text{sp}}}\right) P_m(m); \quad 0 \le z \le \infty$$

$$P_c(x) = \frac{x^{\nu-1}}{\Gamma(\nu)} \left(\frac{\nu}{\sigma_{\text{BW}}}\right)^{\nu} \exp\left(\frac{-\nu x}{\sigma_{\text{BW}}}\right); \quad 0 \le x \le \infty \tag{4.79}$$

where

σ_{sp}: mean spike intensity
σ_{BW}: mean Bragg/Whitecap intensity
m: number of spikes in a range cell
$\bar{N}$: mean number of spikes in each range cell
ν: shape parameter of Bragg/Whitecap component
x: local mean intensity of Bragg/Whitecap component
p_{n}: noise power
$\langle z \rangle$: $\sigma_{\text{BW}} + \bar{N}\sigma_{\text{sp}} + p_{\text{n}}$
cnr: $(\sigma_{\text{BW}} + \bar{N}\sigma_{\text{sp}})/p_{\text{n}}$

It is assumed that the mean number of spikes in each range cell is $\bar{N} \ll 1$, so that the probability of m spikes occurring in a range cell, $P_m(m)$, is given by:

$$
\begin{aligned}
P_m(0) &= 1 - \bar{N} \\
P_m(1) &= \bar{N} \\
P_m(m \geq 2) &= 0
\end{aligned} \tag{4.80}
$$

(The Poisson and negative binomial distributions both tend to this simple limiting form when the mean number of scatterers in a range cell is very small.) The mean clutter intensity is given by $\sigma_{\mathrm{BW}} + \bar{N}\sigma_{\mathrm{sp}}$. The results for P_{FA} and P_d can be normalised to the mean clutter level. Setting σ_{BW} to unity, for convenience, the free parameters are $\nu, \bar{N}, cnr$ and $\rho = \sigma_{\mathrm{sp}}/\sigma_{\mathrm{BW}}$. It should be remembered that in this case the local number of spikes, m, is effectively 0 or 1 only (see equation (4.80)). Considering now the integration of n pulses, the probability of exceeding a threshold, t, is given by:

$$
\begin{aligned}
P_{\mathrm{FA}}(t|n,\rho) = &\int_0^\infty \left(\int_0^\infty P_{\mathrm{FA}}(t|x,s,n) P(s|\rho) P_m(1) ds \right) P_c(x) dx \\
&+ \int_0^\infty P_{\mathrm{FA}}(t|x,s=0,n) P_m(0) P_c(x) dx
\end{aligned} \tag{4.81}
$$

This equation can be recast in a more readily interpreted form

$$
\begin{aligned}
&P_{\mathrm{FA}}(t|n,\rho) \\
&= \int_0^\infty P_{\mathrm{FA}}(t|x,s=0,n) P_c(x) dx \\
&\quad + \bar{N} \int_0^\infty \left(\int_0^\infty \{P_{\mathrm{FA}}(t|x,s,n) - P_{\mathrm{FA}}(t|x,s=0,n)\} P(s|\rho) ds \right) P_c(x) dx
\end{aligned} \tag{4.82}
$$

The first term of (4.82) is the probability of false alarm in the absence of spikes. The second term represents the change in probability of false alarm due to the presence of spikes. The distribution of the spike intensity s, in terms of the parameter ρ defined above is

$$
P(s|\rho) = \frac{1}{\rho} \exp\left(\frac{-s}{\rho}\right); \quad 0 \leq s \leq \infty \tag{4.83}
$$

$P_{\mathrm{FA}}(t|x,s,n)$ is the probability of clutter with local mean intensity x, with clutter-to-noise ratio *cnr*, in the presence of a spike of intensity s, exceeding a threshold t following the integration of n frequency agile pulse returns. The calculations for pulse-to-pulse integration assume that the speckle component of the clutter and the

Table 4.1 Radar and environmental parameters of recorded data

Pol	Height (ft)	Range (km)	Grazing (deg)	Wind direction	Wind speed (kn)	Wave height (ft)	Douglas sea state
V	1343	27.8	0.75	Across-wind	27	6.2	4

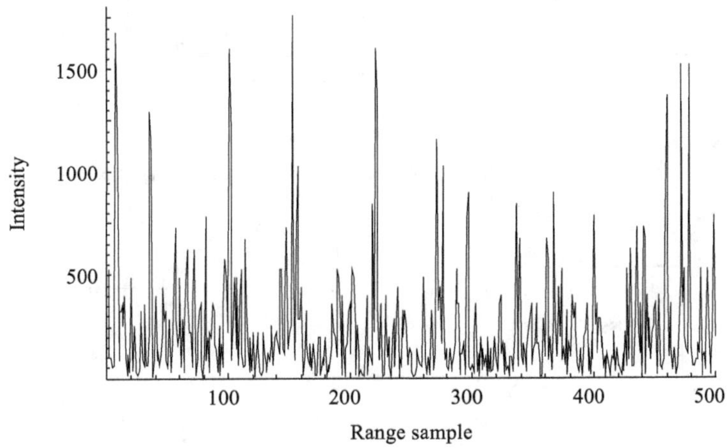

Figure 4.6 Raw intensity data range profile

noise are fully decorrelated from pulse to pulse, whilst the spike and underlying mean intensity components are assumed to be locally constant.

To illustrate these points we examine some very spiky data recorded by an experimental airborne radar, operating in X-band; the pertinent radar and environmental parameters are shown in Table 4.1. The radar had an antenna beamwidth of about 1.5 degrees and a range resolution of 2 m, and employed pulse-to-pulse frequency agility over a 500 MHz bandwidth. The antenna scanned at 40 rpm with a PRF of 2 kHz. These data are analysed to assess how well they fitted to the conventional K distribution and KA distribution (taking account of the radar system noise, and hence the *cnr*).

An important feature of the compound K and KA models is their ability to incorporate pulse-to-pulse correlation for the various components of the clutter (underlying mean component, speckle component, noise and spikes). It is reasonable to assume that the combination of frequency agility and antenna scanning [21] produced 13 independent looks at the speckle and noise over a beam dwell. Therefore the fit to the model was investigated for both the raw data and that resulting from pulse-to-pulse integration over 13 pulses, assuming independent looks at the speckle and noise but with the spikes and underlying mean level components remaining fixed. A plot of the intensity of raw data from a single pulse return is shown in Figure 4.6.

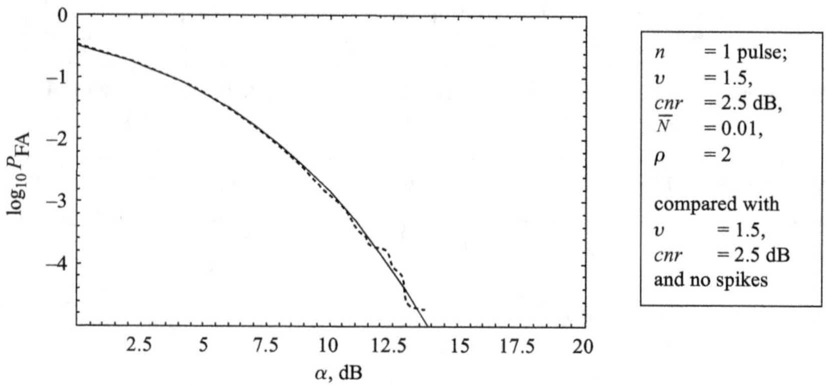

Figure 4.7 *Raw data distribution (dotted line), compared with K and KA distributions (identical solid lines)*

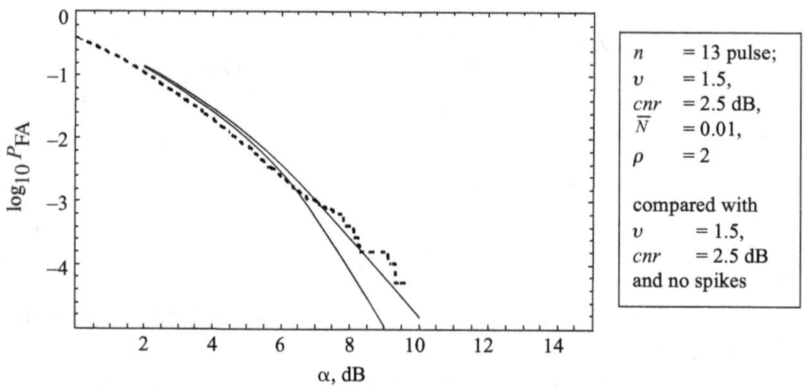

Figure 4.8 *Integrated data distribution (dotted line) compared with K and KA distributions (lower and upper solid lines)*

Figure 4.7 shows the cumulative distribution of this raw data, plotted as the probability of false alarm as a function of threshold multiplier, α, defined as the threshold normalised to the mean data intensity. The clutter-to-noise ratio, determined from the signal level and radar gain settings, was 2.5 dB. It can be seen that the raw data are a reasonable fit to a standard K distribution with $\nu = 1.5$. Figure 4.8 shows the cumulative distribution for the same data following pulse-to-pulse integration over $n = 13$ pulses. A comparison with a K distribution with the same $\nu = 1.5$ and $cnr = 2.5$ dB shows a significant deviation in the tail of the distribution, due to the presence of spikes that are not decorrelated by the frequency agility. A good fit to the data was found using the KA distribution for $\rho = 2$, $\bar{N} = 0.01$ and $cnr = 2.5$ dB. The distribution with the same KA distribution parameters, but for $n = 1$, is also shown in Figure 4.7, where again it can be seen to be a good fit. The fact that the same parameters can fit the data before and after pulse-to-pulse integration is quite striking; it is an indication of the utility of the model and the validity of the physics that underpin it.

It is very desirable that the parameter values to be used can be derived from the data themselves, rather than by trial and error. The value of ν was found for the examples given here by fitting the raw data to a K distribution at a false alarm level of 10^{-2}. At this level the distribution is not significantly affected by the presence of the spikes. The value of $\bar{N}$ determines the point at which the cumulative distribution starts to deviate from the K distribution and was chosen here by inspection. Once the value of clutter-to-noise ratio, ν and $\bar{N}$ have been determined, it remains to establish the value of ρ. This can be found from the increase in the probability of false alarm in the tail of the distribution for a given value of α, compared to that expected for an ideal K distribution. If, for a given value of α, the value of probability of false alarm is $P(\alpha)$ and the expected value for the ideal K distribution (without spikes) is $P_K(\alpha)$, then we can write:

$$\frac{P(\alpha)}{P_K(\alpha)} = \left(1 + \frac{\bar{N}\int_0^\infty \left(\int_0^\infty \{P_{\mathrm{FA}}(t|x,s,n) - P_{\mathrm{FA}}(t|x,s=0,n)\}\, P(s|\rho)ds\right) P_c(x)dx}{P_K(\alpha)}\right) \tag{4.84}$$

This equation can be inverted to find the appropriate value of ρ.

Further examples of the fit of real data to this model can be found in Reference 22.

4.5 The lognormal, Weibull and other non-Gaussian distributions

Our discussion of high resolution sea clutter has been motivated by the random walk model, and its modulation by the large scale structure of the ocean, revealed in part by a high resolution radar system, and has led us to the K distribution. Other non-Gaussian statistical models of sea clutter have been described in the literature, and have also found application in the analysis of radar performance. Foremost among these are the log-normal and Weibull distributions. These models are used principally because they are analytically tractable; they do not show the underlying compound structure that makes the K model so useful. In each a necessarily positive quantity z, such as received power, is considered. The log-normal model [23] then ascribes a normal or Gaussian distribution, with mean m and variance σ^2, to the logarithm of this quantity. Thus we have

$$P(z) = \frac{1}{z\sqrt{2\pi\sigma^2}} \exp\left(-\frac{(\log[z]-m)^2}{2\sigma^2}\right); \quad z \ge 0 \tag{4.85}$$

The moments of this distribution are given by

$$\langle z^n \rangle = \exp(nm + n^2\sigma^2/2) \tag{4.86}$$

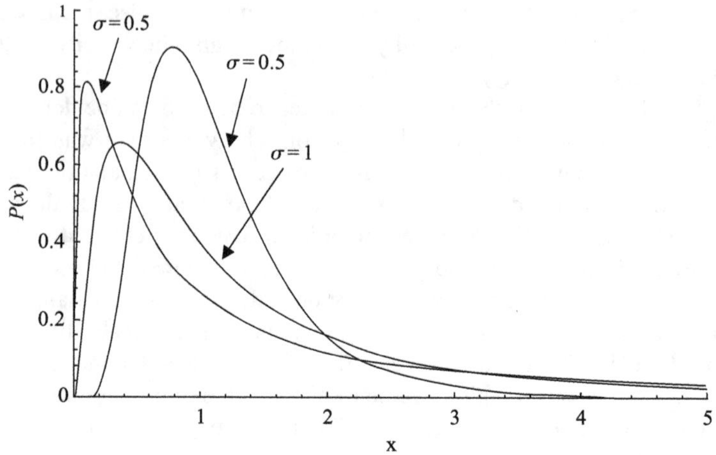

Figure 4.9 Log-normal probability density functions with m = 0

while the cumulative probability is given in terms of the error function (cf. (A1.31))

$$\int_0^z P(z')dz' = \frac{1}{2}\left(1 + \mathrm{erf}\left(\frac{\log z - m}{\sigma\sqrt{2}}\right)\right) \tag{4.87}$$

Examples of the log-normal distribution are shown in Figure 4.9.

The Weibull model [24] assigns the pdf

$$P(z) = \beta \frac{z^{\beta-1}}{a^\beta} \exp(-(z/a)^\beta); \quad z \geq 0 \tag{4.88}$$

to the intensity z; in this case the moments are given by

$$\langle z^n \rangle = a^n \Gamma(n/\beta + 1) \tag{4.89}$$

while the cumulative probability is

$$\int_0^z P(z')dz' = 1 - \exp(-(z/a)^\beta) \tag{4.90}$$

Examples of the Weibull distribution are shown in Figure 4.10, for different values of the shape parameter, β

When $\beta = 1/2$, equation (4.89) reduces to a form reminiscent of equation (4.28)

$$\langle z^n \rangle = n!(4a)^n \frac{\Gamma(n + 1/2)}{\Gamma(1/2)} \tag{4.91}$$

There is a corresponding equivalence between the pdfs of the Weibull and K distribution in this special case. This is best regarded as coincidence and is in no way indicative of a common underlying physical mechanism.

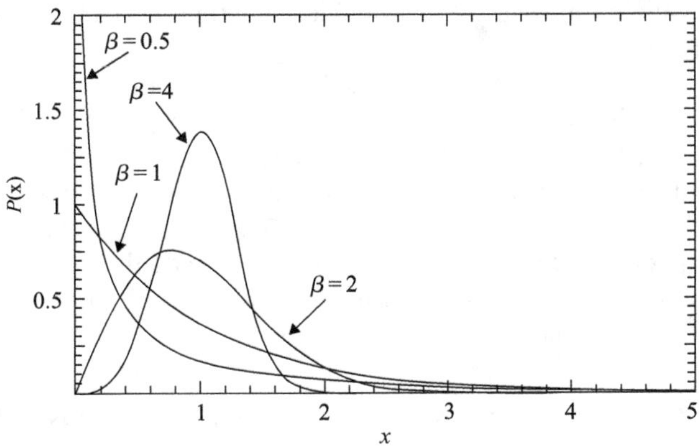

Figure 4.10 Weibull probability density functions

The possibility of constructing a compound model, in which the local power of a Rayleigh distributed speckle component is modulated by a Weibull distributed random variable, has been considered by Bucciarelli *et al.* [25]. In circumstances where the clutter is extremely non-Gaussian, these authors suggest that this model provides an improved description of its single point statistics. Other choices for $P_c(x)$ could be made, and provide potentially useful clutter models that would retain many of the attractive features of the K distribution. However, these would not result in distributions that are as well characterised in terms of tabulated functions, or have the property of infinite divisibility (discussed briefly in Section 4.4.2) possessed by the K distribution.

4.6 Coherent clutter modelling

So far we have applied the compound model to the description of the incoherent intensity of the clutter returns. As we discuss in Section A1.9 of Appendix 1 the two component complex signal detected by a coherent radar system is described in terms of its power (Doppler) spectrum; in this section we will discuss how the compound modelling philosophy can be extended to this case. Capturing the modulation of the local power of a Rayleigh speckle process has played a central role in the modelling of non-Gaussian clutter; in Section 4.3 we saw that its representation by a gamma process led us directly to the K distribution model for the clutter. A Doppler spectrum of a sea clutter return that is derived from a record significantly longer than the decorrelation time of this gamma process will provide an average over these fluctuations in local power. Results obtained in this way are smoothly varying functions of frequency, which can be represented quite effectively as a superposition of two or three Gaussian components. The characteristics (amplitude, centre frequency and width) of the Gaussian components that provide the best fit to Doppler spectrum data can be interpreted explicitly in terms of identifiable scattering mechanisms contributing to the clutter process.

If, however, the Doppler spectrum is determined from a record shorter than or comparable with the decorrelation time of the local power, very variable results are obtained; the Doppler spectrum itself must now be considered as a random variable. An example of the highly variable nature of the Doppler spectrum of sea clutter is shown in Figure 2.15. Thus the power in each Doppler bin fluctuates; how might we attempt to characterise and model these variations? To do this successfully we must first understand how the various scattering mechanisms contributing to the sea clutter are manifest in its Doppler spectrum.

4.6.1 The Doppler signatures of different scattering events

It was noted in Chapter 2 (Section 2.9) that the spectra for horizontally and vertically polarised returns are different; this suggests that the scattering events responsible for sea clutter are polarisation sensitive and contribute to different extents to VV and HH returns. Figure 2.14 identifies the three principal mechanisms from their time evolution. Walker [26,27] investigates the Doppler characteristics in these scattering mechanisms in the controlled environment of the wave tank and through the analysis of real sea clutter data. His discussion both draws on and supports earlier wave tank measurements and clutter analysis [28–30]. Together all this work provides much of the physical understanding we incorporate into our modelling of coherent clutter.

To be consistent with our earlier discussion, we refer to the three mechanisms as Bragg, whitecaps and bursts; each has different characteristics, and so can be distinguished from the others.

Bragg scattering is associated with scattering from resonant capillary waves caused by wind blowing over the sea surface. This mechanism is adequately described by the composite model[2] (see Chapter 3 and Appendix A3 for details) and results in stronger scattering in the VV mode than in HH. These Bragg returns have short decorrelation times (tens of milliseconds), and display the same Doppler shift in both polarisations.

Whitecap scattering, derived from the well-developed surface roughness resulting from a broken wave, contributes in more or less equal measure in VV and HH and is significantly greater in magnitude than Bragg scattering. These events may last for times of the order of seconds but again the scattered amplitudes decorrelate within milliseconds. Their Doppler spectra are broad and centred on a higher Doppler shift than those due to Bragg scattering, as they are usually moving with the crest of a breaking wave.

Burst events are generated by multiple scattering from an incipiently breaking wave travelling along the line of sight of the radar. Detailed electromagnetic scattering calculations, such as those described in Chapter 3, give us considerable insight

[2] Although, as we have shown in Chapter 3, it is important to include extra multipath effects at low grazing angles.

into this mechanism and indicate that its contribution is particularly strong for HH backscatter but is virtually absent for VV. Burst events persist for a much shorter time than do whitecap events, typically for about 0.2 s, but remain coherent over that period.

Walker [27] gives some excellent examples of time histories of these events, their Doppler spectra and autocorrelation functions, for VV and HH polarisations. Further results are presented here for linear and circular polarisations.

4.6.2 Some typical experimental results

Circular polarised results can be obtained by transforming linear 'polarisation scattering matrix' data, as mentioned in Chapter 2. The results presented here were obtained by transforming data recorded with V and H polarisations (using a clifftop radar positioned on Portland Bill, looking out into the Channel) into Left-hand and Right-hand circular polarisation. In order to use the data for polarisation transformation, it is necessary to balance the gain and phase of the transmitter and receiver channels. This was achieved by measuring the returns from a flat plate, which gives no cross-polar (VH or HV) return, to balance the co-polar channels.

Figure 4.11 shows time histories from a single range cell of data. The cell is chosen to exhibit the 'Bragg' scattering behaviour described above. Thus VV is greater than HH, and both the sequences have a noise-like character. The cross-polar channels are small.

When the data transformed to circular polarisation all four channels have similar mean intensities. The details of two cross-polar channels look very similar (which they should, because of reciprocity), but there are differences between the two co-polar channels, and between these and the cross-polar return. In simple terms, for linear polarisation the only significant signal is in VV; in circular therefore, this appears at half the amplitude (a quarter of the intensity) in all four channels.

Figure 4.12 shows the Doppler spectra of all the channels. The VV spectrum is centred close to zero frequency, as expected for 'Bragg' scattering, and all of the circular channels follow suit.

Figure 4.13 is from a range cell with the characteristics of a 'Whitecap'. VV and HH are roughly equal and noiselike. At times where the co-polar signals are large there is a significant amount of cross-polar scattering, although it remains much smaller (more than 10 dB smaller) than the co-polar signals. When the data are transformed to circular there is approximately equal power in all four channels. This implies that the phases of the VV and HH scattering are not correlated.

The Doppler spectra in Figure 4.14 show that the mean Doppler has a larger positive value than the 'Bragg' scattering of Figure 4.12. This implies that the 'Whitecap' has a significant advection velocity, which would be expected from a large breaker. It is also interesting to note the difference between the VV and HH spectra. Although the spectra cover the same Doppler extent, HH peaks at a higher velocity than VV, which implies that the main scattering from HH occurs higher on the wave-crest than VV. This is consistent with a simple multipath illumination model, where forward reflections from the sea surface interfere with direct illumination of the wave-crest.

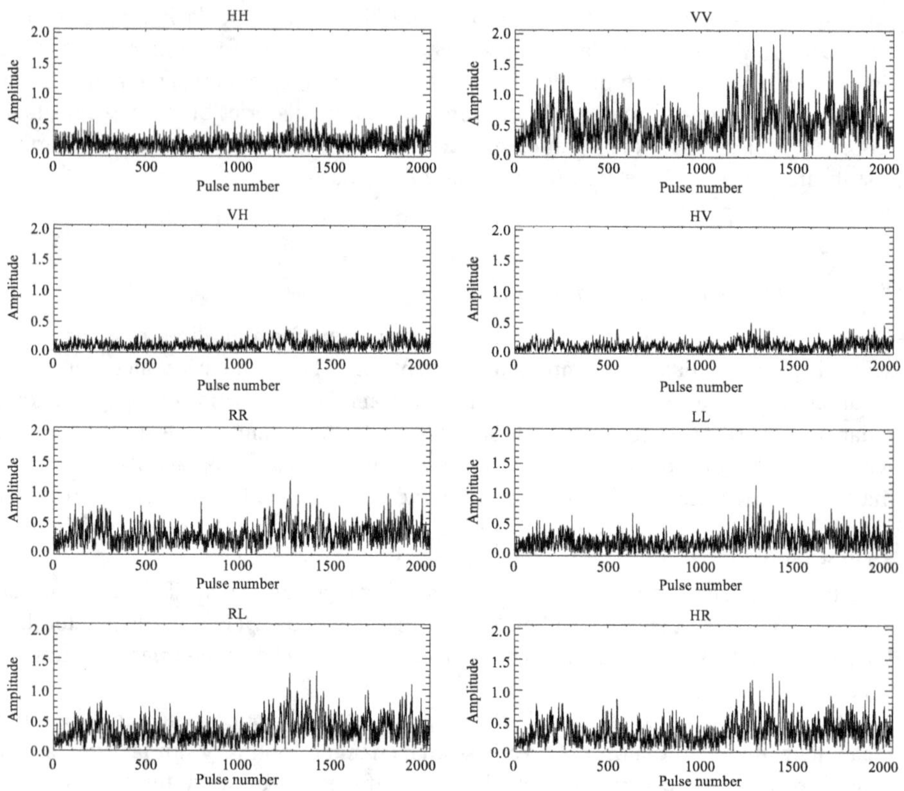

Figure 4.11 Four second time histories showing the calibrated linear channels (top four plots) and the data transformed to circular polarisation (bottom four plots). This range cell has characteristics consistent with Bragg scattering

The different phase changes of the forward scattered wave for VV and HH mean that the constructive interference regions occur at different heights on the sea wave.

All four circular polarisation Doppler spectra in Figure 4.14 look similar, as did the time sequences in Figure 4.13.

Figure 4.15 shows data taken from a range cell containing a discrete HH 'burst'. There is some VV return at the same time as the spike, but this has the character of Bragg scattering. These data are of a fairly rough sea with a wind speed of between 18 and 20 knots. In these conditions discrete spikes on their own are fairly rare. An illustration of clutter collected on the same occasion as these data is shown in Figure 2.13. The Doppler spectra of VV and HH shown in Figure 4.16 confirm that the HH spike is at a different velocity than the VV Bragg scattering is therefore from different parts of the wave. The circular polarisation time sequences and Doppler spectra again have fairly similar power and structure in all four channels.

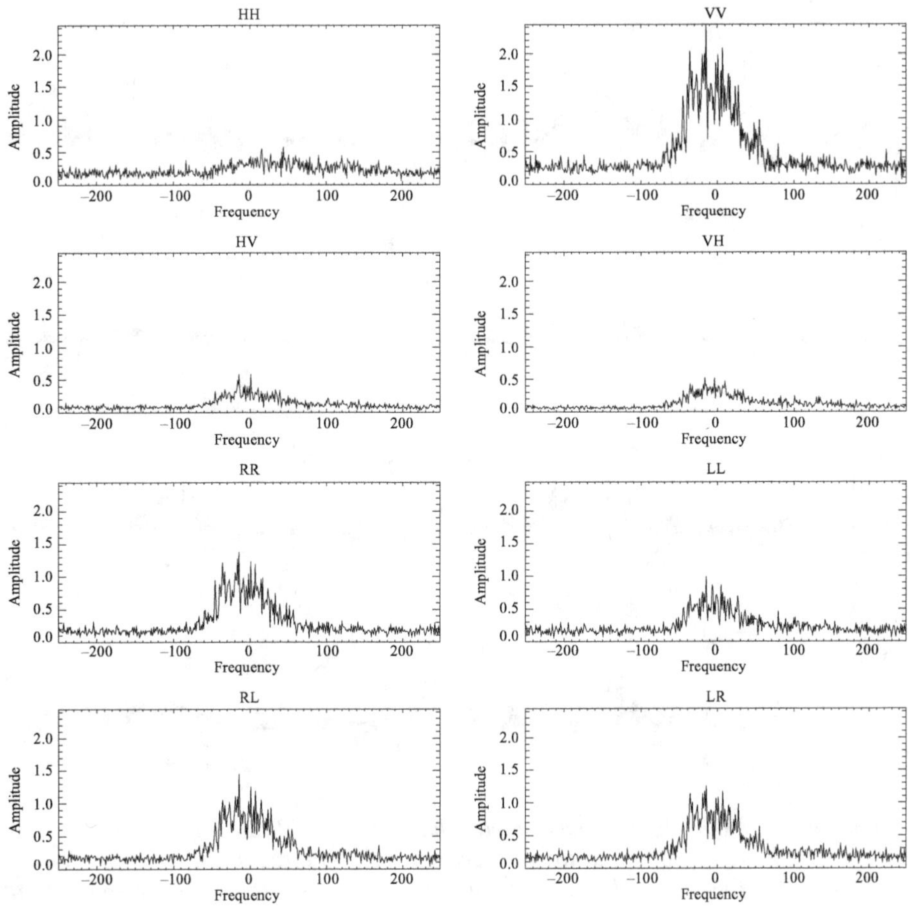

Figure 4.12 Calibrated Doppler spectra of the data in Figure 4.11

It is worth noting that if the spike specular return were free from multipath it would give equal returns in VV and HH. All of the power from this signal would transform into the cross-polar circular channels (LR and RL), with nothing in RR and LL. The fact that the spike is spread across the four circular channels shows the impact of multipath enhancing HH and suppressing VV.

The HH spectrum for the spike specular return in Figure 4.16 is also relatively narrow compared to the spectra of Whitecap or Bragg scattering events. This will be reflected in a longer decorrelation time. Walker [27] measured HH spikes with a decorrelation time of greater than 80 ms, compared with about 10 ms for HH and VV Bragg scattering and about 5 ms for HH and VV Whitecap events.

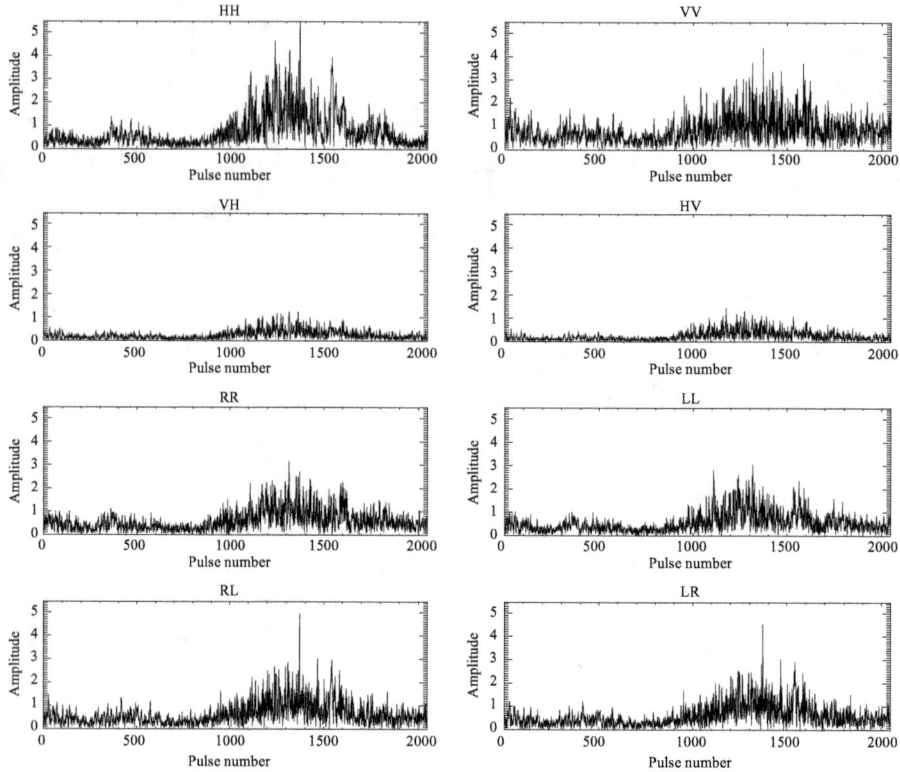

Figure 4.13 *Four second time histories showing the calibrated linear channels (top four plots) and the data transformed to circular polarisation (bottom four plots). This range cell has characteristics consistent with Whitecap scattering*

4.6.3 Models of Doppler spectra

The power spectrum of a complex process, discussed in Section A1.9, is defined in terms of a formal limit of infinite measurement time T. In practice this is taken to be some time that is much greater than the characteristic correlation time of the process under consideration. So, when we analyse sea clutter, the compounded components of which evolve over widely differing time-scales, some care has to be taken over the identification of T. T can be chosen to be much longer than the time-scale typical of the modulation of the local speckle power by the large-scale structure of the sea surface, which will be of the order of many seconds, if not minutes. It is quite appropriate in this case to model the resulting, smoothly varying, power spectrum in terms of a simple functional form that captures the averaged behaviour of the modulation. The modelling of the power spectrum in terms of a sum of Gaussian components, as suggested by Lee and others [28], provides an example of this approach. In this

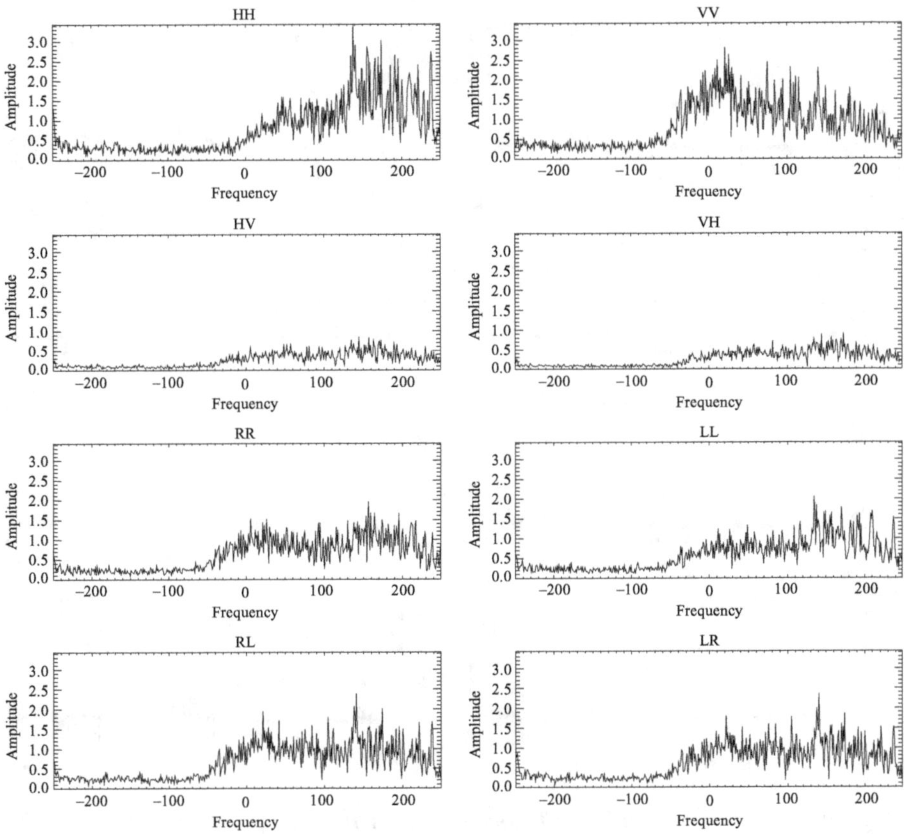

Figure 4.14 Calibrated Doppler spectra of the data in Figure 4.13

the Gaussian function of frequency

$$\Sigma(\omega|a, b, c) = a \exp\left(\frac{-(\omega - b)^2}{c}\right) \tag{4.92}$$

is used as a basic building block from which the power spectrum can be constructed. The parameters a and b control the overall power and position of the centre of this Gaussian unit, while c determines its width. So, for example, the contribution from Bragg scattering would be caricatured by relatively small values of a and b, while a larger value of c would capture its broad spread of frequencies. Similarly, by choosing larger values of a and b, and a smaller c, one could represent the more intense and narrower component typical of a HH burst contribution, suitably displaced from the zero Doppler frequency.

If, however, T is chosen to be shorter than this modulation time scale, and yet to be significantly longer than the decorrelation time of the local speckle process (which may be of the order of a tenth of a second or less), rather different results

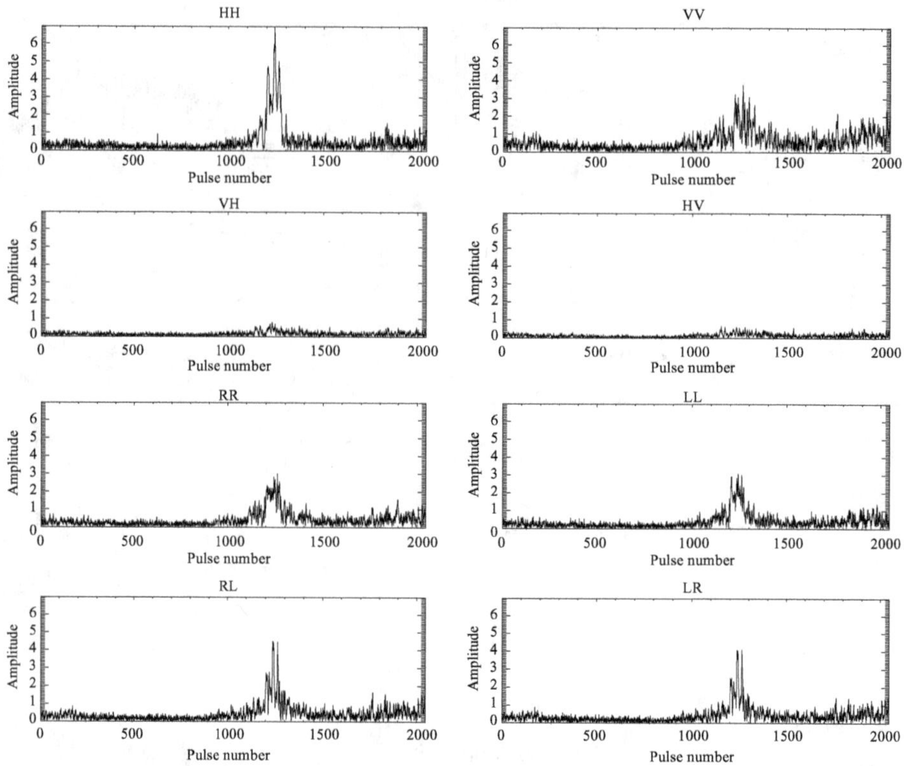

Figure 4.15 *Four second time histories showing the calibrated linear channels (top four plots) and the data transformed to circular polarisation (bottom four plots). This range cell has characteristics consistent with spike scattering*

emerge. A power spectrum, derived in this way, cannot be represented as a sum of Gaussians, or some such simple analytic model. Rather, it can itself be modelled as a random process, varying on the time scale of the slower modulation process. In the K distribution model this local power, x, is represented as a gamma variate. Given a 'short time' power spectrum, determined over an interval T in which x has not changed perceptibly, we can identify this x with the integral of the power spectrum over all frequencies. An attractive way to model this fluctuating Doppler spectrum is to represent it as the product of a gamma variate and suitable function of unit area:

$$S(\omega|x) = x\hat{S}(\omega)$$

$$\int_{-\infty}^{\infty} \hat{S}(\omega)d\omega = 1 \tag{4.93}$$

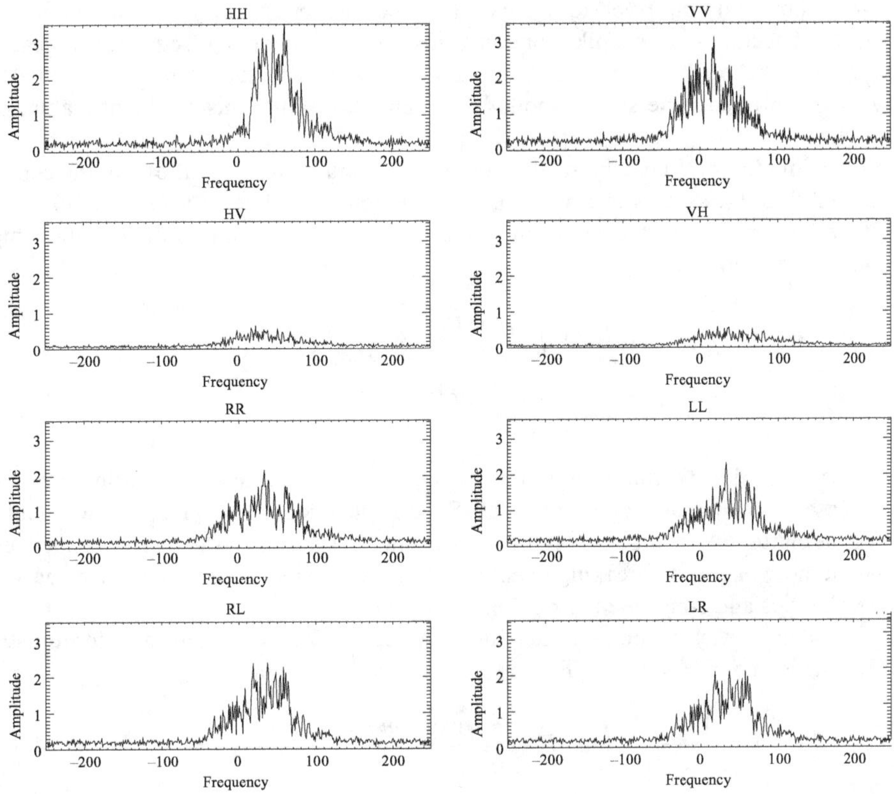

Figure 4.16 Calibrated Doppler spectra of the data in Figure 4.15

Motivated in part by the success of the sum of Gaussians model we choose the form

$$\hat{S}(\omega) = \frac{\exp(-(\omega - \omega_0)^2/2\sigma^2)}{\sqrt{2\pi\sigma^2}} \tag{4.94}$$

The normalised moments of this power spectrum model can be evaluated quite straightforwardly as

$$\frac{\langle S(\omega)^n \rangle}{\langle S(\omega) \rangle^n} = \frac{\langle x^n \rangle}{\langle x \rangle^n} = \frac{\Gamma(\nu + n)}{\nu^n \Gamma(\nu)} \tag{4.95}$$

and are independent of frequency. Thus this simple model predicts that the statistics of the returns in all Doppler bins are identical. Much of the published analysis of Doppler spectra of clutter [26] presents the data in terms of an effective ν parameter that is in general frequency dependent and is defined by

$$\frac{1}{\nu_{\text{eff}}} = \frac{\langle S(\omega)^2 \rangle}{\langle S(\omega) \rangle^2} - 1 \tag{4.96}$$

ν_{eff} is commonly observed to decrease (i.e. the frequency component in the power spectrum becomes more 'spiky' or non-Gaussian) as the Doppler frequency increases (e.g. see Chapter 2, Figures 2.16 and 2.17). It is this feature that we particularly wish to reproduce; the simple model is not able to capture this pertinent feature of the data.

The model (4.93) can be made more general and flexible by allowing the centre and width in (4.94) to vary; we assign a joint conditional density function to these quantities, which may degenerate into delta functions when some explicit relationship is imposed. Thus we have

$$\langle S(\omega)^n \rangle = \frac{1}{(2\pi)^{n/2}} \int dx \int d\omega_0 \int d\sigma \, P_x(x) P(\omega_0, \sigma | x) x^n \times \frac{\exp(-n(\omega - \omega_0)^2/2\sigma^2)}{\sigma^n} \tag{4.97}$$

So for example, we might assume that the width of the Gaussian building block increases with increasing local power x. Such a dependence could result, if we identified the clutter return with a breaking wave event; a stronger return might be associated with a more energetic breaking event which in turn generated a broader distribution of velocities and consequently of Doppler shifts.

A simple way to incorporate this into our model is through a deterministic dependence i.e. $\sigma = \sigma(x)$ or $P(\sigma|x) = \delta(\sigma - \sigma(x))$ so that

$$\langle S(\omega)^n \rangle = \frac{1}{(2\pi)^{n/2}} \int dx \, P_x(x) x^n \frac{\exp(-n\omega^2/2\sigma(x)^2)}{\sigma(x)^n} \tag{4.98}$$

We now make the choice $\sigma(x) = \sqrt{2x}$, primarily to facilitate the analysis; nonetheless the monotonic growth of σ with x captured by this model is sensible. Thus we have

$$\begin{aligned} \langle S(\omega)^n \rangle &= \frac{1}{\pi^{n/2}} \frac{b^\nu}{\Gamma(\nu)} \int dx \exp(-bx) x^{\nu+n/2-1} \exp(-n\omega^2/x) \\ &= \frac{2}{\pi^{n/2}} \frac{b^{\nu/2-n/4}}{\Gamma(\nu)} \omega^{\nu+n/2} n^{(\nu/2+n/4)} K_{\nu+n/2}(2\omega\sqrt{bn}) \end{aligned} \tag{4.99}$$

so that the effective shape parameter can be calculated from

$$\frac{\nu}{\nu_{\text{eff}}} = \nu \left(\frac{\Gamma(\nu)}{\omega^\nu} \frac{2^{(\nu-1)/2}}{b^{\nu/2}} \frac{K_{\nu+1}(2\omega\sqrt{2b})}{K_{\nu+1/2}(2\omega\sqrt{b})^2} - 1 \right) \tag{4.100}$$

As usual K is the modified Bessel function. We see from the results shown in Figure 4.17 that the power spectrum values become more non-Gaussian in its wings, much as in the available experimental data. This enhancement of the non-Gaussian character of the tails is more pronounced for clutter for which the total power itself is more non-Gaussian (i.e. for lower values of ν).

Many Doppler spectra display significant asymmetry; from our earlier discussion we might reasonably associate this with the presence of whitecap and burst

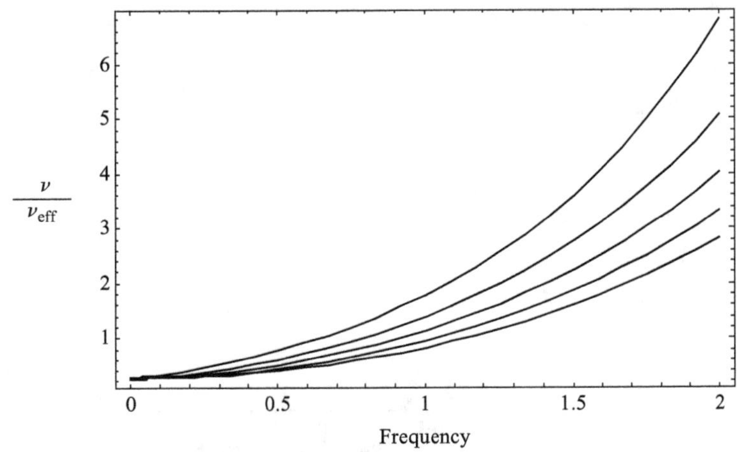

Figure 4.17 *Variation of effective shape parameter with frequency, described by equation (4.98). The parameter b has been set to unity, $\nu = 0.5, 1.0, 1.5, 2.0, 2.5$; ν/ν_{eff} increases with decreasing ν.*

contributions to the scattering. Thus one might set up a spectrum of the form

$$S(\omega) = A[c_1 \exp(-\alpha\omega^2) + c_2 \exp(-\beta(\omega - \omega_1)^2)] \tag{4.101}$$

where A is chosen to maintain a unit total power in the spectrum. Typically this might look like Figure 4.18 and so reproduces the desired asymmetry. Mere multiplication of this 'building block' by a gamma variate, as in (4.20), will not be able to reproduce 'spikiness in the wings'; we must modulate the spectral parameters appropriately if we are to reproduce this experimentally observed behaviour. This can be done in many ways. We will be guided in part by considerations of analytic tractability as the physical foundation of this approach is, at best, phenomenological. Thus we can modulate the power of the central Gaussian, the offset Gaussian or both, or let the centre of either term migrate stochastically. Expressions for general moments of spectral power at any frequency will be rather cumbersome; moments sufficient to determine an effective 'shape parameter' (4.94) should be expressible in useful closed forms.

To illustrate this approach, and indicate how it might be motivated by physical considerations, we suggest, following Walker [27] and our earlier discussion, that the Doppler spectrum of the clutter returns consists of two Gaussian components. One of these, centred about a zero Doppler frequency, represents Bragg scattering by small-scale structure on the sea surface. The other component, whose centre is displaced from zero, represents contributions from breaking wave events. These breaking wave events contribute significantly to the clutter RCS; an increase in local power is attributed to a greater proportion being due to breaking wave events. These

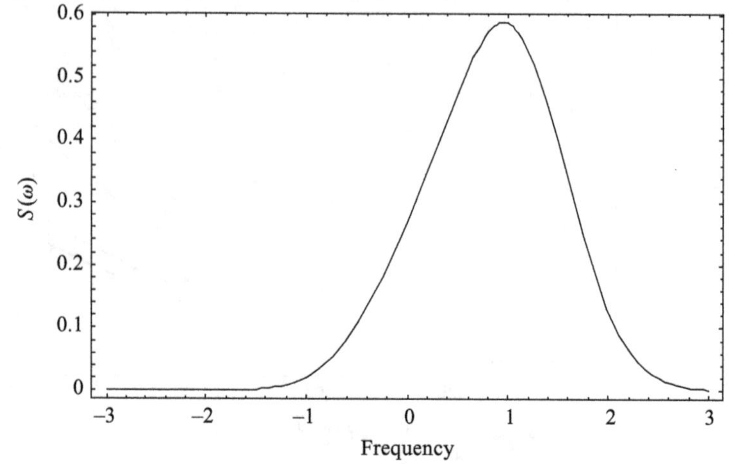

Figure 4.18 *An asymmetric power spectrum modelled by a sum of Gaussians as in (101). Here α, β and ω^{-2} have the same arbitrary units: $\alpha = 2.0$, $\beta = 1.5$, $c_1 = 1$, $c_2 = 4$, $\omega_0 = 1$*

qualitative considerations are captured in the following model for the power spectrum

$$
\begin{aligned}
S(\omega) &= x\frac{a\hat{S}(\omega,\sigma_0,0) + x\hat{S}(\omega,\sigma_1,\Omega_1)}{a+x} \\
\hat{S}(\omega,\sigma,\Omega) &= \frac{\exp(-(\omega-\Omega)^2/2\sigma^2)}{\sqrt{2\pi\sigma^2}}
\end{aligned}
\tag{4.102}
$$

where x is the familiar gamma variate modulating the frequency averaged clutter power. The mean and mean square powers at a given Doppler frequency can now be written as

$$
\begin{aligned}
\langle S(\omega)\rangle &= a\hat{S}(\omega,\sigma_0,0)\left\langle\frac{x}{a+x}\right\rangle_\gamma + \hat{S}(\omega,\sigma_1,\Omega_1)\left\langle\frac{x^2}{a+x}\right\rangle_\gamma \\
\langle S(\omega)^2\rangle &= a^2\hat{S}(\omega,\sigma_0,0)^2\left\langle\frac{x^2}{(a+x)^2}\right\rangle_\gamma + \hat{S}(\omega,\sigma_1,\Omega_1)^2\left\langle\frac{x^4}{(a+x)^2}\right\rangle_\gamma \\
&\quad + 2a\hat{S}(\omega,\sigma_0,0)\hat{S}(\omega,\sigma_1,\Omega_1)\left\langle\frac{x^3}{(a+x)^2}\right\rangle_\gamma
\end{aligned}
\tag{4.103}
$$

Here the angular brackets $\langle\,\rangle_\gamma$ represent an average over the gamma distribution (4.24). A little algebraic reduction shows that these can all be represented in terms of

$\langle x \rangle_\gamma$, $\langle x^2 \rangle_\gamma$ and

$$\left\langle \frac{1}{a+x} \right\rangle_\gamma = b^\nu a^{\nu-1} \exp(ab) \Gamma(1-\nu, ab)$$
$$\left\langle \frac{1}{(a+x)^2} \right\rangle_\gamma = b^\nu a^{\nu-2} \exp(ab) [\Gamma(2-\nu, ab) - ab\Gamma(1-\nu, ab)] \tag{4.104}$$

We also recall the definition from Appendix 1 equation (A1.31) of the incomplete gamma function. In this way we find that the mean value of the Doppler spectrum can be written as

$$\langle S(\omega) \rangle = a\hat{S}(\omega, \sigma_0, 0)(1 - (ab)^\nu \exp(ab)\Gamma(1-\nu, ab)) + \hat{S}(\omega, \sigma_1, \Omega_1) \times \left(\frac{\nu}{b} - a + a(ab)^\nu \exp(ab) \Gamma(1-\nu, ab) \right) \tag{4.105}$$

A similar, but more involved, expression can be derived for the mean square.

Another noteworthy model is described by Miller [31]. In this the central frequency of the Gaussian building block is modulated by a gamma variate that is independent of that modulating the local power; this second gamma variate we denote by y. It is assigned a distinct gamma pdf

$$P(y) = \frac{a^\alpha}{\Gamma(\alpha)} y^{\alpha-1} \exp(-ay) \tag{4.106}$$

If we now form the short-time power spectrum as

$$S(\omega) = x \frac{\exp(-(\omega - y)^2/2\sigma^2)}{\sqrt{2\pi\sigma^2}} \tag{4.107}$$

this, and its higher powers, can be averaged over the separate gamma distributions to give the following result:

$$\begin{aligned} \langle \hat{S}(\omega)^n \rangle = {} & \frac{a^\alpha \sigma^{\alpha-n} \exp(-n\omega^2/2\sigma^2)}{\Gamma(\alpha)(2\pi)^{n/2} n^{\alpha/2}} \frac{\Gamma(n+\nu)}{b^\nu \Gamma(\nu)} \\ & \times \left\{ \Gamma(\alpha/2)\, {}_1F_1(\alpha/2, 1/2; \sigma^2(a - n\omega/\sigma^2)^2/2n) \right. \\ & + \sqrt{\frac{2}{n\sigma^2}} (n\omega - a\sigma^2) \Gamma((\alpha+1)/2) \\ & \left. \times {}_1F_1((\alpha+1)/2, 3/2; \sigma^2(a - n\omega/\sigma^2)^2/2n) \right\} \end{aligned} \tag{4.108}$$

Here we have introduced the confluent hypergeometric function (see Appendix 2, equation A2.14); this direct evaluation provides an alternative to the numerical quadrature suggested by Miller [31]. Once again a closed form such as this may be useful in establishing contact between the model and empirical data; an effective shape parameter can be calculated from equation (4.94).

This brief discussion of the power spectrum of coherent sea clutter has demonstrated how the compound representation, which was initially developed to describe incoherent clutter, again enables us to construct relatively tractable and physically motivated models.

References

1 W. Feller, *'Introduction to probability theory and its applications'*, Vol. 2, Chap. 8, John Wiley & Sons, New York, 1971.
2 S.O. Rice, 'Mathematical analysis of random noise', *Bell Syst. Tech. J.*, **23**, 282, 1944; **25**, 46, 1945.
3 K.D. Ward, 'Compound representation of high resolution sea clutter', *Electron. Lett.*, **17**, 561–65, 1981.
4 G.V. Trunk, 'Radar properties of non-Rayleigh sea clutter', *IEEE Trans.* **AES-8**, 196–204, 1972.
5 G.R. Valenzuela and M.B. Lang, 'On the statistics of sea clutter', NRL Report 7349, 1971.
6 J.M. Blythe *et al.*, 'The CFE clutter model, with application to automatic detection', *The Marconi Rev.*, **32**, 185–206, 1969.
7 K.D. Ward, C.J. Baker and S. Watts, 'Maritime surveillance radar part 1: Radar scattering from the ocean surface', *Proc. IEE*, **137**(part F, No. 2), 51–62, 1990.
8 E. Conte, A. De Maio and C. Galdi, 'Statistical validation of the compound-Gaussian model on clutter data from IPIX radar', Radar 2004 International Conference on Radar Systems, Toulouse, 18–22 October 2004.
9 E. Jakeman, 'On the statistics of K distributed noise', *J. Phys.* A **13**, 31–48, 1980.
10 E. Jakeman and R.J.A. Tough, 'Generalised K distribution: A statistical model for weak scattering', *J. Opt. Soc. AM.*, **4**, 1764, 1987.
11 S. Watts, 'Radar detection prediction for targets in both K-distributed sea clutter and thermal noise', *IEEE Trans.* **AES-23**(2), 40–45, 1987.
12 T. Lamont-Smith, 'Translation to the normal distribution for radar clutter', *IEE Proc. Radar, Sonar Navig.*, **147**(1), 17–22, 2000.
13 T.R. Field and R.J.A. Tough, 'Dynamical models for weak scattering', *J. Math. Phys.*, **46**, 013302, 2005.
14 D. Middleton 'Canonical and non-canonical non-Gaussian noise models of class A interference', *IEEE Trans. Electromag. Compat.*, **EMC-25**, 76–106, 1983.
15 D. Middleton, 'New physical-statistical methods and models for clutter and reverberation: The KA distribution and related probability structures', *IEEE J. Oceanic Eng.*, **24**(3), 261–284, 1999.
16 K.D. Ward and R.J. Tough, 'Radar detection performance in sea clutter with discrete spikes', International Conference Radar 2002, IEE Publication 490, 15–17 October 2002, pp. 253–257.
17 M. Clements and U. Yurtsever, *'The breaking area model for low grazing angle sea clutter'*, Dynamics Technology Inc., Arlington, VA, 1997.

18 R.J. Adler, *'The geometry of random fields'*, John Wiley & Sons, New York, 1981.

19 G.N. Watson *'A treatise on the theory of Bessel functions'*, Section 13.51, University Press, Cambridge, 1966.

20 E. Jakeman and P.N. Pusey, 'A model for non-Rayleigh sea echo', *IEEE Trans. Antennas Propag.*, **AP-24**, 808, 1976.

21 S. Watts and A.G. Stove, *'Non-coherent integration strategies in scanning radar for the detection of targets in sea clutter and noise'*, International Radar Conference Radar 2004, Toulouse, 19–21 October 2004.

22 S. Watts, K.D. Ward and R.J.A. Tough, 'The physics and modelling of discrete spikes in radar sea clutter', IEEE International Radar Conference Radar 2005, Washington, May 2005.

23 G.V. Trunk and S.F. George, 'Detection of targets in non-Gaussian sea clutter', *IEEE Trans* **AES-6**, 620–628, 1970.

24 F.A. Fay, J. Clarke and R.S. Peters, 'Weibull distribution applications to sea clutter', *IEE Conf. Publ.* **155**, 1012, 1977.

25 T. Bucciarelli, P. Lombardo, C.J. Oliver and M. Perrotta, *'A compound Weibull model for SAR texture analysis'*, *Proc. IGARRS*, **1**, 181–183, 1997.

26 D. Walker, 'Experimentally motivated model for low grazing angle radar Doppler spectra of the sea surface', *IEE Proc. Radar, Sonar Navig.*, **147**(3), 114–120, 2000.

27 D. Walker, 'Doppler modelling of radar sea clutter', *IEE Proc. Radar, Sonar Navig.*, **148**(2), 73–80, 2001.

28 P.H.Y. Lee, J.D. Barter, B.M. Lake and H.R. Thompson, 'Lineshape analysis of breaking wave Doppler spectra', *IEE Proc. Radar, Sonar Navig.*, **145**(2), 135–139, 1998.

29 J. Fuchs, D. Regas, T. Waseda, S. Welch and M. Tulin, Correlation of hydrodynamic features with radar backscatter from breaking waves, *IEEE Trans. Geosci. Remote Sens.*, **37**(5), 2442–2460, 1999

30 P.H.Y. Lee *et al.*, 'Power spectral lineshapes of microwave radiation scattered from sea surfaces at low grazing angles', *IEE Proc. Radar, Sonar Navig.*, **142**(5), 252–258, 1995.

31 R.J. Miller, *'Variability in spectra of low-grazing angle sea clutter returns'*, NATO/RTO Publications, Proceedings of SET Symposium on Low Grazing Angle Clutter: Its Characterisation, Measurement and Application, April 2000.

Chapter 5

The simulation of clutter and other random processes

5.1 Introduction

In this chapter we consider simulation techniques that enable us to study various aspects of radar performance, in circumstances where an analytic attack may not be possible, or particularly informative. To complement the analytic clutter modelling discussed in Chapter 4, we will develop methods for the numerical simulation of unwanted radar returns. The clutter models in Chapter 4 are exclusively statistical; this prejudice is still evident in our choice of simulation methods. In essence, we address the problem of generating correlated random numbers with prescribed one and two point statistics (i.e. pdf and correlation function) that incorporate the physical insights developed in Chapters 2–4.

Significant progress can be made in the simulation of Gaussian processes with specified correlation properties; these provide a simple but useful simulation tool. The controlled generation of correlated non-Gaussian processes is more difficult. Fortunately our principal interest is in simulating clutter, for which the compound model has proved to be very effective; the Gaussian simulation techniques allow us to realise the 'speckle' component without too much difficulty. The main difficulty we encounter is in the simulation of the correlated variation in the local power component. Useful progress can be made towards the solution of this problem, adopting a fairly pragmatic approach based on the non-linear transformation of a Gaussian process to one with gamma statistics. The resulting process can be analysed in sufficient detail for the correlation of the output gamma process to be related directly to the more readily controlled correlation of the input Gaussian. Once we are able to simulate a gamma process in this way, we can generate K-distributed clutter, simply by multiplying the local power and speckle components together. When this is done we have a reasonable, physically motivated clutter simulation capability that is potentially very useful in the modelling of radar performance. The implementation of this approach, which has led to most of the simulation results presented elsewhere in the book, will be discussed and illustrated here with some practical examples.

5.2 Generating un-correlated random numbers with a prescribed pdf

Modern computers readily supply us with a succession of un-correlated random numbers x, uniformly distributed between 0 and 1, which we take as the starting point for the development of simulation methods. In Section A1.5 we discuss how the pdf of a function of x can be determined, relatively straightforwardly. In practice, however, one more often wishes to specify the pdf of the variable y (which we will have tailored to model some aspect of the clutter process, in line with the discussion in Chapter 4), and determine the transformation to be applied to the values of x to generate a succession of values of y with this required pdf. Thus, if we write the required pdf as $P_{\mathrm{y}}(y)$ and the transformation of x as

$$y = y(x) \tag{5.1}$$

we can equate the probability that x takes values between x and $x+dx$ with $P_{\mathrm{y}}(y)dy$, the probability that the transformed variable takes values between $y(x)$ and $y(x) + (dy/dx)dx$. This provides us with a first order ordinary differential equation to solve for $y(x)$, subject to the condition $P_{\mathrm{y}}(-\infty) = 0$. In this way we find that

$$x = \int_{-\infty}^{y} P_{\mathrm{y}}(y')\, dy' \tag{5.2}$$

So, given the value x generated by the computer, we solve this equation for the corresponding value of y, which now has the required pdf $P_{\mathrm{y}}(y)$. In some cases the transformation from x to y is relatively straightforward. For example, if y is required to have an exponential distribution, we have

$$P_{\mathrm{y}}(y) = \exp(-y), \quad y \geq 0$$

$$x = \int_{0}^{y} \exp(-y')\, dy' = 1 - \exp(-y)$$

$$y = -\log(1 - x) \tag{5.3}$$

As $1 - x$ and x have the same distribution we can also write this transformation between uniformly and exponentially distributed random numbers as $y = -\log(x)$. The generation of un-correlated Gaussian random numbers is particularly important in practical applications. At first sight it appears that we must solve

$$x = \frac{1}{\sqrt{\pi}} \int_{-\infty}^{y} \exp(-y'^2)\, dy' = \frac{1}{2} + \operatorname{erf}(y) \tag{5.4}$$

at each step in the simulation; fortunately this rather cumbersome procedure can be avoided. Consider two independent Gaussian random variables, y_1, y_2 with the joint pdf (cf. (4.6))

$$P_{y_1 y_2}(y_1, y_2) = \frac{1}{\pi} \exp(-y_1^2 - y_2^2) \tag{5.5}$$

If we now represent these in polar form

$$y_1 = r \cos\theta, \quad y_2 = r \sin\theta \tag{5.6}$$

then the joint pdf of these polar variables is

$$P_{\mathbf{r}\theta}(r,\theta) = P_{\mathbf{r}}(r) P_\theta(\theta)$$
$$P_{\mathbf{r}}(r) = 2r \exp(-r^2), \quad P_\theta(\theta) = \frac{1}{2\pi} \tag{5.7}$$

Thus the polar angle is uniformly distributed, and so can be simulated through

$$\theta = 2\pi x_1 \tag{5.8}$$

We also see that r^2 is exponentially distributed; thus r can be simulated through

$$r = \sqrt{-\log(x_2)} \tag{5.9}$$

So, having generated two independent uniformly distributed random numbers x_1, x_2, we produce two independent Gaussian random variables y_1, y_2 as

$$y_1 = \sqrt{-\log(x_2)} \cos(2\pi x_1), \quad y_2 = \sqrt{-\log(x_2)} \sin(2\pi x_1) \tag{5.10}$$

We note that these y variables each has a zero mean and a variance of $\frac{1}{2}$; a Gaussian random variable with a mean of μ and a variance of σ^2 is obtained when we then form $\mu + \sqrt{2\sigma^2}y$. A Gaussian random variable with zero mean and unit variance is required in many of our simulations; we denote this by g.

Independent random variables with other pdfs can be generated similarly; considerable cunning has been deployed in the efficient evaluation of the transform. Mathematica (and other) software packages implement many of these as built-in functions, so that the generation of un-correlated random variables with a specified pdf presents very few practical problems.

5.3 Generating correlated Gaussian random processes

We recall that the sum of two Gaussian random variables itself has Gaussian statistics; subjecting a Gaussian process to any linear operation (e.g. filtering, integration or differentiation) similarly yields a Gaussian process, albeit with a different mean, variance and correlation function. The qualitative effects of the linear operation are manifest in the correlation properties of the output process, while its single point statistics remain unchanged. A simple, and practically useful, simulation of a correlated Gaussian process can be implemented as follows. Consider the simple recurrence (or feedback) relation

$$x_n = \eta x_{n-1} + \beta g_n \tag{5.11}$$

(g are zero mean, unit variance independent Gaussian random numbers); η is chosen to have magnitude less than 1 so that the variance of the process remains finite. If (5.11) is initiated at $n = 0$ we can write its solution as

$$x_n = \eta^n x_0 + \beta \sum_{r=0}^{n-1} \eta^r g_{n-r} \tag{5.12}$$

Apart from the transient term, this can be viewed as the output of a filter acting on the succession (or time series) of gs. The mean value of x_n, averaged over realisations of the gs consists of the transient term

$$\langle x_n \rangle = \eta^n x_0 \tag{5.13}$$

while its mean square is

$$\langle x_n^2 \rangle = \eta^{2n} x_0^2 + \frac{\beta^2 (1 - \eta^{2n})}{1 - \eta^2} \tag{5.14}$$

Once the transient terms have died down, we see that the output of the process is a zero mean Gaussian process with a modified variance. More importantly, however, correlation has been induced in this output that was absent in the succession of uncorrelated gs that serve as its input. This correlation can be calculated directly:

$$\langle x_n x_{n+m} \rangle = \eta^m \langle x_n^2 \rangle \tag{5.15}$$

Thus we see how the linear feedback process has induced an exponentially decaying correlation in the output Gaussian process, which we can control through the parameter η; β allows us to control the power of the output process. If we choose

$$\beta = \sqrt{1 - \eta^2} \tag{5.16}$$

we ensure that the output has a unit variance, once transients have died down. A simple correlated Gaussian generated in this way can itself be a very useful simulation tool [1].

We see from (5.12) that this method provides an example of the use of a filter to induce correlation in a Gaussian process; as we shall see in Section 5.4 this approach can be connected with the description of the output process in terms of its power spectrum. It is interesting to note that the underlying recurrence (5.11) also makes contact with the stochastic differential equation description of random processes discussed in Section A1.11. There we see that (5.11) provides a realisation of the Ornstein Uhlenbeck (OU) process [2], and can be related directly to the integration of an underlying stochastic differential equation

$$\frac{dx(t)}{dt} = -\alpha x(t) + f(t) \tag{5.17}$$

driven by the white noise f, which has the correlation property

$$\langle f(t_1) f(t_2) \rangle = 2\gamma \delta(t_1 - t_2) \tag{5.18}$$

The emergence of a process with a finite variance as a result of the balance of the damping, characterised by α and random driving term, whose power γ can be interpreted as a diffusion constant, is a simple manifestation of the fluctuation–dissipation theorem. A straightforward generalisation of equation (5.11) allows the process to take complex values and is particularly useful in modelling coherent clutter. Thus

$$\begin{aligned} z_n &= \exp(-(\alpha/2 + i\omega_0)(t_n - t_{n-1}))z_{n-1} \\ &\quad + \sqrt{\frac{(1 - \exp(-\alpha(t_n - t_{n-1})))}{2}}(g_{In} + ig_{Qn}) \end{aligned} \tag{5.19}$$

generates a complex Gaussian process, with a characteristic 'Doppler' frequency ω_0, with a unit mean power and an exponentially decaying intensity ACF.

These simulation techniques can in principle be extended to the multivariate Gaussian case. By doing this we are able to model more complicated correlation properties than the simple exponential decay implicit in (5.11) and (5.19). Each of the components of the multivariate Gaussian will itself have Gaussian single point statistics; in general, however, it will not display the Markov property typical of the OU process. Consequently our modelling capability can be enhanced significantly. To analyse the multi-dimensional case we adopt a vector notation and set up the stochastic differential equation as

$$\frac{d\mathbf{x}(t)}{dt} = \mathbf{A} \cdot \mathbf{x}(t) + \mathbf{f}(t) \tag{5.20}$$

where eigenvalues of $\mathbf{A}$ have negative real parts to ensure stability, and the white noise vector $\mathbf{f}$ has a correlation matrix of the form

$$\langle \mathbf{f}(t_1)\mathbf{f}(t_2)^{\mathrm{T}} \rangle = \delta(t_1 - t_2)\mathbf{G} \tag{5.21}$$

The solution of (5.20) and (5.21) is discussed in Appendix 1, Section A1.11. There we see that the matrix $\mathbf{G}$ is related to the dissipation matrix $\mathbf{A}$ and the covariance matrix $\mathbf{B}$ of $\mathbf{x}$ through

$$\mathbf{AB} + \mathbf{BA}^{\mathrm{T}} = -\mathbf{G} \tag{5.22}$$

To implement the simulation scheme we discretise the evolution equation as

$$\begin{aligned} &\mathbf{x}_n = \mathbf{H} \cdot \mathbf{x}_{n-1} + \mathbf{V}_n \\ &\mathbf{H} = \exp(\mathbf{A}(t_n - t_{n-1})) \\ &\langle \mathbf{V}_n \rangle = \mathbf{0}, \quad \langle \mathbf{V}_n \mathbf{V}_n^{\mathrm{T}} \rangle = \mathbf{B} - \mathbf{H} \cdot \mathbf{B} \cdot \mathbf{H}^{\mathrm{T}} \end{aligned} \tag{5.23}$$

This covariance matrix of $\mathbf{V}_n$ is symmetric and positive definite, and so can be factorised into the product of an upper triangular matrix and its transpose, using the so-called Cholesky decomposition [3] (see also Appendix 1, equations A1.92–A1.96). So, once we have evaluated $\langle \mathbf{V}_n \mathbf{V}_n^{\mathrm{T}} \rangle$, we can effect the decomposition

$$\langle \mathbf{V}_n \mathbf{V}_n^{\mathrm{T}} \rangle = \mathbf{P}\mathbf{P}^{\mathrm{T}} \tag{5.24}$$

This in turn provides us with a realisation of $\mathbf{V}_n$ as

$$\mathbf{V}_n = \mathbf{P} \cdot \mathbf{g}_n \tag{5.25}$$

where $\mathbf{g}_n$ is a vector of zero mean, unit variance, un-correlated Gaussian random numbers. A sequence of vectors of Gaussian random variables with prescribed correlation properties determined by the matrix $\mathbf{H}$ can now be generated through the prescription given in (5.23); if we neglect transient terms we have, much as in (5.15),

$$\langle \mathbf{x}_n \mathbf{x}_{n+m}^{\mathrm{T}} \rangle = \mathbf{B} \cdot (\mathbf{H}^{\mathrm{T}})^m \tag{5.26}$$

Correlated multivariate Gaussian processes generated in this way provide a flexible and well controlled modelling tool; they are not able, however, to reproduce either the single point or correlation properties of sea clutter realistically. In other contexts, however, they can be very useful. If, for example, we take $\mathbf{A}$ to be

$$\mathbf{A} = \mathbf{M} \cdot \Omega \cdot \mathbf{M}^{-1}$$

$$\Omega = \begin{pmatrix} -\omega_1 & 0 & 0 & 0 \\ 0 & -\omega_2 & 0 & 0 \\ 0 & 0 & -\omega_3 & 0 \\ 0 & 0 & 0 & -\omega_4 \end{pmatrix} \quad \mathbf{M} = \begin{pmatrix} 1 & 1 & 1 & 1 \\ -\omega_1 & -\omega_2 & -\omega_3 & -\omega_4 \\ \omega_1^2 & \omega_2^2 & \omega_3^2 & \omega_4^2 \\ -\omega_1^3 & -\omega_2^3 & -\omega_3^3 & -\omega_4^3 \end{pmatrix} \tag{5.27}$$

so that

$$\mathbf{H} = \exp(\mathbf{A}(t_n - t_{n-1})) = \mathbf{M} \cdot \exp(\Omega(t_n - t_{n-1})) \cdot \mathbf{M}^{-1} \tag{5.28}$$

and the covariance of the random driving term as

$$\langle \mathbf{V}_n \mathbf{V}_n^{\mathrm{T}} \rangle = \mathbf{M} \cdot \Psi \cdot \mathbf{M}^{\mathrm{T}}$$
$$(\Psi)_{ij} = \frac{(1 - \exp(-(\omega_i + \omega_j)(t_n - t_{n-1})))}{\omega_i + \omega_j} (\mathbf{M}^{-1})_{i4} (\mathbf{M}^{-1})_{j4} \tag{5.29}$$

a correlated sequence of $\mathbf{x}(n) \equiv \{x_1(n), x_2(n), x_3(n), x_4(n)\}$ can be generated; the power spectrum of the component $x_k(n)$ generated in this way has the form

$$S_{kk}(\omega) = \frac{\omega^{2(k-1)}}{\prod_{j=1}^{4} (1 + \omega^2/\omega_j^2)}; \quad k = 1, 2, 3, 4 \tag{5.30}$$

The more adventurous reader may wish to supply a derivation of the details of this simulation scheme which can be used, for example, in the modelling of effects of uncompensated motions on SAR processing.[1] More generally, models of this type find wide application in the physics literature; Chandrasekhar's [2] discussion of the under-damped harmonic oscillator and the itinerant oscillator and other models described by Coffey [4] provide further examples of the general formalism we have outlined here.

5.4 Fourier synthesis of random processes

We have already seen how the correlated sequence of Gaussian random numbers generated by feedback can be viewed as the output of a simple (one-pole, infinite impulse response) filter. We will now look at how this approach might be extended. Consider the application of a filter to a white noise process f, to give the signal $\phi(t)$

$$\phi(t) = \int_{-\infty}^{\infty} h(t-t')f(t')dt' \tag{5.31}$$

We can evaluate the correlation function of this process as

$$\begin{aligned} \langle \phi(t)\phi(\tau+t)\rangle &= \int_{-\infty}^{\infty} dt_1 \int_{-\infty}^{\infty} dt_2 h(t-t_1)h(t+\tau-t_2)\langle f(t_1)f(t_2)\rangle \\ &= \int_{-\infty}^{\infty} dt_1 h(t-t_1)h(t+\tau-t_1) \end{aligned} \tag{5.32}$$

Alternatively we can Fourier transform the output of the filter to give us

$$\tilde{\phi}(\omega) = \tilde{h}(\omega)\tilde{f}(\omega) \tag{5.33}$$

The power spectra of the input and output processes are therefore related through

$$S_{\phi\phi}(\omega) = |\tilde{h}(\omega)|^2 S_{ff}(\omega) \tag{5.34}$$

noting that the power spectrum of the white noise process is independent of frequency, and is here set to unity. We can now relate the auto-correlation function (ACF) and power spectrum of the output process to the filter function h. This relationship is not unique, as the phase of the filter function is not determined. So, for example, we might identify the positive root of the power spectrum with the FT of the filter function

$$\tilde{h}(\omega) = \sqrt{S_{\phi\phi}(\omega)} \tag{5.35}$$

One can then reconstruct the process $\phi(t)$ by Fourier inversion. In practice this is done discretely, using a FFT type algorithm, which in effect is evaluating a Fourier series (rather than a transform). Each such realisation is of finite length; in practice

[1] This application was described in a private communication by Stuart Ingram and Edward Stansfield of Thales.

the relationship (5.34) is imposed on an average taken over many such realisations of the process. Thus we consider a representation of the Gaussian process of the form

$$\phi(t) = \frac{a_0}{2} + \sum_{n=1}^{\infty}\left(a_n \cos\frac{2\pi nt}{T} + b_n \sin\frac{2\pi nt}{T}\right) \tag{5.36}$$

over a period of length T. The quantities a_n, b_n are independent zero mean Gaussian random variables with a variance given by the power spectrum of the process to be simulated, i.e.

$$\langle a_n^2\rangle = \langle b_n^2\rangle = S\left(\frac{2\pi n}{T}\right)\frac{2\pi}{T} \tag{5.37}$$

If now we wish to realise a sequence of N values of the random variable ϕ we can obtain this by taking the real part of the output of the FFT formed as

$$\phi(m\Delta) = \sum_{n=0}^{N-1}\sqrt{S\left(\frac{2\pi n}{N\Delta}\right)\frac{2\pi}{\Delta}}\exp\left(-i\frac{2\pi nm}{N}\right)(g_{In} + ig_{Qn}) \tag{5.38}$$

The Fourier synthesis method has two advantages over the real time simulation method, in that it can model correlation functions that cannot be expressed as a sum of exponentially decaying and oscillating terms, and can be extended to the simulation of random fields as well as of time series. These features will be illustrated shortly in examples of clutter simulation. Its principal disadvantage is that it produces samples in batches, rather than as a sequence.

5.5 Approximate methods for the generation of correlated gamma distributed random numbers

The simulation methods we have discussed so far have applied only to the generation of correlated Gaussian random variables; much of the discussion in Chapters 2–4 stresses that sea clutter is frequently quite non-Gaussian in its statistical character. Consequently the methods we have developed thus far will be of limited use in the realistic simulation of clutter. We have already seen in Chapter 4 that the gamma distribution

$$\begin{aligned} P_Y(y) &= \frac{b^\nu}{\Gamma(\nu)} y^{\nu-1}\exp(-by) \quad y \geq 0 \\ &= 0 \quad y < 0 \end{aligned} \tag{5.39}$$

plays a central role in clutter modelling. So, if we are to simulate K-distributed clutter we must focus our attention on the generation of gamma-distributed random variables with a specified correlation function

$$\langle y(0)y(t)\rangle = \langle y^2\rangle R_Y(t) \tag{5.40}$$

The form of this correlation function and the requirement that the marginal distributions of $y(0)$ and $y(t)$ have the form (5.39) are not sufficient to specify their joint pdf uniquely. Furthermore, a linear transformation of a gamma-distributed process need not generate a process that itself has gamma single point statistics; this observation is borne out by the results A1.53–A1.55 in Appendix 1. This contrasts with the Gaussian distribution, whose stability ensures that the linear transformation of a Gaussian process yields an output that is itself Gaussian. As a result, the computer generation of a gamma process with a specified correlation function is less straightforward than the corresponding Gaussian simulation and several heuristic and approximate procedures have been developed and described in the literature.

One approach to this problem was developed by Oliver and co-workers [5]. In this an array of uncorrelated gamma variates $y_{\gamma,j}$ is generated (using a variant of the method described in Section 5.2), then weighted to induce the required correlation

$$y_{\Gamma,i} = \sum_{j=1}^{N} w_{ik} y_{\gamma,k} \tag{5.41}$$

Unfortunately the higher order moments of $y_{\Gamma,i}$, which can be characterised in terms of those of $y_{\gamma,j}$ and the weights w_{ik}, are not consistent with a gamma distribution. Furthermore, fundamental restrictions are placed on the correlation properties that can be modelled in this way; in particular it is not able to generate a correlation function that takes values less than the totally decorrelated value $\langle y \rangle^2$. Nonetheless this method is widely used and has been discussed extensively in the literature [6].

An alternative approach to the simulation of a correlated gamma process takes a correlated Gaussian process of zero mean and unit variance as its starting point. This is then mapped onto a gamma process y by the memoryless non-linear transform (MNLT) generated by the solution of the equation (cf. equation (5.2))

$$\frac{1}{\sqrt{2\pi}} \int_{x}^{\infty} \exp\left(\frac{-x'^2}{2}\right) dx' = \frac{1}{\Gamma(\nu)} \int_{y}^{\infty} y'^{\nu-1} \exp(-y')\, dy' \tag{5.42}$$

Used in conjunction with standard methods for generating Gaussian time series and random fields with prescribed correlation properties, this method can generate correlated time series and random fields that exhibit gamma single point statistics. As we shall see, it is possible to establish a direct relation between the correlation functions of the input and output processes. Prior to this development, however, an empirical element entered into the modelling of the correlation in the output gamma process. For example, it was found, on the basis of simulation studies, that a Gaussian process with an exponentially decaying ACF is transformed by the MNLT into a gamma process whose ACF also displays a seemingly exponential decay over several decades. An empirical formula

$$\frac{\tau_G}{\tau_\gamma} = 1 + \frac{0.15}{\nu^{0.7}} \tag{5.43}$$

was devised to relate the observed characteristic decay times of the Gaussian (τ_G) and gamma (τ_γ) processes and the ν parameter of the gamma distribution.

Similar studies of the correlation properties of gamma processes derived from input Gaussian processes with Gaussian and power-law correlation functions revealed the same apparent invariance in functional form of the correlation function under the MNLT. Rules of thumb analogous to (5.43) were devised that encapsulate the results of these numerical studies and were exploited in clutter simulation [7]. However, much of this semi-empirical work has been superseded by a more systematic analysis [8], which we will now describe.

5.6 The correlation properties of non-Gaussian processes generated by MNLT

The MNLT method discussed in the previous section can be adapted to the general non-Gaussian case. The required MNLT or point non-linear transformation is defined by equating the cumulative distribution of a zero mean unit variance Gaussian process, evaluated at the value x taken by this process with the cumulative distribution of the required process, thus determining the latter's value η. So, if the pdf of the values η is $P_{\text{dist}}(\eta)$, we set

$$\int_{\eta}^{\infty} P_{\text{dist}}(\eta')d\eta' = \frac{1}{\sqrt{2\pi}}\int_{x}^{\infty} \exp(-x'^2/2)\,dx' = \frac{1}{2}\text{erfc}(\text{x}/\sqrt{2}) \tag{5.44}$$

where, in the second equality, we have identified the complementary error function. The complementary quantile function $Q_{\text{dist}}(\varsigma)$ of the required distribution is now defined by

$$\int_{Q_{\text{dist}}(\varsigma)}^{\infty} P_{\text{dist}}(\eta)d\eta = \varsigma \tag{5.45}$$

using this we can write the MNLT that takes the input Gaussian random values into the corresponding values of the required non-Gaussian random variable as

$$\eta(x) = Q_{\text{dist}}(\text{erfc}(\text{x}/\sqrt{2})/2) \tag{5.46}$$

The construction of this mapping is in general a non-trivial undertaking. The software package *Mathematica* [9] supplies 'built in' quantile functions for a wide variety of non-Gaussian processes; these are, however, rather slow in execution. A rapidly evaluable representation of (5.46) is a prerequisite if the MNLT approach is to be practically feasible.

We next evaluate the correlation function of the process η. This can be expressed in the form

$$\langle\eta(0)\eta(t)\rangle = \int_{-\infty}^{\infty} dx_1 \int_{-\infty}^{\infty} dx_2\eta(x_1)\eta(x_2)P_{\text{G}}(x_1, x_2, R_{\text{G}}(t)) \tag{5.47}$$

where

$$P_{\mathrm{G}}(x_1, x_2, R_{\mathrm{G}}(t)) = \frac{1}{2\pi\sqrt{1 - R_{\mathrm{G}}(t)^2}} \exp(-(x_1^2 + x_2^2 - 2x_1x_2R_{\mathrm{G}}(t))/2(1 - R_{\mathrm{G}}(t)^2)) \tag{5.48}$$

The two dimensional integration implicit in (5.48) can be avoided by introducing the expansion given in Appendix 2 (Section A2.5)

$$P_{\mathrm{G}}(x_1, x_2, R_{\mathrm{G}}(t)) = \frac{\exp(-(x_1^2 + x_2^2)/2)}{2\pi} \sum_{n=0}^{\infty} \frac{H_n(x_1/\sqrt{2})H_n(x_2/\sqrt{2})}{2^n n!} R_{\mathrm{G}}(t)^n \tag{5.49}$$

Using (5.46) we find that

$$\langle \eta(0)\eta(t) \rangle = \frac{1}{\pi} \sum_{n=0}^{\infty} \frac{R_{\mathrm{G}}(t)^n}{2^n n!} \left(\int_{-\infty}^{\infty} dx \exp(-x^2) H_n(x) Q_{\mathrm{dist}}(\mathrm{erfc(x)/2}) \right)^2 \tag{5.50}$$

Once we have evaluated the integrals

$$\int_{-\infty}^{\infty} dx \exp(-x^2) H_n(x) Q_{\mathrm{dist}}(\mathrm{erfc(x)/2}) \tag{5.51}$$

(5.50) provides us with a power series representation of the mapping between the correlation functions of the input Gaussian and output non-Gaussian processes. This series is rapidly convergent or analytically summable, in most cases of interest. In particular we note that when the series (5.50) contains only two non-negligible terms it reduces to

$$\langle \eta(0)\eta(t) \rangle \approx \langle \eta \rangle^2 + (\langle \eta^2 \rangle - \langle \eta \rangle^2) R_{\mathrm{G}}(t) \tag{5.52}$$

and the correlation functions of the input and output processes are essentially identical. In general the higher order terms in (5.50) characterise the distortion in the correlation function induced by the MNLT that the empirically derived rules of thumb sought to capture in the approximate simulation method described earlier.

So far, we have established a readily evaluable and invertible mapping between the correlation functions of the input Gaussian and output non-Gaussian processes related by the non-linear transformation (5.46). Using this we can now tailor the correlation properties of the input Gaussian process by the Fourier synthesis method, to control the correlation function of the output non-Gaussian process.

5.7 Correlated exponential and Weibull processes

We now consider the correlation properties of MNLT generated non-Gaussian processes that provide useful models of clutter, either as a representation of the envelope of the process or of its local power within the compound model. As a first example we consider the $\nu = 1$ gamma (i.e. negative exponential) distributed variable y with the pdf

$$P(y) = \exp(-y) \tag{5.53}$$

In this case the required MNLT can be written explicitly as

$$y = \log(2/\mathrm{erfc}(x/\sqrt{2})) \tag{5.54}$$

Our general result (5.66) reduces in this case to

$$\langle y(0)y(t)\rangle = \frac{1}{2\pi}\sum_{n=0}^{\infty}\frac{R_{\mathrm{G}}(t)^n}{2^n n!}\left(\int_{-\infty}^{\infty} dx \exp(-x^2/2)H_n(x/\sqrt{2}) \times \log(2/\mathrm{erfc}(x/\sqrt{2}))\right)^2 \tag{5.55}$$

Note that on letting $R_{\mathrm{G}}(t)$ go to zero we recover the totally decorrelated result $\langle y\rangle^2 = 1$.

The numerical integrations over the Hermite polynomials in (5.55) present no real problems and reveal that the series is very rapidly convergent. Thus, once we have calculated and stored the first few of these integrals, (5.55) provides us with a tractable and single valued functional relationship between the correlation functions of the input Gaussian and output negative exponentially distributed processes. This then lets us map the required correlation function of the output process point by point onto that of the corresponding input Gaussian process. Assuming that the latter can be constructed by the Fourier synthesis, we are now able to generate the required correlated, exponentially distributed process by MNLT. Figure 5.1 shows the mapping from R_{G} to R_{exp}. It is reassuring to note that, by setting R_{G} to unity, we recover the fully correlated result $\langle y^2\rangle = 2$. By allowing R_{G} to take negative values the MNLT can generate an exponentially distributed process whose correlation function takes values that, while they are necessarily greater than zero, are less than the fully decorrelated limit $\langle y\rangle^2 = 1$. It is precisely this type of correlation that Oliver's filtering method cannot reproduce [5,6]

We recall (from Section 4.5) that the Weibull process has a pdf of the form

$$\begin{aligned} P_{\mathrm{W}}(\xi) &= \alpha b\xi^{\alpha-1}\exp(-b\xi^{\alpha}); \quad \xi \ge 0 \\ &= 0; \quad \xi < 0 \end{aligned} \tag{5.56}$$

and has been widely used as a model for clutter. The analysis in Section 5.6 can be applied to the correlation properties of a Weibull process generated by an MNLT.

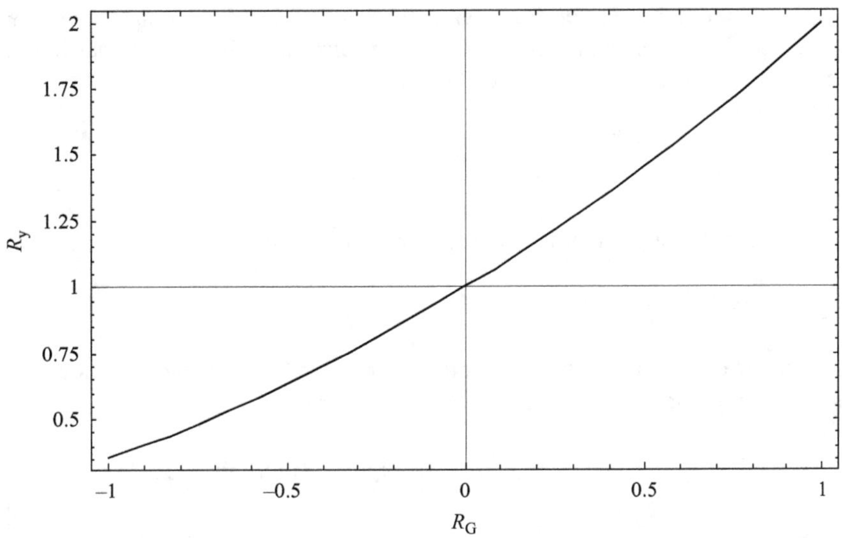

Figure 5.1 *The mapping between the input and output correlation functions under the MNLT (see equation 5.54)*

The required MNLT is given by (we can set b equal to unity, without any real loss of generality)

$$\xi = (\log(2/\mathrm{erfc}(x/\sqrt{2})))^{1/\alpha} \tag{5.57}$$

while the correlation in the input Gaussian process is mapped over into the correlation in the MNLT generated Weibull process by

$$\langle \xi(0)\xi(t)\rangle = \frac{1}{2\pi}\sum_{n=0}^{\infty}\frac{R_G(t)^n}{2^n n!}\left(\int_{-\infty}^{\infty} dx\,\exp(-x^2/2)H_n(x/\sqrt{2})\right.$$
$$\left.\times(\log(2/\mathrm{erfc}(x/\sqrt{2})))^{1/\alpha}\right)^2 \tag{5.58}$$

The coefficients in this expansion can be evaluated by dint of some straightforward numerical integrations. Once this has been done, and the mapping between $\langle \xi(0)\xi(t)\rangle$ and $R_G(t)$ has been inverted, Weibull distributed noise with a prescribed correlation function can be generated.

It is useful to compare the simulation of exponential and Weibull processes just discussed with methods based on a complex Gaussian process [10]. These are simpler to implement, but afford us considerably less freedom in the modelling of the correlation properties of the process.

We consider the complex Gaussian process has real and imaginary parts x, y that are themselves Gaussian processes with zero means and equal (unit) variances; at any one time or place x, y, are un-correlated.

The power

$$z = x^2 + y^2 \tag{5.59}$$

of the process has a pdf that takes the particularly simple form

$$P(z) = \tfrac{1}{2}\exp(-z/2) \tag{5.60}$$

while its correlation function follows directly from the Gaussian factorisation property and the correlations between x, y:

$$\begin{aligned}\langle z_1 z_2\rangle &= \langle (x_1^2 + y_1^2)(x_2^2 + y_2^2)\rangle = 4\langle x^2\rangle + 4(\langle x_1 x_2\rangle^2 + \langle x_1 y_2\rangle^2)\\ &= 4(1 + k_0^2)\end{aligned} \tag{5.61}$$

Thus we see that $\langle z_1 z_2\rangle$ necessarily takes values between those of $\langle z\rangle^2$ and $\langle z^2\rangle$; an undershoot in the correlation function below the value of the square of the mean cannot be modelled through (5.59).

The exponential form taken by (5.60) facilitates the transformation of z into a process with a Weibull pdf (5.56); the required MNTL is simply

$$\xi = \left(\frac{z}{2}\right)^{1/\alpha} \tag{5.62}$$

To relate the correlation function of the output process to that of the input complex Gaussian process, we see from (A2.72), in Appendix 2, that the joint pdf of z_1, z_2 takes the form

$$\begin{aligned}P(z_1, z_2) &= \frac{\exp(-(z_1 + z_2)/2)}{4}\frac{\exp(-k_0^2(t)(z_1 + z_2)/2(1 - k_0^2(t)))}{(1 - k_0^2(t))}\\ &\quad\times I_0(k_0(t)\sqrt{z_1 z_2}/(1 - k_0^2(t)))\\ &= \frac{\exp(-(z_1 + z_2)/2)}{4}\sum_{n=0}^{\infty} L_n(z_1/2)L_n(z_2/2)k_0^{2n}(t)\end{aligned} \tag{5.63}$$

The expansion is quite analogous to (5.49): the L_n are Laguerre polynomials defined as a special case ($\alpha = 0$) of

$$L_n^{\alpha}(z) = \frac{\exp(z)}{z^{\alpha} n!}\frac{d^n}{dz^n}(z^{n+\alpha}\exp(-z)) \tag{5.64}$$

and are discussed in Section A2.5 of Appendix 2.

Arguments paralleling those that lead to (5.50) show that the correlation function of the Weibull variable (5.62) can be expressed as

$$\begin{aligned}\langle \xi(0)\xi(t)\rangle &= \left(\frac{1}{2}\right)^{2/\alpha} \iint dz_1 dz_2 (z_1 z_2)^{1/\alpha} P(z_1, z_2) \\ &= \sum_{n=0}^{\infty} k_0^{2n}(t) \left(\int_0^{\infty} dz z^{1/\alpha} \exp(-z) L_n(z) \right)^2 \\ &= \Gamma(1+1/\alpha)^2 \, {}_2F_1(-1/\alpha, -1/\alpha; 1; k_0^2(t))\end{aligned} \tag{5.65}$$

where we have identified the hypergeometric function through (A2.13) in Appendix 2. The result (5.65) reduces to the expected un-correlated limit if we set $k_0 = 0$, while the fully correlated limit is recovered when $k_0 = 1$.

Much as in the simpler exponentially distributed case, we see that the correlation function of a Weibull process generated from the power of a complex Gaussian process cannot take values less than that of the square of its mean. This constraint is removed when we perform the appropriate MNLT directly on a correlated one-dimensional (1D) Gaussian process. We also note that this method has been adapted to the generation of coherent Weibull clutter, whose power has the distribution (5.56) and whose real and imaginary parts are those of

$$\left(\frac{x^2+y^2}{2}\right)^{1/2\alpha} \frac{x+iy}{\sqrt{x^2+y^2}} \tag{5.66}$$

Thus, in this case, the power spectrum of the input Gaussian process controls the phase of the output process, exactly as it does that of the input process, and the correlation function of the power of the output process. We should also stress that the process generated through (5.66) is not a satisfactory model for coherent sea clutter; it does not incorporate the large separation in time and length scales characteristic of the decays in the correlations in the local speckle and underlying modulation processes present in the clutter. We will discuss the modelling of coherent clutter in more detail shortly.

5.8 The generation of correlated gamma processes by MNLT

The examples discussed in the previous two sections illustrate the principles of the use of the MNLT to give a non-Gaussian noise process with a controlled correlation function. Their application to the generation of a gamma process whose correlation properties model those of variation in the local power characteristic of sea clutter will now be discussed.

In undertaking such simulations our first requirement is that of the quantile function, defined as in (5.45) for the gamma distribution. These are readily available; for example, the *Mathematica* [9] add-on package Statistics 'Continuous Distributions' contains functions that generate quantiles for a wide variety of non-Gaussian

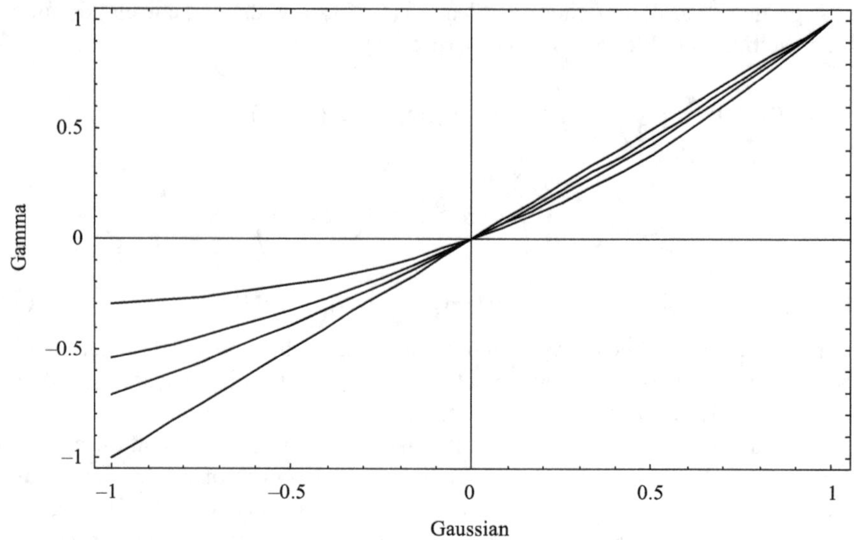

Figure 5.2 *The mapping between the normalised covariances of the input Gaussian process and the gamma processes generated through equation (5.42). Results are shown for $\nu = 0.3, 0, 7, 1.3, \infty$. The deviations from the linear behaviour ($\nu = \infty$) become more marked with decreasing ν*

distributions. As we have already noted, these are rather too slow for use in practical simulations. An alternative approach is to calculate an array of values of probability, ς, for a defined set of threshold values, Q_{dist}, and then interpolate for specific quantile values. The resulting gamma distribution quantile function can be used in the numerical evaluation of the integrals in (5.50). When these results are introduced into the summation of (5.50) with $R_G = 1$, the expected fully correlated value provides a check on its rate of convergence. Once this has been done the series can be used to map out the relationship between the input and output correlation functions in detail. Some typical results are shown in Figure 5.2. Deviations from a linear relation between the input and output correlation functions are more pronounced when $R_G < 0$ and in the extreme non-Gaussian regimes. The apparent 'invariance under MNLT' of the correlation functions in the 'rules of thumb' exploited in the heuristic approach described in Section 5.5 can be understood in terms of these results. We can now replace these empirically derived rules with unambiguous calculations, which need only be made once for each value of the shape parameter ν and stored for subsequent use. The full calculations are also applicable to input correlation functions that take negative values and correspond to values of the output correlation function that are less than the value of the square of the mean of the output process. It is interesting to note that successively larger numbers of terms are required in (5.66) as ν becomes smaller and the gamma pdf exhibits progressively more singular behaviour at the origin.

When the shape parameter ν takes values significantly greater than unity the series (5.50) is increasingly dominated by its first two terms, with the consequence that the functional form of the correlation function was barely affected by the MNLT. Even in the relatively non-Gaussian case of $\nu = 1$, (5.50) reduces to

$$\langle\eta(0)\eta(t)\rangle \approx 1 + 0.816R_G(t) + 0.177R_G(t)^2 + 0.0067R_G(t)^3 + 0.00013R_G(t)^4 + 0.000017R_G(t)^5 \tag{5.67}$$

which demonstrates its rapid convergence in this case. When $\nu = 2$ the first two terms in (5.50) account for 92% of the output correlation function when $R_G = -1$ and 96% when $R_G = 1$; when $\nu = 4.5$ only about a 1% contribution comes from the third and higher order terms in both the extreme correlated and anti-correlated limits. This observation suggests that the MNLT may tend to the form

$$y = \nu + \sqrt{\nu}x \tag{5.68}$$

which establishes a process whose mean and mean square values correspond with those of the required gamma distribution and which takes the correlation function over unchanged. Direct evaluation shows, for example, that in the case where $\nu = 10.0$ the linear relationship (5.68) captures the behaviour of the MNLT for small values of $|x|$ but becomes progressively less satisfactory with increasing $|x| > 1.0$.

Using the method of Fourier synthesis (described in Section 5.4) and the MNLT described here, it is relatively straightforward to generate both time series and random fields with gamma single point statistics and controlled correlation functions. Figures 5.3 and 5.4 show an example of each. Figure 5.3 is a time series with $\nu = 0.3$, a mean amplitude of 1, and a correlation function given by

$$\langle\eta(0)\eta(t)\rangle = 1 + \frac{\exp(-t/10)\cos(t/8)}{\nu} \tag{5.69}$$

Figure 5.4 in a random field of extent 128 in both x and y, with $\nu = 5$ and a correlation function given by

$$\langle\eta(0,0)\eta(x,y)\rangle = 1 + \frac{\exp(-(x+y)/10)\cos(\pi y/8)}{\nu} \tag{5.70}$$

Both examples are chosen to have the property of the correlation function falling below the mean value squared. This shows in the periodic fluctuations in the simulations and is not accessible with other methods of simulation.

An alternative to this procedure was developed by Armstrong and Griffiths [11]. These workers also generated correlated gamma processes with shape parameter ν by MNLT, taking Z, the sum of the powers of N complex Gaussian processes, as their input process. Here N is the integer closest in value to the shape parameter ν. As the pdf of Z takes the form

$$P(Z) = \frac{Z^{N-1}}{2^N(N-1)!}\exp(-Z/2) \tag{5.71}$$

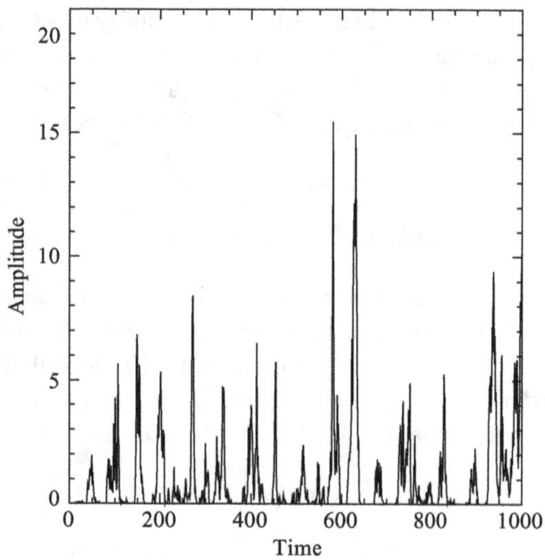

Figure 5.3 *An amplitude time series of a Gamma process with $\nu = 0.3$ and a correlation function given by equation (5.69)*

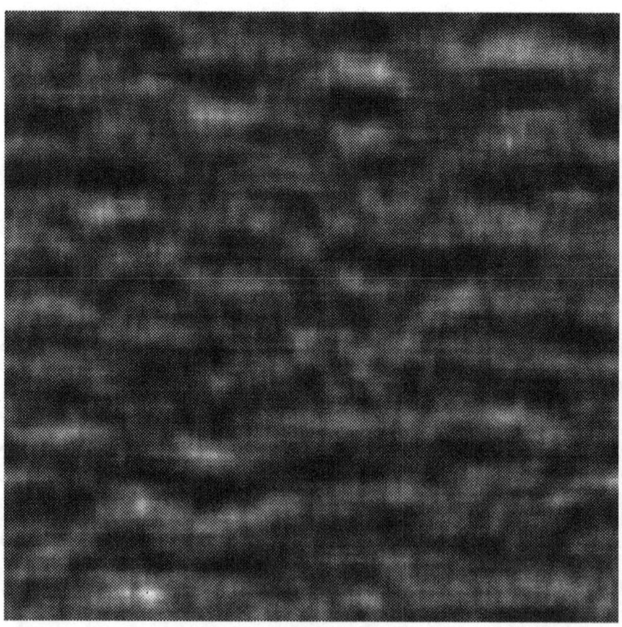

Figure 5.4 *A Gamma distributed random field with $\nu = 5$ and a correlation function given in equation (5.70)*

it approximates quite closely to the required gamma pdf. Armstrong and Griffiths suggest that as the pdfs of the input and output processes are similar, the MNLT

$$\eta(t) \equiv \eta(Z(t)) = Q_{\text{dist}}(p(N, Z(t)/2)) \tag{5.72}$$

where

$$p(N, Z/2) = \frac{\int_{Z/2}^{\infty} z^{N-1} \exp(-z) dz}{(N-1)!} \tag{5.73}$$

will induce relatively little change in the correlation function of the latter relative to that of the former. The analysis we have developed here can be applied directly to the calculation of the correlation function of this output process to both confirm their intuition and to highlight a limitation in their method. To this end we note that the two point statistical properties of Z are determined by the joint pdf (cf. (5.63))

$$P(Z_1, Z_2) = \frac{\exp(-(Z_1 + Z_2)/2)}{4(N-1)!} \frac{\exp(-k_0^2(Z_1 + Z_2)/2(1 - k_0^2))}{(1 - k_0^2)} \times \left(\frac{\sqrt{Z_1 Z_2}}{2k_0}\right)^{N-1} I_{N-1}\left(\frac{k_0\sqrt{Z_1 Z_2}}{(1 - k_0^2)}\right) \tag{5.74}$$

This in turn can be expressed as

$$P(Z_1, Z_2) = \frac{\exp(-(Z_1 + Z_2)/2)}{4(N-1)!} \left(\frac{Z_1 Z_2}{4}\right)^{N-1} \sum_{n=0}^{\infty} \frac{n! k_0^{2n}}{(N+n-1)!} \times L_n^{N-1}(Z_1/2) L_n^{N-1}(Z_2/2) \tag{5.75}$$

Here the associated Laguerre polynomials L_n^{N-1} are obtained from (5.64) by setting $\alpha = N - 1$. The correlation function of Armstrong and Griffiths' output process can now be expressed as

$$\langle \eta(0)\eta(t)\rangle = \frac{1}{(N-1)!} \sum_{n=0}^{\infty} \frac{n!}{(N+n-1)!} k_0^{2n}(t) \left\{ \int_0^{\infty} dq q^{N-1} \times \exp(-q) L_n^{N-1}(q) Q_{\text{dist}}(p(N, q)) \right\}^2 \tag{5.76}$$

This result gives us some insight into the relationship between the correlation functions of the input and output processes of the MNLT (5.72). Their correlation functions are essentially the same

$$\langle \eta(0)\eta(t)\rangle \approx \langle \eta \rangle^2 + (\langle \eta^2 \rangle - \langle \eta \rangle^2) k_0^2(t) \tag{5.77}$$

when only the first two terms in the series (5.76) are non-zero. This is often found to be the case; for example when using an $N = 2$ input Z process to generate a $\nu = 2.5$ gamma process via (5.71) the coefficient of k_0^4 in (5.75) was <0.1% of the value of that of k_0^2 and higher order coefficients were even smaller. These calculations bear out Armstrong and Griffiths' contention that very little violence is done to the correlation function by the MNLT (5.69) applied to an appropriately chosen input Z. These considerations suggest a linear approximation to the MNLT similar in form to (5.68)

$$\eta(Z) = \nu\left(1 - \sqrt{\frac{N}{\nu}}\right) + \frac{1}{2}\sqrt{\frac{\nu}{N}}Z; \quad N - 1 < \nu < N \tag{5.78}$$

Direct evaluation of the MNLT (5.76) shows that this simple expression captures its behaviour for values

$$2(N - \sqrt{N}) \leq Z \leq 2(N + \sqrt{N}) \tag{5.79}$$

but is less satisfactory for values of Zoutside this range and is increasingly inadequate as Z gets very large.

However, if one generates a markedly non-Gaussian process (e.g. $\nu < 0.5$) from a $N = 1$ input Z through (5.72) then terms in (5.76) of orders higher than the second contribute significantly to its correlation function and due allowance must be made for the resulting distortion. We should perhaps also stress that, once again, a method based on the transformation of the power of a complex Gaussian process is unable to model processes whose correlation functions take values less than the squares of their means.

References

1 A.J. Hughes, E. Jakeman, C.J. Oliver and E.R. Pike, 'Photon-correlation spectroscopy: dependence of linewidth error on normalization, clip level, detector area, sample time and count rate', *J. Phys. A: Math. Gen*, **6**, 1327, 1973.

2 S. Chandrasekhar, 'Stochastic problems in physics and astronomy', *Rev. Mod. Phys.*, **15**, 1–89, 1943.

3 G.H. Golub and C.F. Van Loan, '*Matrix computations*', Johns Hopkins University Press, Baltimore, MD, 1989.

4 W.F. Coffey, Y.P. Kalmykov and J.T. Waldon, '*The Langevin equation, with applications to stochastic problems in physics, chemistry and engineering*', World Scientific, Singapore, 2004.

5 C.J. Oliver and R.J.A. Tough, 'On the simulation of correlated K-distributed random clutter', *Optica Acta*, **33**, 223–250, 1986.

6 D. Blacknell, A. Blake, P. Lombardo and C. J. Oliver, 'A comparison of simulation techniques for correlated gamma and K distributed images for SAR applications', Proceedings of the IGARSS 1994, Pasadena CA, 8–12 August 1994.

7 K.D. Ward, R.J.A. Tough and P.W. Shepherd, 'Modeling sea clutter: correlation, resolution and non-Gaussian statistics', Proceedings of the RADAR 97, IEE publication 449, Edinburgh UK, 14–16 October 1997.

8 R.J.A. Tough and K.D. Ward, 'The correlation properties of gamma and other non-Gaussain processes generated by memoryless nonlinear transformation', *J. Phys. D.*, **32**, 3075–3084, 1999.
9 S. Wolfram, '*The Mathematica book*', University Press, Cambridge, 1996.
10 G. Li and K.-B. Yu, 'Modelling and simulation of coherent Weibull clutter', *IEE Proc. F.*, **136**(1), 2–12, 1989.
11 B.C. Armstrong and H.D. Griffiths, 'Modelling spatially correlated K-distributed clutter', *Electron. Lett.*, **27**, 1355–1356, 1991.

Chapter 6
Detection of small targets in sea clutter

6.1 Introduction

One of the main applications of maritime radar is to the detection of small targets, partially obscured by the unavoidable clutter returns from their environment. In this chapter we attempt to identify procedures that allow this detection to be carried out effectively and consider to what extent we can quantify their performance. The statistical models of clutter and target returns developed in earlier chapters provide us with a framework within which we can address these issues in a reasonably systematic fashion and so define and identify optimal detection procedures. In some simple cases these procedures can be implemented in practice; in others they suggest methods that, while not optimal, are nonetheless well-founded in our understanding of the physical processes involved. This approach can be applied usefully to both incoherent and coherent radar operation. The problem of characterising clutter returns in terms of the K and other statistical models, through the systematic estimation of the parameters that define them, can be addressed in much the same way. The parameter estimation techniques that emerge from this discussion form the basis of useful discrimination algorithms that can be applied to Synthetic Aperture Radar (SAR) and other radar imagery. Many of the concepts introduced in this discussion of detection and discrimination are developed making the assumption that independent samples of the target and clutter returns are available; this simplification is appropriate to the analysis of the detection of small targets in clutter. Nonetheless these same principles can be applied in the discrimination between correlated signals and will be considered in the following chapter, where we consider the enhancement and detection of large-scale weak features occurring in radar images of the sea surface. Obviously we cannot give a detailed exposition of detection and estimation theory here. Instead we will try to highlight the principles that underpin their application in maritime radar and emphasise those points where the compound model of clutter, and other aspects of the K distribution approach developed in this book have a particular impact on the discussion.

6.2 Statistical models for probabilities of detection and false alarm

A radar system presents us with a signal whose power (the random variable **z**), is displayed as a function of time or spatial co-ordinates. If this signal changes noticeably as a result of the presence of a target it should then be possible to detect that target. Usually there is a significant increase in the signal in the presence of the target; this implies that we might well achieve successful detections by setting a threshold and identifying any signal in excess of our threshold as a target signature. This procedure need not be fool-proof; mis-attributions of large values of **z** derived from the radar returns from the background (i.e. clutter) to a target return will give rise to false alarms. Obviously if we set the threshold sufficiently high we will tend to avoid such false alarms but only at the cost of missing some 'real' detections. Detection performance calculations attempt to quantify this trade off between detections and false alarms.

To make significant progress we must first characterise the signal **z** in the presence and absence of the target. A statistical description developed in Chapter 4 and Appendix A, in terms of the probability density function $P(z)$ of its value z, is convenient and, given the complexity and uncertain nature of the processes that contribute to the radar return, represents a good compromise between realism and tractability.

We let P_A denote the pdf of the signal derived from the return from the clutter background; P_T will denote the pdf of the signal in the presence of the target. If we now take a threshold Z we can express the probabilities of detection and false alarm as follows

$$P_D = \int_Z^\infty dz P_T(z); \qquad P_{FA} = \int_Z^\infty dz P_A(z) \tag{6.1}$$

A plot of probability of detection vs. probability of false alarm, mapped out as the threshold Z varies, is known as a Receiver Operation Characteristic (ROC) curve [1]; ideally we would like a high probability of detection to be achieved at the expense of a small probability of false alarm. Some typical ROC curves are shown in Figure 6.1.

When plotted linearly the ROC curve corresponding to an inability to distinguish between target and background is just a straight line of unit slope. In practice, due to the usual requirement for a very low probability of false alarm, it is frequently more convenient to plot P_{FA} on a logarithmic scale.

The problem of detection can thus be regarded as that of deciding whether the pdf of z is better described by P_A or P_T, given a value of z. We have already argued that a simple thresholding on z will allow us to make this distinction with some measure of success, which can be quantified in terms of a ROC curve. The question remains as to whether we are able to do better than this. We will now see that there are many circumstances in which we can.

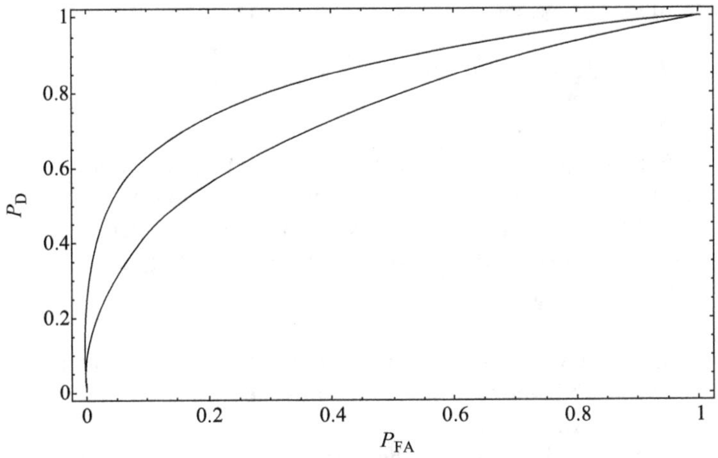

Figure 6.1 *Two ROC curves plotted on linear axes; the detection performance characterised by curve A is significantly better than that associated with curve B*

6.3 Likelihood ratios and optimal detection

Having introduced the concepts of probability of detection and false alarm, we can now consider how we might optimise the detection procedure: is it possible to identify a strategy that allows us to maximise the probability of detection, while keeping the probability of false alarm constant? For the moment we consider this problem in the abstract; we wish to differentiate, on the basis of measurements of the random variable **z**, between two possible pdfs that might describe its single point statistics; these we denote by $P_{\mathbf{z},A}(z)$ and $P_{\mathbf{z},B}(z)$. The appropriate identification of this random variable z and its candidate pdfs allows us to accommodate many practically important radar detection problems within a simple and general framework. So, for example, we might identify z with the intensity of the recorded radar signal and associate $P_{\mathbf{z},A}(z)$ and $P_{\mathbf{z},B}(z)$ with the Rayleigh and Rice pdfs, the former as a simple model for the clutter and the latter for signal plus clutter. Alternatively, we could take z to be a vector of correlated returns that form part or all of a SAR or other two dimensional image; suitably chosen multivariate Gaussians pdfs could then provide appropriate models for $P_{\mathbf{z},A}(z)$ and $P_{\mathbf{z},B}(z)$. Much of the discussion in this and the following chapters will consist of our working through of the implications of this concept. Before we can do this, however, we must identify our optimum procedure. The arguments that guide us to the criterion on which to base this procedure are perhaps a little unfamiliar, and at first sight, deceptively straightforward. Given a measurement z we assign this value on the basis of some test, as yet unspecified, to the set of values Z_A corresponding to the former distribution or Z_B, corresponding to the latter. Between them Z_A and Z_B contain all possible values of z. On the basis of this classification we define a probability of detection (correct assignment to distribution A) and a probability of false alarm (incorrect assignment to distribution A) as

$$P_{\mathrm{D}} = \int_{Z_{\mathrm{A}}} dz P_{\mathbf{z},\mathrm{A}}(z) \quad \text{and} \quad P_{\mathrm{FA}} = \int_{Z_{\mathrm{A}}} dz P_{\mathbf{z},\mathrm{B}}(z) \tag{6.2}$$

We now define the optimum decision rule as that which maximises the probability of detection, while maintaining the false alarm rate at a constant value α. This specification of optimality is known as the Neyman–Pearson criterion [2]. In order to determine the test that satisfies this criterion we consider the quantity

$$F = P_{\mathrm{D}} + \lambda(\alpha - P_{\mathrm{FA}}) \tag{6.3}$$

which we maximise with respect to the choice of decision rule and variations in the parameter λ. The latter condition maintains the false alarm rate at the value α; consequently the maximisation of F with respect to choice of decision rule maximises the probability of detection subject to the constraint of a constant false alarm rate. (This procedure is essentially Lagrange's method of undetermined multipliers.) We now introduce our explicit expressions for the probabilities of detection and false alarm to give

$$F = \lambda\alpha + \int_{Z_{\mathrm{A}}} dz(P_{\mathbf{z},\mathrm{A}}(z) - \lambda P_{\mathbf{z},\mathrm{B}}(z)) \tag{6.4}$$

From this we see that F is maximised if the integral in this expression is carried out over the region where the integrand takes all its positive values. This allows us to identify the optimal decision rule as that in which an observation of z is identified as coming from $P_{\mathbf{z},\mathrm{A}}(z)$ if and only if

$$\Lambda(z) = \frac{P_{\mathbf{z},\mathrm{A}}(z)}{P_{\mathbf{z},\mathrm{B}}(z)} > \lambda \tag{6.5}$$

We conclude that the optimal decision rule should be based on the value of the likelihood ratio $\Lambda(z)$. In practice it is sometimes more convenient to carry out this thresholding on a monotonic function of the likelihood ratio such as its logarithm. The Lagrange multiplier λ can be identified as the threshold on the likelihood ratio that establishes the given false alarm rate α. So, if $P(\Lambda)_{\Lambda,\mathrm{B}}$ is the pdf of the likelihood ratio derived from a measurement of z drawn from distribution B, λ is defined implicitly by

$$\alpha = \int_{\lambda}^{\infty} d\Lambda P(\Lambda)_{\Lambda,\mathrm{B}} \tag{6.6}$$

Here we have introduced the likelihood ratio test on the basis of the Neyman–Pearson criterion; exactly the same test emerges from a consideration of the so-called Bayes risk analysis. In this an intuitively reasonable cost function is constructed in terms of probabilities of detection and false alarm and the optimum test procedure is identified as that which minimises this quantity. Detailed discussions of this approach, which is algebraically more complex but is more readily extendible to the analysis of the testing of multiple, rather than binary hypotheses, are given in the standard textbooks by van Trees [3] and Middleton [4]. Here we merely stress that the two approaches lead to the same, optimal, test procedure.

At this point we have identified the optimum statistic with which to distinguish between signals drawn from two separate distributions. This procedure requires a detailed knowledge of the pdfs of these two distributions, which may not always be available in practice. When this is the case, other sub-optimal test statistics are frequently used, chosen on the basis of convenience and general applicability.

6.4 Some simple performance calculations

As we have seen in Chapter 4 and the Appendix 1, the Gaussian distribution is a much-used statistical model pdf that has the advantages of relative tractability and widespread validity. The former derives in part from the latter; a good model will be studied with sufficient vigour to ensure that it becomes tractable (e.g. through the study of the error function, which characterises the probabilities of detection and false alarm derived from the Gaussian pdf). The wide applicability of the Gaussian model is a consequence of the central limit theorem, which shows that, subject to various conditions, the sum of a 'large' number of random variables has a Gaussian distribution. For these reasons we will use the Gaussian distribution to motivate and illustrate our preliminary discussion of the application of likelihood ratio based analyses.

We recall that a single, one-dimensional Gaussian random variable **y**, takes values y that have a pdf

$$P(y) = \frac{1}{\sqrt{2\pi\sigma^2}} \exp(-(y-m)^2/(2\sigma^2)) \tag{6.7}$$

characterised by two parameters, the mean m and the standard deviation σ. The probability of y exceeding a threshold Y is given in terms of the error function (see A1.31) i.e., $x \to y, X \to Y$:

$$\frac{1}{\sqrt{2\pi\sigma^2}} \int_Y^\infty dy \exp(-(y-m)^2/(2\sigma^2)) = \frac{1}{2}(1 - \operatorname{erf}((Y-m)/\sigma\sqrt{2})) \tag{6.8}$$

This distribution is discussed in more detail in Appendix A1. There we also see that the n-dimensional generalisation of the Gaussian distribution, which includes the effects of correlation between its components, is fairly straightforward. One particularly useful multivariate Gaussian process is the circular 'complex' Gaussian with two independent components of zero mean and equal variance. We recall from Chapter 4 that the I and Q components of a speckle or thermal noise signal have statistics described by this pdf, and provide a simple model of the received signal

$$P(E_I, E_Q) = \frac{1}{\pi x} \exp\left(-\frac{E_I^2 + E_Q^2}{x}\right) \tag{6.9}$$

whose intensity $z = E^2$ has an exponential pdf:

$$P(z) = \frac{1}{x} \exp(-z/x). \tag{6.10}$$

Having recalled these few preliminaries we are in a position to carry through a simple performance calculation that illustrates many of the principles we will employ in more complex problems. We model the target and background returns by complex Gaussians with different mean intensities:

$$P_a(z) = \frac{1}{x_a} \exp(-z/x_a) \tag{6.11}$$

$$P_t(z) = \frac{1}{x_t} \exp(-z/x_t) \tag{6.12}$$

It is quite sensible to assume that $x_t > x_a$; we are in effect modelling the clutter and target returns as complex Gaussian processes of different powers, which we add together to give a resultant complex Gaussian process. Here we are working with solely with intensities and have 'thrown away' any phase-borne information in the signals. We should stress that in this simple analysis we assume that x_t, x_a are known.

The log likelihood ratio (cf. (6.5)) in this case takes the form

$$\log \Lambda = \log\left(\frac{x_a}{x_t}\right) + z\left(\frac{1}{x_a} - \frac{1}{x_t}\right) \tag{6.13}$$

From this result we see that thresholding on the intensity z is the optimal, as well as a convenient, detection procedure. The expressions for the probabilities of detection and false alarm for a given threshold z_{T} are particularly simple in this case:

$$P_{\mathrm{D}} = \exp(-z_{\mathrm{T}}/x_t) \tag{6.14}$$

$$P_{\mathrm{FA}} = \exp(-z_{\mathrm{T}}/x_a) \tag{6.15}$$

If we have N independent measurements of the intensity the appropriate pdfs are

$$P_a(z) = \frac{1}{x_a^N} \exp\left(-\sum_{k=1}^{N} z_k/x_a\right) \tag{6.16}$$

$$P_t(z) = \frac{1}{x_t^N} \exp\left(-\sum_{k=1}^{N} z_k/x_t\right) \tag{6.17}$$

The corresponding log likelihood ratio is

$$\log \Lambda = N(\log x_a - \log x_t) - \left(\frac{1}{x_t} - \frac{1}{x_a}\right)\sum_{k=1}^{N} z_k \tag{6.18}$$

In this case we see that thresholding on the sum of the available intensities gives us the optimum detection procedure. To generate ROC curves we must now calculate the

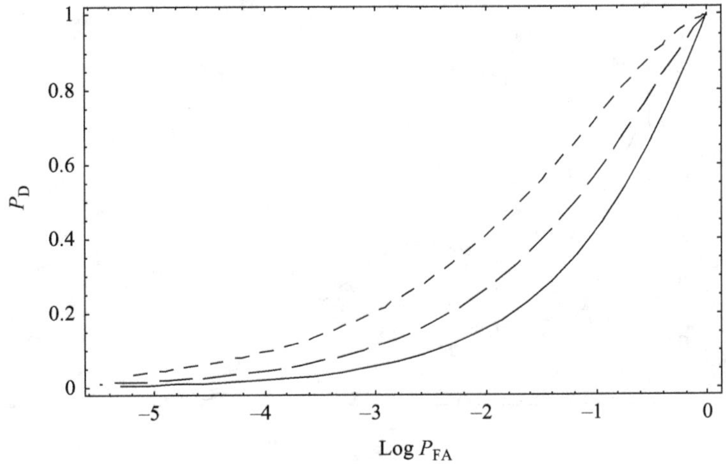

Figure 6.2 *ROC curves characterising discrimination between the pdfs (6.11) with $x_a = 1$, $x_t = 2$ and $N = 1$ (——), 3 (- - - - - -) and 6 (— —)*

pdfs of such sums of intensities drawn from the target and background distributions; this is easily shown to be

$$P(z) = \frac{z^{N-1}}{(N-1)!x^N} \exp(-z/x) \tag{6.19}$$

This result can now be used to plot ROC curves that demonstrate the improvement in detection performance achieved by incoherent averaging in this simple idealised case. Examples are shown in Figure 6.2.

Even this simple foregoing example is of more than didactic interest; it provides a model for the detection of a rapidly fluctuating (Swerling 2) target in thermal noise and as such is quite useful in the assessment of performance of the small target detection.

We can also carry out performance calculations in which the target signal does not fluctuate (Swerling 0 model). In this case the intensity pdf is that given by the Rice model (4.14); we recall that A is the amplitude of the signal

$$P(z) = \frac{1}{x} \exp\left(-\frac{z + A^2}{x}\right) I_0\left(\frac{2A\sqrt{z}}{x}\right) \tag{6.20}$$

The log likelihood ratio appropriate to N independent signals is therefore

$$\log \Lambda = \sum_{k=1}^{N} \log\left[I_0\left(\frac{2A\sqrt{z_k}}{x}\right)\right] - \frac{NA^2}{x} \tag{6.21}$$

In the case of a small signal to noise ratio we can approximate the modified Bessel functions by the first two terms in their Taylor series expansions (cf. (A2.56)) to

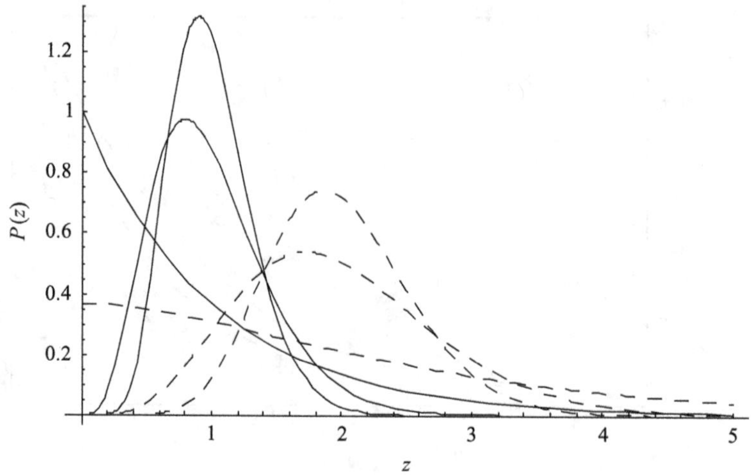

Figure 6.3 *Progressive separation of the clutter (——) and target plus clutter (— —) pdfs for N = 1, 5, and 10 and, in the latter case, a signal to clutter ratio of unity*

obtain

$$\log \Lambda \approx \frac{A^2}{x^2}\left(\sum_{k=1}^{N} z_k - Nx\right) \tag{6.22}$$

This indicates that thresholding on the sum of the received intensities (or squares of voltages) provides an optimal test procedure in the limit of small signal to noise. The pdf of this sum of N independent intensities takes the form

$$P(z)_N = \frac{\exp(-(z + NA^2)/x)}{x}\left(\frac{z}{NA^2}\right)^{(N-1)/2} I_{N-1}\left(\frac{2A\sqrt{Nz}}{x}\right) \tag{6.23}$$

Figure 6.3 shows the progressive separation of the signal plus clutter and clutter pdfs as N increases.

In the limit where the signal to noise ratio is large it is no longer appropriate to invoke the approximation (6.22). Instead we can use (A2.58) to show that

$$\log \Lambda \approx -\frac{N}{2}\log\left(\frac{4\pi A}{x}\right) - \frac{1}{4}\sum_{k=1}^{N}\log z_k + \frac{2A}{x}\sum_{k=1}^{N}\sqrt{z_k} - \frac{NA^2}{x} \tag{6.24}$$

As $\log z_k$ varies slower than $\sqrt{z_k}$ in the large signal limit, thresholding can be carried out on the sum of square roots of detected intensities (i.e. amplitudes of detected voltages). This distinction between 'linear' and 'square law' detection in the high and low signal to noise regimes is much discussed in older radar texts [5], where it is shown that the performance of both detectors is similar. Performance calculations for the linear detector present some difficulties as no tractable analogue of the result

(6.23) can be derived in this case. Therefore it is convenient to assume square law detection for calculations.

The detection strategies developed here require a threshold to be set on a relatively simple discriminant; the expected false alarm rate associated with this threshold will depend on the local clutter power x. Thus, if x is not known or worse still varies significantly, it will be difficult to control the observed false alarms. To overcome this problem we must incorporate some estimate of the local power x, derived from the received data, into our detection processing. One way of doing this is to use clutter data from range and azimuth cells away from the test cell (i.e. the cell being tested for a target). This is the basis for most so-called CFAR (constant false alarm rate) processing schemes, which are discussed in Chapter 9. Alternatively we can use data from the test cell itself. We will see how to do this in principle in the next section, where we consider the generalised likelihood ratio method, more details of whose implementation will be given in Sections 6.6 and 6.7.

6.5 The generalised likelihood ratio method

So far we have been able to deduce useful detection procedures from model clutter and target plus clutter pdfs that contain parameters (e.g. A, x), without specifying what these are. In more complicated cases we have to be more careful. Thus we might write our likelihood ratio as

$$\Lambda = \frac{P_t(z|\{b_1\})}{P_a(z|\{b_0\})} \tag{6.25}$$

As before, P_t and P_a are the pdfs of the 'target and clutter' and 'clutter' signals respectively; these are now characterised explicitly by the sets of parameters $\{b_1\}$ and $\{b_0\}$. Should we know what $\{b_1\}$ and $\{b_0\}$ are, we can make our decision on the basis of a single measurement of the signal. We need only construct the likelihood ratio Λ and compare it with a threshold T; if Λ exceeds this threshold we can ascribe the signal to 'target and clutter', otherwise it is ascribed to 'clutter' alone. The size of T chosen for the test determines the probability of false alarm for the decision process. If however we do not know $\{b_1\}$ and $\{b_0\}$ *a priori* we can still estimate these parameters from the set of received signals and then, using these estimates, form the appropriate likelihood ratio. Once this is done we can base our detection decision on this quantity.

The problem of estimating sets of parameters that define a putative pdf of a set of data is addressed extensively in the statistical literature [6]. In this context the following discussion might appear cursory in the extreme; nonetheless we hope to highlight the principles of its solution that are most relevant to our development of useful detection schemes. For the moment we consider N independent samples $\{z_k\}$ drawn from the clutter signal distribution $P(z|\{b\})$; our ignorance of the values taken by these parameters is manifest in their pdf $P(\{b\})$. The joint pdf of the sample and parameter values can now be written as

$$P(\{z_k\},\{b\}) = P(\{b\})\prod_{k=1}^{N} P(z_k|\{b\}) \tag{6.26}$$

Alternatively we can consider the likelihood of the parameters b taking a particular set of values, given the set of observations $\{z_k\}$; this we write as $P(\{b\}|\{z_k\})$. An equivalent expression for the joint pdf $P(\{z_k\},\{b\})$ can now be constructed:

$$P(\{z_k\},\{b\}) = P(\{b\}|\{z_k\})\prod_{k=1}^{N} P(z_k)$$

$$P(z_k) = \int P(z_k|\{b\})P(\{b\})db \tag{6.27}$$

By equating these two expressions for $P(\{z_k\},\{b\})$ we can express $P(\{b\}|\{z_k\})$ in terms of $P(z|\{b\})$:

$$P(\{z_k\},\{b\}) = P(\{b\})\prod_{k=1}^{N} P(z_k|\{b\}) = P(\{b\}|\{z_k\})\prod_{k=1}^{N} P(z_k)$$

$$P(\{b\}|\{z_k\}) = \frac{P(\{b\})\prod_{k=1}^{N} P(z_k|\{b\})}{\prod_{k=1}^{N} P(z_k)} \tag{6.28}$$

(This provides us with an application of Bayes theorem, discussed briefly in the Appendix A1.) So, given the data set $\{z_k\}$, we can find the set of parameters b that is most likely, by a process of maximising $P(\{b\}|\{z_k\})$. Frequently, in the absence of any prior knowledge, a constant value is assigned to $P(\{b\})$.

Thus, for our given set of signals, we find the values of the parameters $\{b_1\}$ and $\{b_0\}$ that maximise the values of the 'target and clutter' and 'clutter' multivariate pdfs respectively; these correspond to the most likely model parameters, given the set of signals $\{z_i\}$, and no prior knowledge of the model parameters. If we have N independent signals $\{z_i\}$, $i = 1,\ldots,N$, the appropriate multivariate pdfs are

$$P_t^{(N)}(\{z_i\}|\{b_1\}) = \prod_{i=1}^{N} P_t(z_i|\{b_1\})$$

$$P_a^{(N)}(\{z_i\}|\{b_0\}) = \prod_{i=1}^{N} P_a(z_i|\{b_0\}) \tag{6.29}$$

We now estimate the parameters $\{b_1\}$ and $\{b_0\}$, of which there are M_1 and M_0 respectively, from the equations

$$\frac{\partial \log(P_t^{(N)}(\{z_i\}|\{b_1\}))}{\partial b_{1,k}} = 0; \qquad k = 1 \dots M_1$$

$$\frac{\partial \log(P_a^{(N)}(\{z_i\}|\{b_0\}))}{\partial b_{0,k}} = 0; \qquad k = 1 \dots M_0 \tag{6.30}$$

These estimates, which we denote by $\{\hat{b}_1\}$ and $\{\hat{b}_0\}$, are employed to construct the so-called generalised likelihood ratio

$$\hat{\Lambda} = \frac{\prod_{i=1}^{N} P_t(z_i|\{\hat{b}_1\})}{\prod_{i=1}^{N} P_a(z_i|\{\hat{b}_0\})} \tag{6.31}$$

We may now use this quantity, or some suitable approximation to it, as the basis of a detection procedure. As we shall see, this approach can be very effective.

6.6 A Gaussian example

To illustrate these principles, and investigate some consequences of their application, we consider a very simple detection problem, that of distinguishing between independent Gaussian random variables drawn from a distribution with a zero mean and given variance, and from a distribution with the same variance, but having a non-zero mean. Our analysis of this elementary problem nonetheless highlights many of the principles exploited in what follows.

6.6.1 A likelihood ratio based approach

The joint pdf of N independent samples drawn from the zero mean Gaussian distribution is

$$P_1(\{y_k\}) = \frac{1}{(2\pi\sigma^2)^{N/2}} \exp\left(-\frac{1}{2\sigma^2}\sum_{k=1}^{N} y_k^2\right); \tag{6.32}$$

the corresponding pdf for samples drawn from the non-zero mean distribution is

$$P_2(\{y_k\}) = \frac{1}{(2\pi\sigma^2)^{N/2}} \exp\left(-\frac{1}{2\sigma^2}\sum_{k=1}^{N} (y_k - m)^2\right) \tag{6.33}$$

If we know values of the parameters defining these distributions we can form the likelihood ratio

$$\frac{P_1(\{y_k\})}{P_2(\{y_k\})} \tag{6.34}$$

and identify the sufficient statistic

$$\eta = \sum_{k=1}^{N} y_k \tag{6.35}$$

as the optimum discriminant in this case. This is a sum of Gaussian random variables and so is itself a Gaussian random variable. Its pdfs, when constructed from zero mean and non-zero mean Gaussians, are

$$P(\eta) = \frac{1}{\sqrt{2N\pi\sigma^2}} \exp\left(-\frac{\eta^2}{2N\sigma^2}\right)$$
$$P(\eta) = \frac{1}{\sqrt{2N\pi\sigma^2}} \exp\left(-\frac{(\eta - mN)^2}{2N\sigma^2}\right) \tag{6.36}$$

Thus, for a given threshold H we have the probabilities of detection and false alarm (correct assignment to the non-zero mean class, incorrect assignment to the zero mean class) given by

$$P_{\mathrm{D}}(\mathrm{H}) = \frac{1}{2}\left(1 - \mathrm{erf}\left(\frac{\mathrm{H} - Nm}{\sqrt{2N\sigma^2}}\right)\right)$$
$$P_{\mathrm{FA}}(\mathrm{H}) = \frac{1}{2}\left(1 - \mathrm{erf}\left(\frac{\mathrm{H}}{\sqrt{2N\sigma^2}}\right)\right) \tag{6.37}$$

Using these results it is possible to trace out the ROC curves (probability of detection vs. probability of false alarm) characterising the performance of this simple 'detector'.

6.6.2 Generalised likelihood ratio based approach

So far we have assumed that we know the parameters σ, m. If however we assume that one set of samples is drawn from a zero mean distribution and the other from a non-zero mean distribution, neither of whose variances we know, we cannot carry through the foregoing analysis. Instead we have to adopt the generalised likelihood ratio approach described in Section 6.5, in which the data provide us with estimates to be incorporated into the discriminant. In the zero mean case we have a likelihood of the form

$$P_1(\{y_k\}) = \frac{1}{(2\pi\sigma_1^2)^{N/2}} \exp\left(-\frac{1}{2\sigma_1^2}\sum_{k=1}^{N} y_k^2\right) \tag{6.38}$$

Given the data $\{y_k\}$ we can we estimate σ_1^2 as that value which maximises this likelihood; thus we find that

$$\hat{\sigma}_1^2 = \frac{1}{N}\sum_{k=1}^{N} y_k^2 \equiv \langle y^2 \rangle \tag{6.39}$$

If we assume that the data are drawn from the non-zero mean distribution with the pdf

$$P_2(\{y_k\}) = \frac{1}{(2\pi\sigma_2^2)^{N/2}} \exp\left(-\frac{1}{2\sigma_2^2}\sum_{k=1}^{N}(y_k - m)^2\right) \tag{6.40}$$

the mean and variance parameters can then be estimated by likelihood maximisation as

$$\hat{m} = \frac{1}{N}\sum_{k=1}^{N} y_k \equiv \langle y \rangle \tag{6.41a}$$

$$\hat{\sigma}_2^2 = \frac{1}{N}\sum_{k=1}^{N}(y_k - \hat{m})^2 = \langle y^2 \rangle - \langle y \rangle^2 \tag{6.41b}$$

When these parameters are introduced into the likelihood ratio (6.34) we find that the quantity

$$\chi = \frac{\langle y^2 \rangle}{\langle y^2 \rangle - \langle y \rangle^2} \tag{6.42}$$

emerges as a discriminant with which we can distinguish between the zero and non-zero mean distributions (speaking loosely, χ will tend to take larger values in the latter case, especially when the mean is significantly bigger than the variance). To investigate the behaviour of χ more fully we first note that

$$\chi = 1 + \frac{\bar{y}^2}{s^2}$$
$$\bar{y} = \langle y \rangle, \quad s^2 = \langle y^2 \rangle - \langle y \rangle^2 \tag{6.43}$$

where $\bar{y}$, s^2 provide estimators of the mean and variance of the distribution from which the $\{y_k\}$ are drawn. These estimators are themselves random variables; we can determine their joint pdf as follows. The estimate of the variance also incorporates that of the sample mean, which imposes the constraint (6.41a) on the values of $\{y_j\}$ incorporated into (6.41b) and so modifies the form of the pdf of this quantity. To see this we evaluate the characteristic function of the pdf of s^2, explicitly incorporating this constraint

$$\left\langle \exp\left(-\frac{p}{N}\sum_{j=1}^{N}(y_j - \bar{y})^2\right)\delta\left(\bar{y} - \frac{1}{N}\sum_{j=1}^{N} y_j\right)\right\rangle \tag{6.44}$$

It is convenient to represent the delta function as a Fourier integral

$$\delta\left(\bar{y} - \frac{1}{N}\sum_{j=1}^{N} y_j\right) = \frac{1}{2\pi}\int du \exp\left(iu\left(\bar{y} - \frac{1}{N}\sum_{j=1}^{N} y_j\right)\right) \tag{6.45}$$

and then show that

$$\left\langle \exp\left(-\frac{p}{N}\sum_{j=1}^{N}(yx_j-\bar{y})^2\right)\delta\left(\bar{y}-\frac{1}{N}\sum_{j=1}^{N}y_j\right)\right\rangle$$

$$=\frac{1}{(1+2p\sigma^2/N)^{(N-1)/2}}\exp(-N(\bar{y}-m)^2/2\sigma^2)\sqrt{\frac{N}{2\pi\sigma^2}} \tag{6.46}$$

We note that this characteristic function incorporates a factor equal to the probability that the sample mean lies in the interval $\bar{y}$ and $\bar{y}+d\bar{y}$. Laplace inversion then yields the joint pdf of the estimators of the sample mean and variance as

$$\begin{aligned}P(\bar{y},s) &= 2\sqrt{\frac{1}{\pi}}\left(N/2\sigma^2\right)^{N/2}\frac{1}{\Gamma((N-1)/2)}\\ &\quad\times\exp\left(-\frac{1}{2\sigma^2}(Ns^2+N(\bar{y}-m)^2)\right)s^{N-2};\\ &\quad-\infty<\bar{y}<\infty,\quad 0\le s<\infty\end{aligned} \tag{6.47}$$

Here σ, m are the root variance and mean of the underlying Gaussian distribution; $\Gamma(z)$ is the gamma function. We now consider the random variable

$$t=\frac{\bar{y}}{s} \tag{6.48}$$

in terms of which our discriminant takes the form

$$\chi=1+t^2 \tag{6.49}$$

A simple change in variables yields the joint pdf of s and t:

$$\begin{aligned}P(t,s) &= \frac{2}{\Gamma((N-1)/2)\Gamma(1/2)\left(\frac{N}{2\sigma^2}\right)^{N/2}}s^{N-1}\exp\\ &\quad\times\left(-\frac{N}{2\sigma^2}((1+t^2)s^2+m^2-2mst)\right);\quad -\infty<t<\infty,\ 0\le s<\infty\end{aligned} \tag{6.50}$$

from which the marginal pdf and cumulative distribution of t can be obtained by integration. Thus we can write the probability that the discriminant exceeds a

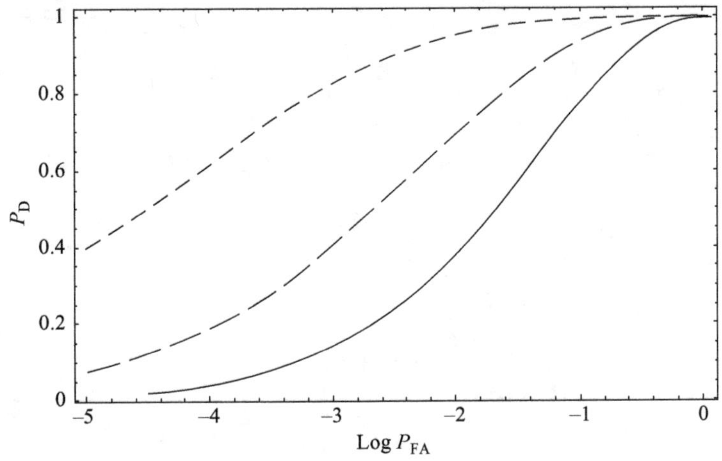

Figure 6.4 ROC curves for likelihood ratio detection, based on 1 (——), 2 (— —) and 4 (- - - - - -) samples. $\sigma = 1.0$, $m = 0$, 2.0

threshold, X

$$\begin{aligned}\text{Prob}(\chi > \text{X}) = \frac{1}{\Gamma\left(\frac{N-1}{2}\right)}\left(\frac{N}{2\sigma^2}\right)^{(N-1)/2}\int_0^\infty ds s^{N-2}\exp\left(-\frac{Ns^2}{2\sigma^2}\right)\\ \times\left(2 - \text{erf}\left(\sqrt{\frac{N}{2\sigma^2}}\left(s\sqrt{(\text{X}-1)} - m\right)\right)\right.\\ \left.- \text{erf}\left(\sqrt{\frac{N}{2\sigma^2}}\left(s\sqrt{(\text{X}-1)} + m\right)\right)\right)\end{aligned} \tag{6.51}$$

Using these results we can plot out the ROC curves for detections based on the likelihood and generalised likelihood ratios; typical results are shown in Figures 6.4 and 6.5.

These figures demonstrate quite strikingly that the detection or discrimination performance based on the likelihood ratio derived from a full knowledge of the candidate pdfs is significantly better than that based on the generalised likelihood ratio. However, the prior knowledge required to implement the former discriminant is rarely, if ever, available and the latter discriminant, or some approximation to it, is usually adopted in practice. We will develop this detection strategy further in the next section where we incorporate more realistic, but less tractable, models for the clutter and target plus clutter pdfs. Nonetheless, many of the qualitative and semi-quantitative conclusions we have drawn here, on the basis of this simple Gaussian model, will be found to remain valid and to provide useful guidelines for the construction of effective detection procedures.

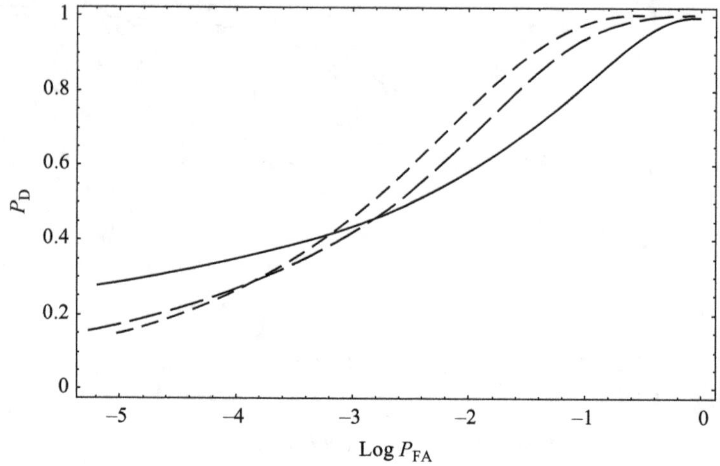

Figure 6.5 *ROC curves for generalised likelihood ratio detection, based on 3 (——), 7 (— —) and 10 (- - - - -) samples. $\sigma = 1.0$, $m = 0, 2.0$*

6.7 The detection of a steady signal in Rayleigh clutter

6.7.1 Generalised likelihood ratio based approach

The generalised likelihood approach demonstrated in the previous section can be extended to more realistic models of target plus clutter and clutter signal distributions; to render the resulting equations tractable one must now almost invariably make some approximation. The behaviour of the discriminants that emerge from these analyses is dominated by that of a 'signal to noise ratio' analogous to that of (6.42) derived in the discussion of the previous section. Thus, for example, we might consider a set of N independent measurements $\{E_k\}$ of the field amplitude. In the case in which these are derived from a clutter return, with a constant but unknown local power x_1, their joint likelihood is given by

$$P(\{E_k\}, x_1) = 2^N \frac{\prod_{k=1}^N E_k}{x_1^N} \exp\left(-\frac{1}{x_1}\sum_{k=1}^N E_k^2\right) \tag{6.52}$$

Should these measurements result from a signal of constant amplitude A, embedded in Rayleigh clutter of power x_2, the corresponding likelihood is built up from the individual Rice pdfs of the signals

$$P(\{E_k\}, x_2, A) = \frac{2^N}{x_2^N} \exp\left(-\frac{NA^2 + \sum_{k=1}^N E_k^2}{x_2}\right) \prod_{k=1}^N (E_k I_0(2E_k A/x_2)) \tag{6.53}$$

Following the generalised likelihood procedure, we first form estimates of the parameters x_1, x_2 and A based on the set of measurements $\{E_k\}$. In the pure clutter case this is a straightforward matter; maximising (6.52) with respect to x_1 yields the result

$$\frac{\partial}{\partial x_1}\log(P(\{E_k\}, x_1)) = -\frac{N}{x_1} + \frac{1}{x_1^2}\sum_{k=1}^{N} E_k^2 = 0$$

$$\hat{x}_1 = \frac{1}{N}\sum_{k=1}^{N} E_k^2 \equiv \langle E^2 \rangle \tag{6.54}$$

The corresponding conditions for maximisation appropriate to the signal plus clutter case are

$$\frac{\partial}{\partial x_2}\log(P(\{E_k\}, x_2, A)) = -\frac{N}{x_2} + \frac{NA^2}{x_2} + \frac{1}{x_2^2} \times \sum_{k=1}^{N} E_k^2 - \frac{2A}{x_2^2}\sum_{k=1}^{N} E_k \frac{I_1(2E_kA/x_2)}{I_0(2E_kA/x_2)} = 0 \tag{6.55}$$

$$\frac{\partial}{\partial A}\log(P(\{E_k\}, x_2, A)) = -\frac{2AN}{x_2} + \frac{2}{x_2}\sum_{k=1}^{N} E_k \frac{I_1(2E_kA/x_2)}{I_0(2E_kA/x_2)} = 0 \tag{6.56}$$

These can be combined to give a simple relationship between the noise power and signal amplitude estimators, rather reminiscent of (6.45),

$$\hat{x}_2 = \langle E^2 \rangle - (\hat{A})^2 \tag{6.57}$$

To make further progress towards the solution of (6.56) some approximations must be made. At first sight we might consider

$$\frac{I_1(2E_kA/x_2)}{I_0(2E_kA/x_2)} \approx 1 \tag{6.58}$$

suggested by the large argument asymptotic behaviour of the modified Bessel function

$$I_\nu(z) \sim \frac{\exp(z)}{\sqrt{2\pi z}}, \quad z \to \infty \tag{6.59}$$

This leads us to

$$\hat{A} = \frac{1}{N}\sum_{k=1}^{N} E_k \equiv \langle E \rangle \tag{6.60}$$

so that (6.54) becomes

$$\hat{x}_2 = \langle E^2 \rangle - \langle E \rangle^2 \tag{6.61}$$

Using these estimators, and the asymptotic form (6.59) of I_0, we can construct the likelihood ratio derived discriminant as

$$\frac{P(\{E_k\}, \hat{x}_2, \hat{A})}{P(\{E_k\}, \hat{x}_1)} \approx \frac{1}{(4\pi)^{N/2}} \frac{\hat{x}_1^N}{x_2^{N/2} \hat{A}^{N/2} \sqrt{\prod_{k=1}^{N} E_k}} \tag{6.62}$$

A suitable test statistic can now be identified as

$$\frac{\langle E^2 \rangle^2}{(\langle E^2 \rangle - \langle E \rangle^2)\langle E \rangle [E]}, \qquad [E] = \left(\prod_{k=1}^{N} E_k \right)^{1/N} \tag{6.63}$$

We note that this is similar to (6.45); (6.63) can also be derived by substituting the asymptotic form into (6.53) and carrying through the appropriate likelihood maximisation process. However, a moment's reflection reveals that each of these analyses is flawed. The condition (6.57) can be recast in the form

$$1 = \frac{\langle E^2 \rangle - x_2}{A^2} \tag{6.64}$$

and, in the large signal limit, is seen to be of order A^{-2}; the condition (6.60), however, is valid only up to order A^{-1}. To obtain a consistent solution to the equations (6.55) (6.56), approximations valid to the same order should be used. Thus we must replace (6.58) by

$$\frac{I_1(2E_k A/x_2)}{I_0(2E_k A/x_2)} \approx 1 - \frac{x_2}{4AE_k} \tag{6.65}$$

(which is readily derived from the large argument asymptotic form (A2.59)) and obtain

$$1 = \frac{\langle E \rangle}{A} - \frac{x_2}{4A^2} \tag{6.66}$$

accurate to order A^{-2}. Combined with (6.65), this yields

$$3A^2 - 4\langle E \rangle A + \langle E^2 \rangle = 0, \tag{6.67}$$

which can be solved to give the estimator

$$\begin{aligned} \hat{A} &= \frac{4\langle E \rangle + 2\langle E \rangle \sqrt{1 - 3(\langle E^2 \rangle - \langle E \rangle^2)/\langle E \rangle^2}}{6} \\ &\approx \langle E \rangle - \frac{1}{2} \frac{(\langle E^2 \rangle - \langle E \rangle^2)}{\langle E \rangle} \end{aligned} \tag{6.68}$$

This in turn, in conjunction with (6.57), gives us

$$\hat{x}_2 = 2(\langle E^2 \rangle - \langle E \rangle^2) \tag{6.69}$$

Equations (6.64), (6.68) and (6.69) can now be used to construct a test statistic of the form

$$\frac{\langle E^2\rangle^2}{(\langle E^2\rangle - \langle E\rangle^2)(3\langle E\rangle^2 - \langle E^2\rangle)} \frac{\langle E\rangle}{[E]} \tag{6.70}$$

This is similar in structure to the result (6.63); its behaviour is again dominated by that of

$$\frac{\langle E^2\rangle}{\langle E^2\rangle - \langle E\rangle^2} \tag{6.71}$$

which itself provides us with another, entirely useful, discriminant. Unfortunately, it is not possible to characterise the statistical properties of any of these, as was done in the Gaussian case discussed in the previous section. Consequently their performance must be assessed by direct simulation, rather than through explicit calculation.

The inconsistencies inherent in the arguments leading to (6.63) have proved to be of little practical consequence; they are nonetheless highlighted when we compare (6.60), (6.61), (6.68) and (6.69) with analytic results derived from the Rice distribution. Thus we have

$$\langle E^2\rangle_{\text{Rice}} = \frac{1}{x_2}\int_0^\infty dE\, E^3 \exp\left(-\frac{E^2 + A^2}{x_2}\right) I_0\left(\frac{2EA}{x_2}\right) = A^2 + x_2 \tag{6.72}$$

which can be compared directly with (6.63), and

$$\begin{aligned}\langle E\rangle_{\text{Rice}} &= \frac{1}{x_2}\int_0^\infty dE\, E^2 \exp\left(-\frac{E^2 + A^2}{x_2}\right) I_0\left(\frac{2EA}{x_2}\right)\\ &= \frac{\sqrt{\pi x_2}}{2}\, {}_1F_1(-1/2, 1, -A^2/x_2) \approx A\left(1 + \frac{x_2}{4A^2}\cdots\right)\end{aligned} \tag{6.73}$$

We have displayed the large signal limiting form of the latter through the large argument behaviour of the confluent hypergeometric function ${}_1F_1$ ([8], see also Appendix A2). Expressed in this form $\langle E\rangle_{\text{Rice}}$ can be compared directly with (6.69), allowing us to make the identifications

$$A \approx \frac{3\langle E\rangle^2_{\text{Rice}} - \langle E^2\rangle_{\text{Rice}}}{2\langle E\rangle_{\text{Rice}}}, \qquad x_2 \approx 2(\langle E^2\rangle_{\text{Rice}} - \langle E\rangle^2_{\text{Rice}}) \tag{6.74}$$

These correspond directly with (6.68) and (6.69), and differ markedly from the estimators (6.60) and (6.61) derived from the approximation (6.58). However the robustness of the general form taken by the various test statistics, and their dominance by the behaviour of (6.71), suggest that only scant attention need be paid to the choice of clutter and target plus clutter models. In many cases, when discussing effects of frequency agility and the construction of CFAR schemes, simple Gaussian based models like that discussed in the previous section are quite adequate.

6.7.2 Peak within interval detection

An alternative, and rather more rough and ready, method of locating a constant signal in clutter, from a sequence of N independent measurements of the received amplitude, is to identify it with the largest of the set of samples. This procedure is frequently referred to as 'peak within interval selection'. Obviously, should there be no target present, this procedure necessarily declares a false alarm. However, if one is able to repeat this a number of times, the false alarms should occur in different, random positions, while a target should follow a deterministic path. If there is a target present we can calculate the probability that the largest signal is the target signal, as a function of signal to noise ratio. This should give us a reasonable figure of merit for the effectiveness of a heuristically constructed, but readily implemented, processing scheme. To start off, we consider N signals, one of which contains target plus clutter (modelled by the Rice distribution); the clutter returns are modelled as identical, independent speckle processes, with Rayleigh distributed amplitudes. The probability that all the amplitude values are less than or equal to E is given by the product of the cumulative probabilities

$$F_{\max}(E) = F_1(E)F_0(E)^{N-1}$$

$$F_1(E) = \frac{2}{x}\int_0^E \exp\left(-\frac{E'^2 + A^2}{x}\right) I_0\left(\frac{2AE'}{x}\right) E'dE'$$

$$F_0(E) = \frac{2}{x}\int_0^E \exp\left(-\frac{E'^2}{x}\right) E'dE' = (1 - \exp(-E^2/x)) \tag{6.75}$$

This in turn provides us with the cumulative probability of the global maximum value; we obtain the associated pdf by differentiation

$$P_{\max}(E) = \frac{dF_{\max}(E)}{dE} = P_1(E)F_0(E)^{N-1} + (N-1)F_1(E)F_0(E)^{N-2}P_0(x)$$

$$P_1(E) = \frac{2E}{x}\exp\left(-\frac{E^2 + A^2}{x}\right) I_0\left(\frac{2AE}{x}\right); \qquad P_0(E) = \frac{2E}{x}\exp\left(-\frac{E^2}{x}\right) \tag{6.76}$$

(This procedure is very similar to that adopted in the analysis of order statistics as shown in Appendix A, Section A1.12 and in Reference 9, in which the contributing independent signals are all assumed to have the same pdf.) The pdf (6.76) consists of two terms. The first is the probability that the target return takes the maximum value, while all the clutter returns take lesser values; the second is the probability that one of the clutter signals takes the maximum value, while the target and remaining clutter signals take lesser values. If we recall Bayes theorem (A1.42) we can write the probability that, given a maximum value E, it comes from the target, as

$$P(\text{target}|E) = \frac{P_1(E)F_0(E)^{N-1}}{P_1(E)F_0(E)^{N-1} + (N-1)F_1(E)F_0(E)^{N-2}P_0(E)} = \frac{P_1(E)F_0(E)^{N-1}}{P_{\max}(E)} \tag{6.77}$$

To find the probability that the peak within interval procedure identifies the target correctly we must average this over the distribution of global maximum values; this leads us to

$$P(\text{target}) = \int_0^\infty P(\text{target}|E')P_{\max}(E')dE' = \int_0^\infty P_1(E')F_0(E')^{N-1}dE' \tag{6.78}$$

It now remains to evaluate this probability of correct target attribution for the simple target and clutter models we are using here. On introducing (6.75) and (6.76) we find that

$$P(\text{target}) = \frac{2}{x}\int_0^\infty \exp\left(-\frac{E^2+A^2}{x}\right) I_0\left(\frac{2AE}{x}\right)(1-\exp(-E^2/x))^{N-1}E\,dE \tag{6.79}$$

This in turn can be expressed in the elementary form

$$P(\text{target}) = \sum_{p=0}^{N-1} \frac{(N-1)!}{p!(N-1-p)!}(-1)^p\frac{1}{(p+1)}\exp\left(-\frac{pA^2}{(p+1)x}\right) \tag{6.80}$$

This is reasonably compact and can be readily evaluated numerically. Some credence is lent to our analysis by the special cases where the SNR, which we identify as A^2/x is zero and very large. In the latter case only the first term in the sum survives and the probability reduces to unity. In the former case the probability of correctly identifying the 'target' reduces to $1/N$, which is again quite reasonable. Figure 6.6 shows the variation of this probability of correct target identification as a function signal to clutter ratio for several values of N.

6.8 Applications to coherent detection

The approach just taken to the discrimination between incoherent clutter and signal plus clutter returns can be extended quite straightforwardly to the case of coherent signals, with separately measured in-phase and quadrature components. In this case

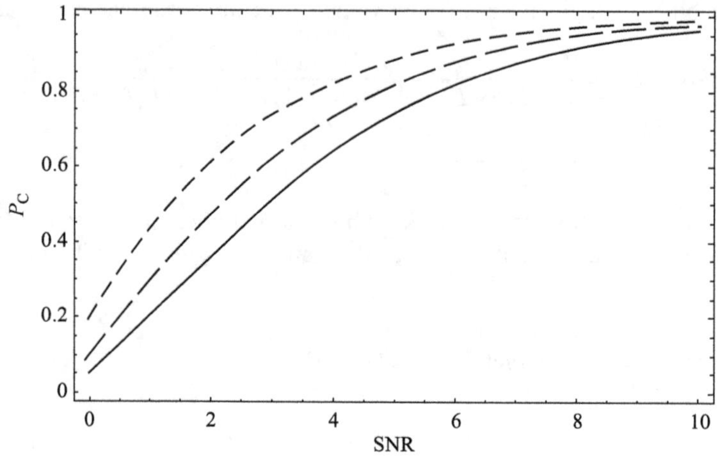

Figure 6.6 Probability of correct target identification as a function of SNR for N = 5 (——),10 (— —) and 20 (- - - - - -)

the circular complex Gaussian process provides us with a useful model of the clutter signal into which a signal of constant amplitude can be introduced straightforwardly.

Each of the received signals has I and Q components $\{E_{I,k}, E_{Q,k}\}$, $k = 0, \ldots, N-1$ which we represent as the real and imaginary parts of a complex number

$$\varepsilon_k = E_{I,k} + i E_{Q,k} \tag{6.81}$$

In the absence of a target return the pdf of these quantities is, assuming that the clutter samples are un-correlated,

$$P(\{\varepsilon_k\}, x_1) = \frac{1}{(\pi x_1)^N} \exp\left(-\frac{1}{x_1} \sum_{k=0}^{N-1} |\varepsilon_k|^2\right) \tag{6.82}$$

In the presence of a target return we have

$$P_1(\{\varepsilon_k\}, x_2, A) = \frac{1}{(\pi x_2)^N} \exp\left(-\frac{1}{x_2} \sum_{k=0}^{N-1} |\varepsilon_k - A \exp(ik\Delta)|^2\right) \tag{6.83}$$

Here A is a complex number characteristic of the target return and Δ is the inter-pulse phase shift in the target signal which, for simplicity, we assume to be constant.

Just as in our discussion of incoherent detection, we use likelihood maximisation to estimate the parameters characterising these pdfs, given a set of data. We then substitute these parameter estimates into a likelihood ratio and obtain a discriminant that takes the form

$$\chi = \frac{\hat{x}_1}{\hat{x}_2} \tag{6.84}$$

The carets again denote estimates derived from the received data.

If we assume that there is no target return present we make the estimate

$$\hat{x}_1 = \frac{1}{N}\sum_{k=0}^{N-1} |\varepsilon_k|^2 = \frac{1}{N}\sum_{k=0}^{N-1} (E_{I,k}^2 + E_{Q,k}^2) \tag{6.85}$$

If, however, it is assumed that there is a target present we must estimate s_1, A, Δ. The likelihood maximisation condition for x_2 then yields the estimator

$$\hat{x}_2 = \frac{1}{N}\sum_{k} |\varepsilon_k - A\exp(ik\Delta)|^2 \tag{6.86}$$

This can be used with (6.85) to construct the discriminant (6.84) once we have estimated the remaining parameters A, Δ.

The condition satisfied when the likelihood is maximised with respect to A gives us the estimator

$$\hat{A} = \frac{1}{N}\sum_{k=0}^{N-1} \varepsilon_k \exp(-ik\Delta) = \Phi(\Delta) \tag{6.87}$$

This we recognise as a discrete Fourier transform of the data, which is readily evaluated numerically [10]. Finally we estimate the phase shift between successively returned pulses from the condition that the likelihood (6.83) is maximised with respect to this quantity

$$\frac{\partial \log P_1}{\partial \Delta} = 0 \Rightarrow A^* \sum_{k=0}^{N-1} k\varepsilon_k \exp(-ik\Delta) - A \sum_{k=0}^{N-1} k\varepsilon_k^* \exp(ik\Delta) = 0 \tag{6.88}$$

We see straight away that this condition is equivalent to

$$\frac{\partial(\Phi(\Delta)^*\Phi(\Delta))}{\partial \Delta} = 0 \tag{6.89}$$

Thus we can estimate Δ by identifying the defining condition with that satisfied by a maximum in the power spectrum of the data. A suitably fine-grained array of Δ values can be constructed by 'zero-padding' the data; should there be more than one local maximum in the power spectrum, intuition dictates that we choose the global maximum to identify our estimator $\hat{\Delta}$.

Armed with this estimator we can then calculate $\hat{x}_2$ as

$$\hat{x}_2 = \frac{1}{N}\sum_{k=0}^{N-1} |\varepsilon_k - \exp(ik\hat{\Delta})\Phi(\hat{\Delta})|^2 \tag{6.90}$$

recalling that $\Phi(\hat{\Delta})$ has already been obtained as a by-product of an earlier calculation. The discriminant (6.78) can now be constructed as

$$\chi = \frac{\sum_{k=0}^{N-1} |\varepsilon_k|^2}{\sum_{k=1}^{N} \left|\varepsilon_k - \exp(ik\hat{\Delta})\Phi(\hat{\Delta})\right|^2} \tag{6.91}$$

This discussion of the construction of the coherent discriminant provides another example of the utility of the generalised likelihood ratio approach. The use of numerical Fourier transformation techniques allows us to search for a solution of the likelihood maximisation condition (6.89) quite straightforwardly. The underlying philosophy is the same as that of the incoherent case. A set of data is analysed to provide an estimate of the signal; this is then subtracted from the data and a discriminant is constructed as the ratio of the power in the signal plus noise to that in the noise alone.

6.9 The estimation of clutter parameters

6.9.1 Maximum likelihood estimators for gamma and Weibull parameters

We will now discuss the way in which the maximum likelihood method can be used to estimate parameters occurring in the clutter models discussed in Chapter 4. First we will demonstrate how this is done for data with gamma and Weibull distributions. Consider a set of independent data $\{x_j\}$, each of which we assume to be gamma distributed, with the likelihood

$$P(\{x_j\}|b,\nu) = \frac{b^{N\nu}}{\Gamma(\nu)^N}\left(\prod_{j=1}^{N} x_j^{\nu-1}\right)\exp\left(-b\sum_{j=1}^{N} x_j\right) \tag{6.92}$$

Arguing just as before, we identify the most probable values of b, ν consistent with the given data as those that satisfy the maximum likelihood conditions

$$\frac{\partial \log(P(b,\nu|\{x_j\}))}{\partial b} = \frac{N\nu}{b} - \sum_{j=1}^{N} x_j = 0$$

$$\frac{\partial \log(P(b,\nu|\{x_j\}))}{\partial \nu} = N\log(b) + \sum_{j=1}^{N}\log(x_j) - N\psi(\nu) = 0 \tag{6.93}$$

Here $\psi(\nu)$ is the logarithmic derivative of the gamma function

$$\psi(\nu) = \frac{d\log(\Gamma(\nu))}{dz} = \frac{1}{\Gamma(\nu)}\frac{d\Gamma(\nu)}{d\nu} \tag{6.94}$$

whose properties are reviewed in Appendix A2. These maximum likelihood conditions provide us with an equation satisfied by ν

$$\log \nu + \frac{1}{N}\sum_{j=1}^{N} \log x_j - \log\left(\frac{1}{N}\sum_{j=1}^{N} x_j\right) - \psi(\nu) = 0 \tag{6.95}$$

Once this has been solved numerically to give our estimate $\hat{\nu}$ of ν this can be used to give the estimate of b as

$$\hat{b} = \frac{\hat{\nu}}{\langle x \rangle} \tag{6.96}$$

In the case where the data $\{x_j\}$ are assumed to have a Weibull distribution their likelihood is given by

$$P(\{x_j\}|a, \beta) = \frac{\beta^N}{a^{\beta N}} \prod_{j=1}^{N} x_j^{\beta-1} \exp\left(-\sum_{j=1}^{N} x_j^{\beta}/a^{\gamma}\right) \tag{6.97}$$

from which we derive the maximum likelihood conditions

$$\frac{\partial \log(P(a, \beta|\{x_j\}))}{\partial a} = -\frac{N\beta}{a} + \frac{\beta}{a^{\beta+1}} \sum_{j=1}^{N} x_j^{\beta} = 0$$

$$\frac{\partial \log(P(a, \beta|\{x_j\}))}{\partial \beta} = \frac{N}{\beta} - N \log a + \sum_{j=1}^{N} \log x_j + \frac{1}{a^{\beta}} \sum_{j=1}^{N} x_j^{\beta}(\log a - \log x_j) = 0 \tag{6.98}$$

These can be simplified to

$$\frac{1}{\beta} + \frac{1}{N}\sum_{j=1}^{N} \log x_j - \frac{\sum_{j=1}^{N} x_j^{\beta} \log x_j}{\sum_{j=1}^{N} x_j^{\beta}} = 0$$

$$a = \left(\frac{1}{N}\sum_{j=1}^{N} x_j^{\beta}\right)^{1/\beta} \tag{6.99}$$

the first of which can be solved numerically to give $\hat{\beta}$, which can then be substituted into the second to give $\hat{a}$.

6.9.2 *Tractable, but sub-optimal, estimators for K and Weibull parameters*

Attempts to set up a maximum likelihood scheme for estimating the parameters of a K distribution directly quickly come to grief, foundering on analytically intractable quantities such as $\partial K_\nu(x)/\partial \nu$. As we have already mentioned, the analysis of K distributed data frequently reduces to that of the underlying gamma distribution of the integrated data, to which the ML estimation procedure embodied in (6.95) and (6.96) can be applied. Oliver has discussed these and related approximate parameter estimation schemes in great detail [11]; invariably they require the solution of a transcendental equation involving the digamma function. Thus for example, the relation appropriate to a K distributed intensity

$$\langle \log(z) \rangle - \log(\langle z \rangle) = \psi(\nu) - \log(\nu) - \gamma \tag{6.100}$$

provides an equation that can be solved for the shape parameter ν, when estimators for $\langle \log(z) \rangle$ and $\langle z \rangle$ are formed from the data. (We recognise γ as Euler's constant (A2.22)). If the intensities are integrated incoherently over N looks prior to analysis (6.94) is replaced by

$$\langle \log(z) \rangle - \log(\langle z \rangle) = \psi(\nu) - \log(\nu) - \psi(N) - \log(N) \tag{6.101}$$

and a similar procedure can be adopted. As each of these equations involves the digamma function, their numerical solution is a little time consuming. It is possible, however, to avoid this difficulty.

We note that for an N look K-distributed intensity, the following relation holds:

$$\frac{\langle z \log z \rangle}{\langle z \rangle} - \log\langle z \rangle = \psi(N+1) - \psi(N) + \psi(\nu+1) - \psi(\nu) \tag{6.102}$$

which, at first sight, looks quite as intractable as (6.100) or (6.101), as it involves digamma functions. However, if we recall the fundamental property of the gamma function (A2.2), take its logarithm

$$\log(\Gamma(z+1)) = \log(\Gamma(z)) + \log(z) \tag{6.103}$$

and differentiate, we find that

$$\psi(z+1) = \psi(z) + \frac{1}{z} \tag{6.104}$$

Using this (6.102) can be reduced to

$$\frac{\langle z \log z \rangle}{\langle z \rangle} - \langle \log z \rangle = \frac{1}{N} + \frac{1}{\nu} \tag{6.105}$$

This is easily solved for the shape parameter, once we have introduced estimators for the average values $\langle z \rangle, \langle \log z \rangle$ and $\langle z \log z \rangle$. A detailed analysis presented in Reference 12 shows that this simple estimation technique performs quite as well as those discussed by Oliver. We also note that in the limit $N \to \infty$ (6.105) reduces to the estimator for the shape parameter obtained from a similar analysis based on the gamma distribution of the local power x, evaluated by averaging over a large number of looks.

An alternative to the maximum likelihood analysis of Weibull distributed data is provided by the solution of the equation

$$\langle \log(z) \rangle - \log(\langle z \rangle) = -\frac{\gamma}{\beta} - \Gamma\left(\frac{1}{\beta} + 1\right) \tag{6.106}$$

6.10 Implications of the compound form of non-Gaussian clutter

In Chapter 4 we have seen that the spiky non-Gaussian character of high resolution sea clutter can be captured very effectively through its representation as a Rayleigh speckle process whose local power is itself a gamma distributed random variable. We are now able to estimate the parameters characterising this distribution and so refine our understanding of the behaviour of the clutter. Here we consider briefly the implications of this prior knowledge for the detection methodology we have developed. In deriving (6.70) we saw how we might construct a discriminant to detect a signal of constant amplitude, embedded in Rayleigh speckle with an unknown local power. The latter we were able to estimate from the processed data and incorporate within the discriminant, assuming that this local power did not change during the period over which the detection took place. This is a reasonable and necessary assumption which, up until now, has been made in the absence of any prior knowledge of the distribution of the local power. How might we exploit the information that local power x is gamma distributed, with a pdf specified by shape and scale parameters ν, b?

6.10.1 Modified generalised likelihood ratio based detection

Once again the generalised likelihood ratio approach provides us with a framework within which this question can be addressed. In the case where the constant signal is absent we can write the joint pdf of the field amplitudes $\{E_k\}$ and local power x in the form

$$P(\{E_k\}, x_1, \nu, b) = P(\{E_k\}, x_1)\frac{b^\nu}{\Gamma(\nu)}x_1^{\nu-1}\exp(-bx_1) \tag{6.107}$$

where $P(\{E_k\}, x_1)$ is as given in (6.5). In doing this we are effectively introducing prior knowledge of the distribution of the parameter x, just as we discussed in Section 6.5. To estimate x_1 we again form the maximum likelihood condition, whose solution will provide us with our estimator. Thus we find that

$$\frac{\partial \log(P(\{E_k\}, x_1, \nu, b))}{\partial x_1} = \frac{\partial \log(P(\{E_k\}, x_1))}{\partial x_1} + \frac{\nu - 1}{x_1} - b \tag{6.108}$$

and (cf. (6.48))

$$\frac{(\nu - N - 1)}{\hat{x}_1} - b + \frac{N\langle E^2\rangle}{\hat{x}_1^2} = 0 \tag{6.109}$$

This gives us a quadratic equation to solve for $\hat{x}_1$

$$\hat{x}_1 = \frac{\sqrt{(1-(\nu-1)/N)^2 + 4\langle E^2\rangle b/N} - (1-(\nu-1)/N)}{2b/N} \tag{6.110}$$

(We have chosen the root that reduces to (6.54) in the separate limits of large N and non-fluctuating, but unknown, x_1 (i.e. $\nu \to \infty$; $b = \nu/x_1$)).

If we assume that a constant signal of amplitude A is present, the joint pdf of the field amplitudes $\{E_k\}$ and local power x is now

$$P(\{E_k\}, x_2, A, \nu, b) = P(\{E_k\}, x_1, A)\frac{b^\nu}{\Gamma(\nu)}x_1^{\nu-1}\exp(-bx_1) \tag{6.111}$$

while the maximum likelihood conditions take the form

$$\begin{aligned}\frac{\partial \log(P(\{E_k\}, x_2, A, \nu, b))}{\partial x_2} &= \frac{\partial \log(P(\{E_k\}, x_2, A))}{\partial x_2} + \frac{\nu - 1}{x_1} - b = 0\\ \frac{\partial \log(P(\{E_k\}, x_2, A, \nu, b))}{\partial A} &= \frac{\partial \log(P(\{E_k\}, x_2, A))}{\partial A} = 0\end{aligned} \tag{6.112}$$

We note that the condition derived from the variation in A is identical with (6.56); as a consequence we may carry through much of our earlier analysis without modification. In this way we find that

$$\frac{(\nu - 1 - N)}{x_2} - b + \frac{N}{x_2^2}(\langle E^2\rangle - A^2) = 0 \tag{6.113}$$

The large signal limit analysis again provides us with the estimator (6.74) for A, which in turn leads to the quadratic whose solution is the estimator $\hat{x}_2$

$$\begin{aligned}&\frac{(\nu - 1 - N)}{x_2} - b + \frac{2N}{x_2^2}(\langle E^2\rangle - \langle E\rangle^2) = 0\\ &\hat{x}_2 = \frac{\sqrt{(1-(\nu-1)/N)^2+8(\langle E^2\rangle-\langle E\rangle^2)b/N} - (1-(\nu-1)/N)}{2b/N}\end{aligned} \tag{6.114}$$

A useful discriminator can now be constructed from (6.107) and (6.114) as (cf. (6.71)) $\hat{x}_1/\hat{x}_2$.

6.10.2 *Modified peak within interval detection*

The results of our earlier analysis of the performance of the 'peak within interval' technique can be modified quite straightforwardly to take account of the compound representation of non-Gaussian clutter. In essence, the probability of correct identification of the signal plus clutter sample (6.80) need only be averaged over a gamma distribution of the local power x. In this way we find that

$$\begin{aligned}P(\text{target}) = &\frac{2}{\Gamma(\nu)}\sum_{p=0}^{N-1}\frac{(N-1)!}{(p+1)!(N-1-p)!}\\ &\times(-1)^p\left(\frac{A^2bp}{p+1}\right)^{\nu/2}K_\nu\left(2\sqrt{\frac{A^2bp}{p+1}}\right)\end{aligned} \tag{6.115}$$

which is again quite easy to evaluate. In Figure 6.7 we plot out the probability of correct identification for 10 samples, as a function of signal to clutter power ratio.

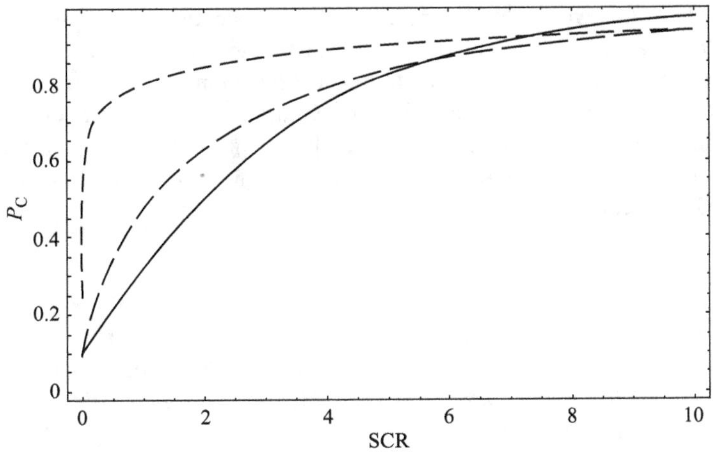

Figure 6.7 Plots of probability of correct target identification, as a function of SCR, for 10 samples and $\nu = 10$, (——), *1* (— —) *and 0.1* (- - - - - -)

From this we see that there is a higher probability of correct identification at low SCR for smaller ν. This is to be expected, as this spiky clutter is characterised by significant periods of quiescence, during which an added signal would readily exceed a modest threshold. For large SCR however, the technique yields a lower probability of correct identification at higher thresholds, in progressively more spiky clutter whose occasional excursions can be mistaken for signal returns. It is interesting to compare these findings with those of the discussion of the performance of CFAR in non-Gaussian clutter presented in Chapter 9, where qualitatively similar conclusions emerge from a rather different analysis.

6.11 Concluding remarks

In this chapter we have established a well-defined framework, provided by the likelihood ratio concept and the statistical models developed in Chapter 4, within which useful detection schemes can be developed in a systematic way. The closely related problem of parameter estimation is also considered: maximum likelihood techniques derived from Bayes' theorem prove to be quite tractable for simple Gaussian clutter models, and can be incorporated into generalised likelihood ratio based detection methods. Some relatively simple examples of the application of these principles have been discussed in detail, to the point where contact is made with the small target detection strategies discussed in Chapters 8 and 9. The principles demonstrated here can then be applied to other detection scenarios. The estimation of parameters characterising non-Gaussian clutter is more problematic: the maximum likelihood derived equations are now much less easy to solve. Nonetheless, useful estimation methods have been derived that can be applied to gamma, Weibull and K-distributed data. The compound representation of clutter developed in Chapter 4 plays a central, but rather subtle, role in this work. Small target detection procedures are applied to localised samples of data, to which the relatively simple and tractable Gaussian derived methods can be applied. Consequently we should expect the methods described in Section 6.7 to be relatively effective even in spiky, non-Gaussian clutter; their performance can then be improved incrementally and relatively straightforwardly, when prior knowledge of the gamma distribution of local power can then be brought to bear, as is described in Section 6.10.

References

1 W.W Peterson, T.G. Birdsall and W.C. Fox, 'The theory of signal detectability', Section 2.7, *IRE Trans. Inform. Theory*, **PGIT-6**, 204–221, 1956.

2 E.S. Pearson, 'The Neyman–Pearson story: 1926–34', in F.N. David (Ed.), '*Research papers in Statistics*', John Wiley & Sons, London, 1966.

3 H.L. van Trees, *'Detection, Estimation and Modulation Theory, Part 1'*, John Wiley & Sons, New York, 1968.

4 D. Middleton, '*Statistical Communication Theory*', McGraw Hill, New York, 1960.

5 M.I. Skolnik, '*Introduction to Radar Systems*', Section 9.6, McGraw Hill, New York, 1962.

6 M.G. Kendall, '*The Advanced Theory of Statistics*', Vol. 2, Chaps. 17–20, Griffin, London, 1946.

7 M.G. Kendall, '*The Advanced Theory of Statistics*', Vol. 1, Chap. 10, Griffin, London, 1943.

8 E.T. Copson, '*Introduction to the Theory of Functions of a Complex Variable*', Section 10.62, Clarendon Press, Oxford, 1970.

9 R.V. Hogg and A.T. Craig, '*Introduction to Mathematical Statistics*', Section 4.6, Collier Macmillan, London, 1978.

10 W.H Press, S.A. Teukolsky, W.T. Vetterling and B.P. Flannery, '*Numerical recipes in FORTRAN; The art of scientific computing.*', 2nd Edn., University Press, Cambridge, 1992.

11 C.J. Oliver, 'Optimum texture estimators or SAR clutter', *J. Phys. D*, **26**, 1824–1835, 1994.

12 D. Blacknell and R.J.A. Tough, 'Parameter estimation for the K distribution based on [zlog(z)]', *IEE Proc. Radar, Sonar, Navig.*, **148**, 309–312, 2001.

Chapter 7

Imaging ocean surface features

7.1 Introduction

In the previous chapter we considered how we might best detect small, localised targets in a background of sea clutter. The identification of the likelihood ratio as an optimum discriminant, and of its more practically useful approximations, provided us with a unifying framework for this discussion. However, small target returns are not the only features of interest in maritime radar imagery. Large-scale correlated structures arising from surface currents, ship wakes, the presence of surfactants and other sources can frequently be discerned and are a valuable source of information in many circumstances. In this chapter we will discuss how such ocean surface features might best be enhanced and detected. Once again the likelihood ratio concept is a very useful guiding principle, which leads us to methods that enable us to both enhance these features and exploit our prior knowledge of their structure to detect them more effectively. So, paradoxically, a discussion of the processing of images that are frequently interpreted and assessed in qualitative terms, will involve us in a fair amount of detailed formal analysis. Much of this will be based on the multivariate Gaussian distribution; we commence this chapter with a review of its pertinent properties, which are discussed in more detail in Appendix 1.

7.2 The analysis of correlated Gaussian data

So far we have considered the analysis of un-correlated data sets, whose joint pdf can be decomposed into a product of those of the individual data values (e.g. (6.26)). In general the extension of these arguments to correlated sets of data is rather intractable. There is one case however in which considerable progress can be made and which provides a realistic model of many detection scenarios. The joint pdf of a set of n correlated Gaussian random variables $\mathbf{x}$ can be written in the form

(see also Appendix 1)

$$P(\mathbf{x}) = \frac{\exp(-\frac{1}{2}(\mathbf{x}-\boldsymbol{\mu})^{\mathrm{T}} \cdot \mathbf{K}^{-1} \cdot (\mathbf{x}-\boldsymbol{\mu}))}{(2\pi)^{n/2}\sqrt{\det \mathbf{K}}} \tag{7.1}$$

The vector **x** might consist of values of signals in a clique of pixels within a SAR image, for example. Here **K** and $\boldsymbol{\mu}$ are the covariance matrix and vector of mean values of **x**. The relatively simple form of this likelihood facilitates the implementation of methods of detection and discrimination based on the principles outlined in Section 6.3. In subsequent sections we will see how these concepts can also be applied in the exploitation of multi-channel radar systems, such as polarimetric and interferometric SAR, which provide us with effective techniques for the enhancement of surface features in radar imagery of the sea. We will also find that the compound modelling of non-Gaussian clutter discussed in Chapter 4 can be extended to the description of correlated processes, and results in a formulation equivalent to that provided by spherically invariant random processes (SIRPs).

7.2.1 χ processing

We can acquire a great deal of insight into the processing of correlated data of this type simply by considering the quadratic form that contributes to the logarithm of the likelihood (7.1) i.e.

$$\chi_a = \frac{1}{2}\mathbf{x}^{\mathrm{T}} \cdot \mathbf{K}_a^{-1} \cdot \mathbf{x} \tag{7.2}$$

Given the matrix $\mathbf{K}_a^{-1}$ it is an easy matter to evaluate χ_a. Our experience in the previous chapter suggests that quantities of this type should provide us with a useful discriminant; we refer to data analysis based on (7.2) and its analogues as χ processing. Here we have set the vector of means $\boldsymbol{\mu}$ to zero; this is a particularly benign simplification for stationary and correlated Gaussian processes.

7.2.2 χ_a processing and the whitening filter

The covariance matrix $\mathbf{K}_a$ of the ambient signal (and consequently its inverse) is necessarily symmetric

$$K_{ij} = K_{ji} = \langle x_i x_j \rangle \tag{7.3}$$

To ensure that the pdf (7.1) is integrable we must also require that the quadratic form (7.2) is greater than zero, save when all the elements of **x** vanish, i.e. χ_a is positive definite. Consequently it is possible to factorise $\mathbf{K}_a^{-1}$ into the product of an upper triangular matrix **Q** and its transpose

$$\mathbf{K}_a^{-1} = \mathbf{Q}^{\mathrm{T}} \cdot \mathbf{Q} \tag{7.4}$$

(This representation is known as a Cholesky decomposition, and is discussed in Section A1.6 of Appendix 1). If the vector of data **x** is subject to the linear transformation

$$\mathbf{v} = \mathbf{Q} \cdot \mathbf{x} \tag{7.5}$$

the elements of **v** are necessarily un-correlated

$$\langle \mathbf{v}\mathbf{v}^{\mathrm{T}} \rangle = \mathbf{Q} \cdot \mathbf{K}_a \cdot \mathbf{Q}^{\mathrm{T}} = \mathbf{Q} \cdot (\mathbf{Q}^{\mathrm{T}} \cdot \mathbf{Q})^{-1} \mathbf{Q}^{\mathrm{T}} = \mathbf{Q} \cdot (\mathbf{Q}^{-1} \cdot (\mathbf{Q}^{\mathrm{T}})^{-1}) \cdot \mathbf{Q}^{\mathrm{T}} = \mathbf{1} \tag{7.6}$$

For this reason, processing that utilises the quadratic form (7.2) is sometimes referred to as a 'whitening transform'. To determine the pdf of the value taken by χ_a we first construct the characteristic function

$$\begin{aligned} C(\omega) &= \langle \exp(i\omega\chi_a) \rangle_a \\ &= \frac{1}{(2\pi)^{n/2}\sqrt{\det \mathbf{K}_a}} \int d^n x \exp\left(-\frac{(1-i\omega)}{2} \mathbf{x}^{\mathrm{T}} \cdot \mathbf{K}_a^{-1} \cdot \mathbf{x} \right) \\ &= \frac{1}{(1-i\omega)^{n/2}} \end{aligned} \tag{7.7}$$

The pdf of χ_a can now be recovered by Fourier inversion; while the form it takes in this case is quite simple, its derivation is instructive. Thus we have

$$P(\chi_a) = \frac{1}{2\pi} \int_{-\infty}^{\infty} \frac{\exp(-i\omega\chi_a) d\omega}{(1-i\omega)^{n/2}} \tag{7.8}$$

and note that the integrand is an analytic function in the upper half of the complex plane. When χ_a takes negative values, we must evaluate (7.8) by closing the contour of integration in the upper half plane [1]; Cauchy's theorem tells us that $P(\chi_a)$ therefore vanishes in this case. For positive χ_a we 'close' in the lower half plane and obtain

$$\begin{aligned} P(\chi_a) &= \frac{\exp(-\chi_a)}{\Gamma(n/2)} \chi_a^{n/(2-1)}, \quad \chi_a \geq 0 \\ &= 0, \quad \chi_a < 0 \end{aligned} \tag{7.9}$$

This has the gamma distribution form typical of the pdf of a sum of squares of zero mean Gaussian random variables with equal variance, and so includes (6.19), with $n = 2N$.

So far we have assumed χ_a to be constructed from the covariance matrix $\mathbf{K}_a$ that describes the statistics of the data **x** i.e.

$$P(\mathbf{x}) = \frac{\exp\left(-\frac{1}{2} \mathbf{x}^{\mathrm{T}} \cdot \mathbf{K}_a^{-1} \cdot \mathbf{x} \right)}{(2\pi)^{n/2}\sqrt{\det \mathbf{K}_a}} \tag{7.10}$$

(The subscript a might thus be taken to denote a statistic incorporating knowledge of *ambient* signals.) If, however, the data are described by a different covariance matrix $\mathbf{K}_t$,

$$P(\mathbf{x}) = \frac{\exp\left(-\frac{1}{2}\mathbf{x}^{\mathrm{T}} \cdot \mathbf{K}_t^{-1} \cdot \mathbf{x}\right)}{(2\pi)^{n/2}\sqrt{\det \mathbf{K}_t}} \tag{7.11}$$

then the statistics of the resulting quadratic form are rather different; in this case the subscript t might connote '*target*'. Once again the characteristic function provides us with a useful handle on the problem

$$\begin{aligned} C(\omega) &= \langle \exp(i\omega\chi_a)\rangle_t \\ &= \frac{1}{(2\pi)^{n/2}\sqrt{\det \mathbf{K}_t}} \int d^n x \exp\left(-\frac{1}{2}\mathbf{x}^{\mathrm{T}} \cdot (-i\omega\mathbf{K}_a^{-1} + \mathbf{K}_t^{-1}) \cdot \mathbf{x}\right) \\ &= \frac{1}{\sqrt{\det(\mathbf{1} - i\omega\mathbf{K}_t \cdot \mathbf{K}_a^{-1})}} \end{aligned} \tag{7.12}$$

Proceeding with the evaluation of the pdf of χ_a we recognise that the determinant in the denominator of (7.12) can be written as a polynomial, the locations of whose zeros will tell us a great deal about the behaviour of χ_a

$$\det(\mathbf{1} - i\omega\mathbf{K}_t \cdot \mathbf{K}_a^{-1}) = \prod_{j=1}^{n} (1 - i\omega\mu_j^{-1}) \tag{7.13}$$

Here the μ_j are solutions to the eigenvalue problem

$$\det(\mathbf{K}_t - \mu\mathbf{K}_a) = 0 \tag{7.14}$$

and are necessarily positive. The characteristic function (7.12) might yet be invertible in terms of known functions (we will consider a case where this is so in Section 7.4); even when this is not the case it can still furnish us with useful information. We see in Section A1.4 of Appendix 1 that expectation values can be read off from a series expansion of a characteristic function; in this case we find that

$$\langle\chi_a\rangle = \frac{1}{2}\sum_j \mu_j^{-1}, \quad \langle\chi_a^2\rangle = \frac{1}{2}\left(\sum_j \mu_j^{-2} + \frac{1}{2}\left(\sum_j \mu_j^{-1}\right)^2\right) \tag{7.15}$$

The normalised variance of χ_a now provides us with a measure of the spikiness of the output of a χ_a processor

$$\frac{\langle\chi_a^2\rangle - \langle\chi_a\rangle^2}{\langle\chi_a\rangle^2} = 2\frac{\sum_j \mu_j^{-2}}{\left(\sum_j \mu_j^{-1}\right)^2} \tag{7.16}$$

By setting the derivatives of this expression with respect to each of the μ_j to zero we see that this measure of spikiness (and so of a propensity for the declaration of false alarms) is minimised when the eigenvalues μ_j all take the same value. This condition holds when

$$\mathbf{K}_a \equiv \mathbf{K}_t \tag{7.17}$$

This observation accounts, at least in part, for the utility of 'whitening filter' techniques used to extract target features from a correlated background [2]. If the covariance matrix characterising the ambient background is known *a priori*, the χ_a that incorporates this covariance will, loosely speaking, generate as few false alarms as are compatible with this prior knowledge, when not applied to 'target' data.

7.2.3 χ_o processing

So far, our discussion of the exploitation of χ_a processing has taken no account of any knowledge we might have of the correlation properties of the target signal, accommodated within the Gaussian model (7.11). The likelihood ratio criterion developed in Section (6.3) provides us with the means with which to do this; the quadratic form

$$\chi_o = \frac{1}{2}\mathbf{x}^{\mathrm{T}} \cdot (\mathbf{K}_a^{-1} - \mathbf{K}_t^{-1}) \cdot \mathbf{x} \tag{7.18}$$

should provide us with the optimal test statistic with which to discriminate between target and ambient data sets, provided that we know the covariance matrices $\mathbf{K}_a$ and $\mathbf{K}_t$. (Here we might associate the subscript o with optimality.) The characteristic function again leads us quite directly to the pdf of the test statistic; in the case of χ_o we will see that it can, in some cases, take both positive and negative values. For the moment we consider the case where the data $\mathbf{x}$ are drawn from the ambient distribution (7.10); the characteristic function is constructed as

$$\begin{aligned} C(\omega) &= \langle \exp(i\omega\chi_o)\rangle_a \\ &= \frac{1}{(2\pi)^{n/2}\sqrt{\det \mathbf{K}_a}} \int d^n x \exp\left(-\frac{1}{2}\mathbf{x}^{\mathrm{T}} \cdot (\mathbf{K}_a^{-1} - i\omega(\mathbf{K}_a^{-1} - \mathbf{K}_t^{-1}))\right) \\ &= \frac{1}{\sqrt{\det(\mathbf{1} - i\omega(\mathbf{1} - \mathbf{K}_a \cdot \mathbf{K}_t^{-1}))}} \end{aligned} \tag{7.19}$$

Once again we can develop the determinant as a polynomial, whose zeros can be expressed in terms of the eigenvalues of (7.14):

$$\det(\mathbf{1} - i\omega(\mathbf{1} - \mathbf{K}_a \cdot \mathbf{K}_t^{-1})) = \prod_{j=1}^{n}\left(1 - i\omega\left(\frac{\mu_j - 1}{\mu_j}\right)\right) \tag{7.20}$$

The pdf of the optimal test statistic formed from ambient data can now be generated by Fourier inversion.

$$P(\chi_o)_a = \frac{1}{2\pi}\int_{-\infty}^{\infty} \frac{\exp(-i\omega\chi_A)}{\sqrt{\prod_{j=1}^{n}(1 - i\omega((\mu_j - 1)/\mu_j))}} d\omega \tag{7.21}$$

In much the same way, the pdf of χ_o formed from data drawn from the target distribution (7.11) can be derived as

$$P(\chi_o)_t = \frac{1}{2\pi} \int_{-\infty}^{\infty} \frac{\exp(-i\omega\chi_A)}{\sqrt{\prod_{j=1}^{n}(1 - i\omega(\mu_j - 1))}} d\omega \tag{7.22}$$

Arguing much as we did in the derivations of (7.9) we see that the optimal test statistic χ_o will take positive values when all the positive eigenvalues μ_j are greater than unity, and that it can take both positive and negative values when they take values both greater and less than one. Should all the eigenvalues be less than unity, χ_o takes only negative values.

As χ_o is constructed as the logarithm of a likelihood ratio, discrimination or detection techniques based on its measurement should have an optimal performance. In particular they should perform at least as well as those based on the ambient statistic χ_a. Processing schemes based on both χ_o and χ_a have been proposed in the literature, in a variety of contexts. Obviously they depend on fidelity of the prior knowledge incorporated in the covariance matrices $\mathbf{K}_a$ and $\mathbf{K}_t$; the former is doubtless easier to determine from an appropriate analysis of a separate ambient signal. In the special case where the ambient and test signals have the same correlation structure, and differ only in their power i.e.

$$\mathbf{K}_t^{-1} = \alpha \mathbf{K}_a^{-1} \tag{7.23}$$

χ_o and χ_a, derived from both target and ambient signals, have pdfs that take the same gamma form as (7.9), albeit with different scale parameters. A simple calculation shows [3] that, under these circumstances, the performance of χ_o and χ_a based discrimination is identical; this finding perhaps accounts for the observation that χ_a discriminators frequently perform as well as χ_o discriminators [2]. In practice, χ processing is exploited to enhance anomalous features, whose correlation structure is different from that of an ambient background, and has been applied particularly extensively to SAR imagery.

7.3 The Wishart distribution

In Section 6.5 of the previous chapter we discussed how the parameters which specify the pdf of a set of independent, identically distributed random variables could be determined by the process of likelihood maximisation. Here we will address the analogous problem for correlated Gaussian data: the estimation of the vector of their means and their covariance matrix. The solution of this problem proceeds, in principle, very much as did that in the uncorrelated variable case. We will also briefly discuss the pdf of these estimators, the so-called Wishart distribution.

7.3.1 The real Wishart distribution

We consider a set of p correlated Gaussian random variables, with a vector of means $\boldsymbol{\mu}$ and a covariance matrix $\mathbf{K}$, of which we have N samples. From (7.1) we see that the joint pdf of these sample values can be written as

$$P(\{\mathbf{x}_j, j = 1, \ldots, N\}|\mathbf{K}) = \frac{1}{(2\pi)^{Np/2}|\det(\mathbf{K})|^{N/2}} \times \exp\left(-\frac{1}{2}\sum_{j=1}^{N}(\mathbf{x}_j - \boldsymbol{\mu})^{\mathrm{T}} \cdot \mathbf{K}^{-1} \cdot (\mathbf{x}_j - \boldsymbol{\mu})\right) \tag{7.24}$$

If we introduce the matrix $\mathbf{A}$ derived from the data as

$$\mathbf{A} = \frac{1}{N}\sum_{j=1}^{N}(\mathbf{x}_j - \boldsymbol{\mu})(\mathbf{x}_j - \boldsymbol{\mu})^{\mathrm{T}} \tag{7.25}$$

we can re-write this as

$$P(\{\mathbf{x}_j, j = 1, \ldots, N\}|\mathbf{K}) = \frac{1}{(2\pi)^{Np/2}|\det(\mathbf{K})|^{N/2}} \exp\left(-\frac{N}{2}\mathrm{Tr}(\mathbf{K}^{-1} \cdot \mathbf{A})\right) \tag{7.26}$$

The logarithm of the joint pdf can now be expressed as

$$\log P(\{\mathbf{x}_j, j=1, \ldots, N\}|\mathbf{K}) = C + \frac{N}{2}\log(|\det \mathbf{K}^{-1}|) - \frac{N}{2}\sum_{k,l=1}^{p}(\mathbf{K}^{-1})_{kl}(\mathbf{A})_{lk} \tag{7.27}$$

Arguing much as we did in Section 6.6 we maximise this log likelihood to obtain equations satisfied by the estimators. In this way we find that

$$\frac{\partial \log P}{\partial \boldsymbol{\mu}} = -\sum_{j=1}^{N}\mathbf{K}^{-1} \cdot (\mathbf{x}_j - \boldsymbol{\mu}) = 0; \quad \mathbf{m} = \frac{1}{N}\sum_{j=1}^{N}\mathbf{x}_j \tag{7.28}$$

which yields the maximum likelihood estimator $\mathbf{m}$ of the vector of means. Minimising (7.27) with respect to elements of the inverse of the covariance matrix $\mathbf{K}$ leads to

$$\frac{\partial \log P}{\partial (\mathbf{K}^{-1})_{kl}} = \frac{N}{2}(\mathbf{K})_{lk} - \frac{N}{2}(\mathbf{A})_{lk} = 0 \tag{7.29}$$

where we have used the result (A1.83) from Appendix 1. (7.29) can be solved immediately to yield an estimate of the covariance matrix

$$(\hat{\mathbf{K}})_{lk} = (\mathbf{A})_{lk} \tag{7.30}$$

where the estimator $\mathbf{m}$ of the vector of mean values (7.28) must now be introduced into $\mathbf{A}$ in place of $\boldsymbol{\mu}$.

We note that the exponent in the pdf (7.24) can be written solely in terms of these estimators

$$\sum_{j=1}^{N} (\mathbf{x}_j - \boldsymbol{\mu})^{\mathrm{T}} \cdot \mathbf{K}^{-1} \cdot (\mathbf{x}_j - \boldsymbol{\mu})$$

$$= \sum_{j=1}^{N} (\mathbf{x}_j - \mathbf{m} + \mathbf{m} - \boldsymbol{\mu})^{\mathrm{T}} \cdot \mathbf{K}^{-1} \cdot (\mathbf{x}_j - \mathbf{m} + \mathbf{m} - \boldsymbol{\mu})$$

$$= N\mathrm{Tr}(\hat{\mathbf{K}} \cdot \mathbf{K}^{-1}) + N(\mathbf{m} - \boldsymbol{\mu})^{\mathrm{T}} \cdot \mathbf{K}^{-1} \cdot (\mathbf{m} - \boldsymbol{\mu}) \tag{7.31}$$

as a consequence of the identity

$$\sum_{j=1}^{N} (\mathbf{x}_j - \mathbf{m}) = 0 \tag{7.32}$$

The exponential term in the pdf decomposes into a product of a Gaussian characterising the distribution of the estimated mean, and a term that depends only on the estimator of the covariance matrix

$$\exp\left(-\frac{N}{2}(\mathbf{m} - \boldsymbol{\mu})^{\mathrm{T}} \cdot \mathbf{K}^{-1} \cdot (\mathbf{m} - \boldsymbol{\mu})\right) \exp\left(-\frac{N}{2}\mathrm{Tr}(\mathbf{K}^{-1} \cdot \hat{\mathbf{K}})\right) \tag{7.33}$$

The pdf of the estimators can be derived on the basis of this observation; the vector $\mathbf{m}$ of estimated means itself has a multivariate Gaussian distribution, with a mean of $\boldsymbol{\mu}$ and a covariance matrix $\mathbf{K}/N$. In the case of the covariance matrix element estimators, their joint pdf consists of the second exponential term shown in (7.33), supplemented by a more complicated expression, containing the analogues of s^{N-2} and normalisation terms in (6.47). The explicit derivation of this factor will not be considered here; a careful exposition can be found in Chapters 14 and 28 of Kendall's classic work [4]. The resulting form of the pdf of this so-called Wishart distribution [5] is

$$P(\hat{\mathbf{K}}) = \left(\frac{2}{N}\right)^{pN/2} \frac{|\det \hat{\mathbf{K}}|^{(N-p-1)/2} \exp(-(N/2)\mathrm{Tr}(\mathbf{K}^{-1} \cdot \hat{\mathbf{K}}))}{\pi^{(p(p-1))/4} \prod_{m=1}^{p} \Gamma((N-p+m)/2) |\det \mathbf{K}|^{N/2}} \tag{7.34}$$

7.3.2 The complex Wishart distribution

Thus far we have considered estimators of real multivariate Gaussian processes, whose covariance matrices $\mathbf{K}$ are arbitrary, save for the requirements that they are symmetric and generate a positive definite quadratic form. A practically important special case is provided by the so-called p-variate complex Gaussian process. This consists of $2p$ Gaussian random variables $\{x_j, y_j\}$, $j = 1, \ldots, p$, which can be thought of as the real and imaginary parts of the complex numbers

$$z_j = x_j + iy_j \tag{7.35}$$

whose correlation properties can be summarised through

$$\langle x_j x_j \rangle = \langle y_j y_j \rangle = \sigma_j^2; \quad \langle x_j y_j \rangle = 0$$
$$\langle x_j x_k \rangle = \langle y_j y_k \rangle = \alpha_{jk}\sigma_j\sigma_k; \quad \langle x_j y_k \rangle = -\langle x_k y_j \rangle = \beta_{jk}\sigma_j\sigma_k; \qquad j \neq k \tag{7.36}$$

We also assume, as is almost invariably the case in practice, that x_j, y_j have zero means. Consequently the $2p \times 2p$ covariance matrix formed from $\{x_j, y_j\}, j = 1, \ldots, p$ is specified by $p + 2((p(p-1))/2) = p^2$ parameters, rather than the $p(2p+1)$ required in the general case. The complex representation (7.35) can be used to good effect to present such processes in a relatively compact form. As

$$\langle z_j{}^* z_k \rangle = \langle (x_j - iy_j)(x_k + iy_k) \rangle = 2\sigma_j\sigma_k(\alpha_{jk} + i\beta_{jk}); \quad j \neq k$$
$$\langle |z_k^2| \rangle = 2\sigma_k^2 \tag{7.37}$$

a p times p Hermitian matrix can be constructed

$$C_{jk} = \langle z_j{}^* z_k \rangle = (C_{kj})^* \tag{7.38}$$

that captures the correlation properties implicit in a $2p \times 2p$ real symmetric covariance matrix $\mathbf{R}$ matrix incorporating the results (7.36). In much the same way the vector of the $2p$ real variables $\{x_j, y_j\}, j = 1, \ldots, p$ can be replaced by the p complex random variables $\{z_j, j = 1, \ldots, p\}$.

The pdf of these can then be written in the more compact form if we identify the real and complex random variables and their covariance matrices through

$$\zeta = \{x_1, y_1, \ldots, x_p, y_p\}$$
$$\eta = \{z_1, z_2, \ldots, z_p\}$$
$$\mathbf{R} = \langle \eta\eta^{\mathrm{T}} \rangle, \quad \mathbf{C} = \langle \mathbf{z}\mathbf{z}^{*\mathrm{T}} \rangle \tag{7.39}$$

The pdfs of the real and complex variables can then be written as

$$P(\zeta) = \frac{1}{(2\pi)^p\sqrt{\det \mathbf{R}}} \exp\left(-\frac{1}{2}\zeta^{\mathrm{T}} \cdot \boldsymbol{R} \cdot \zeta\right) \tag{7.40}$$

and

$$P(\eta) = \frac{1}{\pi^p \det \mathbf{C}} \exp\left(-\eta^{*\mathrm{T}} \cdot \mathbf{C} \cdot \eta\right) \tag{7.41}$$

and are equivalent. The theory of the estimators of the elements of the Hermitian matrix $\mathbf{C}$ can be developed just as in the real variable case and their pdf derived (a detailed account of this analysis is given by Goodman [6]). In this way we can identify the estimator of the covariance matrix as

$$\hat{\boldsymbol{\Sigma}} = \frac{1}{N}\sum_{j=1}^{N} \mathbf{z}_j \mathbf{z}_j^{\mathrm{T}*} \tag{7.42}$$

and find that

$$P(\hat{\mathbf{\Sigma}}) = \frac{|\det \hat{\mathbf{\Sigma}}|^{N-p} \exp(-N\mathrm{Tr}(\mathbf{C}^{-1} \cdot \hat{\mathbf{\Sigma}}))}{\pi^{p(p-1)/2} \prod_{m=1}^{p} \Gamma(N+1-m) |\det \mathbf{C}|^{N} N^{Np}} \tag{7.43}$$

7.4 Polarimetric and interferometric processing

The χ processing techniques discussed in Section 7.2 effectively enhanced anomalous features by exploiting the correlations present in a single radar image. Some features discernible within a SAR image of the ocean can be enhanced by appropriately combining two or more SAR images of the same scene. An interferometric SAR system consists of two antennae, separated by a distance ℓ along the fuselage of its moving platform. Of these, the front antenna transmits and receives while the back antenna receives. The received signals from the two antennae are processed to form two SAR images of the same scene, separated in time by $\tau = \ell/2v$ where v is the velocity of the platform. (The reader might wish to consider why the delay takes this value.) Each pixel in the SAR image carries a complex signal with amplitude and phase: this can be modelled as a circular Gaussian process with a local power and temporal correlation properties determined by a power spectrum. Typically τ will be of the order of 5–50 ms, dependent on the radar operating frequency. On this time scale the resolved features of the image barely change; the only significant change is the phases of the pixel signals. Thus if we register (superimpose) the two images, measure the phase increments in the various pixel signals and display these in some suitable colour coding we generate an image that highlights specific surface features. It would be expected that objects moving along the SAR line of sight should be picked out by this procedure which, in some ways, acts like a large-scale MTI (moving target indication) radar. (More details of the implementation of this interferometric mode can be found in the paper by Goldstein and Zebker [7]). Figure 7.1, taken from Reference 21, shows an intensity SAR image and a simultaneous interferometric SAR image of a region of the South Falls sandbank. The modulations are caused by a tidal current flow over the sandbank, which induces straining and a consequential modulation of the sea wave spectrum (as described in Section 3.4). The intensity image is similar to Figure 3.36 and shows how the wave spectrum is suppressed as the tidal flow speeds up to pass over the sandbank, and then is amplified as the flow slows on the other side. The interferometric image, where the phase difference is plotted as intensity, shows that this process also affects the sea clutter Doppler spectrum as would be expected. The interferometric technique thus provides a useful alternative method for detecting this type of feature.

Another useful set of enhancement methods exploits the correlations between two images of the same scene, obtained using different radar polarisations [8]. Once again, the signal in each pixel can be modelled by a complex Gaussian process, which is correlated with that in the corresponding pixel in the second image.

The analysis of this pair of correlated circular complex Gaussian random variables allows us to present many of the ideas developed in Section 7.2 in a more

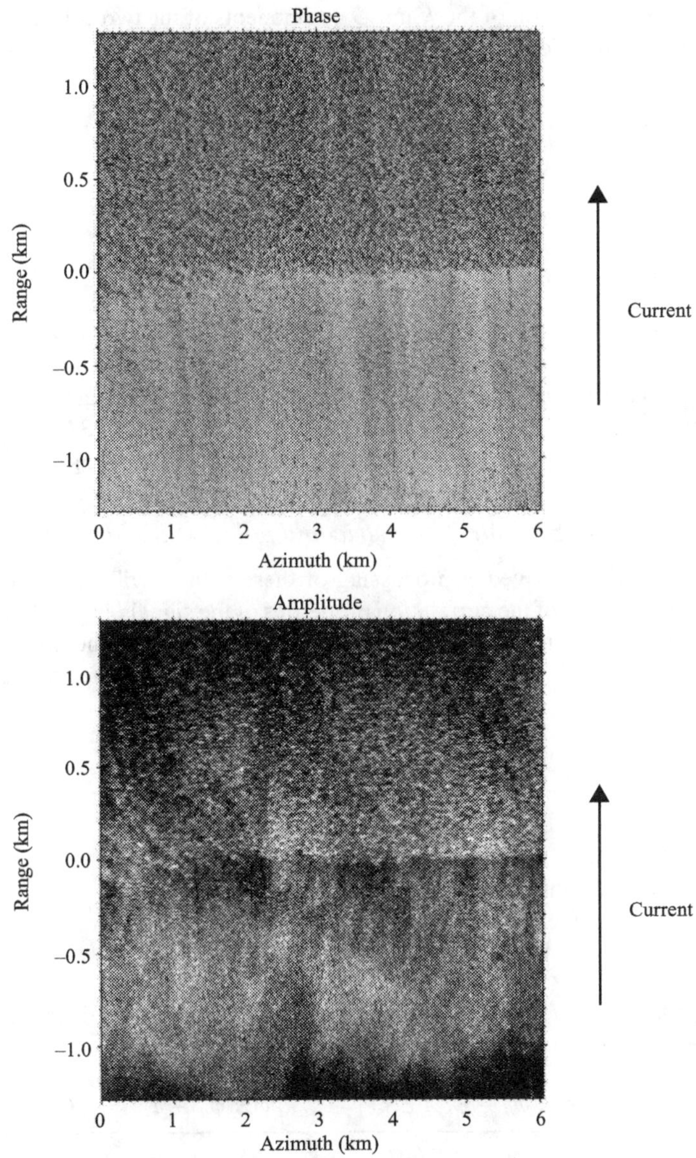

Figure 7.1 Comparison of SAR (lower) and Along Track Interferometric (ATI) SAR (upper) images of the South Falls sandbank [21] (images courtesy of QinetiQ)

concrete setting. Some of the results we obtain are relevant to the polarimetric and interferometric processing of SAR data and, perhaps surprisingly, are intimately related to the generalised K distribution introduced as a model of weak scattering in Section 4.

The covariance matrix of the I and Q components of the two correlated complex Gaussian quantities takes the form (cf. A1.126–128)

$$\mathbf{K} = \psi \begin{pmatrix} 1 & 0 & \rho & \lambda \\ 0 & 1 & -\lambda & \rho \\ \rho & -\lambda & 1 & 0 \\ \lambda & \rho & 0 & 1 \end{pmatrix} \tag{7.44}$$

Here

$$k^2 = \rho^2 + \lambda^2 \leq 1 \tag{7.45}$$

is a measure of the decorrelation of the process; where appropriate we will use subscripts a and t to distinguish between ambient and target characteristics. For convenience we have assumed that the powers of the two correlated signals are identical. This simplified analysis is directly applicable to interferometric SAR processing; a simple re-scaling allows it to accommodate polarimetric processing [9].

7.4.1 χ processing of interferometric and polarimetric data

The likelihood ratio derived χ processing of these signals will be considered first. Given full knowledge of the correlation properties of the signals, χ_o processing yields the optimum discriminant between the ambient and targets. The solutions to the eigenvalue problem (7.14) are doubly degenerate, as a consequence of the statistical equivalence of I and Q components, and take the values

$$\mu_\pm = \frac{\psi_t}{\psi_a} \frac{(1 - \rho_a\rho_t - \lambda_a\lambda_t) \pm \sqrt{(1 - \rho_a\rho_t - \lambda_a\lambda_t)^2 - (1 - k_a^2)(1 - k_t^2)}}{(1 - k_a^2)} \tag{7.46}$$

If we now introduce the phases

$$\phi_a = \tan^{-1}\left(\frac{\lambda_a}{\rho_a}\right), \quad \phi_t = \tan^{-1}\left(\frac{\lambda_t}{\rho_t}\right) \tag{7.47}$$

the eigenvalues (7.46) can be written in a form that displays its dependence on the amplitude, coherence and phase associated with each signal

$$\mu_\pm = \frac{\psi_t}{\psi_a} \frac{(1 - k_a k_t \cos(\phi_a - \phi_t)) \pm \sqrt{(1 - k_a k_t \cos(\phi_a - \phi_t))^2 - (1 - k_a^2)(1 - k_t^2)}}{(1 - k_a^2)} \tag{7.48}$$

From this we see explicitly that $\mu_\pm$ are both positive and that χ_o will take exclusively positive (negative) values when the amplitude of the target signal is significantly larger (smaller) than that of the ambient signal. If the target and ambient signals have similar amplitudes, χ_o may take both positive and negative values. The characteristic function of the distribution of a sum of N independent χ values is obtained by raising the appropriate single look characteristic function to its Nth power. An analysis of multi-look processing can be carried through on the basis of this result [3]; the pdfs from which probabilities of detection and false alarm are calculated by Fourier inversion. The results

$$
\begin{aligned}
\frac{1}{2\pi}\int_{-\infty}^{\infty}&\frac{\exp(-i\omega\chi)}{(1-(i\omega/\gamma))^N(1-(i\omega/\delta))^N}d\omega \\
&= \frac{(\gamma\delta)^N}{(\gamma-\delta)^N}\frac{\chi^{N-1}}{(N-1)!}\exp\left(-\frac{(\gamma+\delta)}{2}\chi\right) \\
&\quad\times\sum_{n=0}^{N-1}\frac{(N+n-1)!}{n!(N-n-1)!}\frac{1}{(\chi(\gamma-\delta))^n} \\
&\quad\times\left\{(-1)^n\exp\left(-\frac{(\delta-\gamma)}{2}\chi\right)+(-1)^N\exp\left(-\frac{(\gamma-\delta)}{2}\chi\right)\right\} \\
&= \frac{(\gamma\delta)^N}{(\gamma-\delta)^{N-1}}\frac{\chi^N}{(N-1)!}\exp\left(-\frac{(\gamma+\delta)\chi}{2}\right)i_{N-1}\left(\frac{(\gamma-\delta)\chi}{2}\right), \quad \chi\geq 0
\end{aligned} \tag{7.49}
$$

and

$$
\begin{aligned}
\frac{1}{2\pi}\int_{-\infty}^{\infty}&\frac{\exp(-i\omega\chi)}{(1-(i\omega/\gamma))^N(1+(i\omega/\delta))^N}d\omega \\
&= \left(\frac{\gamma\delta}{\gamma+\delta}\right)^N\exp(\delta\chi)(-\chi)^{N-1}\sum_{n=0}^{N-1}\frac{(N+n-1)!}{n!(N-n-1)!}\frac{1}{(-\chi(\gamma+\delta))^n}, \quad \chi<0, \\
&= \left(\frac{\gamma\delta}{\gamma+\delta}\right)^N\exp(-\gamma\chi)\chi^{N-1}\sum_{n=0}^{N-1}\frac{(N+n-1)!}{n!(N-n-1)!}\frac{1}{(\chi(\gamma+\delta))^n}, \quad \chi>0, \\
&= \frac{(\gamma\delta)^N}{(\gamma+\delta)^{N-1}}\frac{|\chi|^N}{\pi(N-1)!}k_{N-1}\left(\frac{\gamma+\delta}{2}|\chi|\right)
\end{aligned} \tag{7.50}
$$

lead to expressions for the required pdfs, in terms of elementary functions and of the modified spherical Bessel functions

$$
i_n(z)=\sqrt{\frac{\pi}{2z}}I_{n+1/2}(z), \quad k_n(z)=\sqrt{\frac{\pi}{2z}}K_{n+1/2}(z) \tag{7.51}
$$

Appropriate values of the constants γ, δ can be identified by reference to equations (7.13), (7.21) and (7.22). Details of the analysis of multi-look χ processing, based on these results, are given in Reference 3; readers should note that (7.49) and (7.50) correct minor typographical errors in the referenced paper.

7.4.2 *Phase increment processing of interferometric data*

In modelling Goldstein and Zebker's interferometric processing we assume that each pixel signal is independent; this simplifies the analysis. The interferometric SAR (InSAR) displays the increments in the phases of these signals that develop over the delay time τ; we represent each of these signals as a circular complex Gaussian process. Signals in the same pixel form a correlated pair, described by a covariance matrix much like (7.44). It will also be instructive to compare interferometric processing with the χ based techniques discussed earlier; identifying a common basis for the modelling of the two techniques provides a first step in this direction.

The signal consists of I and Q components (in phase and quadrature) whose temporal correlation properties are given by

$$\begin{aligned}
&\langle E_I(t)E_I(t+\tau)\rangle = \langle E_Q(t)E_Q(t+\tau)\rangle = \psi\rho(\tau)\\
&\langle E_I(t)E_Q(t+\tau)\rangle = -\langle E_Q(t)E_I(t+\tau)\rangle = \psi\lambda(\tau)\\
&\langle E_I^2\rangle = \langle E_Q^2\rangle = \psi\\
&\rho(\tau) = \frac{\int d\omega\, \cos(\omega\tau)S(\omega)}{\int d\omega\, S(\omega)}; \quad \lambda(\tau) = \frac{\int d\omega\, \sin(\omega\tau)S(\omega)}{\int d\omega\, S(\omega)}
\end{aligned} \tag{7.52}$$

These results allow us to identify the appropriate parameters in (7.44) while $\mathbf{x}$ is the column vector

$$\mathbf{x} = \begin{pmatrix} E_I(t) \\ E_Q(t) \\ E_I(t+\tau) \\ E_Q(t+\tau) \end{pmatrix} \tag{7.53}$$

As can be seen from Appendix 1, Section A1.9, the joint distribution of the amplitudes and phases at two different times takes the form

$$\begin{aligned}
&P(E(t), E(t+\tau), \theta(t), \theta(t+\tau))\\
&\quad = \frac{E(t)E(t+\tau)}{(2\pi\psi)^2(1-k^2)} \exp\left\{ -\frac{1}{2\psi(1-k^2)}[E(t)^2 + E(t+\tau)^2\right.\\
&\qquad \left. - 2kE(t)E(t+\tau)\cos(\theta(t+\tau) - \theta(t) - \phi_0)]\right\}
\end{aligned} \tag{7.54}$$

where

$$\phi_0 = \tan^{-1}(\lambda/\rho), \quad k^2 = \rho^2 + \lambda^2 \tag{7.55}$$

The result (7.55) now allows us to give a useful interpretation of the phases introduced in (7.47).

In the InSAR processing the signal in a given pixel is multiplied by the complex conjugate of the (complex) signal in that pixel occurring at a time τ later. The phase of this quantity is identical with the phase increment to be displayed in the image.

Thus if we write

$$\varsigma(\tau) = E(t)E(t+\tau), \quad \phi(\tau) = \theta(t+\tau) - \theta(t) \tag{7.56}$$

we can find the joint pdf of these quantities by making a change of variables in (7.54) and integrating over those that are not required. In this way we find that

$$P(\varsigma, \phi) = \frac{\varsigma}{2\pi\psi^2(1-k^2)} \exp\left(\frac{k\varsigma\cos(\phi - \phi_0)}{\psi(1-k^2)}\right) K_0\left(\frac{\varsigma}{\psi(1-k^2)}\right) \tag{7.57}$$

In the case of the interferometer attention is focused on the distribution of phase increments, obtained from (7.57) by integrating over ς. In this way we find that

$$P(\phi) = \frac{1-k^2}{2\pi}\left\{\frac{\beta(\pi/2 + \sin^{-1}\beta)}{(1-\beta^2)^{3/2}} + \frac{1}{(1-\beta^2)}\right\} \tag{7.58}$$

where $\beta = k\cos(\phi - \phi_0) \cdot \phi_0$, defined in (7.55), is the most probable value of the phase increment, i.e. the mode of the distribution. The corresponding pdf of the quantity ς is obtained from (7.57) as

$$P(\varsigma) = \int_0^{2\pi} d\phi\, P(\varsigma, \phi) = \frac{\varsigma}{\psi^2(1-k^2)} I_0\left(\frac{k\varsigma}{\psi(1-k^2)}\right) K_0\left(\frac{\varsigma}{\psi(1-k^2)}\right) \tag{7.59}$$

whose form is identical with that of the generalised K distribution (cf. (4.38)).

In essence the interferometer distinguishes between signals that have different power (Doppler) spectra by comparing the phase increments that develop in those signals in a given time. The statistical approach we are developing allows us to make a quantitative assessment of the effectiveness of this technique. The phase increments derived from signals with different power spectra will have different pdfs. If these have a significant separation then InSAR will discriminate between them effectively; if they lie one on top of the other discrimination will not be possible. Thus if m_1, m_2 are the means of the two distributions and their variances are σ_1, σ_2 then a quantitative measure of the separation of the distributions will be given by

$$\Delta = \frac{|m_1 - m_2|}{\sqrt{\sigma_1^2 + \sigma_2^2}} \tag{7.60}$$

This quantity provides a useful figure of merit for the InSAR system. The analysis we have carried out allows us to evaluate Δ and trace its dependence on the different Doppler spectra and the lag time, which in turn depends on the antenna separation and the aircraft velocity. Maximising Δ will identify the optimum parameters for the operation of the interferometer. The evaluation of the mean and variance of the distribution of phase increments is described in the paper of Tough *et al.* [9]. The mean of the phase difference distribution can be calculated as

$$\langle\phi\rangle = \frac{k \sin \phi_0}{\sqrt{1 - k^2 \cos^2 \phi_0}} \cos^{-1}(k \cos \phi_0) \tag{7.61}$$

while

$$\langle\phi^2\rangle = \frac{\pi^2}{3} - \pi \sin^{-1}(k \cos \phi_0) + [\sin^{-1}(k \cos \phi_0)]^2 - \frac{1}{2}\sum_{n=1}^{\infty}\frac{k^{2n}}{n^2} \tag{7.62}$$

The infinite sum in (7.63) is known as Euler's dilogarithm; its behaviour when k is close to 1 is captured by

$$F_2(z) = \sum_{n=1}^{\infty}\frac{z^n}{n^2}; \quad F_2(z) = \frac{\pi^2}{6} - F_2(1-z) - \log(z)\log(1-z) \tag{7.63}$$

Thus for each candidate Doppler spectrum and delay time we calculate λ and ρ from (7.52) and hence the modal phase increment ϕ_0 and the correlation coefficient k from (7.55). This then allows us to calculate the means and variances of these distributions through (7.61) to (7.62) and thence construct appropriate figures of merit Δ. In practice, significant discrimination is not possible without some reduction in the spread of the distributions of phase increments.

7.4.3 *Coherent summation and discrimination enhancement*

Goldstein and Zebker [7] suggested that InSAR images could be enhanced by a coherent summation procedure, which in effect reduces the variance of the measured phase increments. In this, neighbouring pixels are grouped into cliques, a coherent sum of the complex products formed from the time separated pixel signals within each clique is constructed and its phase is measured. This is then displayed through some suitable colour coding over the area of the clique. In doing this we reduce the variances of the distributions of phase increments, while preserving their modes. This in turn improves the figure of merit Δ; as a consequence better imagery can be formed in this way.

Figure 7.2 (from Reference 21) shows this process applied to the imagery of Figure 7.1. The graphs show the amplitude and interferometric phase difference as a function of vertical distance on the images. The small pixel-to-pixel variation about the mean trend is the result of intensity averaging (for the intensity image) and complex product averaging (in the case of the interferometric image).

Analysis of this 'interferometric averaging' process can be carried out, just as in the single look case, using the infinite divisibility of the generalised K distribution (which is much like that of the more familiar K distribution mentioned in Section 4.4.2). If we label pixels by a suffix k, represent the complex signal in each as

$$E_k(t)\exp(i\theta_k(t)) \tag{7.64}$$

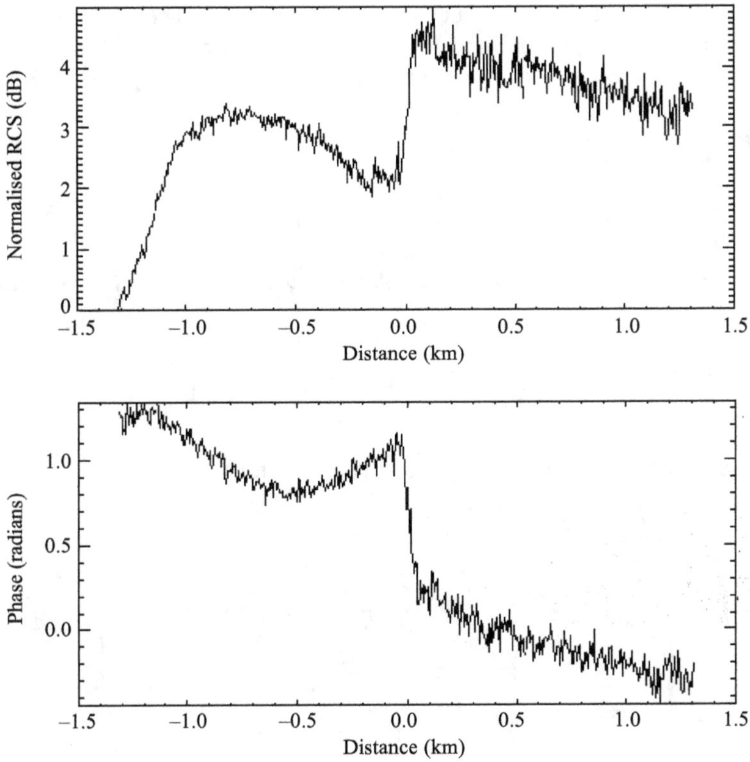

Figure 7.2 *Comparison of average RCS and average phase across sandbank corresponding to images shown in Figure 7.1 [21] (courtesy of QinetiQ)*

The product formed from two temporally separated signals derived from the same pixel can be written as

$$\xi_k(\tau) = \varsigma_k(\tau)\exp(i\phi_k(\tau)) = E_k(t)E_k(t+\tau)\exp(i(\theta(t+\tau) - \theta(t))) \quad (7.65)$$

The coherently summed signal is given by

$$\Omega(\tau) = \sum_{k=1}^{N} \xi_k(\tau) = \Xi(t)\exp(i\Phi(\tau)) \quad (7.66)$$

We note that in forming this coherent sum we are, in effect, constructing a maximum likelihood estimator of the off-diagonal element of the coherence matrix characterising the correlated complex signals (cf. (7.42)). Thus the Wishart distribution introduced in Section 7.3 is expected to be of relevance in the analysis of output of multi channel radar systems, of which the interferometric and polarimeteric SARs are two-channel special cases. A discussion of the interferometric SAR, which deals only with the phases of the signals, can be carried through in terms of the generalised

K distribution approach we have been developing here. As a consequence of the infinite divisibility of the generalised K distribution the joint pdf of the amplitude and phase of the coherent sum (7.66) can be written in the relatively simple form

$$P(\Xi, \Phi) = \frac{1}{2\pi \psi^{N+1}(1-k^2)2^{N-1}(N-1)!} \times \exp\left[\frac{k\Xi \cos \Phi}{\psi(1-k^2)}\right] \Xi^N K_{N-1}\left(\frac{\Xi}{\psi(1-k^2)}\right) \tag{7.67}$$

Finally we note that the pdf of Ξ is obtained by integrating (7.67) over the phase variable:

$$P(\Xi) = \frac{1}{\psi^{N+1}(1-k^2)2^{N-1}(N-1)!} I_0\left[\frac{k\Xi}{\psi(1-k^2)}\right] \Xi^N K_{N-1}\left(\frac{\Xi}{\psi(1-k^2)}\right) \tag{7.68}$$

while the distribution of phase increment takes the form

$$P_N(\Phi) = \frac{(1-k^2)^N}{2\pi} \times \left\{ \begin{aligned} &\frac{(2N-2)!}{(N-1)!2^{2(N-1)}}\left[\frac{(2N-1)}{(1-\beta^2)^{N+1/2}}\beta\left(\frac{\pi}{2}+\sin^{-1}\beta\right)+\frac{1}{(1-\beta^2)^N}\right] \\ &+\frac{1}{2(N-1)!}\sum_{r=0}^{N-2}\frac{\Gamma(N-1/2)}{\Gamma(N-1/2-r)}(N-r)!\,\frac{1+(2r+1)\beta^2}{(1-\beta^2)^{r+2}} \end{aligned} \right\} \tag{7.69}$$

where now $\beta = k \cos(\Phi - \phi_0)$. As we see from Figure 7.3 the distribution of the phase increment becomes progressively narrower as N increases, while its mode is unchanged. Consequently the figure of merit Δ is enhanced and a more effective discrimination between different features in the InSAR image is possible.

This analysis is discussed in rather more detail in, amongst others, papers by Tough *et al.* [9], Barber [10], and Lee *et al.* [11]. In Reference 9 its applications to both interferometric and polarimetric processing and its relationship to the analysis in terms of the Wishart distribution are considered. Blacknell and Tough [3] also give a comparison of the performance of discriminators based on the χ_a and χ_o statistics, and the amplitude and phase variables Ξ and Φ. As we would expect, χ_o based discrimination always performs the best. Ξ based discrimination can perform almost as well when there is a significant difference between the powers of the two signal pairs, while the performance Φ based interferometric discrimination can approach that of χ_o when the two signals are alike in power, but have significantly different correlation properties.

7.5 Feature detection by matched filtering

So far we have discussed methods by which regions within a radar image of the sea that differ significantly in their correlation properties can be distinguished by appropriate processing. This has in the main been based on a correlated Gaussian model for the

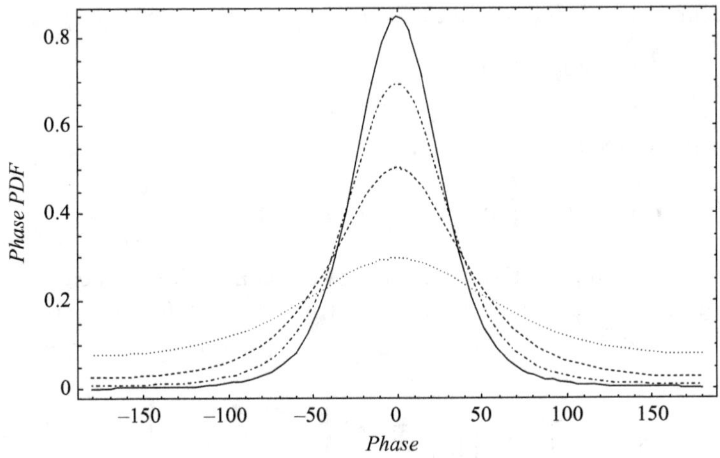

Figure 7.3 *Reduction of the variance in phase increment by coherent averaging; plots of $P_N(\Phi)$ with $k = 0.4$ and $N = 1$ (······), 4 (------), 8 (-·-·-·-) and 12 (——)*

signals in a small number of pixels, and has been motivated by the likelihood ratio principles we developed in the previous chapter. No mention has been made of the large-scale structure of the image, and no attempt has been made to exploit any prior knowledge we might have of this structure. In this section we consider how this might be done; once again the likelihood ratio concept proves to be extremely valuable in motivating the discussion.

Initially, we assume that the image has been pre-processed in such a way that, in the absence of a large scale feature, the signal at a point $\mathbf{r}$ in the two dimensional image can be described by a stationary, correlated complex Gaussian process y. Pre-processing of this type is frequently referred to as de-trending.

The assumption of stationarity requires that the spatial auto-correlation function of the process $\langle y(\mathbf{R}+\mathbf{r})y(\mathbf{R})\rangle$ is independent of $\mathbf{R}$. We assume that y can be represented in terms of spatial Fourier components $\tilde{y}(\mathbf{k})$ which are themselves complex Gaussian random variables:

$$y(\mathbf{r}) = \frac{1}{(2\pi)^2}\int d^2k \exp(-i\mathbf{k}\cdot\mathbf{r})\tilde{y}(\mathbf{k}) \tag{7.70}$$

Thus we see that the spatial autocorrelation function of y is given by

$$\langle y(\mathbf{R}+\mathbf{r})y^*(\mathbf{R})\rangle = \frac{1}{(2\pi)^4}\int d^2k \int d^2q \times \exp(-i((\mathbf{k}-\mathbf{q})\cdot\mathbf{R}+\mathbf{k}\cdot\mathbf{r}))\langle\tilde{y}(\mathbf{k})\tilde{y}^*(\mathbf{q})\rangle \tag{7.71}$$

For this to be independent of $\mathbf{R}$ the Fourier components must be uncorrelated i.e.

$$\langle\tilde{y}(\mathbf{k})\tilde{y}^*(\mathbf{q})\rangle = \langle|\tilde{y}(\mathbf{k})|^2\rangle\delta(\mathbf{k}-\mathbf{q}) \tag{7.72}$$

The autocorrelation function of the noise process is then related to its power spectrum

$$S(\mathbf{k}) = \frac{\langle|\tilde{y}(\mathbf{k})|^2\rangle}{4\pi^2 A} \tag{7.73}$$

(A is the area of the image) through

$$\langle y(\mathbf{R}+\mathbf{r})y^*(\mathbf{R})\rangle = \int d^2k \, \exp(-i\mathbf{k}\cdot\mathbf{r})S(\mathbf{k}) \tag{7.74}$$

Thus the joint pdf of the set of signals in the Fourier transform of a background noise image is given by (introducing labels i, j to denote the N pixel positions and adopting a discrete notation)

$$P(\{\tilde{y}_{i,j}\,|\,\sigma_{i,j}\}) = \frac{1}{(2\pi)^N \prod_{i,j}\sigma^2_{i\cdot j}} \exp\left(-\sum_{i,j}\frac{\tilde{y}_{i,j}\tilde{y}^*_{i,j}}{2\sigma^2_{i,j}}\right) \tag{7.75}$$

The statistical independence of the Fourier components $\tilde{y}(\mathbf{k})$ simplifies the form of this pdf significantly, compared with that of the pixel signals $y(\mathbf{r})$. Thus, in the χ processing of an image, we wished to exploit the correlation between the signals in pixels in close proximity, and so worked explicitly in $\mathbf{r}$ space. Here we wish to focus our attention on the large-scale features in the image and so simplify the description of the background noise as much as possible. We see that this can be done very effectively working in the Fourier domain.

We now consider a signature $s(\mathbf{r}\,|\,\{\alpha\})$, characterised by the parameters $\{\alpha\}$, superimposed additively on the noise background. The construction of this template incorporates our prior knowledge of the large scale structure of, for example, a wake or a surface current. On Fourier transformation we will obtain an image consisting of

$$\tilde{s}(\mathbf{k}\,|\,\{c_p\}) = \int d^2r \, \exp(i\mathbf{k}\cdot\mathbf{r})s(\mathbf{r}\,|\,\{c_p\}) \tag{7.76}$$

added to the uncorrelated noise described by the pdf (7.75); we assume that the values of $\sigma_{i,j}$ can be determined from measurements taken on ambient ocean images. The likelihood of the signature plus noise signal achieving a set of values $\{\tilde{y}_{i,j}\}$ is given by

$$P(\{\tilde{y}_{i,j}\,|\,\sigma_{i,j};\{\alpha\}\}) = \frac{1}{(2\pi)^N \prod_{i,j}\sigma^2_{i,j}} \times \exp\left(-\sum_{i,j}\frac{(\tilde{y}_{i,j}-\tilde{s}_{i,j}(\{\alpha\}))(\tilde{y}_{i,j}-\tilde{s}_{i,j}(\{\alpha\}))^*}{2\sigma^2_{i,j}}\right) \tag{7.77}$$

The likelihood ratio that tests between the hypotheses of signature plus noise and noise alone is thus given by

$$\Lambda = \frac{P(\{\tilde{y}_{i,j}\,|\,\sigma_{i,j};\{\alpha\}\})}{P(\{\tilde{y}_{i,j}\,|\,\sigma_{i,j}\})} \tag{7.78}$$

The behaviour of this is determined by that of the quantity

$$\Re\left(\sum_{i,j}(\tilde{y}_{i,j}\tilde{s}_{i,j}{}^{*}(\{\alpha\}))/2\sigma_{i,j}^{2}\right) \tag{7.79}$$

If the parameters $\{\alpha\}$ that determine the form of the large-scale feature are known *a priori* this quantity can be calculated from the data and used as the basis of a detection procedure. Should the $\{\alpha\}$ not be known however, they can be estimated from the data by finding those values that maximise $P(\{\tilde{x}_{i,j} \mid \sigma_{i,j}; \{\alpha\}\})$ and consequently the likelihood ratio. In this way a quantity defined by

$$\chi = \max_{\{\alpha\}} \Re\left(\sum_{i,j}(\tilde{y}_{i,j}\tilde{s}_{i,j}{}^{*}(\{\alpha\}))/2\sigma_{i,j}^{2}\right) \tag{7.80}$$

can be formed and used as the basis of our detection procedure. Image enhancement can then be achieved by reconstructing an image of the feature based on the parameters $\{\alpha\}$ that effect the maximisation of χ.

This processing is made much simpler by the complete absence of correlation between the complex Gaussian processes in each pixel in the transformed image. This in turn is a consequence of the assumed statistical stationarity of the pre-processed image. Should this image have spatially varying statistical properties, there can be significant inter-pixel correlation in the transformed image. If the form of this correlation is known, a modification of the matched filter through a 'whitening transformation' quite analogous to that discussed in Section 7.2. Thus, if the correlation in the Fourier transformed noise background is described by a covariance matrix $\mathbf{K}$ and we represent the known signature by the vector $\mathbf{s}$, we find that the quantity analogous to (7.81) is

$$\Re(\mathbf{y}^{\mathrm{H}} \cdot \mathbf{K}^{-1} \cdot \mathbf{s}) \tag{7.81}$$

In principle, the estimation of the unknown parameters occurring in this expression through a maximum search procedure may be possible; should the elements of $\mathbf{K}$ have to be estimated in this way the whole process will be prohibitively involved. An alternative approach to the problem of non-stationarity is to divide the pre-processed image into sub-images, each of whose statistical properties is effectively stationary within its limited spatial extent.

7.6 False alarm rates for matched filter processing

The matched filter technique described in the previous section provides us with the optimum method for the recovery of a known signal, corrupted by the addition of Gaussian noise [12], and as such, is very widely used in the analysis of time series and two dimensional images. In particular the presence or absence of the signal can be inferred by thresholding the output of the filter. Nonetheless, this output may exceed a prescribed threshold in the absence of the signal, and a false alarm be declared.

The probability of false alarm as a function of threshold is a useful measure of the effectiveness of the matched filter detection method. In the absence of the 'target' signal, the output of the matched filter is itself a correlated Gaussian process; the statistical properties of its local maxima can provide us with significant insight into the performance of the filter. For example, the classic work of Rice [13] can be applied directly to the analysis of time series. In the present context of large scale features in radar imagery, however, attention is focused on the two dimensional case for which the corresponding calculations are much less tractable. Longuet-Higgins [14] has discussed some aspects of this problem in detail; it is only recently that Oakley [15] has extended his analysis to provide an assessment of the performance of matched filter detection. In this, each local extremum exceeding a threshold is identified with a false alarm. In some circumstances it is more reasonable to threshold on the global maximum; this would be the case if one were trying to detect a single weak large-scale feature in the image. We will consider this problem, of the rate of false alarms triggered by the global maximum in the output of a matched filter, in more detail, following the treatment presented in Reference 16 quite closely.

7.6.1 A simple model for the global maximum single point statistics

To make significant progress towards its solution we adopt a simple and non-rigorous approach that combines elementary order statistics (discussed in Section A1.12 of Appendix 1) with the properties of the local maxima in the Gaussian process. This provides us with a simple expression for the probability that the global maximum of a one-dimensional (1D) correlated Gaussian random process of finite support exceeds a given threshold. Furthermore, this probability of false alarm turns out to be completely insensitive to the finer details of the correlation function of the process itself. Once the general principles of this approach have been established, it is relatively straightforward to generalise the analysis to a two-dimensional (2D) random field, generated as the output of a matched filter such as that discussed in Section 7.5.

Our discussion of the statistics of the global maximum of a correlated Gaussian random process is couched in terms of the probability of false alarm. The corresponding probability density function can be obtained from this by differentiation. This approach proves to be calculationally convenient and establishes direct contact with matched filter performance.

A simplified representation of this global maximum thresholding process and its false alarm characteristics can be constructed as follows. The normalised Gaussian random field representing the filter output in the absence of a target signal has the familiar pdf

$$P_0(x) = \frac{1}{\sqrt{2\pi}} \exp(-x^2/2) \tag{7.82}$$

As a consequence of the central limit theorem it is reasonable to adopt this process as a model for the filter output, even when it is applied to an image with non-Gaussian statistics. The output will have several local maxima, whose individual values have a pdf $\hat{P}(x)$ that is in general different from P_0. The global maximum is the largest of these local maxima and has a pdf $P_{\max}$ that we must determine. The probability of false alarm is given by

$$\mathrm{P}_{\mathrm{FA}}(X) = \int_X^{\infty} P_{\max}(x)\,dx \tag{7.83}$$

To model this false alarm rate we first consider N independent samples of the random variable; any one of these can take the maximum value, subject to the condition that all the other variables independently take values less than this maximum value. Thus we have

$$P_{\max}(x, N) = NF(x)^{N-1}P_0(x) \tag{7.84}$$

Here

$$F(x) = \int_{-\infty}^{x} P_0(x')\,dx' \tag{7.85}$$

is the cumulative distribution of the signal values. The result (7.84) is an example of an 'order statistics' calculation of the type discussed in Section A1.12 of Appendix 1. Within this simple model the false alarm rate for the matched filter detection is

$$P_{\mathrm{FA}}(X) = 1 - F(X)^N \tag{7.86}$$

To extend this independent sample analysis to correlated processes we note that one effect of correlation is to reduce the number of effectively independent samples in the image from N to some smaller N_{eff}. The correlation will also determine the form of $\hat{P}(x)$. Taking these effects into account we form an estimate of the pdf of the global maximum as

$$\hat{P}_{\max}(x, N_{\mathrm{eff}}) = N_{\mathrm{eff}}\hat{F}(x)^{N_{\mathrm{eff}}-1}\hat{P}(x) \tag{7.87}$$

with

$$\hat{F}(x) = \int_{-\infty}^{x} \hat{P}(x')\,dx' \tag{7.88}$$

from which the false alarm rate is calculated as

$$P_{\mathrm{FA}}(x) = 1 - \hat{F}(x)^{N_{\mathrm{eff}}} \tag{7.89}$$

The result (7.89) provides us with our general prescription for the calculation of the false alarm curve. To implement this we must calculate N_{eff} and $\hat{F}(x)$ for 1D time series and 2D spatial images.

7.6.2 *The global maximum of a 1D Gaussian process and the matched filter false alarm curve for a time series*

If we specialise first to the case of a time series, we can make consistent estimates of $\hat{P}(x)$ and N_{eff} for a record of a 1D Gaussian process representing the normalised filter output. This has a power spectrum $S(\omega)$ with moments

$$\int_{-\infty}^{\infty} d\omega\, S(\omega) = 1$$

$$\int_{-\infty}^{\infty} d\omega\, \omega^2 S(\omega) = m_2$$

$$\int_{-\infty}^{\infty} d\omega\, \omega^4 S(\omega) = m_4 \tag{7.90}$$

Rice [13] shows that the number of local maxima expected per unit time in the record is

$$N_1 = \frac{1}{2\pi}\sqrt{\alpha m_2}$$

$$m_4 = \alpha m_2^2 \tag{7.91}$$

Provided the process decorrelates significantly between maxima we can make the identification

$$N_{\text{eff}} = N_1 T \tag{7.92}$$

where T is the duration of the record. The pdf of x, the locally maximum value, is [13]

$$\hat{P}(x) = \frac{1}{\sqrt{2\pi}}\left[\varepsilon \exp(-x^2/2\varepsilon^2) + \sqrt{(1-\varepsilon^2)}x \exp(-x^2/2) \times \int_{-\infty}^{x\sqrt{1-\varepsilon^2}/\varepsilon} \exp(-x'^2/2)\, dx'\right]$$

$$\varepsilon^2 = 1 - \frac{1}{\alpha} \tag{7.93}$$

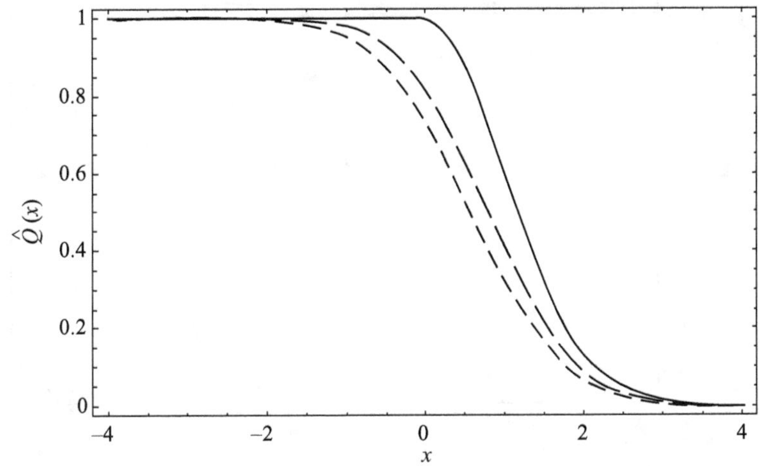

Figure 7.4 $\hat{Q}(x)$ *for* $\alpha = 1.01$ *(solid line), 2.5 (long dash) and 5.0 (short dash)*

Integration by parts shows that

$$\hat{Q}(x) = 1 - \hat{F}(x) = \int_x^\infty P(x')\,dx' = \frac{1}{\sqrt{2\pi}}\left[\int_{x/\varepsilon}^\infty \exp(-x'^2/2)\,dx' + \sqrt{(1-\varepsilon^2)}\exp(-x^2/2)\int_{-\infty}^{x\sqrt{1-\varepsilon^2}/\varepsilon} \exp(-x'^2/2)\,dx'\right] \tag{7.94}$$

So, in going from a noise process with significant structure on a whole range of the timescales ($\alpha \gg 1$) to the narrow band process ($\alpha \to 1(+)$) the maximum value pdf goes from a Gaussian to a Rayleigh form, and the false alarm curve is displaced to larger threshold values. This is illustrated in Figure 7.4, showing the variation of $\hat{Q}(x)$ with threshold for several values of α. However, N_{eff} decreases progressively as this transition is made so that, implicit in the general formula (7.89), there are two counteracting influences on the global maximum false alarm curve.

We now calculate the corresponding false alarm curves, keeping the value of m_2 constant, for a segment of record T such that $N_{\text{eff}} = 50\sqrt{\alpha}$. We see that the false alarm curve barely changes with α and what is in effect a single curve emerges as shown in Figure 7.5.

These calculations demonstrate that the false alarm curves, and thus the pdfs of the global maxima, are effectively independent of α. Consequently we can capture the behaviour of the whole family of false alarm curves by the simple expression

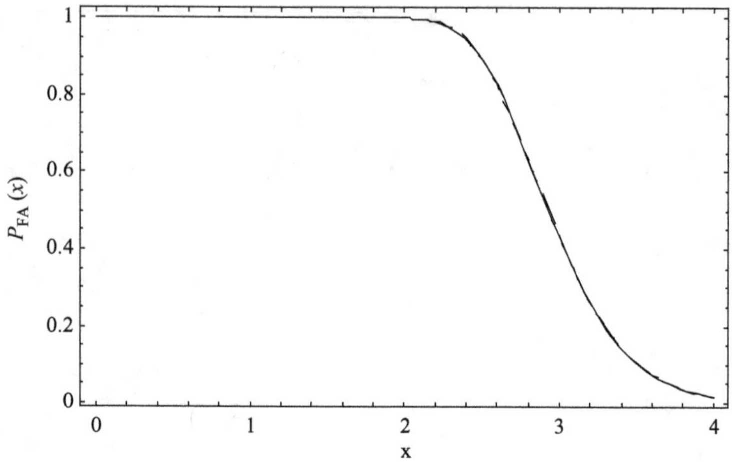

Figure 7.5 *False alarm curves for α = 1.01 (solid line), 2.5 (long dash), 5.0 (short dash) and constant m_2. The curves are barely distinguishable by eye*

$(\alpha = 1)$

$$P_{FA}(x) = 1 - (1 - \exp(-x^2/2))^{m_2^{1/2}T/2\pi} \tag{7.95}$$

The corresponding approximation to the pdf of the global maximum is

$$P_{max}(x) \approx \frac{\sqrt{m_2}}{2\pi}T(1 - \exp(-x^2/2))^{m_2^{1/2}T/2\pi-1}x\exp(-x^2/2) \tag{7.96}$$

To gain some insight into this finding we observe that, provided N_{eff} is reasonably large, the form of the false alarm curve is determined by the large argument behaviour

$$\begin{aligned}\hat{F}(x) &\sim 1 - \sqrt{\frac{m_2^2}{m_4}}\exp(-x^2/2)\\ &= 1 - \frac{\exp(-x^2/2)}{\sqrt{\alpha}}\end{aligned} \tag{7.97}$$

so that

$$P_{FA}(x) = 1 - (1 - \frac{1}{\sqrt{\alpha}}\exp(-x^2/2))^{(\sqrt{\alpha m_2}/2\pi)T} \tag{7.98}$$

This expression shows that, given a large N_{eff}, $P_{FA}(x)$ is effectively independent of α. Thus (7.98) can be recast in the form (7.95). Our analysis demonstrates explicitly how the effects of the fine structure of the time series, captured by the parameter α, cancel out in the construction of the false alarm curve and accounts quite straightforwardly for its remarkable robustness. Furthermore we note that, while the approximate pdf of the globally maximal value (7.96) is derived from the large x asymptotic result (7.97), it is valid for all values of x for which it is not vanishingly small. This 1D special case gives us a first indication that the statistics of the global maximum of a Gaussian process can be described by simple analytical expressions such as (7.95) and (7.96), displaying their dependence on T and m_2. The latter can be identified with the second derivative or curvature of the correlation function measured at zero time separation:

$$\frac{d^2\langle f(0)f(t)\rangle}{dt^2}\Big|_{t=0} = -m_2 \tag{7.99}$$

7.6.3 Extension to 2D matched filters

Equation (7.98) can be written in the slightly different form

$$P_{\mathrm{FA}}(x) = 1 - \left(1 - \frac{N(x)}{D_{\max}}\right)^{T D_{\max}} \tag{7.100}$$

Here $N(x)$ is the number of local maxima per unit area exceeding the threshold x and $D_{\max}$ is the total number of local maxima per unit time. Equation (7.94) shows that $N(x)$ is a complicated function of the second and fourth moments of the power spectrum of the matched filter output. The observed robustness of the false alarm curve is a consequence of cancellations and simplifications that occur when the number of local maxima in the image $T D_{\max}$ is reasonably large. Under these circumstances

$$\left(1 - \frac{N(x)}{D_{\max}}\right)^{T D_{\max}} \approx (1 - N(x))^{\mathrm{T}} \tag{7.101}$$

and the false alarm rate is independent of $D_{\max}$. Furthermore $(1 - (N(x)/D_{\max}))^{T D_{\max}}$ will be negligibly small, and the probability of false alarm will be effectively unity, unless $N(x)/D_{\max}$ is itself small. This will be the case when the threshold x is large; under these circumstances $N(x)$ reduces to a simple function of x alone, pre-multiplied by a root mean square frequency characteristic of the power spectrum of the matched filter output.

It is expected that corresponding simplifications will occur in an evaluation of the false alarm curve for a random field in two dimensions. Any calculation of the analogue of $D_{\max}$ is rendered redundant by the cancellation implicit in (7.101). (This is particularly fortunate as the evaluation of $D_{\max}$ in two dimensions is a difficult problem, solved by Longuett-Higgins [14], and discussed, from a different standpoint, in References 15 and 16.) The determination of the false alarm curve therefore reduces to the asymptotic evaluation of $N(x)$ in the high threshold limit. This calculation has been discussed in detail by Adler [17], who shows that

$$N(x) \sim \frac{\sqrt{\det \Lambda_2}}{(2\pi)^{3/2}} x \exp(-x^2/2) \tag{7.102}$$

where Λ_2 is the symmetric matrix of derivatives of the correlation function of the random field, evaluated at the origin. Substituting (7.102) into (7.101) gives

$$P_{FA}(x) \approx 1 - \left(1 - \frac{\sqrt{\det \Lambda_2}}{(2\pi)^{3/2}} x \exp(-x^2/2)\right)^{V_M}$$
$$\approx 1 - (1 - x \exp(-x^2/2))^{(\sqrt{\det \Lambda_2}/(2\pi)^{3/2})A} \quad (7.103)$$

where A is the area of the support of the random field. The corresponding approximation to the pdf of the global maximum of the pdf is

$$P_{max}(x) \approx \frac{\sqrt{\det \Lambda_2}}{(2\pi)^{3/2}} A(x^2 - 1)$$
$$\times \exp(-x^2/2)(1 - x \exp(-x^2/2))^{(\sqrt{\det \Lambda_2}/(2\pi)^{3/2})A-1} \quad (7.104)$$

We stress once more that, while our analysis is based on the high threshold limiting behaviour of the pdfs of the individual local maxima in the field, it yields a pdf of the global maximum over all values of x for which it takes effectively non-zero values.

These results have been derived on the basis of an approximate analysis. Consequently they should be verified in some independent way; in Reference 16 this is done by computer simulation. This demonstrates that our simple model successfully describes the single point statistics of the global maximum of a two dimensional Gaussian random field of reasonably large support.

We see from (7.103) that the probability of false alarm for a detection procedure based on the thresholding of the global maximum of the output of a two dimensional matched filter is relatively insensitive to the details of its correlation structure. Furthermore, as this structure is determined to a significant extent by the image preprocessing and the form of the matched filter, the false alarm curve obtained using a given filter will be effectively independent of other operating conditions. Consequently this detection procedure has an effectively 'universal' false alarm curve that can be understood in terms of the relatively simple concepts we have introduced here, and has been widely observed in practice. We should also note that, while we have carried through this argument in terms of a real Gaussian field, very similar results can be derived when the threshold is applied to the envelope of a complex Gaussian random field [18].

7.7 A compound model for correlated signals

So far our discussion has centred on the Gaussian model introduced in Section 7.1. In Chapter 4 we saw that the corresponding Gaussian model for the clutter intensity, introduced in Section 4.2, could be adapted to the non-Gaussian regime by allowing its local power x to vary. The gamma distribution provided a useful and realistic model of this variation, which led directly to the compound K description of sea clutter that has met with considerable success since its inception. In light of this, we might consider to what extent this compound model can be usefully extended to correlated signals.

Much as in our discussion of small target detection, we see that the role of the gamma distribution modulation of local power in the processing of large-scale image features is quite limited. The χ and phase increment based feature enhancement methods are applied locally to many small subsets of the overall image, within any one of these the approximation of Gaussian statistics on which these methods are based holds. Conversely, when large-scale feature detection methods such as matched filtering are applied to the image as a whole, any large-scale modulation in the power of the signals is suppressed by pre-processing whose intention is to reduce the ambient, featureless image to as good an approximation to a Gaussian noise process as is possible. Consequently the compound nature of the clutter process has relatively little impact on the image enhancement and detection techniques considered in this chapter.

Nonetheless, correlated processes have been discussed in terms of a compound pdf by several authors. Tough *et al.* [9], whose work was motivated by the analogy provided by the compound representation of the K distribution, allow the quantity ψ to be a gamma distributed random variable. Quite reasonably, this has no effect on the statistics of the phase increment variable ϕ, while the distribution of the amplitude ς has a pdf of the form

$$P(\varsigma) = \left(\frac{\nu}{\langle\psi\rangle}\right)^{\nu} \frac{1}{\Gamma(\nu)} \frac{\varsigma}{(1-k^2)} \int_0^{\infty} I_0\left(\frac{k\varsigma}{\psi(1-k^2)}\right) K_0\left(\frac{\varsigma}{\psi(1-k^2)}\right) \psi^{\nu-3} \times \exp(-\nu\psi/\langle\psi\rangle)\, d\psi \tag{7.105}$$

If it is assumed that ψ does not vary over the extent of the clique within which coherent averaging described in Section 7.4.3 is carried out then the pdf of the resultant phase increment (7.69) is unchanged, while that of the amplitude Ξ is modified in the same way as is (7.105). The consequences of this so-called product model are discussed in some detail in Reference 19.

To motivate another approach to the modelling of correlated clutter processes we now consider a n-dimensional multivariate Gaussian with the pdf

$$P_G(\mathbf{x}_{(n)}) = \frac{\exp(-(1/2s)\mathbf{x}_{(n)}^{\mathrm{T}} \cdot \mathbf{K}_{(n)}^{-1} \cdot \mathbf{x}_{(n)})}{(2\pi s)^{n/2}\sqrt{\det \mathbf{K}_{(n)}}} \tag{7.106}$$

Here the subscripts n emphasise the dimensionality of the process while the subscript G reminds us that the pdf is of Gaussian form. If we now integrate this pdf over one of the components x_1 of this multivariate Gaussian, we are left with the pdf of a $n-1$ dimensional multivariate Gaussian. This procedure, and its extension to the integration over an arbitrary subset of the components of $\mathbf{x}_{(n)}$, are discussed in

Section A1.6 of Appendix 1 (equations (A1.97–A1.105)), and entail no more than a succession of completions of the square in the exponent. The result of this integration over x_1 can be written as

$$\int dx_1 P_G(\mathbf{x}_{(n)}) = \frac{\exp(-(1/2s)\mathbf{x}^{\mathrm{T}}_{(n-1)} \cdot \mathbf{K}^{-1}_{(n-1)} \cdot \mathbf{x}_{(n-1)})}{(2\pi s)^{(n-1)/2}\sqrt{\det \mathbf{K}_{(n-1)}}} = P_G(\mathbf{x}_{(n-1)}) \tag{7.107}$$

where $\mathbf{K}_{(n-1)}$ is now the covariance matrix formed from the random variables $\mathbf{x}_{(n-1)} \equiv \{x_2, \ldots, x_n\}$. Is it possible, in some way, to extend this property of the multivariate Gaussian to a non-Gaussian multivariate pdf? If we now assign a gamma distribution to the quantity s (which we use here in preference to our more customary x, to avoid confusion with the components of the multivariate Gaussian $\mathbf{x}$) in (7.106), we see that pdf of the set of variables $\mathbf{x}_{(n)}$ becomes

$$\begin{aligned} P_{NG}(\mathbf{x}_{(n)}) &= \frac{b^{\nu}}{\Gamma(\nu)(2\pi)^{n/2}\sqrt{\det \mathbf{K}_{(n)}}} \int_0^{\infty} \exp(-bs)s^{\nu-1-n/2} \\ &\quad \times \exp\left(-\frac{1}{2s}\mathbf{x}^{\mathrm{T}}_{(n)} \cdot \mathbf{K}^{-1}_{(n)} \cdot \mathbf{x}_{(n)}\right) \\ &= \frac{b^{(n/4)+(\nu/2)}2^{1-(\nu/2)+(n/4)}\left(\mathbf{x}^{\mathrm{T}}_{(n)} \cdot \mathbf{K}^{-1}_{(n)} \cdot \mathbf{x}_{(n)}\right)^{(2\nu-n)/4}}{\Gamma(\nu)(2\pi)^{n/2}\sqrt{\det \mathbf{K}_{(n)}}} \\ &\quad \times K_{\nu-(n/2)}\left(\sqrt{2b\mathbf{x}^{\mathrm{T}}_{(n)} \cdot \mathbf{K}^{-1}_{(n)} \cdot \mathbf{x}_{(n)}}\right) \end{aligned} \tag{7.108}$$

Here the subscript *NG* emphasises that the pdf is now of non-Gaussian form; as usual K is a modified Bessel function. Nonetheless, we see from (7.107) that a multivariate pdf of this form does have an invariance property that can be written compactly as

$$\int dx_1 P_{NG}(\mathbf{x}_{(n)}) = P_{NG}(\mathbf{x}_{(n-1)}) \tag{7.109}$$

This property gives the name spherically invariant random process (SIRP) to a multivariate process with the pdf of the form (7.108). Considerable attention has been paid to these processes as a model for coherent and other correlated clutter processes, in part as a result of the interesting mathematical properties which motivated much of their development [20–21]. Here we merely stress that, when s is interpreted as a randomly varying local power, the analogy with the compound from of the K distribution is quite striking, and should be borne in mind when SIRPs are used in the analysis of clutter data.

References

1 E.T. Copson, '*An introduction to the theory of functions of a complex variable*', Clarendon Press, Oxford, 1970.

2 L.M. Novak, M.C. Burl and W.W. Irving, 'Optimal polarimetric processing for enhanced target detection', *IEEE Trans. Aerospace Electr. Syst.*, **29**, 234–244, 1993.

3 D. Blacknell and R.J.A. Tough, 'Clutter discrimination in polarimetric and interferometric synthetic aperture radar imagery', *J. Phys. D, Appl. Phys.*, **30**, 551–566, 1997.

4 M.G. Kendall, 'Advanced theory of statistics: Vol.1' Sects. 14.12–14.14, '*Advanced theory of statistics: vol.2*', Sects. 28.4–28.10, Charles Griffin, London, 1943, 1946.

5 J. Wishart, 'The generalised product-moment distribution in samples from a normal multi variate population', *Biometrica*, **20**, 32–43, 1928.

6 N.R. Goodman, 'Statistical analysis based on a certain multivariate complex Gaussian distribution (an introduction)', *Ann. Math. Stat.*, **34**, 152–177, 1963.

7 R.M. Goldstein and H.A. Zebker, 'Interferometric radar map of ocean currents', *Nature*, **328**, 707–709, 1987.

8 F.T. Ulaby and C. Elachi, 'Radar polarimetry for geoscience applications', Artech House, New York, 1990.

9 R.J.A. Tough, D. Blacknell, and S. Quegan, 'A statistical description of polarimetric and interferometric synthetic aperture radar data', *Proc. Roy. Soc.*, **A449**, 567–589, 1995.

10 B.C. Barber, 'The non-isotropic random walk' and 'The phase statistics of a multi channel radar interferometer', *Waves Random Media*, **3**, 243–256 and 257–266, 1993.

11 J.S. Lee, A.R. Miller, and K.W. Hoppel, 'Statistics of phase difference and product magnitude of multi-look processed Gaussian signals', *Waves Random Media*, **4**, 307–317, 1994.

12 D.O. North, 'An analysis of factors which determine signal/noise discrimination in pulsed-carrier systems', RCA Technical Report, PTR-6C, 1943.

13 S.O. Rice, 'Mathematical analysis of random noise', *Bell Syst. Tech. J.*, **23**, 282–332, 1944 and, **24**, 46–156, 1945.

14 M.S. Longuet-Higgins, 'Statistical analysis of a random moving surface', *Phil. Trans. Roy. Soc.*, **A249**, 321–388, 1957.

15 J.P. Oakley, 'Statistical properties of local extrema in two-dimensional Gaussian random fields', *IEEE Trans. Sig. Proc.*, **46**, 130–140, 1998.

16 R.D. Hill, R.J.A. Tough, and K.D. Ward, 'The distribution of the global maximum of a Gaussian random field and the performance of matched filter detectors', *IEE Proc. Vis. Image Signal Process*, **147**, 297–903, 2000.

17 R.J. Adler, '*The geometry of random fields*', John Wiley, New York, 1981.

18 R. D. Hill, R.J.A. Tough, and K.D. Ward, 'The false alarm curve for the envelope of a Gaussian random field', *Electron. Lett.*, **37**(4), 258–259, 2002.

19 B. Picinbono, 'Spherically invariant and compound Gaussian stochastic processes', *IEEE Trans. Information Theory*, 77–79, 1970.
20 E. Conte and M. Longo, 'Characterisation of radar clutter as a spherically invariant random process', *IEE Proceedings-F*, **134**(2), pp. 191–197, April 1987.
21 R.D. Hill, 'Along-track interferometric SAR measurements of the South Falls sandbank', submitted for publication.

Chapter 8

Radar detection performance calculations

8.1 Introduction

An important application of the understanding of sea clutter developed in Chapters 2–4 is the prediction of radar detection performance. While many of the phenomena associated with sea clutter are not understood in detail in physical terms, empirically derived statistical models do provide us with the means with which we might calculate detection performance and show how observed clutter characteristics can affect radar performance. This modelling of detection performance is the subject of the current chapter. Here we address the 'end-to-end' problem, which takes as its input radar system parameters and yields estimates of detection probability, but postpone the discussion of the maintenance of a constant false alarm rate (CFAR), and the associated losses, until the next chapter. The confirmation that real radar systems do in fact perform in a manner consistent with these calculations is presented in Chapter 10; thus this and the following two chapters now justify, in a practical context, much of the analysis presented in the rest of the book.

We consider just one system in detail; an airborne, scanning, sea surface surveillance radar, which operates at I-Band (9–10 GHz), is non-coherent, uses frequency agility and has pulse-to-pulse and scan-to-scan integration. The problem is that of detecting small, point sea-surface targets. Having worked through and understood this example, the reader should be well placed to carry out similar analyses of other types of radar. A few pointers towards how this may be done are given towards the end of the chapter.

We cannot emphasise too strongly that the effects of sea clutter can have a very significant impact on detection performance. If these effects are not modelled realistically in the design process, it is unlikely that a radar system will fully meet its operational requirements.

8.2 Radar equation and geometry

The starting point for most performance calculations is the Radar equation. In its simplest form it gives the received signal power, p_r, as

$$p_r = \frac{p_t G_t}{4\pi R_1^2} \times \frac{1}{L_a} \times \frac{\sigma}{4\pi R_2^2} \times A_e \tag{8.1}$$

We have followed Skolnik [1] in writing the right hand side as the product of factors representing the physical processes taking place. The first term is the power density (power per unit area) at a range R_1 from an antenna of gain G_t, which is transmitting a power of p_t. The second term is the two-way propagation loss. The third term accounts for the effective isotropic scattering of the power collected from an area σ, which is the radar cross section (RCS) of the scattering object. The product of the first three terms gives the power density of the scattered signal at a range R_2 from the scattering object. This power is collected by a receiving antenna, whose effective area is A_e. Using basic antenna theory, the gain of a lossless antenna is related to the antenna effective aperture A_e by the expression

$$G = \frac{4\pi}{\lambda^2} A_e \tag{8.2}$$

where λ is the radar wavelength. Thus, for a monostatic radar with a common transmitting and receiving antenna, we get

$$p_r = \frac{p_t G^2 \lambda^2 \sigma}{(4\pi)^3 R^4 L_a} \tag{8.3}$$

The propagation loss, L_a, accounts for absorption due to the atmosphere and any weather (e.g. rain, mist and clouds).[1] Whilst the average effect of atmospheric refraction is modelled by using a 'four-thirds' Earth radius in the curved Earth geometry calculations below, unusual refraction (such as ducting) and associated multipath propagation are included within the propagation loss.[2] For our purposes in this chapter we use a clear air, two-way propagation loss of 0.02 dB km^{-1} (appropriate to the I-band operation of a low or medium altitude radar detecting targets on the sea surface).

In order to calculate the detection performance, we need to evaluate the average signal, clutter and noise powers (p_s, p_c and p_n) at the output of the radar receiver

[1] The propagation losses to the target and clutter patch are taken to be the same because we are considering a target on the sea surface.

[2] Multipath interference between local reflections from the sea surface and the direct, plane wave illumination are included within the sea clutter and target fluctuation statistical models.

matched filter. This introduces some extra terms as follows:

$$p_s = \frac{p_t \mu_c G^2 \lambda^2 \sigma_t}{(4\pi)^3 R^4 L_a L_\mu} \tag{8.4}$$

$$p_c = \frac{p_t \mu_c G^2 \lambda^2 \sigma^0 A_c}{(4\pi)^3 R^4 L_a L_\mu} \tag{8.5}$$

$$p_n = kTBF_n \tag{8.6}$$

σ_t is the average target RCS, σ^0 is the normalised sea clutter RCS (i.e. the average RCS divided by the illuminated area) and A_c is the area of sea illuminated (to be consistent with our clear-air model for propagation, we have not included weather clutter, which would require an 'illuminated volume' term). μ_c is the pulse compression gain, which accounts for any coding present within the radar pulse to enhance the range resolution to less than the pulse length. The loss, L_μ, comprises all of the microwave losses and filter mismatch losses within the radar, antenna and radome. The noise power equation (8.6) consists of Boltzmann's constant ($k = 1.38 \times 10^{-23}$), the ambient reference temperature (T_0 = 290 K), the matched filter bandwidth (B) and the noise figure (F_n, which accounts for noise generated within the receiver above the theoretical thermal noise minimum).

Calculation of the illuminated area, A_c, in equation (8.5) takes account of the viewing geometry, the radar antenna beamshape and for pulsed radars, the range resolution of the pulse (determined, for a matched filter, by its bandwidth B). In most cases, as shown in Figure 8.1, the area of the clutter patch is defined by the azimuth beamwidth and the pulse length and is

$$A_c = \alpha \rho R \theta_{az} \sec(\phi_{gr}), \tag{8.7}$$

where θ_{az} is the antenna azimuth beamwidth and ϕ_{gr} is the local grazing angle. The range resolution, ρ, is related to the radar pulse bandwidth, B, by $\rho = c/2B$. The factor α accounts for the actual compressed pulse shape and the azimuth beamshape,

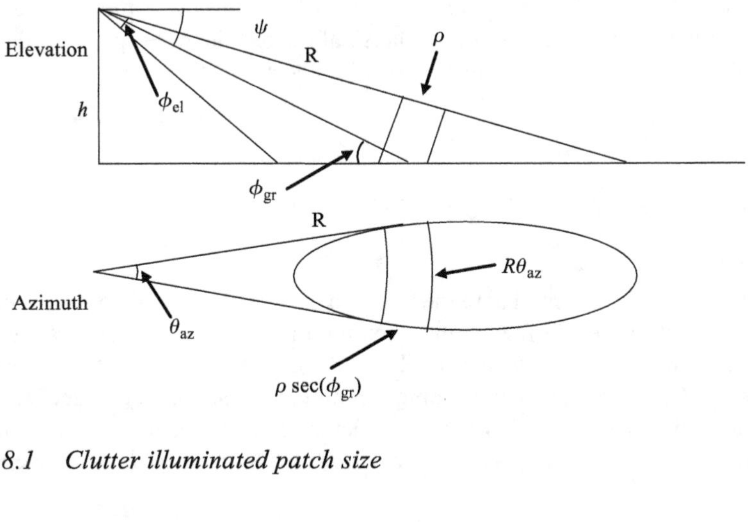

Figure 8.1 Clutter illuminated patch size

including the range and azimuth sidelobes ($\alpha = 1$ corresponds to a rectangular pulse and beamshape). Thus, for small grazing angles,

$$\alpha\rho\theta_{az} = \frac{1}{v^2(0)G^2(0,0)} \int_{-R/2}^{R/2} v^2(r)dr \int_{-\pi}^{\pi} G^2(\theta,0)d\theta \tag{8.8}$$

where $v(r)$ is the pulse shape amplitude as a function of range and $G(\theta,\phi)$ is the one-way antenna gain as a function of azimuth angle, θ, and elevation angle, ϕ, and elevation angle, π. For a narrow band radar, the illuminated area may be limited by the antenna elevation beamshape, rather than the compressed pulse length (see Figure 8.1). For a nominally conical beamshape, the antenna beam illuminates an ellipse on the ground of area:

$$A_c = \alpha\pi R^2 \frac{\sin^2\phi_{gr}\cos^2(\phi_{el}/2)}{(\cos^2(\phi_{el}/2) - \cos^2\phi_{gr})^{3/2}} \sin(\phi_{el}/2)\tan(\theta_{az}/2) \tag{8.9}$$

where ϕ_{el} is the elevation beamwidth, α now accounts for the azimuth and elevation beamshapes, and ϕ_{gr} is the grazing angle at the centre of the elevation beam. (Equation 8.9 assumes that the Earth is flat over the illuminated area and is only valid if $\phi_{gr} > \phi_{el}/2$. It should be used with care in (8.5) because the grazing angle, and therefore σ^0, varies over the illuminated patch). In the calculations presented in this section we use equation (8.7) and assume a rectangular pulse and a Gaussian azimuth beam shape (with θ_{az} defined to be the one-way 3 dB beamwidth). This gives a value for α of 0.753 from equation (8.8).

In order to apply the above equations we need to calculate some geometric values. Using a curved Earth geometry, the grazing angle, ϕ_{gr}, of the radar beam at the sea surface is

$$\phi_{gr} = \sin^{-1}\left(\frac{h}{R} + \frac{h^2}{2r_eR} - \frac{R}{2r_e}\right) \tag{8.10}$$

h is the height (altitude) of the radar, r_e is the Earth's radius (multiplied by $\frac{4}{3}$ to account for atmospheric refraction), and R is the slant range. The corresponding depression angle of the radar beam from the horizontal at the radar is

$$\psi = \sin^{-1}\left(\frac{R^2 + h^2 + 2r_eh}{2R(r_e + h)}\right) \tag{8.11}$$

and the slant range to the radar horizon is

$$R_{horizon} = (h^2 + 2hr_e)^{1/2} \tag{8.12}$$

In the calculations presented here the beam scans in azimuth and points the elevation beam-centre at the sea surface at the maximum range of the selected range scale or the horizon, whichever is the shorter. Thus, the depression angle of the beam-centre is calculated using the maximum range in equation (8.11). For targets and clutter at shorter range the depression angle is calculated (again using 8.11) and the offset from the beam-centre value is used to calculate the antenna gain. A 'cosec squared' elevation beam pattern may be used in order to maintain detection over a wide range interval. Thus the elevation pattern is

$$G(0,\phi) = G(0,0)2^{-(2\phi/\phi_{\rm el})^2} \quad \text{for } \phi \geq -\frac{\phi_{\rm el}}{2}$$
$$G(0,\phi) = G(0,0)\frac{\sin^2(\phi_{\rm el}/2)}{2\sin^2(\phi)} \quad \text{for } \phi < -\frac{\phi_{\rm el}}{2} \tag{8.13}$$

8.3 Normalised sea clutter RCS models

The value of σ^0, the normalised clutter RCS, in equation (8.5) depends on the sea conditions and the viewing geometry (i.e. the grazing angle and the aspect angle with respect to the wind direction). As discussed in Chapters 2 and 3, there is a significant body of experimental data relating to this variation, but physical modelling is at an early stage. Thus it is necessary to use a model derived from experimental data.

In the 1970s at the Royal Radar Establishment (RRE) in the United Kingdom [2], data from a number of sources (principally [3] and [4]) were used to produce an expression at 9–10 GHz radar frequency for σ^0, the normalised clutter RCS (expressed in dB), versus grazing angle, polarisation and sea state. The expression, which is for grazing angles of less than 10°, is:

$$\sigma^0 = a[s] + b[s]\log_{10}\left(\frac{180\phi_{\rm gr}}{\pi}\right) \quad \text{if } \phi_{\rm gr} \leq \frac{\pi}{180} = 1^\circ$$
$$= a[s] + c[s]\log_{10}\left(\frac{180\phi_{\rm gr}}{\pi}\right) \quad \text{if} \phi_{\rm gr} > \frac{\pi}{180} = 1^\circ \tag{8.14}$$

where s is the sea state ($s = 1$ to 6), $\phi_{\rm gr}$ is the grazing angle, and

$$a_{\rm HH}[s] = [-52, -46, -42, -39, -37, -35.5]$$
$$b_{\rm HH}[s] = [21, 17.5, 12.5, 10.5, 7, 3.5]$$
$$c_{\rm HH}[s] = [1.015, 3.39, 2.03, 1.35, 2.03, 2.37]$$

$$a_{\rm VV}[s] = [-51.5, -45.5, -41, -38.5, -36, -34.5]$$
$$b_{\rm VV}[s] = [15, 12, 11.5, 11, 9.5, 8]$$
$$c_{\rm VV}[s] = [8.2, 9.5, 8, 7.5, 7, 6.5] \tag{8.15}$$

Data were used from all wind directions, and so the model averages over all aspect angles. Figures 8.2 and 8.3 show equation (8.14) for sea states 1–5, and for Vertical and Horizontal polarisations. The model has been used extensively in the UK over the past 30 years for airborne radar performance specification and evaluation. Sometimes an additional cosine term was added to equation (8.14) to increase σ^0 by 3 dB in the upwind direction and to decrease it by 3 dB in the downwind direction.

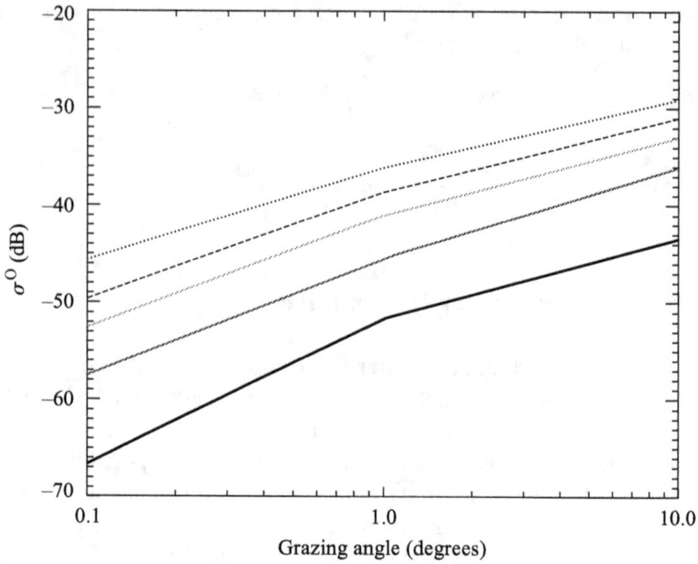

Figure 8.2 *VV sea clutter normalised RCS at 9.5 GHz as a function of grazing angle from the RRE model [2] for sea states 1 to 5 (black: 1, dark grey: 2, light grey: 3, dashed line: 4, dotted line: 5)*

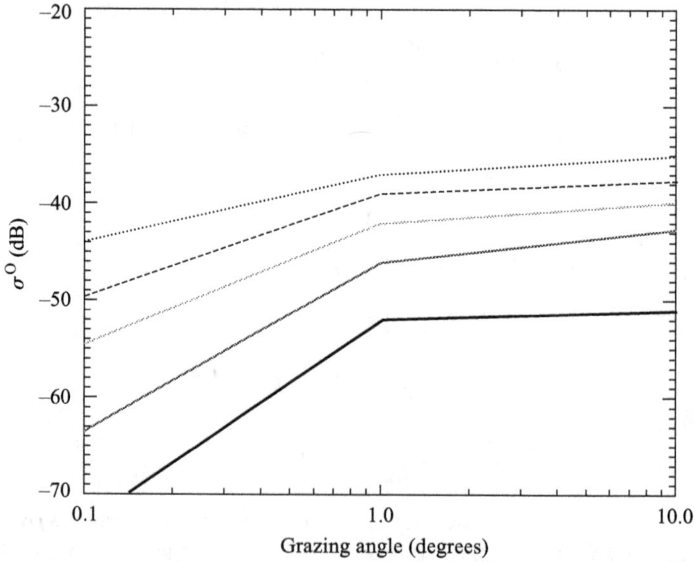

Figure 8.3 *HH sea clutter normalised RCS at 9.5 GHz as a function of grazing angle from the RRE model [2] for sea states 1 to 5 (black: 1, dark grey: 2, light grey: 3, dashed line: 4, dotted line: 5)*

Also in the 1970s workers at the Georgia Institute of Technology developed a σ^0 sea clutter model [5] (often now referred to as the GIT model). This model covers radar frequencies from 1 to 100 GHz (although the equations presented here relate only to 1–10 GHz) and is based on an underlying multipath model as well as other more general trends observed in experimental data sets. Thus, the wind velocity, U, is derived from the sea state, s, using

$$U = 3.16s^{0.8} \tag{8.16}$$

The 'average wave height', h_{av}, is given by

$$h_{\text{av}} = 0.00452U^{2.5} \tag{8.17}$$

A roughness parameter is defined as

$$\sigma_\phi = (14.4\lambda + 5.5)\frac{\phi_{\text{gr}} h_{\text{av}}}{\lambda} \tag{8.18}$$

where λ is the radar wavelength (in metres). This is used to evaluate a multipath or interference parameter

$$A_i = \frac{\sigma_\phi^4}{1 + \sigma_\phi^4} \tag{8.19}$$

The wind direction dependence is introduced through

$$A_u = \exp(0.2\cos(\theta_{\text{w}})(1 - 2.8\phi_{\text{gr}})(\lambda + 0.015)^{-0.4}) \tag{8.20}$$

where θ_{w} is the wind direction relative to the radar look direction and ϕ_{gr} is the grazing angle in radians. The variation on sea state is contained within the last factor

$$A_{\text{w}} = \left(\frac{1.94U}{1 + (U/15.4)}\right)^{1.1/(\lambda+0.015)^{0.4}} \tag{8.21}$$

These factors are put together to give σ^0 for the two polarisations as follows:

$$\sigma^0_{\text{HH}} = 65.91 + 10\log_{10}(\lambda\phi_{\text{gr}}^{0.4} A_i A_u A_{\text{w}}) \tag{8.22}$$

$$\begin{aligned}\sigma^0_{\text{VV}} = \sigma^0_{\text{HH}} &- 1.05\log_e(h_{\text{av}} + 0.015) + 1.09\log_e(\lambda)\\ &+ 1.27\log_e(\phi_{\text{gr}} + 0.0001) + 9.7\end{aligned} \tag{8.23}$$

Plots of σ^0 versus grazing angle and sea state are shown in Figures 8.4 (VV polarisation) and 8.5 (HH polarisation). Comparing these with Figures 8.2 and 8.3 for the RRE model shows that there are there are considerable differences between the models, which may seem surprising given that both models were derived from data at around the same period. However, it just shows how variable clutter results are for apparently similar sea conditions. A discussion of the differences is given in Reference 6.

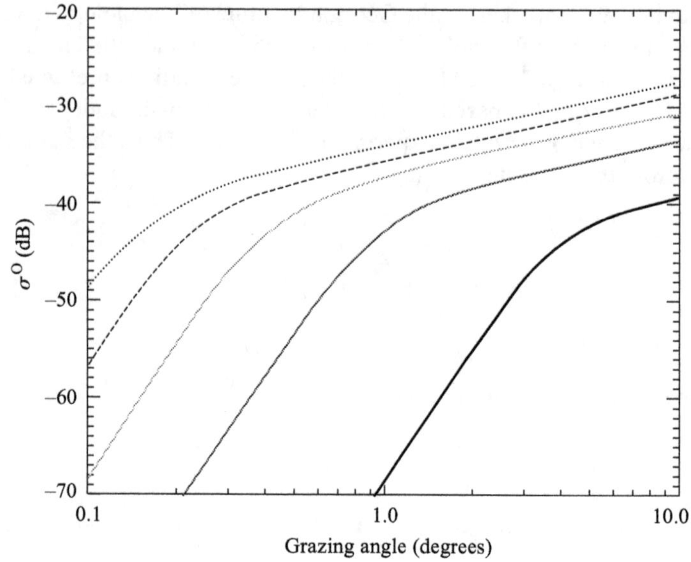

Figure 8.4 *VV clutter normalised RCS at 9.5 GHz as a function of grazing angle from the GIT model [5] for sea states 1 to 5 (black: 1, dark grey: 2, light grey: 3, dashed line: 4, dotted line: 5)*

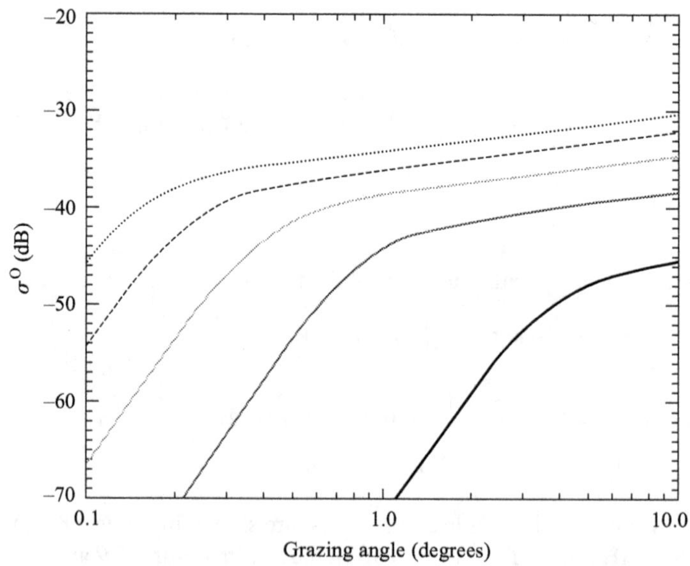

Figure 8.5 *HH clutter normalised RCS at 9.5 GHz as a function of grazing angle from the GIT model [5] for sea states 1 to 5 (black: 1, dark grey: 2, light grey: 3, dashed line: 4, dotted line: 5)*

8.4 Sea clutter fluctuations and false alarms

Having established the signal, clutter and noise powers from equations (8.4) to (8.6), we now need to calculate the detection performance. The first step in this process is the calculation of false alarm rate. The system to be modelled has pulse-to-pulse and scan-to-scan integration. We assume that a threshold is applied after pulse-to-pulse integration and there is a CFAR (constant false alarm rate) system to adjust the threshold for a specified false alarm rate after scan-to-scan integration. Details of CFAR processing are given in Chapter 9. There is a loss associated with this adaptive method of setting the threshold, known as a CFAR loss; we include it in the detection loss described in the next section, and model the threshold here as being fixed. After thresholding, binary 'm out of n', scan-to-scan integration is performed.

In order to obtain a realistic assessment of the false alarm rate, it is necessary to use a model for the fluctuations of the clutter signal that faithfully represents the characteristics described in Chapter 2. If the processing were just a simple threshold applied to a single pulse return, it would be sufficient to use any realistic model for the single-point distribution of sea clutter. A number of appropriate models of this type are described in Chapter 4; for example the Weibull, Log-normal and K distributions. In our case, however, we need to model the partial correlation of frequency agile sea clutter for pulse-to-pulse integration, and only the compound form of the K distribution allows this to be done. It turns out, for both coherent and non-coherent radar, that once the processing moves beyond a single-pulse fixed-threshold, the correlation properties are important and therefore should be modelled.

As described in Chapters 2 and 4, the compound form of the K distribution assumes that the clutter power, z, may be described by an exponential distribution of mean power, x,

$$P(z|x) = \frac{1}{x}\exp\left(-\frac{z}{x}\right) \tag{8.24}$$

where x itself fluctuates with a Gamma distribution

$$P(x) = \frac{b^{\nu}}{\Gamma(\nu)}x^{\nu-1}\exp(-bx) \tag{8.25}$$

Here, ν is the shape parameter, and $\langle x\rangle = \nu/b$ is the sea clutter mean power (and is equal to p_c from equation (8.5)). The model has been developed through the analysis of experimental data, which has led to an empirical model at I-band (9–10 GHz) for the dependence of ν on radar, environmental and geometric parameters:

$$\log_{10}(\nu) = \frac{2}{3}\log_{10}(\phi_{gr}^{o}) + \frac{5}{8}\log_{10}(A_c) - k_{pol} - \frac{\cos(2\theta_{sw})}{3} \tag{8.26}$$

where

ϕ_{gr}^{o} is the grazing angle in degrees
A_c is the radar resolved area
k_{pol} is a polarisation dependent parameter (1.39 for VV and 2.09 for HH)

θ_{sw} is the aspect angle with respect to the swell direction.
(The last term is omitted if there is no swell)

As discussed in Chapter 4 and elsewhere, the importance of separating the clutter power z into the two components ($z|x$, the speckle, and x, the modulation) lies in their different correlation properties. Speckle decorrelates in a few milliseconds at I-band, and samples separated spatially by more than the radar resolution are independent. The modulation carries all the correlation associated with the sea wave structure, and can therefore be correlated for over a second and over distances of many tens of metres. Furthermore, the speckle tends to be decorrelated by radar frequency agility whilst the modulation is unaffected. Thus for a scanning radar, with pulse to pulse frequency agility, x remains approximately constant over a beam dwell time while independent samples of $z|x$ are obtained from pulse to pulse.

Adding noise to the clutter is performed as described in Chapter 4. If we assume that frequency agility provides independent samples of speckle, noise is added by offsetting x by the noise power p_n in equation (8.24),

$$P(z|x) = \frac{1}{x + p_n} \exp\left(-\frac{z}{x + p_n}\right) \tag{8.27}$$

We assume that the radar has a 'square law detector'. Therefore, writing the sum of N pulses as

$$\mu = \sum_{i=1}^{N} z_i \tag{8.28}$$

we get the pdf of $\mu|x$ to be

$$P(\mu|x) = \frac{\mu^{N-1}}{(x + p_n)^N (N-1)!} \exp\left(\frac{-\mu}{x + p_n}\right) \tag{8.29}$$

The probability of false alarm, given x, for a threshold Y is

$$P_{FA}(Y|x) = \int_{Y}^{\infty} \frac{\mu^{N-1}}{(x + p_n)^N (N-1)!} \exp\left(\frac{-\mu}{x + p_n}\right) d\mu = \frac{\Gamma(N, Y/(x + p_n))}{(N-1)!} \tag{8.30}$$

where $\Gamma(\bullet, \bullet)$ is the incomplete Gamma function. The overall probability of false alarm after pulse-to-pulse integration is therefore

$$P_{FA}(Y) = \frac{1}{(N-1)!} \int_{0}^{\infty} \Gamma\left(N, \frac{Y}{x + p_n}\right) P(x) dx \tag{8.31}$$

where $P(x)$ is given in equation (8.25). As shown in Chapter 4, the clutter pdf, $P(z)$ (which is derived by the combination of equations (8.24) and (8.25) and integration over x), may be expressed in terms of the K Bessel function – hence the name 'K distribution'. The same is true of the probability of false alarm, P_{FA}, in equation (8.31).

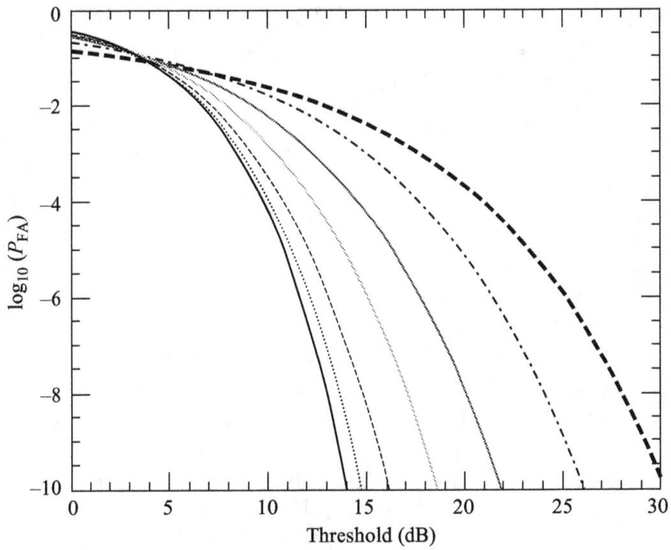

Figure 8.6 *Probability of false alarm (log to the base 10) versus threshold (in dB relative to mean clutter power) for a single pulse of K-distributed clutter. The lines correspond to difference values of shape parameter ν (black dashed line: 100, dotted line: 30, dashed line: 10, light grey: 3, dark grey: 1, dashed dotted line: 0.3, black: 0.1)*

However, in computational terms it is often found to be convenient to integrate over the Gamma distribution of x numerically.

The figures on this and subsequent pages show the probability of false alarm versus threshold, calculated using the equations above. Figure 8.6 is for a single pulse of K-distributed clutter with no (radar system) noise, i.e. $p_n = 0$. The different curves are for different degrees of 'spikiness' or shape parameter, ν. The curve on the left is for $\nu = 100$, which gives results close to the limit of infinite ν. In this limit, the Gamma distribution of equation (8.25) approaches a Delta function and the K distribution is noise-like (i.e. it approaches the negative exponential distribution of equation (8.24) with x being constant). The other curves to the right in Figure 8.6 are for decreasing values of ν, and thus represent 'spikier' clutter. As ν decreases, the threshold required for any particular P_{FA} (below about 10^{-1}) increases. This means that the 'spikier' clutter tends to obscure targets and thus tends to reduce detection performance.

Figure 8.7 shows P_{FA} for the sum of 10 pulses of K-distributed clutter ($N = 10$ in equation (8.28)). Here we assume that the speckle component of clutter, i.e. $z|x$, is independent from pulse to pulse (this may be due to the use of frequency agility or the time delay between pulses), and that the modulation, x, is completely correlated. In this case equation (8.31) applies, with $p_n = 0$. The results show that the effect of pulse-to-pulse integration is to move the curves to the left and to make them somewhat steeper. Thus, P_{FA} at a given threshold is reduced, by the averaging of the speckle

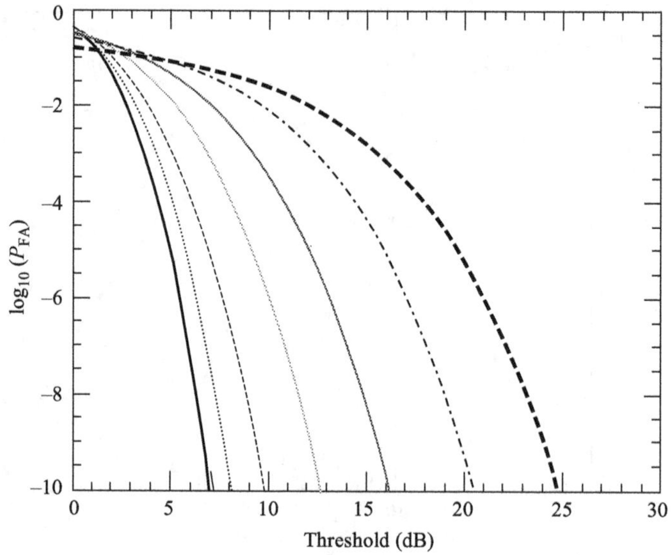

Figure 8.7 Probability of false alarm (log to the base 10) versus threshold (in dB relative to mean clutter power) for the sum of 10 pulses of K-distributed clutter. The lines correspond to difference values of shape parameter ν (black dashed line: 100, dotted line: 30, dashed line: 10, light grey: 3, dark grey: 1, dashed dotted line: 0.3, black: 0.1)

fluctuations. However, the spacing of the curves remains approximately the same for 1 and 10 pulses. Hence the degradation in performance due to the 'spikiness' of clutter is largely unaffected.

Figure 8.8 shows the effect of adding noise to the clutter. The conditions are the same as Figure 8.7, except that noise of equal power to the clutter has been added. The effect of the addition is to make the combination of noise and clutter less spiky than the clutter alone. As the threshold is normalised by the total interference (i.e. noise plus clutter), the effect of the decreased spikiness is to decrease the false alarm rate for a given threshold. The shape of the curves is nonetheless very similar to the case of clutter without noise, and this has led to the concept of an 'effective ν' parameter for clutter plus noise (Section 9.3.3.1 and Reference 7). This approach provides approximate results without the need to calculate results with clutter plus noise explicitly.

As described earlier, Figures 8.7 and 8.8 (and the formulae derived above) assume that the speckle is completely decorrelated from pulse-to-pulse. This is not always the case for a number of reasons, for instance:

- The radar uses fixed frequency and the time between pulses is not sufficient for temporal decorrelation.
- The agile frequencies are not spaced apart by at least the pulse bandwidth.
- The clutter RCS is not uniformly distributed across the radar resolution range cell.

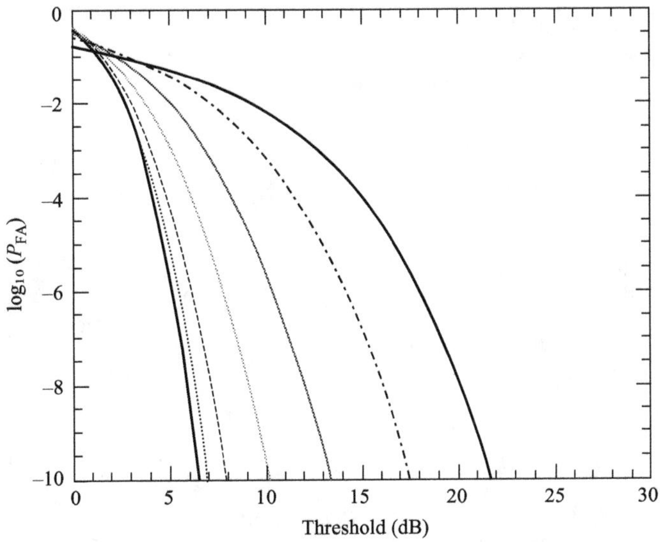

Figure 8.8 Probability of false alarm (log to the base 10) versus threshold (in dB relative to mean clutter plus noise power) for the sum of 10 pulses of K-distributed clutter plus noise, with a clutter to noise ratio of 0 dB. The lines correspond to difference values of shape parameter ν (black dashed line: 100, dotted line: 30, dashed line: 10, light grey: 3, dark grey: 1, dashed dotted line: 0.3, black: 0.1)

In these cases there is pulse-to-pulse correlation of clutter speckle, and the effective number of samples of speckle is less than that of noise. In order to calculate P_{FA}, it is necessary to include clutter within the target distribution. The means to do this are described in the next section on probability of detection. In the mean time, we show results of the effect in Figure 8.9, where P_{FA} is plotted versus threshold for the $\nu = 0.1$ curve of Figure 8.8, but with the number of independent speckle samples varying from 1 to 10. The clutter to noise ratio is 0 dB. As the number of speckle samples decreases, the false alarm rate increases; but only for threshold values that give a P_{FA} of less than 10^{-2}.

Another mechanism that causes the clutter speckle to be correlated is the presence of discrete spikes. This is discussed in Chapters 2 and 4, where the KA model is used to describe the effect. The clutter is modelled as a mixture of two distributions: the K distribution plus a spike (with P_{FA} calculated using a Swerling type 1 target model for the spike); and a K distribution alone. Results are shown in Figure 8.10, where it can be seen (for the KA parameters chosen) that the discrete spikes produce a well defined 'kink' in the curve for large values of ν, but have less effect for low values of ν.

The results above all relate to a single radar azimuth scan, which occurs over a time that is short compared with the correlation time of the clutter modulation. Many radar systems perform scan-to-scan integration to improve detection performance. Often the time between scans is long enough for independent samples of modulation

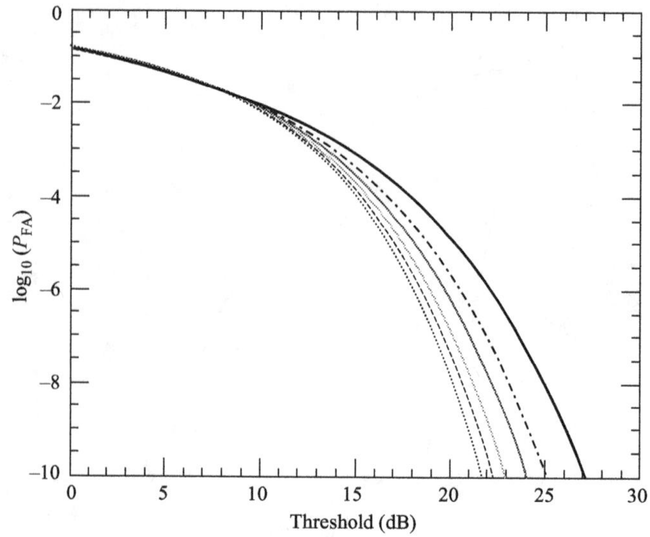

Figure 8.9 Probability of false alarm (log to the base 10) versus threshold (in dB relative to mean clutter plus noise power) for the sum of 10 pulses of K-distributed clutter plus noise, with a clutter to noise ratio of 0 dB and a ν value of 0.1. The lines correspond to difference numbers of independent samples of clutter speckle (black: 1, dashed dotted line:2, dark grey: 3, light grey: 5, dashed line: 7, dotted line: 10)

to be obtained from scan-to-scan. In these cases and for 'm out of n' scan-to-scan integration the probability of false alarm becomes[3]

$$P_{FA}^{(s-s)} = \sum_{i=m}^{i=n} \frac{n!}{(n-i)!i!} P_{FA}^{i}(Y)(1 - P_{FA}(Y))^{n-i} \tag{8.32}$$

As discussed above, the radar is assumed to have a CFAR and a specified output false alarm time, τ_{FA}. If we define the false alarm time to be the reciprocal of the average number of false alarms per second, we obtain

$$\tau_{FA} = \frac{1}{P_{FA}^{(s-s)} f_{opp}} \tag{8.33}$$

where f_{opp} is the number of independent opportunities for false alarms.[4] Furthermore, if the number of pulses integrated (N) covers a beamwidth of the radar scan, f_{opp} is

[3] Often targets are able to move through a number of resolution cells during scan-to-scan integration. If this is the case it is necessary either to collapse resolution cells together before the scan-to-scan integration, or to search through all of the possible target tracks in the scan-to-scan process. In some situations both options may be used together. Whatever solution is chosen, it will alter equation (8.32).

[4] This analysis is somewhat idealised because it does not take account of the duration of false alarms and the dependency of this on threshold. A more detailed account is given in Section 10.3.3.

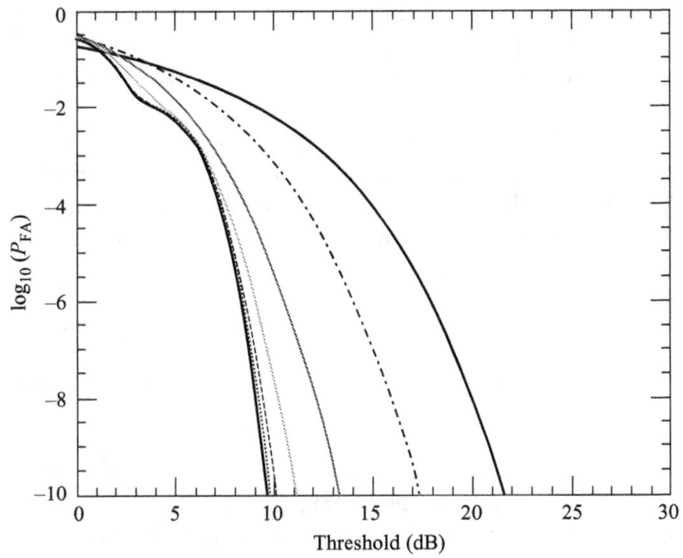

Figure 8.10 *Probability of false alarm (log to the base 10) versus threshold (in dB relative to mean clutter plus noise power) for the sum of 10 pulses of KA distributed clutter plus noise, with a clutter to noise ratio of 0 dB. The KA spike parameters (see Section 4.8) are $\bar{N} = 0.01$ and $\rho = 5$. The lines correspond to different values of K shape parameter ν (black: 100, dotted line: 30, dashed line: 10, light grey: 3, dark grey: 1, dashed dotted line: 0.3, black: 0.1)*

given by

$$f_{opp} = \frac{R_{scale}}{\rho} \frac{f_{prf}}{N} \tag{8.34}$$

where R_{scale} is the instrumented range interval, ρ is the range resolution (as used in equation (8.7)), and f_{prf} is the radar pulse repetition interval. Equations (8.33) and (8.34) are used to derive the output false alarm rate, $P_{FA}^{(s-s)}$, from τ_{FA} and then equations (8.32) and (8.31) are inverted numerically to give the threshold Y. This threshold is now used in the detection calculations.

8.5 Target RCS models and detection probability

Radar targets are often complex objects, which produce a wide variety of reflections. It is generally not practical to model the detailed EM scattering properties of the targets and so statistical methods are often employed. The most widely used statistical target models were introduced by Swerling [8], who along with Marcum [9], provided methods to calculate detection performance in noise. Later it became clear that the constant target model (often referred to as Swerling 0) and the four Swerling models

(1–4) were not sufficient to describe the range of target behaviours observed in practice. Thus Swerling [10] generalised his models to the 'Generalised Chi-Squared' or 'Gamma' distribution model. Shnidman [11] has derived numerical schemes to evaluate the detection performance of Gamma distributed targets in noise.

We are interested in the detection of targets in K-distributed clutter and noise. For the case where the speckle has identical correlation properties to the noise, this may be calculated from detection results in noise by adding the speckle to noise and integrating over the Gamma distribution for the clutter modulation. Thus (by analogy with equations (8.30) and (8.31) for P_{FA}),

$$P_{\mathrm{D}}(Y) = \int_0^\infty P_{\mathrm{D}}(Y|x)P(x)dx \tag{8.35}$$

where $P_{\mathrm{D}}(Y|x)$ is derived using the methods for detection in noise, and $P(x)$ is given by equation (8.25). For a single pulse from a target of amplitude A, the pdf of the received intensity, $P(z|A, x)$, is a Rice distribution [9] as follows

$$P(z|A,x) = \frac{1}{x+p_{\mathrm{n}}}\exp\left(-\frac{z+A^2}{x+p_{\mathrm{n}}}\right) I_0\left(\frac{2A\sqrt{z}}{x+p_{\mathrm{n}}}\right) \tag{8.36}$$

and $P_{\mathrm{D}}(Y|x)$ for equation (8.35) is given by

$$P_{\mathrm{D}}(Y|x) = \int_Y^\infty \frac{1}{x+p_{\mathrm{n}}}\exp\left(-\frac{z+A^2}{x+p_{\mathrm{n}}}\right) I_0\left(\frac{2A\sqrt{z}}{x+p_{\mathrm{n}}}\right)dz \tag{8.37}$$

The numerical evaluation of equation (8.37) can be simplified by approximating equation (8.36) by a Gamma distribution; this is achieved by matching the first two moments. Pulse-to-pulse integration is then derived from the convolution properties of the Gamma distribution, and target fluctuations are incorporated by numerical integration over the target pdf. The details and some results from this approach are given in References 7, 12 and 13. A further extension to the case of fixed frequency (where the clutter speckle is constant from pulse-to-pulse) is achieved by integrating numerically over the distribution of target plus constant speckle; results are given in Reference 14.

An alternative method for the evaluation of $P_{\mathrm{D}}(Y|x)$ is to extend the method adopted by Shnidman in References 11 and 15. This is based on the probability density function (pdf) of the sum of N square law detected radar returns from a target in noise and speckle (i.e. the N fold convolution of equation (8.36)), which is a multi-look Rice distribution [9]

$$P(\mu|s,N) = \left(\frac{\mu}{s}\right)^{(N-1)/2} e^{-(\mu+s)} I_{N-1}(2\sqrt{\mu s}) \tag{8.38}$$

where μ is the sum of the N pulse returns, z, normalised by the noise and local speckle power

$$\mu = \frac{1}{x + p_{\mathrm{n}}} \sum_{i=1}^{N} z_i \tag{8.39}$$

and s is the sum of the target powers, A_i^2, from the N pulses,[5] again normalised by the noise and local speckle power

$$s = \frac{1}{x + p_{\mathrm{n}}} \sum_{i=1}^{N} A_i^2 \tag{8.40}$$

$I_{\mathrm{n}}(.)$ is the modified Bessel function of the first kind of order n. A remarkable feature of equation (8.38) is that the pdf of the sum of the N normalised pulse returns, μ, depends only upon the sum, s, of the normalised target powers from all of the pulses. Therefore, in order to calculate P_{D}, it is only necessary to know the distribution of s, rather than the distributions of the individual A_i^2. This means that scan-to-scan and pulse-to-pulse target fluctuations (and all things in-between) can be evaluated using the same equation. By adopting the Gamma distribution for s, all of the original Marcum and Swerling models become special cases of this family. Thus

$$P(s|S,k) = \frac{s^{k-1}}{\Gamma(k)} \left(\frac{k}{S}\right)^k e^{-ks/S} \tag{8.41}$$

where k is the shape parameter, and S is given by

$$S = \frac{N\langle A^2 \rangle}{x + p_{\mathrm{n}}} \tag{8.42}$$

The following values of k deliver the standard Swerling models:

- Swerling 1: $k = 1$
- Swerling 2: $k = N$
- Swerling 3: $k = 2$
- Swerling 4: $k = 2N$

It is possible to extend this approach to situations where the effective number of independent clutter speckle samples, L, is less than the number of pulses, N. This is achieved by setting the speckle power, x, to zero in equations (8.39), (8.40) and (8.42), and by incorporating the clutter speckle with the target. For an individual pulse return, the sum of the target and clutter speckle is a Rice distribution. We can therefore represent[6] the pdf of the sum, β, of the target-plus-speckle returns as an L pulse Rice

[5] Note that the target power can vary from pulse-to-pulse.

[6] This is an approximation taken in the context of the approximate concept of 'effective number of speckle samples'.

distribution

$$P(\beta|s,\alpha,L) = \frac{1}{\alpha}\left(\frac{\beta}{s}\right)^{(L-1)/2} \exp\left(-\frac{s+\beta}{\alpha}\right) I_{L-1}\left(\frac{2\sqrt{s\beta}}{\alpha}\right) \tag{8.43}$$

where α in equation (8.43) is related to the local clutter speckle power x by

$$\alpha = \frac{xN}{Lp_\text{n}} \tag{8.44}$$

s and S are now given by

$$s = \frac{1}{p_\text{n}}\sum_{i=1}^{N} A_i^2, \qquad S = \frac{N\langle A^2\rangle}{p_\text{n}} \tag{8.45}$$

and the pdf of the sum of target, speckle and noise is now given by (cf. (8.38))

$$P(\mu|\beta,N) = \left(\frac{\mu}{\beta}\right)^{(N-1)/2} e^{-(\mu+\beta)} I_{N-1}(2\sqrt{\mu\beta}) \tag{8.46}$$

The efficient numerical evaluation of $P_\text{D}(Y|x)$ from equations (8.46), (8.43) and (8.41) may seem to be an impossible task. It can, however, be simplified by expanding the Bessel function in equation (8.46) in terms of its argument, and expressing $P_\text{D}(Y|x)$ as

$$\begin{aligned} P_\text{D}(Y|x) &= \int_Y^\infty P(\mu|\beta;N)P(\beta|s;\alpha,L)P(s|S;k)d\beta\, ds\, d\mu \\ &= \sum_{m=0}^{N-1} e^{-Y}\frac{Y^m}{m!} + \sum_{m=N}^{\infty} e^{-Y}\frac{Y^m}{m!}\left(1 - \sum_{i=0}^{m-N}\left\langle e^{-\beta}\frac{\beta^i}{i!}\right\rangle\right) \end{aligned} \tag{8.47}$$

where

$$\left\langle \exp(-\beta)\frac{\beta^i}{i!}\right\rangle = \int \exp(-\beta)\frac{\beta^i}{i!} P(\beta|s;\alpha,L)P(s|S,k)\, d\beta\, ds \tag{8.48}$$

A similar expansion of the Bessel function in equation (8.43) leads, after some manipulation, to

$$\begin{aligned} \left\langle \exp(-\beta)\frac{\beta^i}{i!}\right\rangle &= \frac{(L-1+i)!}{i!(L-1)!}\frac{\alpha^i}{(\alpha+1)^{L+i}}\left(\frac{k(\alpha+1)}{k(\alpha+1)+S}\right)^k \\ &\quad \times \Psi\left(i, L, k, \frac{S}{\alpha(k(\alpha+1)+S)}\right) \end{aligned} \tag{8.49}$$

where $\Psi(i, L, k, z)$ is given by the recurrence relation

$$
\begin{aligned}
&(L+i)\Psi(i+1, L, k, z) \\
&\quad = (2i+L+z(i+k))\Psi(i, L, k, z) - i(z+1)\Psi(i-1, L, k, z) \\
&\Psi(0, L, k, z) = 1, \quad \Psi(1, L, k, z) = 1+\frac{kz}{L}
\end{aligned}
\tag{8.50}
$$

In the limit of k approaching infinity, the target is non-fluctuating. In this limit a more useful expression for (8.48) is

$$
\left\langle \exp(-\beta)\frac{\beta^i}{i!} \right\rangle = \frac{(L-1+i)!}{i!(L-1)!}\frac{\alpha^i}{(\alpha+1)^{L+i}} \exp\left(-\frac{S}{\alpha+1}\right)\Phi\left(i, L, \frac{S}{\alpha(\alpha+1)}\right)
\tag{8.51}
$$

where $\Phi(i, L, z)$ is given by the recurrence relation

$$
\begin{aligned}
(L+i)\Phi(i+1, L, z) &= (2i+L+z)\Phi(i, L, z) - i\Phi(i-1, L, z) \\
\Phi(0, L, z) = 1, &\qquad \Phi(1, L, z) = 1+\frac{z}{L}
\end{aligned}
\tag{8.52}
$$

The overall probability of detection can now be evaluated by substituting equation (8.48) into (8.35); the integration over the gamma distribution for the local speckle power, x, is then carried out numerically. It has been found that this technique is very reliable and that its accuracy is easy to control[7]. Thus special cases do not require special formulae as they can all be evaluated straightforwardly using this one method; for example, in order to calculate the probability of false alarm for Figure 8.9 above, the signal S is set to zero.

Example detection results using the method described above are presented in Figures 8.11–8.18. Figure 8.11 shows the probability of detection versus signal to interference ratio (SIR, signal to clutter-plus-noise ratio) for K-distributed clutter with $\nu = 10$ and a non-fluctuating target. The two graphs compare detection from a single pulse and detection using the sum of ten pulses; the integration improvement is approximately 6 dB, but depends in detail on P_D and P_{FA}. Figure 8.12 shows similar graphs with the clutter $\nu = 0.1$. The performance in this 'spikier' clutter is worse than for Figure 8.11, which is expected from the threshold results in Figures 8.6 and 8.7 above. The other significant effect of the 'spikiness' is the steepening of the P_D versus SCR curves. This may be understood from the observation that, in this type of clutter, the modulation value is usually much lower than the threshold, which has to be high to limit the false alarms caused by the relatively rare, large excursions

[7] A bound may be put on the error incurred by stopping the summation (8.47) at $m = m'$, using the results $\sum_{m=0}^{\infty} \exp(-Y)Y^m/m! = 1$ and $\sum_{m=0}^{\infty} \langle \exp(-\beta)\beta^m/m! \rangle = 1$. Thus, $P_D - P_D(m') < 1 - \sum_{m=0}^{m'} \exp(-Y)Y^m/m!$ if $m < N$, and $P_D - P_D(m') < (1 - \sum_{m=0}^{m'} \exp(-Y)Y^m/m!)(1 - \sum_{m=0}^{m'-N} \langle \exp(-\beta)\beta^m/m! \rangle)$ if $m \geq N$. All of the terms needed to evaluate the accuracy are calculated as part of the expression for $P_D(m')$.

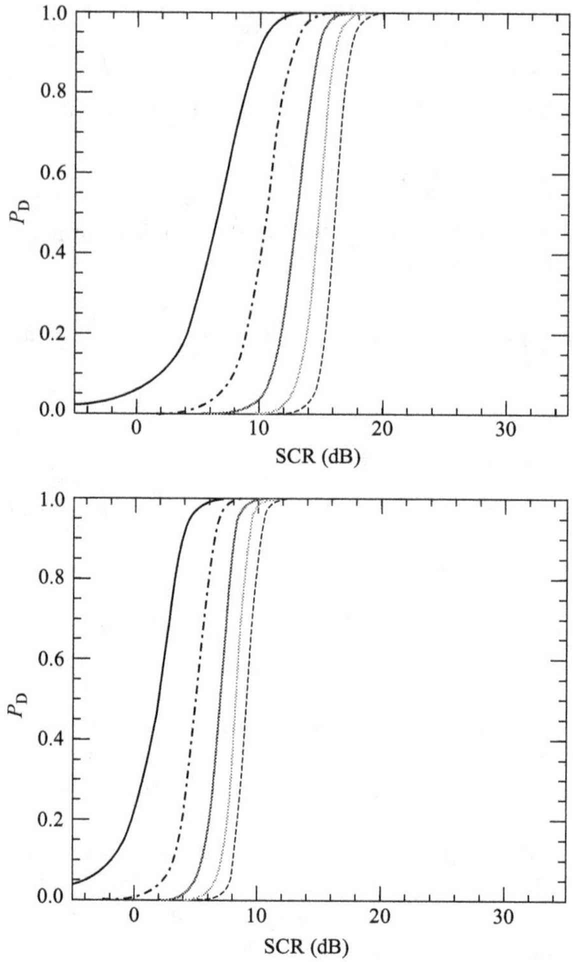

Figure 8.11 *Probability of detection versus signal to clutter level (in dB) for a non-fluctuating target in K-distributed clutter with $\nu = 10$. The top graph is a single pulse and the lower graph is the non-coherent sum of 10 pulses. The lines correspond to different false alarm probabilities (black: 10^{-2}, dashed dotted line: 10^{-4}, dark grey: 10^{-6}, light grey: 10^{-8}, dashed line: 10^{-10})*

(i.e. the spikes). Thus P_D for a target below the threshold is very low, P_D for a target above the threshold is high, and there is an abrupt transition between the two states.

Figures 8.13 and 8.14 show the results for the same settings as Figures 8.11 and 8.12, except the target is fluctuating as a Swerling type 1, i.e with $k = 1$ in equation (8.41). The target fluctuations cause P_D to be averaged over a range of values of signal to clutter ratio, and so the averaged curves are less steep than those from a non-fluctuating target.

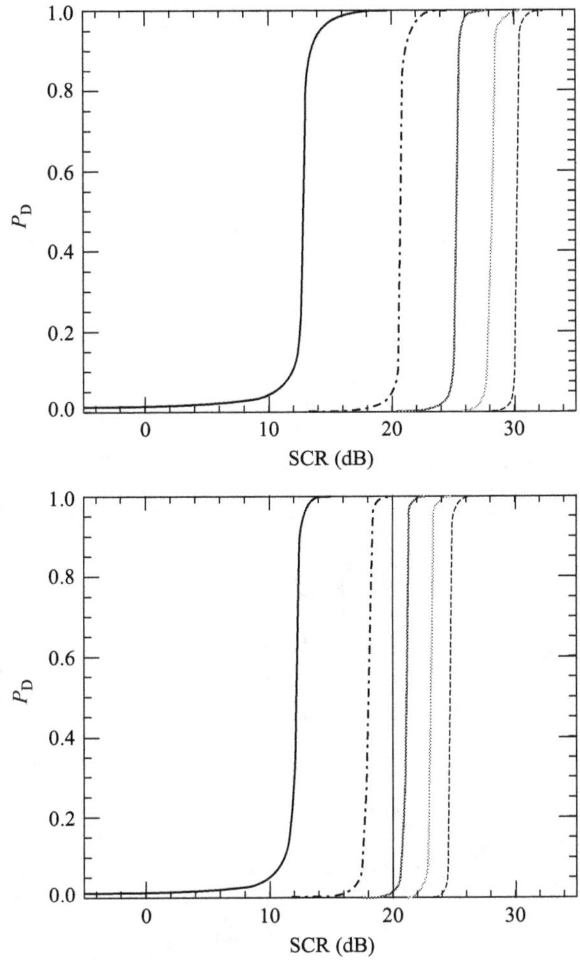

Figure 8.12 *Probability of detection versus signal to clutter level (in dB) for a non-fluctuating target in K-distributed clutter with $\nu = 0.1$. The top graph is a single pulse and the lower graph is the non-coherent sum of 10 pulses. The lines correspond to different false alarm probabilities (black: 10^{-2}, dashed dotted line: 10^{-4}, dark grey: 10^{-6}, light grey: 10^{-8}, dashed line: 10^{-10})*

Figures 8.15 and 8.16 show some effects of fixed frequency operation in thermal noise. Detection curves are plotted for a Swerling type 1 target in K-distributed clutter and noise, following the pulse-to-pulse integration of 10 pulses. The clutter is assumed to be fully correlated and $P_{FA} = 10^{-4}$. The top plot of Figure 8.15 shows the results for a clutter to noise ratio (CNR) of 30 dB and shape parameter of $\nu = 0.3$, 1, 3, 10 and 100. Because of the high CNR and the target and clutter being fully correlated, this result is nearly the same as for single pulse detection in

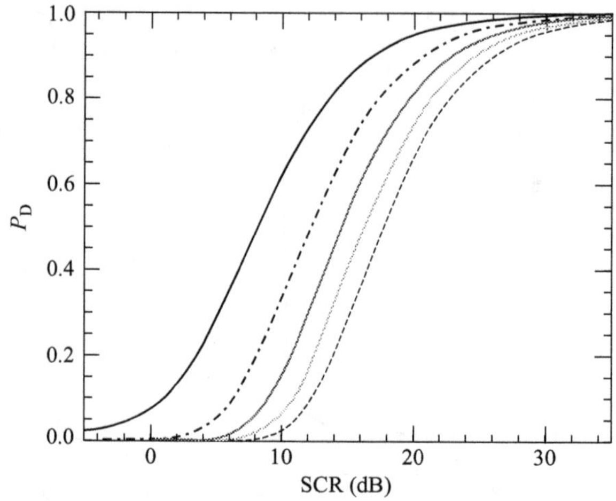

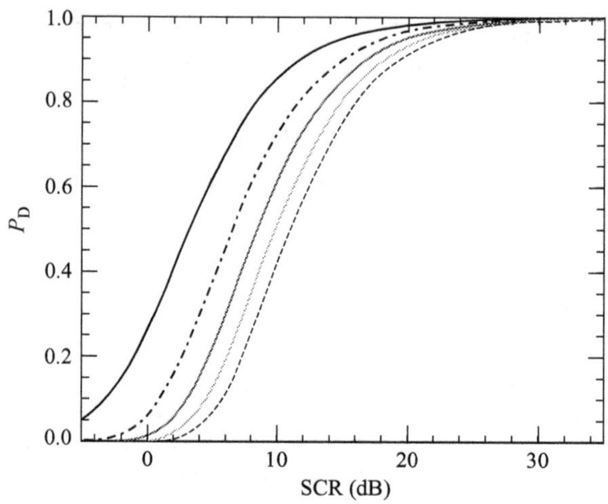

Figure 8.13 *Probability of detection versus signal to clutter level (in dB) for a Swerling 1 target in K-distributed clutter with* $\nu = 10$. *The top graph is a single pulse and the lower graph is the non-coherent sum of 10 pulses. The lines correspond to different false alarm probabilities (black:* 10^{-2}, *dashed dotted line:* 10^{-4}, *dark grey:* 10^{-6}, *light grey:* 10^{-8}, *dashed line:* 10^{-10}*)*

clutter alone. The lower plot of Figure 8.15 and the two plots in Figure 8.16 show CNR of 10, 0 and -10 dB, respectively. As the CNR reduces, the results approach those of noise alone, following the integration of 10 pulses. It may be noted that even with CNR $= -10$ dB, the presence of very spiky clutter introduces a significant detection loss (e.g. for $\nu = 0.3$ an increase in SIR of about 5 dB is required to achieve $P_D = 0.5$ compared with $\nu = \infty$).

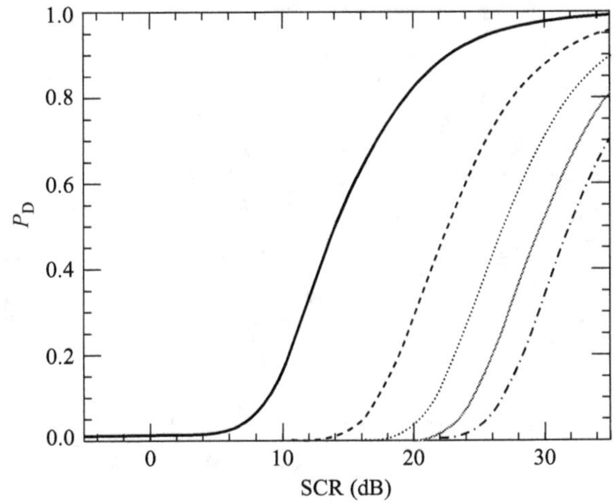

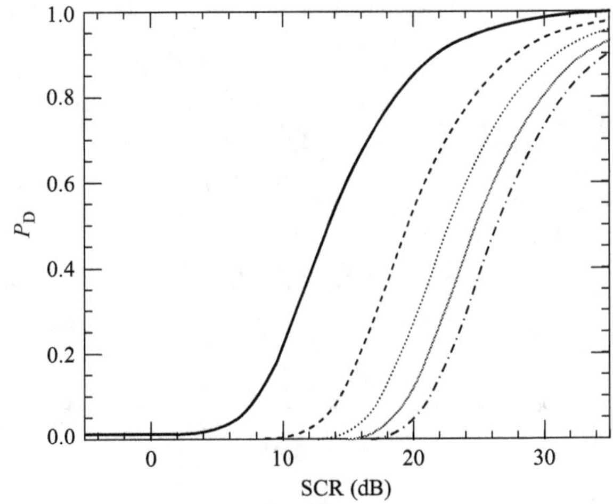

Figure 8.14 *Probability of detection versus signal to clutter level (in dB) for a Swerling 1 target in K-distributed clutter with $\nu = 0.1$. The top graph is a single pulse and the lower graph is the non-coherent sum of 10 pulses. The lines correspond to different false alarm probabilities (black: 10^{-2}, dashed dotted line: 10^{-4}, dark grey: 10^{-6}, light grey: 10^{-8}, dashed line: 10^{-10})*

The final set of plots show the effect of target model on detection. The interference is clutter alone, the speckle is independent from pulse-to-pulse, 16 pulses are integrated and $P_{FA} = 10^{-4}$. Figure 8.17 shows the standard Swerling target types. The top plot is for noiselike clutter ($\nu = \infty$) and the lower plot is for more spiky clutter ($\nu = 1$). The general Gamma target model allows more wildly fluctuating targets than the Swerling types to be used, as discussed in Reference 10. Figure 8.18 shows the effect of using $k = 0.5$, 0.3, 0.2 and 0.1 in equation (8.41). In these cases

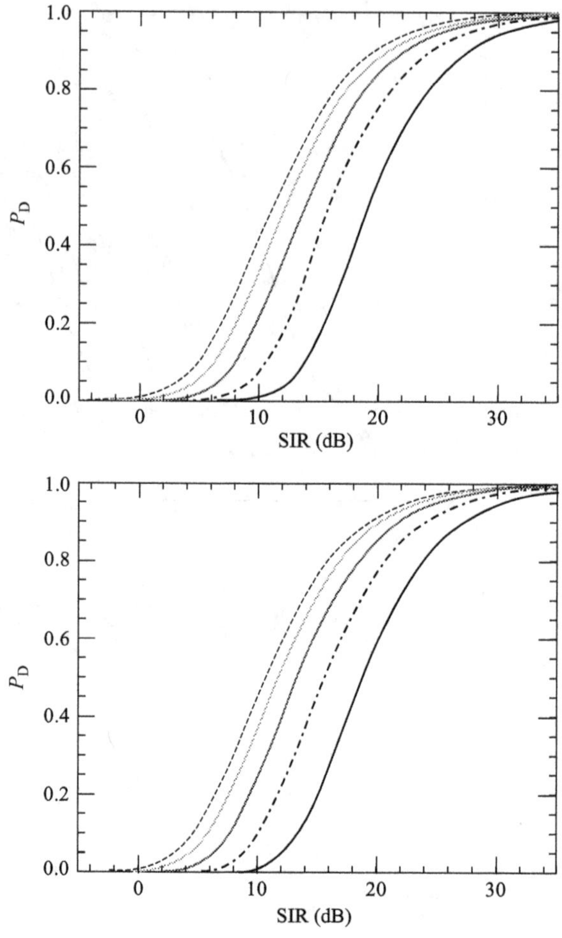

Figure 8.15 Probability of detection versus signal to interference level (in dB) for a Swerling 1 target in fixed frequency clutter and noise following the noncoherent integration of 10 pulses, with $P_{FA} = 10^{-4}$*. The top plot is for CNR = 30 dB, the lower is for CNR = 10 dB. The lines correspond to different shape parameter ν (dashed line: 100, light grey: 10, dark grey:3, dashed dotted line: 1, black:0.3)*

the fluctuations make it very hard to obtain a high P_D. This is often the case for real targets in complex environments.

All of the results above are at the output of pulse-to-pulse integration, before the scan-to-scan integration. For the output of the scan-to-scan integrator we need to apply a binomial function similar to equation (8.32). Thus, for a fluctuating target, the final result is

$$P_D^{(s-s)} = \sum_{i=m}^{i=n} \frac{n!}{(n-i)!i!} P_D^i(Y)(1 - P_D(Y))^{n-i} \tag{8.53}$$

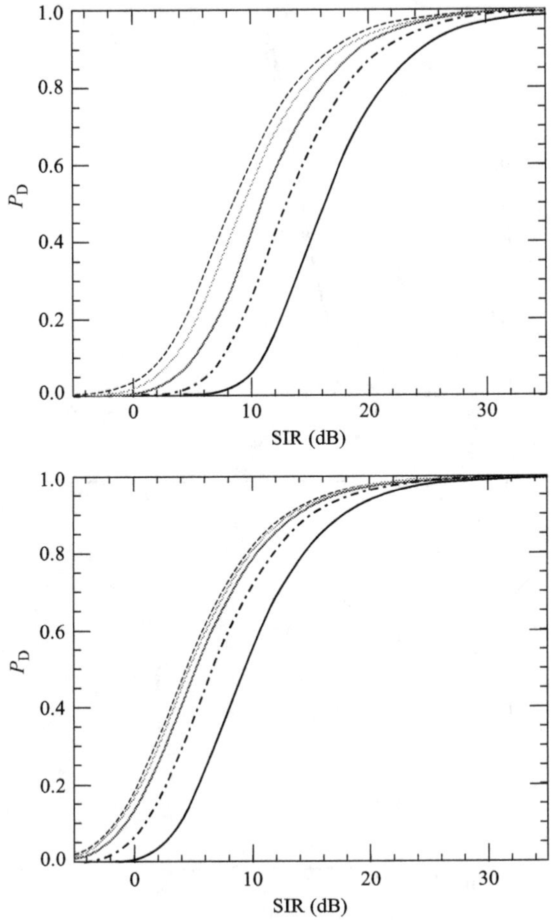

Figure 8.16 *Probability of detection versus signal to interference level (in dB) for a Swerling 1 target in fixed frequency clutter and noise, following the non coherent integration of 10 pulses with $P_{FA} = 10^{-4}$. The top plot is for CNR = 0 dB, the lower is for* CNR $= -10$ *dB. The lines correspond to different shape parameter ν (dashed line: 100, light grey: 10, dark grey: 3, dashed dotted line: 1, black: 0.3)*

Equation (8.53) provides us with the means to calculate the probability of detection from the threshold derived in the previous section. This applies to the target radar cross section, σ_t, specified back in equation (8.4). It is often difficult to interpret results such as this. For instance, if 50% P_D is required and a 1 m^2 target is found to have 30% P_D, how far short are we? Will a 2 m^2 target reach the required detection performance? In order to answer these types of question it is often useful to specify the required P_D and then to calculate the minimum detectable target size. This can be done numerically by using a root finding algorithm. The results in the next section are obtained in this way.

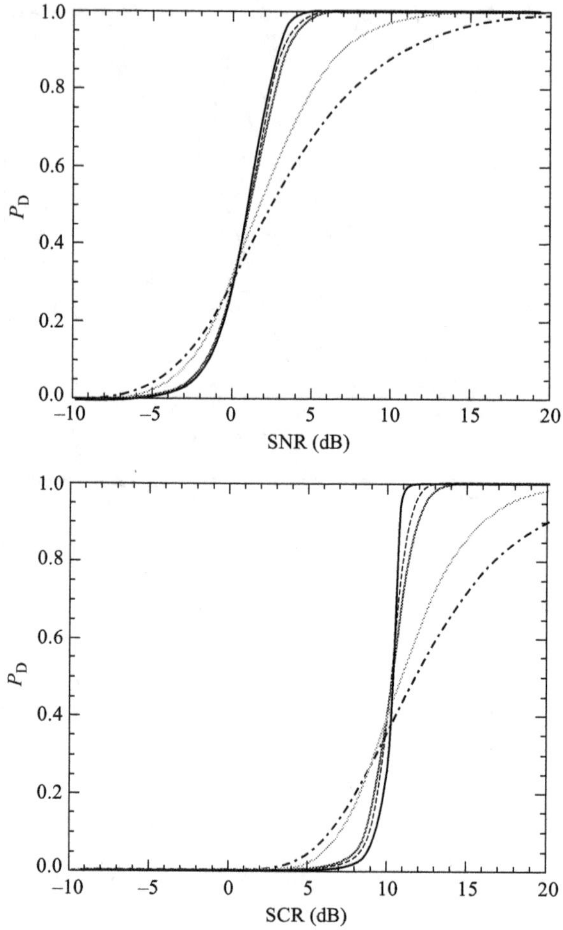

Figure 8.17 Probability of detection versus signal to interference level for Swerling target models following non-coherent integration of 16 pulses (black: Sweling 0, dashed dotted line: Swerling 1, dark grey: Swerling 2, light grey: Swerling 3, dashed line: Swerling 4). For the top plot, the interference is noise. For the lower plot, the interference is K-distributed clutter with $\nu = 1$

8.6 Detection performance

Before deriving performance results it is worthwhile reviewing the radar application we are going to consider. A common requirement for maritime surveillance radar is to be able to detect a small slow surface target in a few seconds. In medium and high sea states the small target is competing with sea clutter for detection. Thus in order to minimise the clutter, a high resolution is preferred (i.e. a narrow antenna beam-width and a short compressed pulse-width). Unfortunately this leads to non-Gaussian (or spiky)

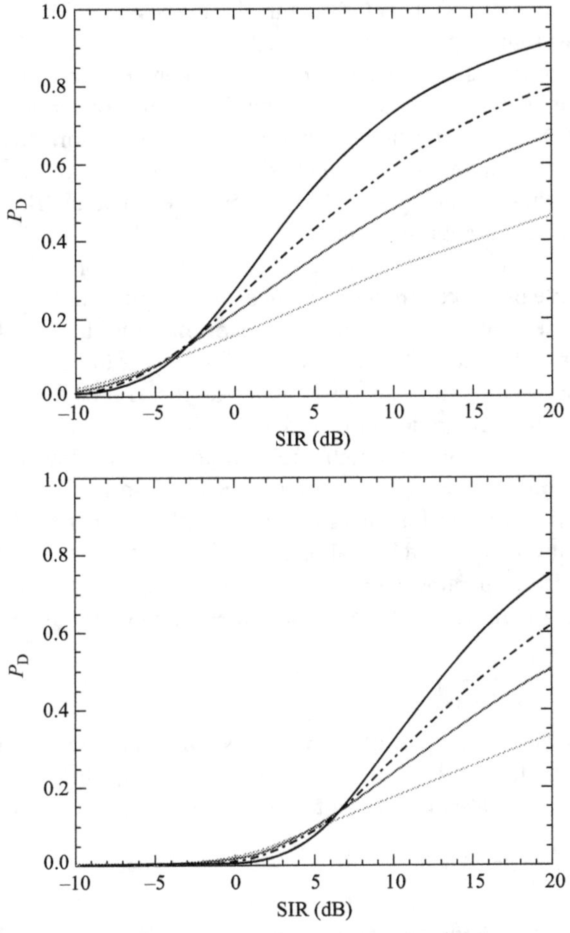

Figure 8.18 Probability of detection versus signal to interference level for Gamma distributed target models following non-coherent integration of 16 pulses (black: k = 0.5, dashed dotted line: k = 0.3, dark grey: k = 0.2, light grey: k = 0.1). For the top plot, the interference is noise. For the lower plot, the interference is K-distributed clutter with ν = 1

clutter, which tends to reduce performance. A scanning antenna is generally required to give all round coverage, and the scan rate is adjusted to give a number of pulses within the beam dwell time and a number of scans within the overall allowed detection spikiness interval. The radar returns may be integrated from pulse to pulse and from scan to scan to improve the detection performance. As the clutter spikiness varies over the field of spikiness view (due to changes of aspect, grazing angle and across range resolution and also the reduction of the 'effective' spikiness in the presence of system noise), it is necessary to adapt the detection threshold applied to the signal in order to maintain a CFAR. This CFAR threshold is usually derived from range and azimuth cells surrounding the target test cell.

The small target's slow velocity often results in the target Doppler frequency being within the sea clutter Doppler spectrum. Therefore conventional Moving Target Detection (MTD) processing is not effective. Also, the short beam dwell time of the narrow scanning antenna beam results in the resolution of conventional Pulsed Doppler processing not being sufficient to provide a useful Clutter Improvement Factor. Thus non-coherent processing is generally used.

With non-coherent processing, the beam dwell-time is not sufficient to decorrelate the clutter speckle on a fixed frequency. Thus pulse to pulse integration is ineffective, and a fast scan rate (many scans a second) to decorrelate the clutter, followed by scan to scan integration is desirable. An alternative to fast scan is the use of frequency agility within the beam dwell time. The agility tends to decorrelate the clutter speckle and makes pulse to pulse integration more effective. However, using the radar bandwidth for frequency agility reduces the range resolution, and the slower scan rate reduces the scan to scan integration. A discussion of these trade-offs is given in Reference 16.

We will define a radar for this application with parameters, losses and default detection conditions as listed in Table 8.1 and refer to it as the 'baseline radar'. Various other derived parameters are required for performance assessment. The radar processing is assumed to have within beam integration over n_{wbi} pulses given by

$$n_{\text{wbi}} = \text{int}\left(\frac{\theta_{\text{bw}} f_{\text{prf}}}{2\pi f_{\text{revs}}}\right) \tag{8.54}$$

(where int () is the integer part of the value) when this number exceeds the number of independent agile frequencies, i.e. $n_{\text{wbi}} > B_{\text{agile}}/B_{\text{pulse}}$, then an effective number of integrated clutter pulses is defined to be the number of agile frequencies,

$$n_{\text{eff}} = \text{int}\left(\frac{B_{\text{ag}}}{B_{\text{pulse}}}\right) \tag{8.55}$$

The number of scans in the target exposure time, T_{exp}, is given by

$$n_{\text{scans}} = \text{int}(T_{\text{exp}} f_{\text{revs}}) \tag{8.56}$$

This leads to a scan-to-scan integration binomial threshold of m out of n_{scans}, where

$$m = \text{nint}\left(\frac{2n_{\text{scans}}}{3}\right) \tag{8.57}$$

(where nint() is the nearest integer) except for the special case of $n_{\text{scans}} = 2$, where m is also set to 2. Further discussion of the choice of m may be found in Reference 12. For the parameter values given in Table 8.1, we have $n_{\text{wbi}} = 10$; $n_{\text{eff}} = 5$; $n_{\text{scans}} = 10$ and $m = 7$; these are varied in later sections.

Now, using the methods described in this chapter, the baseline performance may be calculated in terms of the minimum target size, σ_{min}, which can be detected with the required probability of detection, P_{D}, and false alarm rate (1 per T_{FA}) in the defined target exposure time, T_{exp} (all specified in Table 8.1). Figure 8.19 shows σ_{min} as a function of range and sea-state, at a radar height of 600 ft. Figure 8.20 shows a

Table 8.1 Baseline radar parameters and losses and default detection conditions

Parameter	Value	Symbol
Transmitter frequency	9.5 GHz	f_{tx}
Frequency agility range	500 MHz	B_{agile}
Peak radiated power	20 kW	P_{tx}
Transmit antenna gain	33 dB	G_{tx}
Receive antenna gain	33 dB	G_{rx}
Azimuth 3 dB beamwidth	2.0°	θ_{bw}
Elevation 3 dB beamwidth	8.0°	ϕ_{bw}
Azimuth and elevation beam shape	Gaussian	
Antenna polarisation	HH	
Mechanical scan sector	360°	
Mechanical scan rate	60 rpm	f_{revs}
Ambient/Antenna temperature	300°K	
Pulse repetition frequency	1.8 kHz	f_{prf}
Pulse width	25 μs	
Pulse bandwidth (3 dB)	100 MHz	B_{pulse}
Range resolution	1.5 m	
System noise figure[a]	4.0 dB	
Randome loss (2-way)	1.0 dB	
Target processing losses[b]	5.0 dB	
Clutter processing losses[c]	1.0 dB	
Range scale	40 n miles	
Probability of detection	50%	
False alarm time	60 s	T_{FA}
Maximum radial target speed	10 ms^{-1}	
Target exposure time	10 s	T_{exp}
Radar target model	Swerling 2	
Sea state	4	
Swell	none	
Radar altitude	600 feet	

[a]The noise figure includes all the microwave losses that are not accounted for in the peak radiated power value.
[b]Target losses consist of: CFAR, beam-shape, range-straddling, pulse-compression, IF-mismatch, and pulse-droop losses.
[c]Clutter losses consist of: pulse-compression and pulse droop losses.

similar set of plots for a radar height of 3000 ft. In both figures there are two graphs corresponding to the GIT and RRE σ^0 clutter models. At both heights the two models give results differing by in the region of 10 dB. In most situations the GIT model gives more pessimistic predictions. However, at very low grazing angles (low height, long range) when the multipath factor in the GIT model becomes effective (i.e. below the

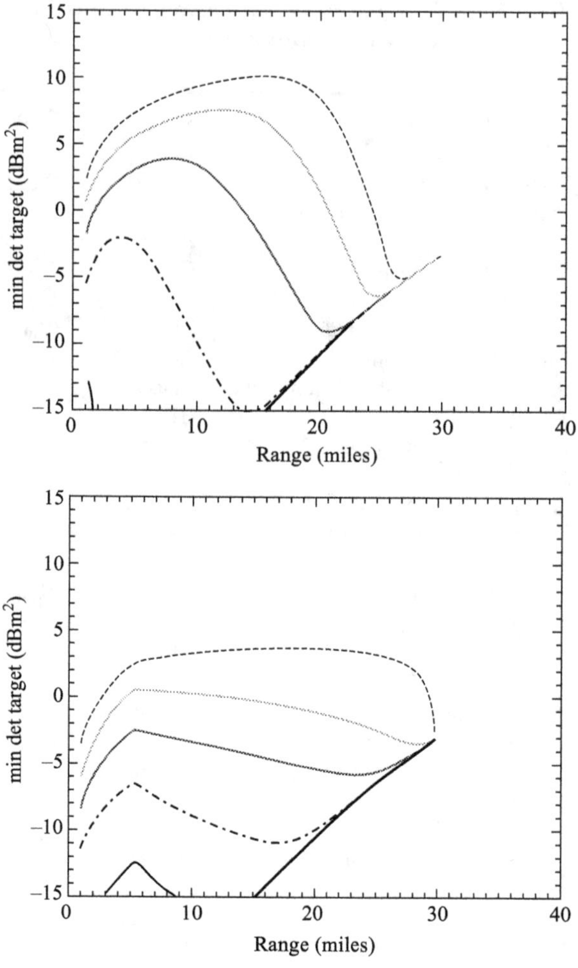

Figure 8.19 Minimum detectable target versus range for a radar with the parameters of Table 8.1 at 600 ft altitude for sea states 1 to 5 (black = 1, dashed dotted line = 2, dark grey = 3, light grey = 4, dashed line = 5). The top plot uses the GIT σ^0 clutter model and the lower plot uses the RRE σ^0 clutter model

critical angle of equation (8.18)), the situation reverses and the RRE model is more pessimistic.

This effect is more clearly illustrated in plots of the radar performance versus height. Figure 8.21 shows the minimum detectable target at a range of 20 nautical miles for radar altitudes from 300 to 3000 ft. In most cases the minimum detectable target does not vary much with height because the increase of clutter power with grazing angle is offset, to some extent, by the decrease in spikyness. At grazing angles lower than the 'critical angle' in the GIT model however, there is a significant

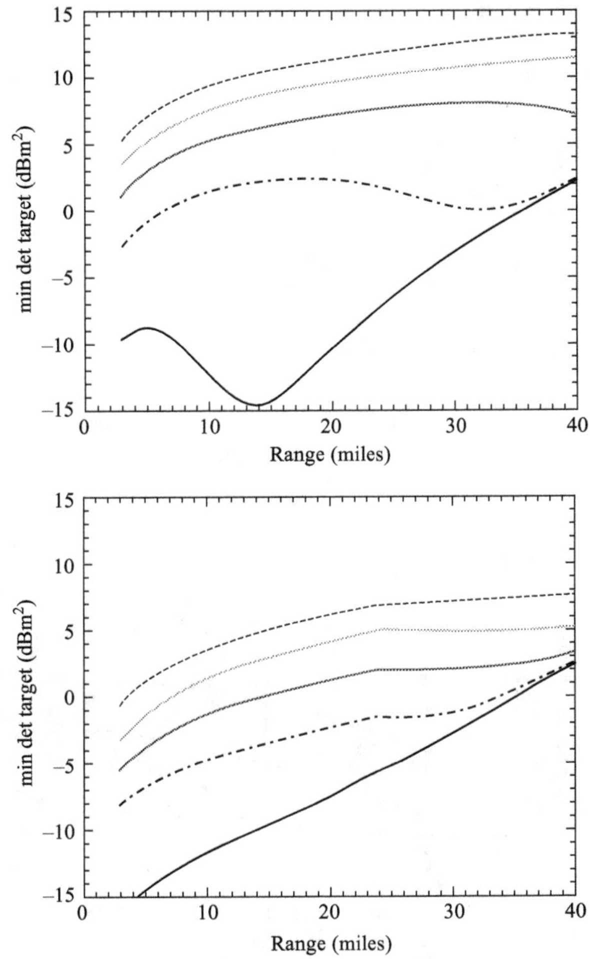

Figure 8.20 Minimum detectable target versus range for a radar with the parameters of Table 8.1 at 3000 f altitude for sea states 1–5 (black = 1, dashed dotted line = 2, dark grey = 3, light grey = 4, dashed line = 5). The top plot uses the GIT σ^0 clutter model and the lower plot uses the RRE σ^0 clutter model

variation with height. Thus, in sea-state 5 the GIT model predicts that there is a 20 dB loss associated with increasing the radar altitude from 300 to 700 ft. The equivalent loss using the RRE model is 4 dB.

Radar polarisation: In general, the choice of polarisation is an unresolved issue for a radar system of this type, and depends on the clutter model in use. Figures 8.19–8.21 are for HH polarisation. Figure 8.22 is for the same conditions as Figure 8.21, but

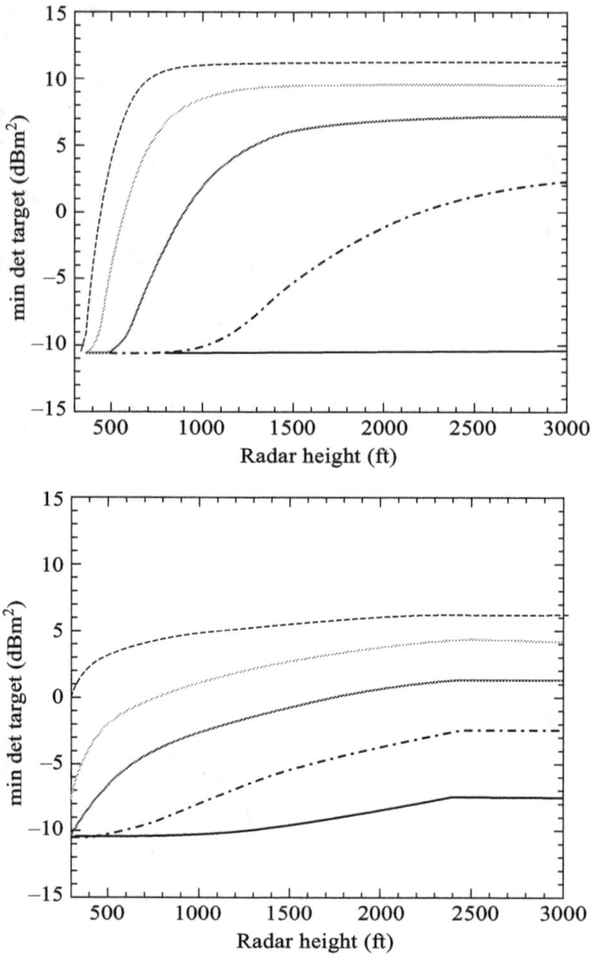

Figure 8.21 *Minimum detectable target versus radar height for a radar with the parameters of Table 8.1 at 20 miles range for sea states 1 to 5 (black = 1, dashed dotted line = 2, dark grey = 3, light grey = 4, dashed line = 5). The top plot uses the GIT σ^0 clutter model and the lower plot uses the RRE σ^0 clutter model*

with the use of VV polarisation. These figures show that VV is predicted to be better than HH.

The difference in performance between the two polarisations comes from the empirical model for ν (HH being spikier than VV) and the model for σ^0 (VV having a higher radar cross section, RCS, than HH). These two effects almost cancel each other, and the resulting difference is probably too small to be significant given the large spread of the original data used to develop the models for ν and σ^0. Further work is needed to refine both of the models so that a direct polarisation comparison is

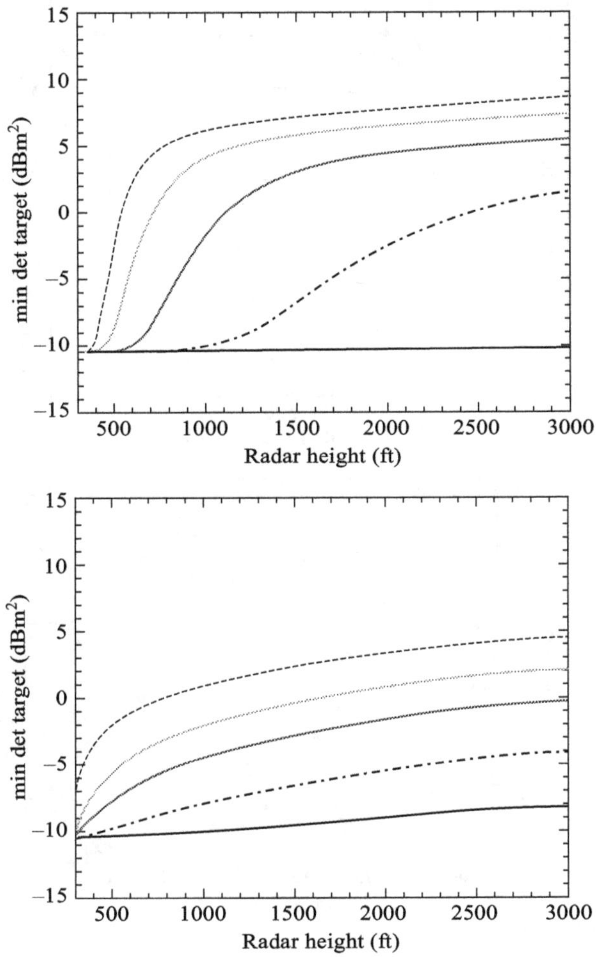

Figure 8.22 *Radar performance as plotted in Figure 8.21, but with vertical polarisation*

more meaningful. In the meantime experience suggests that in some conditions HH is better than VV, but in other conditions the situation is reversed.

Target models: The results so far in this section are all for a Swerling 2 target model. Figure 8.23 shows the effect on performance of varying the k parameter in the Gamma distribution model of equation (8.41). The sea-state is 4 and the polarisation is HH. The lines are for $k = 5$ (corresponding to Swerling 2 in this case), $k = 0.5$ and $k = 0.3$. This range of values for k corresponds to realistic targets on the sea surface, and results in a variation of about 7 dB in performance for the cases considered.

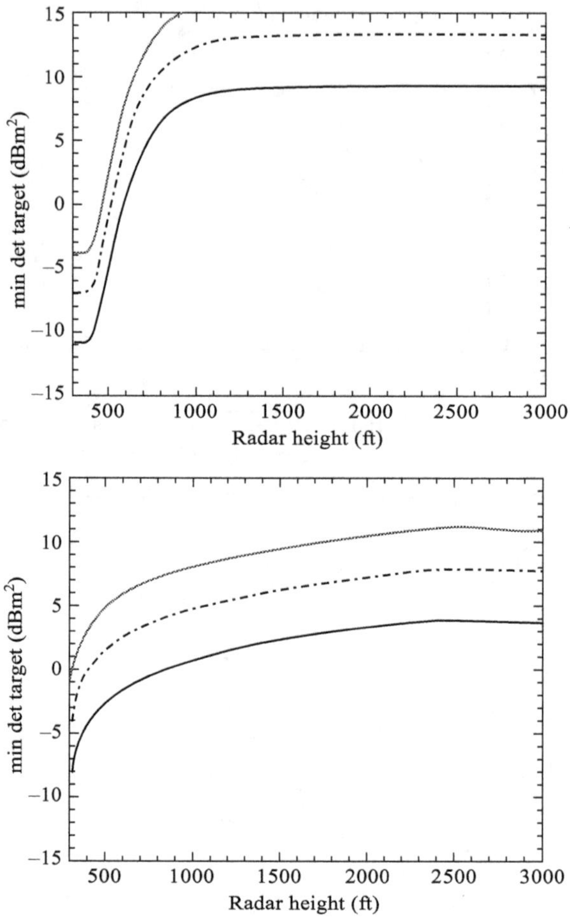

Figure 8.23 *Minimum detectable target versus radar height for a radar with the parameters of Table 8.1 at 20 miles range for sea state 4. The lines correspond to different Gamma distributed target fluctuation models (black; $k = 5$, dashed dotted line: $k = 0.5$, dark grey: $k = 0.3$). The top plot uses the GIT σ_0 clutter model and the lower plot uses the RRE σ_0 clutter model*

Target exposure time: The results so far have considered a target exposure time of 10 s and a scan rate of 60 rpm. Thus 10 scans have been available for scan-to-scan integration and a '7 out of 10' binary threshold has been applied. Figure 8.24 shows the effect of varying the target exposure time between 2 and 12 s. Equations (8.56) and (8.57) are used to derive the scan-to-scan integration, and the GIT clutter RCS model is used. The black and dashed lines show the results for 600 and 3000 ft altitude, respectively. Reducing the exposure time from 10 to 2 s results in a loss of between 7 and 8 dB at both heights.

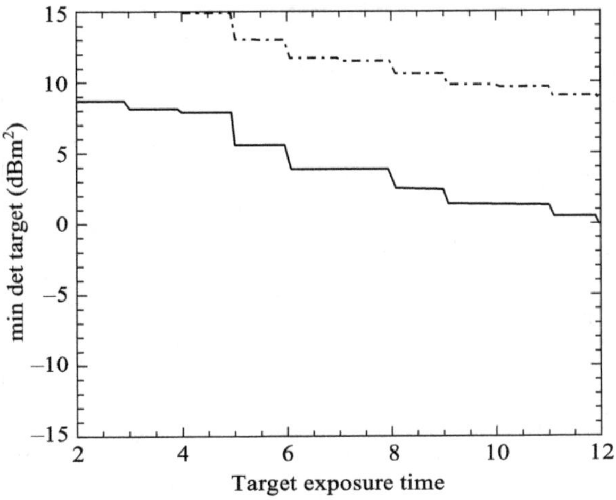

Figure 8.24 *Minimum detectable target versus target exposure time for scan rates of 60 rpm (black line: 600 ft altitude, dashed dotted line: 3000 ft). The other radar parameters are as in Table 8.1, the range is 20 miles, and the sea-state is 4*

Radar resolution: When all of the effects of resolution are taken into account, it is found that performance generally improves as the resolution is increased. However, because the overall bandwidth is constrained and the lower resolution systems are optimised through the use of frequency agility, the improvements are fairly small. This is due to an increase in signal to clutter ratio obtained through high resolution being offset partly by the increase in spikiness of the clutter. Frequency agility at lower resolution helps to decorrelate the clutter speckle, which further offsets the high-resolution advantage.

Figure 8.25 shows the effect of varying the resolution from 100 m down to 0.3 m. The performance improves as the resolution cell is reduced from 100 m down to 2 m. Further reduction down to 0.3 m results in fairly level performance with steps as the number of agile frequencies is reduced due to the constraint of overall bandwidth. The calculations assume that frequency agility completely decorrelates the speckle component of the clutter. However, the reflectivity is not always uniformly distributed across the resolution cell, an effect which accounts for the increase in spikyness as the cell size is reduced. It also means that the speckle has a residual correlation when frequency agility is applied. This phenomenon is investigated in more detail in Reference 17.

Scan rate: The choice of scan rate is a trade-off between pulse-to-pulse and scan-to-scan integration. At high resolution, where the bandwidth constraint does not allow frequency agility, there is little point of doing pulse-to-pulse integration in clutter (because it is correlated). Therefore, increasing the scan rate shows a significant performance advantage, as it increases the number of independent clutter samples.

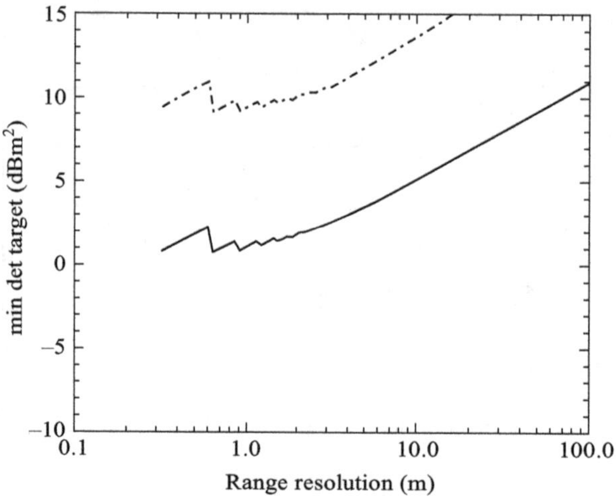

Figure 8.25 Minimum detectable target versus range resolution. The other radar parameters are as in Table 8.1, the range is 30 miles, the radar altitude is 600 ft (black line) and 3000 ft (dashed dotted line). The sea-state is 4. HH polarisation and the GIT clutter RCS model are used

Figure 8.26 shows this effect for a range resolution of 1.5 m. Increasing the scan rate from 30 to 300 rpm improves the calculated performance by 13 dB for HH and 6 dB for VV. These calculations, however, assume that the clutter spikes are independent from scan to scan and this assumption is increasingly in error as the scan rate increases. We do not yet have a good model for the lifetime of spikes, but data show that most last for between 100 ms and 2 s. Thus we might expect the performance improvement with fast scan to be in reality about 5 dB or so, less than shown in Figure 8.26. (A further discussion on the optimisation of scan rate may be found in Reference 16).

8.7 Modelling other types of radar

The discussions in this chapter have concentrated on a specific radar type: an airborne, scanning, sea surface surveillance radar, which operates at I-Band (9–10 GHz), is non-coherent, uses frequency agility and has pulse-to-pulse and scan-to-scan integration. This choice has been to allow end-to-end performance calculations to be performed, and for it to be demonstrated that sea clutter modelling has a significant effect on the results.

There are, of course, many other types of radar system, and the performance calculations for these are different in many details. In particular, coherent radar systems, such as MTI (moving target indication), Pulsed Doppler, FMCW (frequency modulation, continuous wave), and SAR (synthetic aperture radar), all described in Reference 1, use the complex amplitude of the received signal and perform coherent integration. The evaluation of the effect of this processing is generally incorporated

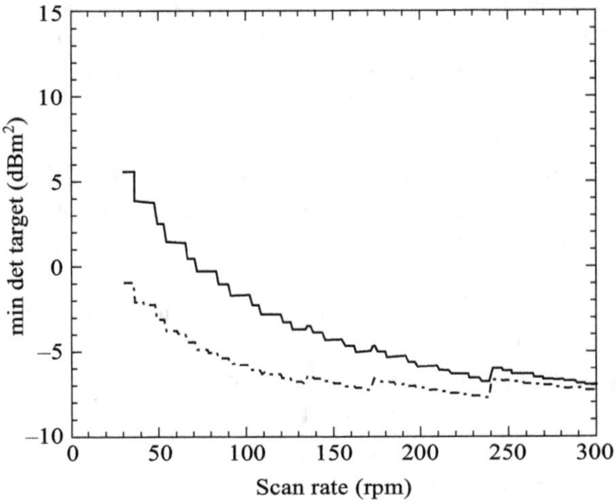

Figure 8.26 Minimum detectable target versus scan rate and polarisation (HH – black line and VV – dashed dotted line) for a range resolution of 1.5 m. The other radar parameters are as in Table 8.1, the range is 20 miles, radar altitude is 600 ft and the sea-state is 4

within the radar equation in terms of gains and losses, which depend upon the spectrum of the received signal and imperfections in the radar and the processing performed [18]. However, at some point all of these systems calculate the magnitude of the processed signal and then perform some form of non-coherent processing, such as integration, CFAR and thresholding.

The information in this book on sea clutter and the compound K distribution model may be used in performance calculations for these other radar types as follows. First, the descriptions of the sea clutter Doppler spectrum given in Chapters 2, 3, and 4 may be used to model coherent integration effects. Also, the magnitude signal, when it is derived, often turns out to be K distributed. It is therefore possible to apply the detection calculation methods derived here. This latter application is not limited to sea clutter; SAR images of land are often described in terms of K distribution, with speckle and modulation in the form of image texture. A detailed discussion of the use of the compound K distribution in SAR image exploitation may be found in Reference 19.

References

1 M.I. Skolnik, *'Radar handbook'*, McGraw-Hill, second edition, 1990.
2 *'RRE σ_0 sea clutter model'*, Royal Radar Establishment, 1974 (unpublished).
3 F.E. Nathanson, *'Radar design principles'*, McGraw-Hill, New York, 1969.
4 J.C. Daley, J.T. Ransome, J.A. Burkett and J.R. Duncan, *'Sea clutter measurements on four frequencies'*, Naval Research Lab report 6806, 1968.

5 M. Horst, F. Dyer and M. Tuley, *'Radar sea clutter model'*, International IEEE AP/S URSI Symposium (IEEE Antennas and Propagation), 1978.
6 T.P. Leonard, I. Antipov and K.D. Ward, *'A comparison of radar sea clutter models'*, Proceedings of the Radar 2002, IEE International Radar Conference, 2002.
7 S. Watts, 'Radar detection prediction for targets in both K-distributed sea clutter and thermal noise', IEEE AES-23, No.2, pp. 40–45, January 1987.
8 P. Swerling, 'Probability of detection for fluctuating targets', IRE Trans IT-6, pp. 269–308, 1960.
9 J.L. Marcum, 'A statistical theory of target detection by pulsed radar', IRE Trans IT-6, pp. 59–144, 1960.
10 P. Swerling, 'Probability of detection for some additional fluctuating target cases', IEEE AES-33, No. 2, 698–709, April 1997.
11 D. Shnidman, 'Radar detection probabilities and their calculation', IEEE AES-31, No. 3, pp. 928–950, July 1995.
12 S. Watts, 'Radar detection performance in sea clutter using the compound K-distribution model', *IEE Proc. Part F*, 7, pp. 613–620, 1985.
13 S. Watts and D.C. Wicks, *'Empirical models for detection prediction in K-distributed radar sea clutter'*, Proceedings of the IEEE 1990 International Radar Conference, pp. 189–194, 1990.
14 S. Watts, *'Radar performance in K-distributed sea clutter'*, SET Panel Symposium on 'Low grazing angle clutter: its characterisation, measurement and application', Laurel, Maryland, USA, 25–27 April 2000.
15 D. Shnidman, *'Expanded swerling target models'*, IEEE AES-39, No. 3, pp. 1059–1068, 2003.
16 S. Watts and A. G. Stove, *'Non-coherent integration strategies in scanning radar for the detection of targets in sea clutter and noise'*, Proceedings of the International Radar Conference Radar 2004, Toulouse 19–21 October 2004, paper 10P-SP-229.
17 K.D. Ward, R.J.A. Tough and P.W. Shepherd, *'Modelling sea clutter: correlation, resolution and non-Gaussian statistics'*, IEE Conference Publ. 449, Radar 97, pp. 95–99, 1997.
18 J.A. Scheer and J.L. Kurtz, *'Coherent radar performance estimation'*, Artech House, Boston, 1993.
19 C.J. Oliver and S. Quegan, *'Understanding synthetic aperture radar images'*, Artech House, Boston, 1995.

Chapter 9

CFAR detection

9.1 Introduction

A radar is often required to detect targets against a changing background of clutter and thermal noise, the clutter reflectivity and statistics varying with range and look direction, dependent on the chosen radar parameters, operating height and the prevailing weather conditions. For a scanning radar over the sea, this change is continuous and usually unpredictable.

The problem faced by the radar signal processing is that of reliably detecting targets, while at the same time rejecting the returns from the sea and thermal noise. To do this, we must monitor features in the radar returns that allow us to discriminate between real and spurious target returns. The final discrimination is usually done on the basis of amplitude at a detection threshold. The choice of the quantity that is thresholded depends on the signal processing and the features chosen to discriminate between real and false targets. A wide range of such features may be exploited in this way, and are associated with different types of signal processing.

Typical signal processing schemes include:

- pulse-to-pulse integration
- scan-to-scan integration
- polarisation discrimination
- Doppler processing

Pulse-to-pulse integration exploits differences in amplitude statistics and pulse-to-pulse correlation between targets and clutter or noise. Against thermal noise the gain can be quantified in term of an effective increase of power on the target.

Scan-to-scan integration is often used to exploit the longer term correlation characteristics of targets and clutter. In particular, discrete clutter spikes may not be affected by pulse-to-pulse integration, and appear to be target-like. However, such spikes usually only persist over periods of the order of a second or so. Longer term integration, from scan-to-scan over several seconds can be used to distinguish them from real persistent targets. One approach to such integration is 'track before detect' [1].

Polarisation has often been proposed as a discriminant [2]. This can be exploited in a radar that adapts its polarisation dynamically to minimise the clutter returns, whilst hopefully still detecting the target.[1] A simpler approach is to select from different fixed polarisation (e.g. horizontal, vertical or circular).

The Doppler shifts of moving targets relative to the sea clutter provide another potential discriminator. However, as discussed in Chapter 2, the clutter spectrum can vary considerably over time and space and is also dependent on the polarisation, viewing geometry and prevailing sea conditions. An example of a time history of the clutter spectrum in a single range cell is shown in Figure 2.15. Similar spatial variations between range cells are also observed. For fast targets, with a Doppler frequency that is well separated from the clutter, discrimination from clutter should be relatively straightforward. However, as discussed in Chapter 2, the clutter statistics in the edges on the clutter spectrum are particularly non-Gaussian and spiky. Control of false alarms and detection of small targets in this Doppler region is difficult.

In all these cases, targets are eventually discriminated from clutter by some form of acceptance threshold. In this chapter, we discuss the methods for setting such a detection threshold.

As discussed in Chapter 6, the standard detection strategy is to try and maintain a fixed probability of false alarm, whilst also attempting to maximise the probability of detection. The requirement to maintain a constant false alarm rate (CFAR) is central to the practical operation of many radar systems. The actual false alarm rate that can be tolerated depends on the radar application. Typically, the false alarm rate on a radar display might be specified at, say, one false alarm per scan or one false alarm per minute. Dependent on the amount of processing after the first detection threshold (scan-to-scan integration etc., as discussed in Chapter 8), such a displayed false alarm rate might equate to a probability of false alarm, P_{FA}, in the range, say, 10^{-4}–10^{-6} at the first threshold. In other applications, values as high as 10^{-2} or as low as 10^{-8} may be encountered. False alarms and their effect on performance are discussed in more detail in Chapter 10.

Given the probability density function, pdf, of the interference (i.e. clutter-plus-noise) and target-plus-interference, Chapter 8 has shown how to calculate the probability of false alarm, P_{FA}, and probability of detection, P_D. In Chapter 8, it was assumed that that the various parameters (clutter-to-noise ratio, target-to-noise ratio, clutter shape parameter etc.) are known *a priori*. We must now consider how a radar can adapt its detection threshold to provide a CFAR, when the relevant parameters are not known but must be estimated from the data.

[1] At the time of writing this book and as far as we are aware, radars that are able to measure and adapt to the coherent polarisation scattering matrix are still only available as experimental systems.

9.2 Adaptation to changing clutter amplitude

In Rayleigh clutter or noise with mean power x, the pdf of the amplitude, E, of the radar returns may be expressed (as shown in Chapter 4) as:

$$P(E) = \frac{2E}{x}\exp\left(-\frac{E^2}{x}\right); \qquad 0 \le E \le \infty \tag{9.1}$$

For a fixed threshold, t, the P_{FA} is given by

$$P_{\mathrm{FA}} = \exp\left(-\frac{t^2}{x}\right) \tag{9.2}$$

In other words, the required threshold is given by

$$t^2 = -x\log_e(P_{\mathrm{FA}}) \tag{9.3}$$

The radar detection processing must be able to estimate x, the mean power of the interference, in order to set a threshold. In general, the mean power is unknown and may also be varying temporally and spatially, requiring a continuously adapting estimate of its value.

In this section of the chapter, methods for adapting to unknown clutter amplitude are described. The first requirement is to maintain the received signal level within the dynamic range of the radar receiver, to prevent distortion of the signals. Then, to set a detection threshold, the local mean level of the interference (i.e. clutter-plus-noise) must be estimated. The use of the cell-averaging CFAR system, with its many variants, is developed in this context.

9.2.1 Control of received signal dynamic range

Adaptation to the amplitude of the received signals is required to prevent limiting in the receiver and to maintain a reference level against which detections may be made. Figure 9.1 shows how the signal-to-interference ratio (i.e. signal to clutter-plus-noise ratio) may vary as a function of range and azimuth for an airborne radar in a specific scenario. Plots such as this can be prepared using the radar range equation and the models for clutter reflectivity described in Chapter 8. The dynamic range of the receiver is often achieved by swept gain or sensitivity time control (STC). A selection of different laws may be available to the operator to cope with different conditions. It can be seen from Figure 9.1 that ideally the law should also vary with azimuth. The variation of signal level from the surface return depends on the elevation beamshape of the antenna. Airborne maritime surveillance radars sometimes employ an approximation to a cosec^2 shaping in elevation to control the dynamic range of the signals and to ensure continuous coverage in elevation. However, maximum control of dynamic range is achieved by using automatic gain control (AGC) that adapts to the local signal level. AGC operates a closed loop control system that varies the RF attenuation as a function of range and bearing, to maintain a given average level at the detector output. The time constant of the AGC must be short enough to

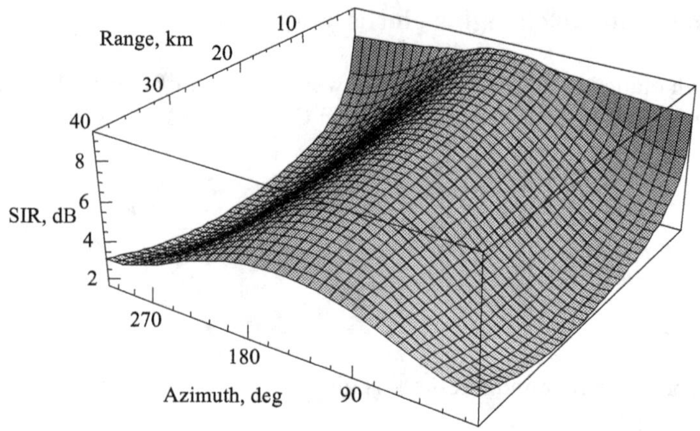

Figure 9.1 An example of the variation of signal-to-interference ratio, as a function of range and azimuth for an airborne radar

adjust quickly to large changes, such as land-sea interfaces, but not significantly to large targets. However, even with a good AGC, the received signal will occasionally saturate on very large targets and good recovery from limiting is an essential design feature.

Even with AGC and STC, the mean level of the clutter or noise is not usually controlled exactly. If automatic detection is required, an adaptive threshold must be set to maintain a CFAR.

9.2.2 Log FTC receiver for Rayleigh clutter

If the clutter is known to be Rayleigh distributed, one method widely used in older radars for setting a detection threshold in an unknown clutter power level, is the Log FTC receiver. In this, a logarithmic video amplifier is followed by a fast time constant circuit (FTC) or differentiator (essentially a high-pass filter). It is found that a logarithmic receiver gives an output with a constant variance, independent of the clutter power. There is an unknown mean level (i.e. a slowly varying offset on the video signal), which is removed by differentiating the output with the FTC circuit. This works satisfactorily, provided that the offset varies sufficiently slowly compared with the differentiator time-constant. However, problems are encountered with range extensive targets, which may be suppressed by the differentiator, and clutter edges where the differentiator produces target-like spikes.

The effect of this CFAR system (for Rayleigh clutter or noise, following a square law detector) may be described as follows. After square law detection, the clutter variates, z, will have an exponential pdf $P(z)$, so that $P(z) = 1/x \exp(-z/x)$, where x is the mean clutter power. After the logarithmic amplifier, the data are transformed to y, where now $y = a \log_e bz$, where a and b are parameters of the logarithmic amplifier.

The mean and mean square values of value of y are given by

$$\bar{y} = \int_0^\infty \frac{a}{x} \log_e(bz) \exp\left(-\frac{z}{x}\right) dz \quad \text{and}$$

$$\overline{y^2} = \int_0^\infty \frac{(a \log_e(bz))^2}{x} \exp\left(-\frac{z}{x}\right) dz \tag{9.4}$$

Using these results (see, e.g. Appendix 2 equations A2.25–A2.27):

$$\int_0^\infty e^{-\mu z} \log z \, dz = -\frac{1}{\mu}(\gamma + \log \mu)$$

$$\int_0^\infty e^{-\mu z} (\log z)^2 \, dz = \frac{1}{\mu}\left(\frac{\pi^2}{6} + (\gamma + \log \mu)^2\right) \tag{9.5}$$

where Euler's constant $\gamma = 0.5772167$, we obtain

$$\bar{y} = a(\gamma + \log(xb))$$

$$\overline{y^2} = a^2\left(\frac{\pi^2}{6} + (\gamma + \log(xb))^2\right) \tag{9.6}$$

Thus the variance of y is given by $\text{var}(y) = a^2\pi^2/6$, which is independent of the clutter power, x, and the gain of the amplifier, b. The mean level, $\bar{y}$, is removed by the FTC.

9.2.3 Cell-averaging CFAR detector

In thermal noise or clutter an adaptive estimation of the mean level can be achieved by a cell-averaging (CA) CFAR system [3] that estimates the expected mean level of the cell under test in the absence of a target, as an average of the surrounding cells in range and/or azimuth.

Figure 9.2 shows the operation of a double-sided CA CFAR system. The cell under test is compared with a threshold, t, which is estimated from the mean level of the surrounding range cells, with clutter values E_i. In the example in Figure 9.2, $M/2$ cells either side of the cell under test are used to estimate the mean clutter level. To reduce contamination from range-extensive targets in the cell under test, the cell-averaging cells are separated by a gap of G cells, as shown in Figure 9.2. A threshold multiplier, α, is used to scale the estimate of mean level to give the required threshold, t:

$$t = \frac{\alpha}{M}\left(\sum_{i=-M/2-G}^{i=-G-1} E_i + \sum_{i=G+1}^{i=M/2+G} E_i\right) \tag{9.7}$$

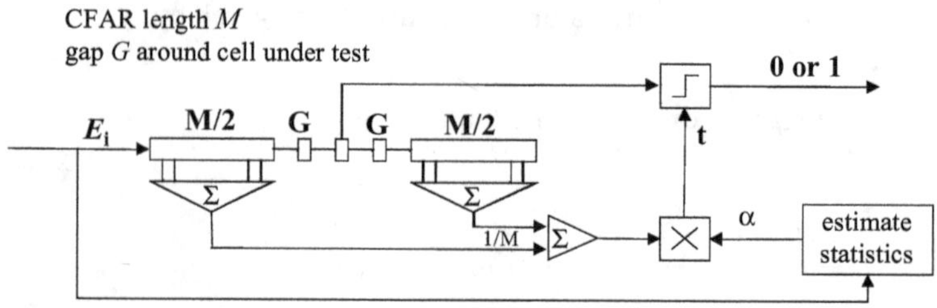

Figure 9.2 Double-sided cell-averaging CFAR detector

The value of threshold multiplier, α in Figure 9.2, to achieve a given value of P_{FA} is dependent on the amplitude statistics of the clutter and, as will be seen, the configuration of the cell-averaging system. For clutter or noise with a known constant pdf, this can be calculated in advance for a given value of probability of false alarm (P_{FA}) at the threshold. In unknown or varying clutter statistics, the clutter pdf must be estimated directly or indirectly so that in turn the threshold multiplier can be estimated. This is discussed later in the chapter. For the remainder of this section, the performance of the CA CFAR systems are assessed on the assumption that the amplitude statistics are known and it is then shown how to calculate the appropriate value of α.

A number of points must be noted concerning the operation of a CA CFAR. First, a CA CFAR does not estimate the clutter mean level exactly, as it only uses a finite data sample. The estimate of mean level has a distribution of its own giving a fluctuating threshold instead of the fixed value that would result from ideal knowledge of the clutter or noise statistics. The effect of this threshold fluctuation is that the target signal-to-noise ratio, SNR, required for a given probability of detection, P_D, and probability of false alarm, P_{FA}, is larger than for the ideal case. This increase in SNR required is defined as a CFAR loss.

As will be discussed later, the CFAR loss is also observed as an increase in the value of the threshold multiplier, α, to achieve the desired value of P_{FA}. For $P_D = 0.5$, this is very close to the CFAR loss calculated from the change in signal-to-noise ratio for a given P_{FA}.

A number of assumptions are implicit in the basic cell-averaging systems described above. In particular it is assumed that the clutter is spatially uniform and that the surrounding cells are representative of the cell under test. The larger the number of surrounding cells that can be used, the better the estimate of the mean clutter level, provided that the more distant cells are still representative of the cell under test. In the analysis given below, it is also assumed that the averager cells are independent of the cell under test and the clutter distribution is stationary over the extent of the cell-averaging. For rapidly varying clutter a short cell-averager must be used, giving a poor estimate, quantified as a CFAR loss.

Some environments present particular problems for CA CFAR systems. For example, clutter edges can cause false alarms or excessive detection loss. This is illustrated in Figure 9.3, which shows in idealised form the variation in P_{FA} at a

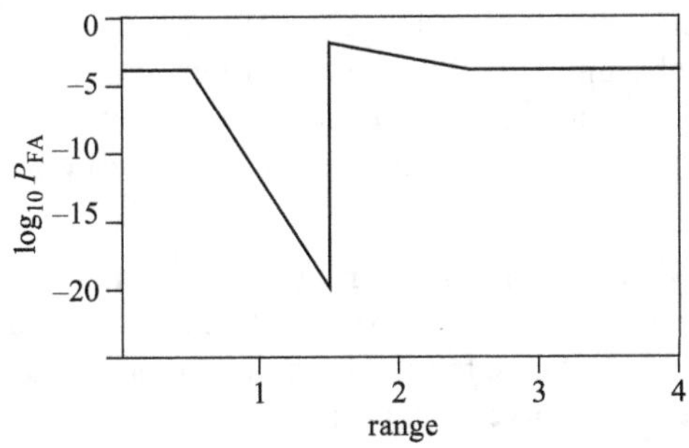

Figure 9.3 *Variation in probability of false alarm for a double-sided CFAR operating over a 10 dB step in clutter intensity, at range position 1.5; cell-averager length 1 range unit either side of the cell under test and nominal* $P_{FA} = 10^{-4}$

clutter edge in response to the threshold set by a double sided CA CFAR, with cell-averager lengths of 1 range unit either side of the cell under test (assumed to estimate the local mean level exactly). The clutter edge comprises a step in amplitude of 10 dB at a range of 1.5 units in Figure 9.3 and the nominal false alarm rate is 10^{-4}. As the cell-averager encounters the higher clutter level, its estimate of the locally lower level becomes too high and the false alarm rate drops. Immediately after the edge the opposite effect occurs and a high false alarm rate is observed before the threshold finally settles to the correct value again. The increased false alarms at such an edge persists from scan to scan but may not concern the radar operator if their source is obvious (e.g. the land–sea interface). However, alarms due to unexpected edges, such as clouds or unusual sea clutter conditions, can pose more of a problem, requiring operator intervention to investigate them.

Another problem occurs in high target densities. If there are targets in the cell-averager cells the system no longer maintains a true CFAR performance. This may be acceptable in very sparse target environments but degrades performance in high target densities.

In addition to the standard CA CFAR, a large number of variants have been proposed (e.g. Reference 4) to mitigate some of these specific problems in practical scenarios. The following paragraphs summarise briefly some of the more common variants, taking the basic CA CFAR configuration as the baseline for comparison. It should be noted that all the techniques introduce additional losses compared with the double-sided CA CFAR system when used in spatially uniform clutter or noise, with a single target present.

9.2.3.1 CFAR variants

Cell-averaging CFAR: This configuration estimates the mean level from cells surrounding the cell under test in range and/or azimuth and uses that to set a threshold in the cell under test. It is susceptible to increased false alarms from clutter edges and desensitisation from targets or impulsive interference in the CA window. Its operation assumes spatially uniform clutter or variations that occur slowly compared to the length of the cell window. A double-sided configuration is shown in Figure 9.2. Single-sided configurations may also be used for specific applications or ease of implementation.

'Greater of' CFAR: This configuration uses the greater of (GO) the mean level estimates either side of the cell under test. This technique is widely used to limit false alarms during clutter transitions, but is susceptible to interfering targets in the CFAR window. It cannot easily resolve closely spaced targets.

'Smaller of' CFAR: This configuration uses the smaller of (SO) the mean level estimates either side of the cell under test. It is able to resolve closely spaced targets but gives a high false alarm rate around clutter edges.

Ordered statistics (OS) CFAR: The M range cells in the cell-averager window $(z(1), z(2), \ldots, z(M))$ are ranked according to magnitude to give ordered samples, $z(1) < z(2) < \cdots < z(M)$. The noise power is estimated from the magnitude of the kth largest cell. The optimum value of k is found to be around $3M/4$. This approach eliminates up to $M - k$ interfering targets (assuming they all have magnitudes significantly greater than the clutter) with no degradation in performance.

Trimmed mean (TM) CFAR: Here the samples are ranked according to magnitude and $T1$ samples are trimmed from the lower end and $T2$ samples trimmed from upper end. If $T1 = k - 1$ and $T2 = M - k$ the detector becomes an ordered statistics CFAR and for no trimming it becomes a CA CFAR. Ideally $T2$ should correspond to the number of interfering targets whist $T1$ should be as small as possible. However, to minimise alarms at clutter boundaries $T1$ should be large and $T2$ should be small.

Censored mean (CM) CFAR: Another approach to reducing the effects of interfering targets is to censor the largest n samples of the window samples ranked according to magnitude. The remaining $M - n$ samples are averaged to estimate the clutter mean level as for the normal CA CFAR.

Variations on a theme: There are many other variations [4] including excision CFAR, generalised censored mean level CFAR, greater of order statistics estimator CFAR, censored greater-of CFAR and so on.

9.2.3.2 CFAR loss in noise

Earlier in this section the concept of a CFAR loss was introduced to account for the use of an estimate of the mean interference level, as opposed to its exact value. In this section, the CFAR loss is quantified for thermal noise or spatially uniform Rayleigh clutter.

Referring back to the CA CFAR circuit in Figure 9.2, the threshold, t, is formed from the average of M surrounding clutter cells and a threshold multiplier, α. The estimate of the mean level is not exact and fluctuates from cell to cell. This fluctuation means that a higher value of threshold multiplier must be used to achieve the required average value of probability of false alarm.

The threshold has its own pdf, which is calculated here for a cell-averager in noise following a square-law detector and pulse-to-pulse integration of N pulses.

For a square law detector, the pdf of a Swerling 2 target with signal-to-noise ratio S, in thermal noise (mean level unity) following the integration of N pulses is given by

$$P(z) = \frac{1}{(1+S)^N} \frac{z^{N-1}}{\Gamma(N)} \exp\left(-\frac{z}{1+S}\right), \qquad 0 \le z \le \infty \tag{9.8}$$

If the cell-averager uses M samples, then the CFAR threshold, t, has a distribution given by

$$P(t) = \left(\frac{M}{\alpha}\right)^{MN} \frac{t^{MN-1}}{\Gamma(MN)} \exp\left(-\frac{Mt}{\alpha}\right), \qquad 0 \le t \le \infty \tag{9.9}$$

As the threshold fluctuates, so P_{FA} and P_D vary. The average values of P_{FA} and P_D are calculated by averaging over all possible values of t.

The P_D following this threshold is therefore given by

$$\overline{P_D} = \int_0^\infty \left(\int_t^\infty P(z)\,dz\right) P(t)\,dt \tag{9.10}$$

This can be solved by a change of variables to $y = z/t$ and changing the order of integration to give

$$\begin{aligned}\overline{P_d} &= \int_1^\infty \frac{\Gamma(N+MN)}{\Gamma(N)\,\Gamma(MN)(1+S)^N} \\ &\quad \times \left(1 + \frac{y\alpha}{M(1+S)}\right)^{-N-MN} y^{N-1} \left(\frac{\alpha}{M}\right)^N dy \\ &= \left(\frac{M(1+S)}{\alpha + M(1+S)}\right)^{MN} \sum_{p=0}^{N-1} \frac{(MN-1+p)!}{(MN-1)!\,p!} \\ &\quad \times \left(\frac{\alpha}{\alpha + M(1+S)}\right)^p \end{aligned} \tag{9.11}$$

It may be noted that the integrand above is the distribution of the signal plus noise after normalisation by the cell-averager threshold. The value of α required for a given P_{FA} is determined from equation (9.11) with $S = 0$.

The CFAR loss is defined as the increase in signal-to-noise ratio to maintain the same P_{FA} and P_D as a fixed threshold. A close approximation to the CFAR loss for $P_D = 0.5$ can be obtained from the increase in the threshold multiplier, α, to achieve

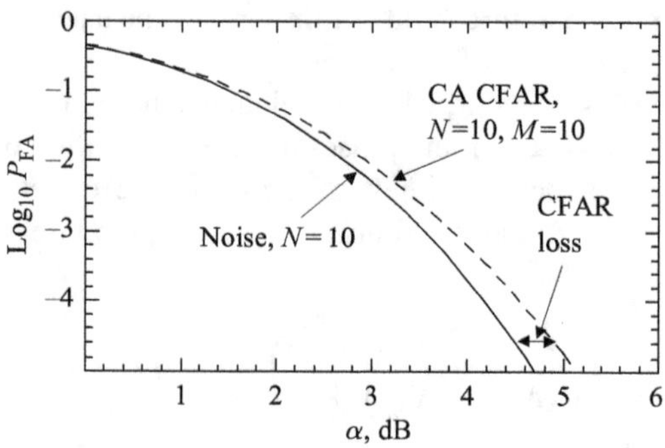

Figure 9.4 Distribution of noise after normalisation by the CA CFAR threshold, illustrating CFAR loss as a function of P_{FA}

a constant average P_{FA}. The fixed threshold P_D is

$$P_D = \frac{\Gamma(N, \alpha N/(1+S))}{\Gamma(N)} \tag{9.12}$$

where Γ(a,b) is the incomplete Gamma function [5]. P_{FA} is again found by setting $S = 0$.

For the case when $N = 1$, equation (9.11) can be simplified to give

$$\overline{P_D} = \left(1 + \frac{\alpha}{M(1+S)}\right)^{-M} \tag{9.13}$$

and equation (9.12) simplifies to

$$P_D = \exp\left(-\frac{\alpha}{1+S}\right) \tag{9.14}$$

Setting $S = 0$, we obtain the approximate CFAR loss (for $P_D = 0.5$ and $N = 1$) from the increase in the value of α:

$$\text{CFAR loss} \cong \frac{M(1 - P_{FA}^{-1/M})}{\log_e(P_{FA})} \tag{9.15}$$

As discussed above, the effect of the fluctuating threshold can also be analysed by normalising the noise to the threshold (see equation (9.11)). Figure 9.4 shows the cumulative distribution of the sum of $N = 10$ independent samples of thermal noise, following square-law detection. Also shown is the distribution of the noise following normalisation by the running estimate of mean level from a cell-average with $M = 10$ independent cells. Since the resulting increase in α is approximately equal to the CFAR loss (for $P_D = 0.5$) it can be read off here as a function of P_{FA}.

The above results are for the Swerling Case 2 target. Results for other target models in thermal noise can be found in Reference 6.

Finally, Figure 9.5 shows the CFAR loss as a function of cell-averager length for single pulse detection, $P_D = 0.5$ and various values of P_{FA}. For typical values

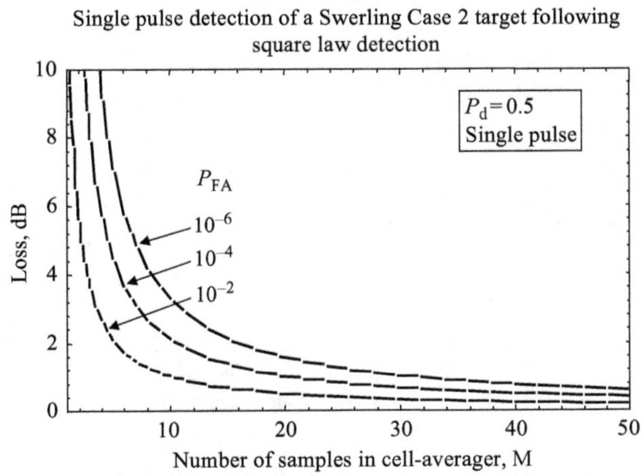

Figure 9.5 Example of CFAR loss for single pulse detection

of P_{FA}, the loss is less than 1 dB for large values of M. For small values of M the loss becomes very large. The use of pulse-to-pulse integration ($N > 1$) reduces the CFAR loss.

9.2.3.3 Greater-of CFAR in noise

Similar estimates can be made of the losses associated with other cell-averaging variants. For the GO-CFAR, the threshold is formed using the greater of the two cell-averaging regions either side of the cell under test. For single pulse detection of a Swerling 2 target and a square law detector, assume that each side of the cell-averager has $M/2$ samples (clearly M must be an even integer) and produces two estimates, z_1 and z_2, respectively. Let each z have a pdf $P(z)$ and probability distribution $F(z)$. For two independent samples z_1 and z_2 let the maximum be m; the pdf of m is given (from Appendix 1 equation (A1.193) or [7]) by $P_2(m) = 2[F(m)]P(m)$. Now:

$$P_2(m) = 2F(m)P(m)$$

$$= 2\frac{m^{M/2-1}}{\Gamma(M/2)}\exp(-m)\int_0^m \frac{z^{M/2-1}}{\Gamma(M/2)}\exp(-z)\,dz$$

$$P_D = \int_0^\infty P_2(m)\int_{2\alpha m/M}^\infty \frac{1}{1+S}\exp\left[-\frac{y}{1+S}\right]dy\,dm$$

$$= \int_0^\infty P_2(m)\exp\left(-\frac{2\alpha m}{M(1+S)}\right)dm \tag{9.16}$$

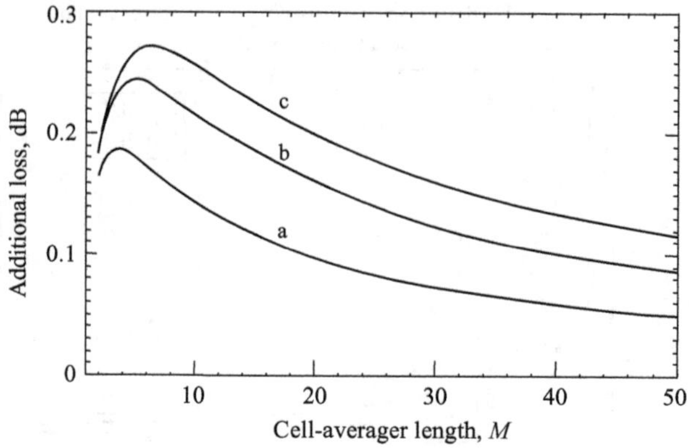

Figure 9.6 Additional loss of GO CFAR compared to a standard CA CFAR; Swerling Case 2 target, $P_D = 0.5$.
a: $P_{FA} = 10^{-2}$; b: $P_{FA} = 10^{-4}$; c: $P_{FA} = 10^{-6}$

This expression for P_D can be expressed in a closed form:

$$P_D = \frac{2}{\Gamma(M/2)^2} \int_0^\infty m^{M/2-1} \exp(-m) \exp\left(-\frac{2\alpha m}{M(1+S)}\right) \times \int_0^m z^{M/2-1} \exp(-z)\, dz\, dm$$

$$= \frac{2\Gamma(M)}{\Gamma(M/2)^2} \int_0^1 p^{M/2-1} \frac{dp}{(1 + (2\alpha/(M(1+S))) + p)^M}$$

$$= \frac{2}{(M/2-1)!}\left[(M/2-1)!\left(\frac{M(1+S)}{M(1+S)+2\alpha}\right)^{M/2} - \sum_{p=0}^{M/2-1} \frac{(M/2-1+p)!}{p!\,2^{p+M}}\left(\frac{M(1+s)}{\alpha+M(1+s)}\right)^{M+p}\right] \quad (9.17)$$

Figure 9.6 shows the additional loss associated with a GO CFAR compared with a standard double-sided CA CFAR, with $M/2$ cells either side of the cell under test.

9.2.3.4 OS CFAR in noise

In an OS CFAR, the M range cells in the cell-averager window $(z(1), z(2), \ldots, z(M))$ are ranked according to magnitude to give ordered samples, $z(1) < z(2) < \cdots < z(M)$. The noise power is estimated from the magnitude of the kth largest cell.The optimum value of k is found to be around $3M/4$. This approach eliminates up to $M-k$ interfering targets (assuming they all have magnitudes significantly greater than the clutter) with no degradation in performance.

Let $Z_1, Z_2, \ldots, Z_M$ be a sequence of statistically independent, identically distributed random variables, with pdf $P(z)$ and distribution function $F(z)$. The pdf of the kth value of the ordered statistic is given by Appendix 1 equation (A1.196) or Reference 8:

$$F_{Z_{(k)}}(z) = P_k(z)$$

$$= k\binom{M}{k}(1 - F_Z(z))^{M-k}(F_Z(z))^{k-1}P_Z(z) \tag{9.18}$$

with

$$P_1(z) = M(1 - F_Z(z))^{M-1}P_Z(z) \quad \text{for } k = 1$$

and

$$P_M(z) = M(F_Z(z))^{M-1}P_Z(z) \quad \text{for } k = M$$

For a square law detector $P_Z(z) = (1/x)(\exp(-z/x))$ and then

$$P_k(z) = \frac{k}{x}\binom{M}{k}\left(\exp\left(-\frac{z}{x}\right)\right)^{M-k+1}\left(1 - \exp\left(-\frac{z}{x}\right)\right)^{k-1} \tag{9.19}$$

If the CFAR threshold multiplier is α, then the P_{FA} is given by

$$P_{\text{FA}} = \int_0^\infty \exp\left(-\frac{\alpha z}{x}\right) P_k(z)\,dz$$

$$= k\binom{M}{k}\int_0^\infty e^{-(\alpha+M+1-k)y}(1 - e^{-y})^{k-1}dy \tag{9.20}$$

and P_{D} is given by:

$$P_{\text{D}} = k\binom{M}{k}\int_0^\infty e^{-((\alpha/(1+S))+M+1-k)y}(1 - e^{-y})^{k-1}dy$$

$$= \frac{M!}{(M-k)!}\left(\prod_{r=1}^{k}\left(1 + M + \frac{\alpha}{1+S} - r\right)\right)^{-1} \tag{9.21}$$

The expression for P_{FA} is found by putting $S = 0$.

It can be seen that the threshold multiplier, α, controlling P_{FA} does not depend on the clutter mean level x and so the method is a CFAR under changing mean [8].

As with other configurations, the CFAR loss for the OS CFAR is calculated as the additional SNR required to achieve a given P_{D} and P_{FA} compared with a fixed

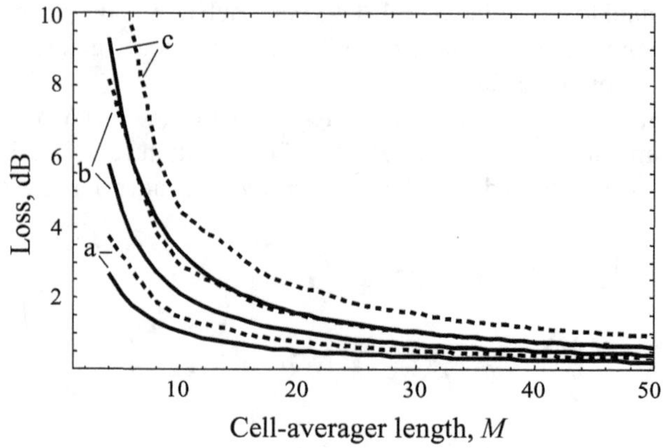

Figure 9.7 *Examples of CFAR loss of OS CFAR, compared with CA CFAR; Swerling Case 2 target, $P_D = 0.5$*

——: *CA-CFAR*

------: *OS-CFAR, $k = 3M/4$*

a: $P_{FA} = 10^{-2}$; b: $P_{FA} = 10^{-4}$; c: $P_{FA} = 10^{-6}$

threshold. It is shown in Figure 9.7 for single pulse detection and $k = 3M/4$ following a square law detector, with a Swerling Case 2 target and $P_D = 0.5$. The equivalent results for the CA CFAR are shown for comparison.

It should be noted that in the cases of the GO-CFAR and OS-CFAR, the use of the threshold multiplier, α, as a guide to CFAR loss is not applicable, as the quantities being multiplied by α are not comparable estimates of the mean level.

9.2.3.5 CFAR loss in K-distributed clutter

The calculations of CFAR loss above have assumed operation in noise or spatially uniform Rayleigh clutter. The CFAR loss in K-distributed clutter can be calculated using a simple extension to the method for thermal noise. For the discussion in this section, it is assumed that each of the M cells of the cell-averager provides an independent sample of the overall clutter distribution.

It should be noted that the calculations given here are for a linear detector, whilst those for noise in the earlier section were for a square law detector. The difference in values will be very small (typically <0.1 dB), especially for large value of M and low values of P_{FA}.

For a linear detector and single pulse detection, the pdf of the K distribution is given (see (4.25)) by

$$P(E) = \frac{4b^{(\nu+1)/2}E^{\nu}}{\Gamma(\nu)}K_{\nu-1}\left(2E\sqrt{b}\right); \qquad o \le E \le \infty \tag{9.22}$$

where the mean square clutter level is given by $\langle E^2 \rangle = \nu/b$ and the mean clutter level is

$$\langle E \rangle = \sqrt{\frac{\pi}{b}} \frac{\Gamma(\nu + 1/2)}{2\Gamma(\nu)}$$

As in equation (9.7) the CA CFAR threshold is given by

$$t = \frac{\alpha}{M} \left(\sum_{i=-M/2-G}^{i=-G-1} E_i + \sum_{i=G+1}^{i=M/2+G} E_i \right) \tag{9.23}$$

The pdf of the threshold $P(t)$ is assumed to be the pdf of the sum of M independent K-distributed samples. As with noise, the CA CFAR probability of false alarm is given by

$$P_{\text{FA}} = \int_0^\infty \left(\int_t^\infty P(E)\, dE \right) P(t)\, dt \tag{9.24}$$

The value of α required to achieve a given P_{FA} can be compared with the ideal fixed threshold where

$$P_{\text{FA}} = \frac{2}{\Gamma(\nu)} \alpha^\nu \left(\frac{\sqrt{\pi}}{2} \frac{\Gamma(\nu + 1/2)}{\Gamma(\nu)} \right)^\nu K_\nu \left(\alpha \frac{\sqrt{\pi}\Gamma(\nu + 1/2)}{\Gamma(\nu)} \right) \tag{9.25}$$

where now

$$\alpha = \frac{\text{threshold}}{\text{mean}} = 2t\sqrt{b/\pi}\,\Gamma(\nu)/\Gamma(\nu + 1/2)$$

Semi-analytic solutions can be found for $\nu = 0.5$ and 1.5.

For $\nu = 0.5$:

$$P(E) = 2\sqrt{b} \exp\left(-2\sqrt{b}\, E\right) \tag{9.26}$$

and

$$P(t) = \frac{2\sqrt{b}M}{\alpha\, \Gamma(M)} \left(\frac{2\sqrt{b}Mt}{\alpha} \right)^{M-1} \exp\left(-2\sqrt{b}Mt/\alpha\right) \tag{9.27}$$

For the ideal fixed threshold:

$$P_{\text{FA}} = \exp(-\alpha) \tag{9.28}$$

Similarly for $\nu = 1.5$

$$P(E) = 4bE \exp\left(-2\sqrt{b}\, E\right) \tag{9.29}$$

and

$$P(t) = \frac{t^{2M-1}}{\Gamma(2M)} \left(\frac{2\sqrt{b}M}{\alpha} \right)^{2M} \exp\left(-2\sqrt{b}Mt/\alpha\right) \tag{9.30}$$

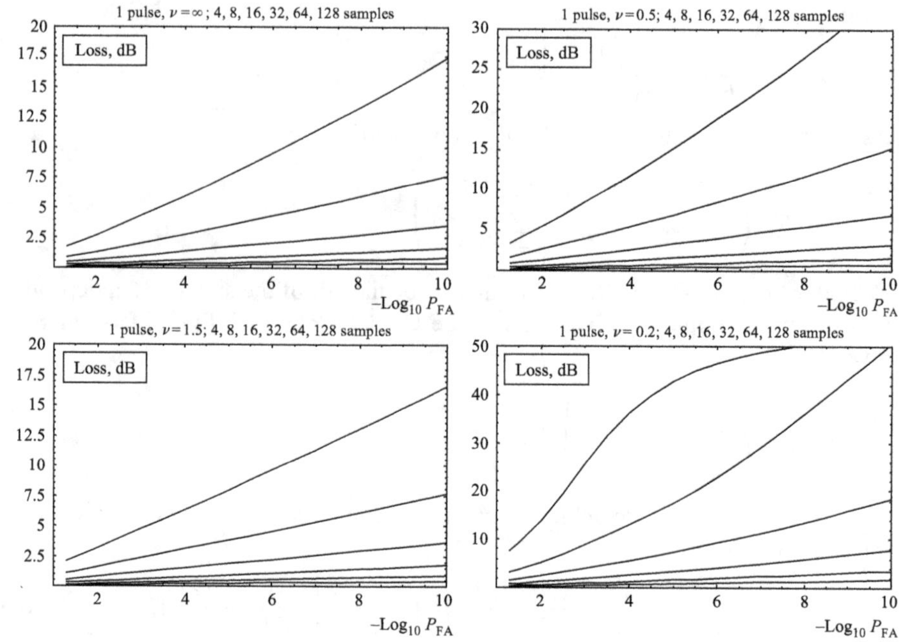

Figure 9.8 CA CFAR loss for single pulse detection in K-distributed clutter (linear detector)

and for the ideal fixed threshold

$$P_{FA} = (1 + \alpha) \exp(-\alpha) \tag{9.31}$$

A similar approach is taken for other values of ν and for CA CFAR operation following pulse-to-pulse integration of N pulses. For these cases semi-analytic solutions are not available and now $P(E)$ must be calculated numerically and $P(t)$ is obtained by M-fold numerical convolution of $P(E)$.

Figure 9.8 shows the values of CFAR loss calculated for single pulse detection and assuming M independent samples in the cell-averager. In this case, the CFAR loss is approximated (as we did earlier) by the increase in threshold multiplier needed to maintain a given P_{FA}. The values for $\nu = \infty$ (Gaussian clutter, Rayleigh envelope) are included for comparison. It can be seen that for small values of ν and M, the losses are very large. For example, the loss for $\nu = 0.2$ and $M = 4$ exceeds 50 dB for low values of P_{FA}. (It should be noted that the unexpected shape of this particular curve is probably due to numerical errors in the convolution and integration calculations.)

Figure 9.9 shows equivalent results following pulse-to-pulse integration of the clutter returns over 10 pulses. It is assumed that the speckle component of the clutter is independent from pulse-to-pulse but the underlying mean level is constant. The losses are less than for single pulse detection but again it can be seen that very large

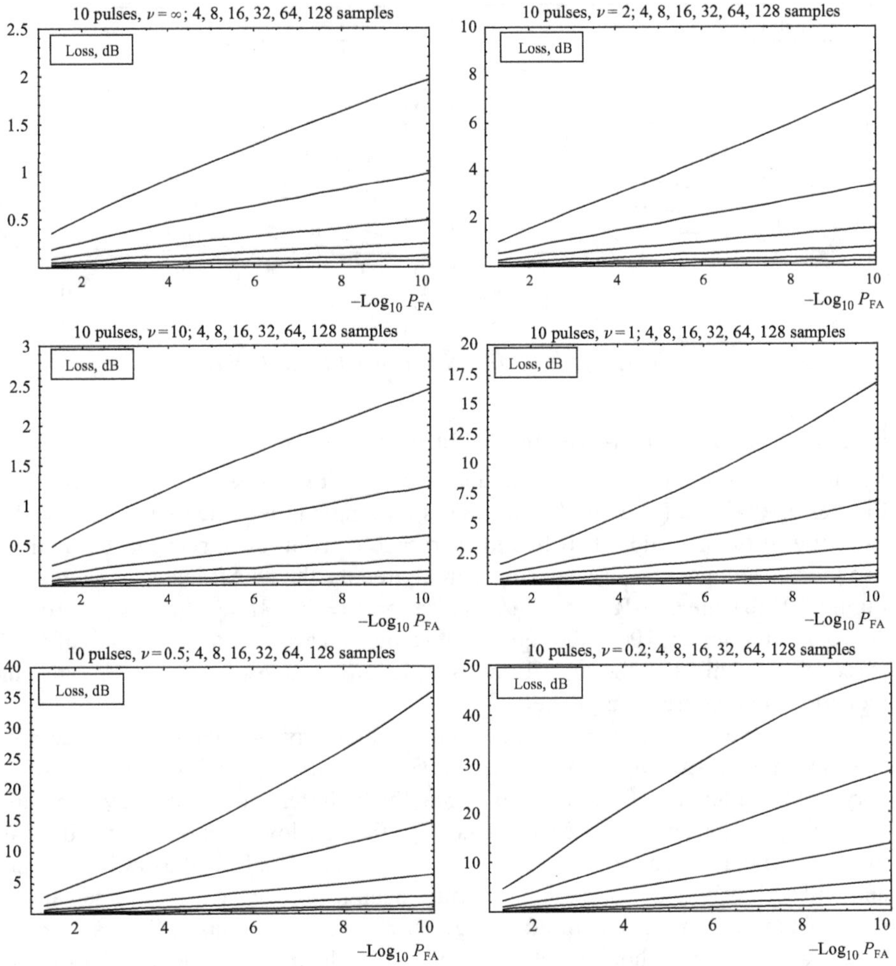

Figure 9.9 *CA CFAR loss for detection in K-distributed clutter following integration over 10 pulses (linear detector, assuming independent speckle from pulse to pulse)*

values are observed for small values of ν and M. Loss values above about 40 dB may also be subject to numerical errors, as in Figure 9.8.

The very large values of loss are only observed for very short averager lengths and very spiky clutter. For the cell-averager configurations typically employed in practical radar systems and for typical clutter conditions, the losses incurred are quite acceptable. Also, as described in Section 9.2.3.7, when spatial correlation is present, short averager lengths can be beneficial. Therefore a trade-off is necessary to optimise the averager length, depending on clutter conditions.

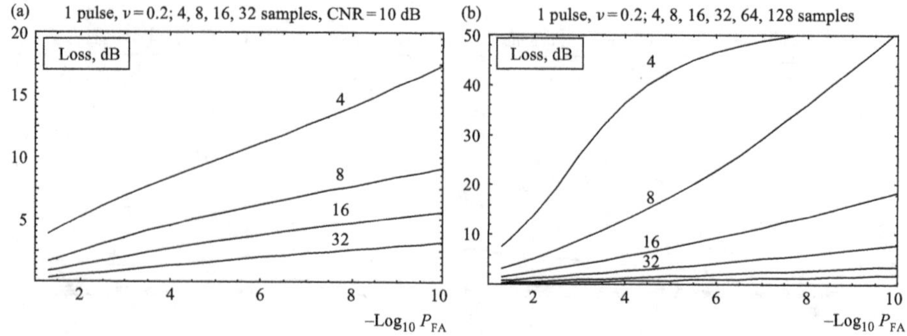

Figure 9.10 *CFAR loss, 1 pulse, (a) $CNR = 10\,dB$; (b) $CNR = \infty$*

9.2.3.6 CFAR loss in K-distributed clutter plus noise

A similar analysis has also been undertaken for CFAR loss in clutter plus noise. Figures 9.10(a) and (b) compare the results for single pulse detection with $\nu = 0.2$, and clutter-to-noise ratios of 10 dB and ∞, respectively. It can be seen that the addition of a modest amount of noise can considerably reduce the CFAR loss. For example in Figure 9.10(b), the CFAR loss with no noise for $\nu = 0.2$, $M = 8$ and $P_{FA} = 10^{-6}$ is 24 dB. From Figure 9.10(b), the equivalent value with a clutter-to-noise ratio of 10 dB is reduced to 6 dB. For shorter cell-averager lengths and lower false alarm rates, the reductions in loss are even greater.

That such a small amount of added noise should have such a marked effect may at first sight be surprising. However, this observation can be explained in terms of the very spiky clutter ($\nu = 0.2$) in this example. Such clutter is dominated by very large isolated spikes, set against a background significantly lower than the overall mean. For much of the time, a short cell-averager will only see data from the low background and will seriously underestimate the overall mean level. Hence a very large value of threshold multiplier, α, is required for a given value of P_{FA} compared to that required for an ideal fixed threshold which is set relative to the true mean level. The addition of even a small amount of thermal noise significantly increases the average clutter-plus-noise level surrounding the spikes and hence the cell-averager achieves a better estimate of the overall mean and the CFAR loss is reduced. As the noise increases the effect of the clutter spikes is reduced until in the limit the CFAR loss is that for noise alone, which is always less than or equal to the equivalent loss for K-distributed clutter.

9.2.3.7 Ideal CFAR detection and CFAR gain in K-distributed clutter

In the results presented so far, it has been assumed that the cell-averager uses independent samples of the overall clutter distribution to estimate the cell under test. If the cell under test is correlated with the surrounding data, a better estimate may be achieved in some circumstances. Figure 9.11 shows an example of the underlying mean level of clutter and the threshold achieved by a short double-sided cell averager

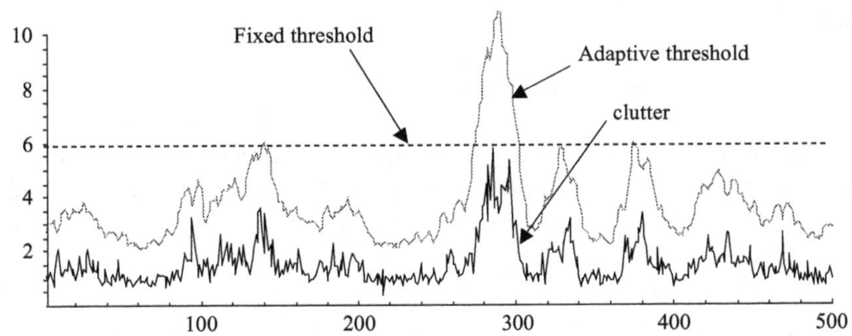

Figure 9.11 Fixed and CA CFAR thresholds in spatially correlated clutter

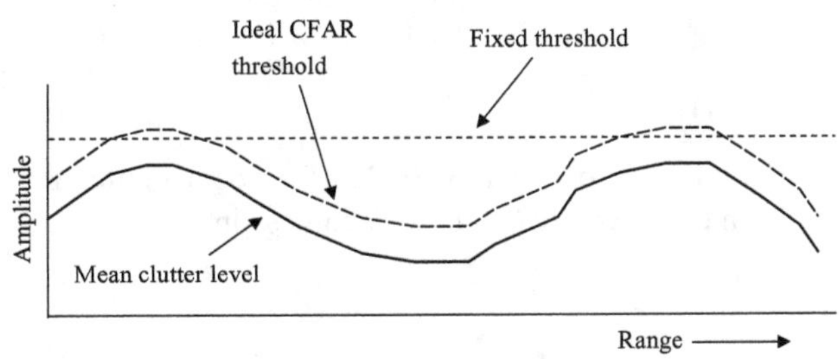

Figure 9.12 Ideal CFAR threshold

($M = 10$ in Figure 9.2). Also in Figure 9.11 is the equivalent fixed threshold required to achieve the same P_{FA}, averaged over all range cells. It can be seen that, on average, the CA CFAR threshold is much lower than the fixed threshold. In the CA CFAR this is observed as a lower value of threshold multiplier, α, to achieve a given value of P_{FA}. Now, rather than a CFAR loss, it appears that there is a CFAR gain, relative to the 'ideal' fixed threshold.

The limiting case, where the threshold follows the underlying mean level exactly, is shown, in simplified form, in Figure 9.12. This is known as 'ideal CFAR' performance [9]. Ideal CFAR performance in K-distributed clutter plus noise can be calculated in a similar fashion to fixed threshold detection.

Equations (9.32)–(9.34) show the performance calculations for a fixed threshold. (Note that we are using a square law detector, for analytic convenience, rather than the linear detector assumed in Sections 9.2.3.5 and 9.2.3.6).

Following the sum of N square-law detected pulse returns, assuming that independent samples of the speckle component are obtained for each pulse, it is found (equations (8.25) and (8.31)) for a fixed threshold Y, that

$$P_{FA}(Y) = \int_0^\infty P(x) \frac{1}{\Gamma(N)} \Gamma\left(N, \frac{Y}{x + p_n}\right) dx \tag{9.32}$$

where

$$P(x) = \frac{x^{\nu-1} b^{\nu}}{\Gamma(\nu)} \exp(-bx) \tag{9.33}$$

The notation here is the same as in Chapter 8, so that x is the local clutter power, ν/b is the overall mean clutter power and p_n is the noise power. The fixed threshold Y is given by

$$Y = \alpha N \left(\frac{\nu}{b} + p_n \right) \tag{9.34}$$

where α is the threshold multiplier. The value of α is derived by solving equation (9.32) to (9.34) for a given value of P_{FA}. The threshold multiplier for a CA CFAR tends to this value as the length of the cell-averager, M, becomes large.

For ideal CFAR detection it is assumed that the threshold can adapt exactly to the local mean intensity of the clutter, x. The ideal CFAR threshold, Y, is thus:

$$Y = \alpha N(x + p_n) \tag{9.35}$$

When this is substituted into equation (9.32) the local clutter power plus noise $(x + p_n)$, cancels and the overall P_{FA} now becomes simply

$$P_{\text{FA}}(Y) = \frac{\Gamma(N, \alpha N)}{(N-1)!} \tag{9.36}$$

which is the same as the fixed threshold result in noise or spatially uniform Rayleigh clutter. The probability of detection is calculated in a manner similar to that described in Chapter 8, but again with the threshold following the local clutter mean. Here, however, an integral over the local mean, x, is required, as the local SCR changes as a function of x. Figure 9.13 shows examples of ideal CFAR detection curves for different values of clutter-to-noise ratio and clutter shape parameter. There are several points to note from these results. First, for ideal CFAR detection, the best performance is achieved in the spikiest clutter (i.e. the lowest value of ν). This is to be expected, as in very spiky clutter the clutter has a very low local mean level between very large spikes. In these regions a target is more easily detected, provided that the threshold adapts appropriately. Performance in Rayleigh clutter (i.e. $\nu = \infty$) is identical to the fixed threshold performance, since there is no spatial variation of the local mean level. As might be expected, reducing the clutter-to-noise ratio also reduces the performance in spiky clutter, tending to the case of noise (Rayleigh clutter) only. However, it should be noticed that, even for low clutter to noise ratios, ideal CFAR performance in spiky clutter is still better than would be achieved with a fixed detection threshold.

Figure 9.14 shows a comparison of ideal CFAR and fixed threshold detection curves. The difference between the curves represents the CFAR gain.

It can be seen that the CFAR gain is very large in spiky clutter (i.e. small values of ν); the ideal CFAR gain for $\nu = 0.1$, with $P_{\text{D}} = 0.6$ and $P_{\text{FA}} = 10^{-4}$ is about 30 dB. Of course, this ideal performance is predicated on knowing exactly the local mean level of the clutter. In very spiky clutter with little spatial correlation between range

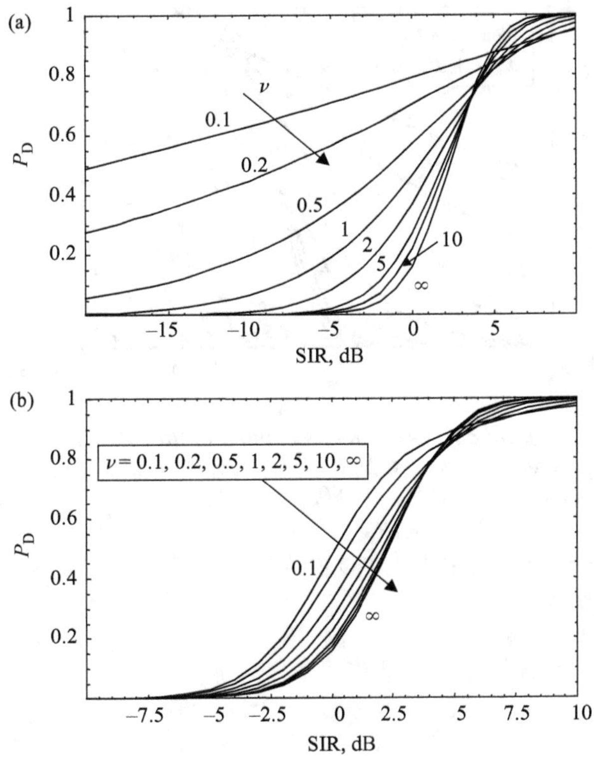

Figure 9.13 P_D *v SIR for ideal CFAR detection, Swerling 2 target,* $N = 10$ *pulses integrated (independent speckle),* $P_{FA} = 10^{-4}$*.*
a: $CNR = 30\,dB$*; b:* $CNR = 0\,dB$*.*

cells it is unlikely that a CA CFAR would be able to estimate the local mean level. However, where significant spatial correlation is present, as illustrated in Figure 9.11, it may be possible to achieve some CFAR gain, if not ideal CFAR performance. This is discussed in the following section.

9.2.3.8 CFAR gain with a CA CFAR

The performance of a CA CFAR in spatially correlated clutter can be assessed in terms of the value of threshold multiplier α needed to achieve a given value of the P_{FA}, averaged over all ranges. Referring to Figure 9.14, some insight can be gained into this use of α as a measure of CFAR gain. For ideal CFAR performance, the value of α is fixed at that required for the speckle component of the clutter (i.e. Gaussian clutter, with shape parameter $\nu = \infty$). For fixed threshold detection, α is given approximately by the SCR required for $P_D = 0.5$. As the CA CFAR performance approaches ideal CFAR performance, the change in α clearly underestimates the CFAR gain in spiky clutter (say, $\nu < 1$) when the P_D is less than, say, 0.8. For $P_D > 0.8$, the change in α

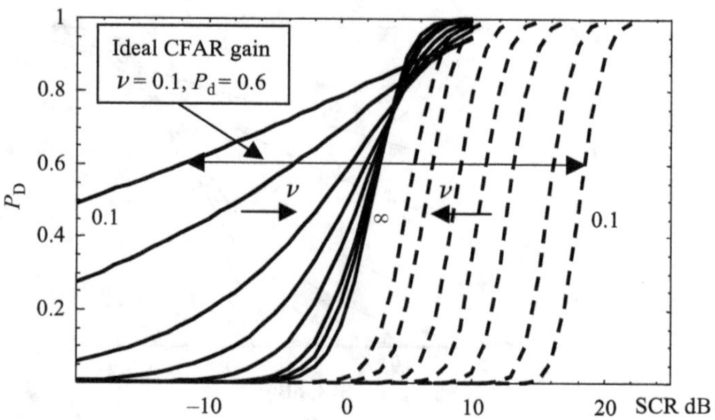

Figure 9.14 Detection curves for K-distributed clutter, 10 pulses integrated, independent speckle; $CNR = \infty$; Swerling 2 target; $\text{Log}\,(P_{FA}) = -4$; *$\nu = 0.1, 0.2, 0.5, 1.0, 2.0, 5.0, 10.0, \infty$*
———: Ideal CFAR
- - - - - - - - -: Fixed threshold.

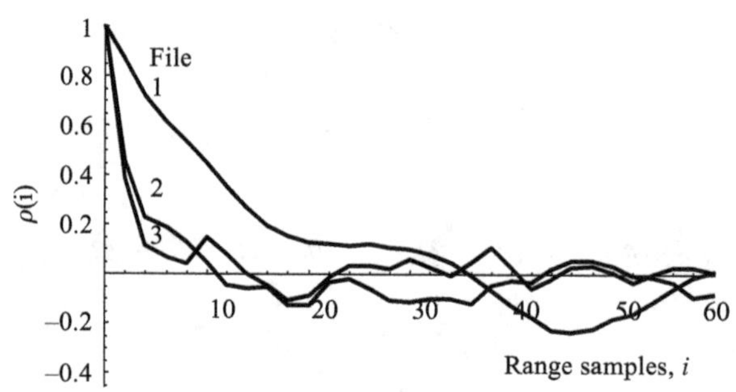

Figure 9.15 Clutter range correlation coefficient ρ for recorded data and examples of simulated range profiles (see also Figure 2.40)

tends to overestimate the CFAR gain. However, in practice ideal CFAR performance is not achieved, especially in very spiky clutter, and in most circumstances the change in α is a reliable guide to relative CFAR performance for P_D values of about 0.5.

The actual value of α required for a given P_{FA} is dependent on the clutter statistics, the spatial correlation of the clutter and the configuration of the cell-averager. It is difficult to calculate explicitly and so performance is best assessed using Monte Carlo simulations. Figure 9.15 shows the range correlation functions of three samples of real clutter data (labelled Files 1, 2 and 3) taken from airborne radar trials (see Section 2.10.1). We have generated spatially correlated compound K-distributed clutter samples from these correlation functions using the method described in Chapter 5 (Section 5.8), and used these to assess CFAR loss or gain.

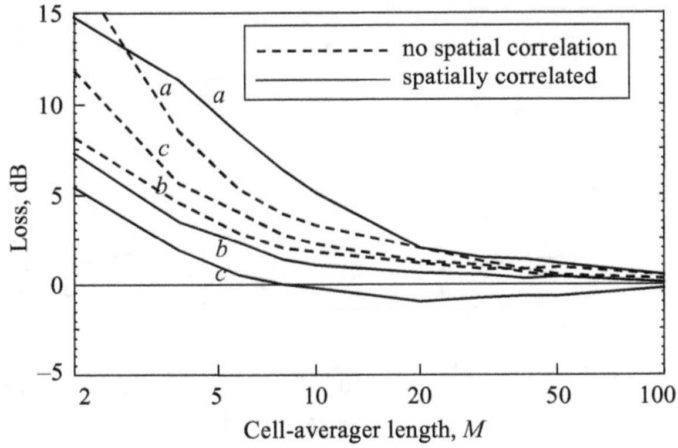

Figure 9.16 *CFAR loss v. M for a single pulse, N = 1, and G = 2, P_{FA} = 0.002. a: File 3 (ν = 0.5); b: File 2 (ν = 5.0); c: File 1 (ν = 1.0)*

Figures 9.16 and 9.17 show the CFAR loss in K-distributed clutter as a function of cell-averager length, M, and with a gap $G = 2$ either side of the cell under test (see Figure 9.2). Also shown for comparison are the values of CFAR loss for clutter with no spatial correlation and the same values of shape parameter. The results in Figure 9.16 are for single pulse detection ($N = 1$). Figure 9.17 is for $N = 10$ pulse returns integrated prior to operation of the CA CFAR and the speckle component of the clutter decorrelated from pulse to pulse (e.g. by using frequency agility). The change in value of threshold multiplier, α, is used here to give a guide to CFAR loss (see discussion above). A number of different behaviours are observed. The data from File 3 have a short but finite correlation length. This has the effect of reducing the number of independent samples in the cell-averager but leaving the cell-averager output decorrelated from the cell-under-test. The CFAR loss in this case is then greater than would be expected for totally uncorrelated clutter samples in the cell-averager. At the other extreme, File 1 shows a negative CFAR loss (i.e. a CFAR gain); this type of behaviour was originally reported in Reference 10. The cell-averager output and the cell-under-test are highly correlated for short cell-averager lengths. The increased fluctuation of the threshold due to small values of M is offset by the ability of the cell-averager to follow local variations in the mean clutter level. This is illustrated in Figure 9.18, which shows some of the clutter from File 1 used in this simulation and the CA CFAR threshold compared with the fixed threshold achieving the same value of P_{FA}.

A similar simulation has been undertaken to assess the effects of noise on CFAR loss and gain. It is shown in Section 9.2.3.6 that the addition of thermal noise considerably reduces the CFAR loss in this case, where the clutter samples across the cell-averager are uncorrelated. This was illustrated in Figure 9.10, where the addition of only a small amount of noise leads to a very marked reduction in loss for short cell-averager lengths. To assess the effect of added noise when the clutter is correlated

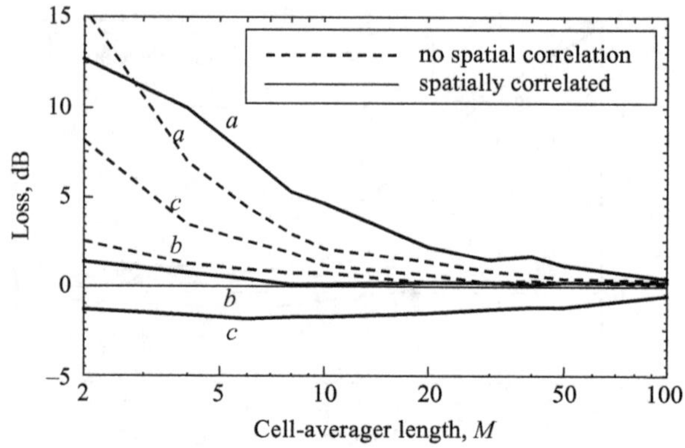

Figure 9.17 *CFAR loss v. M for 10 pulses integrated, N = 10 (assuming independent speckle), and G = 2, $P_{FA} = 0.002$.*
a: File 3 ($\nu = 0.5$); b: File 2 ($\nu = 5.0$); c: File 1 ($\nu = 1.0$)

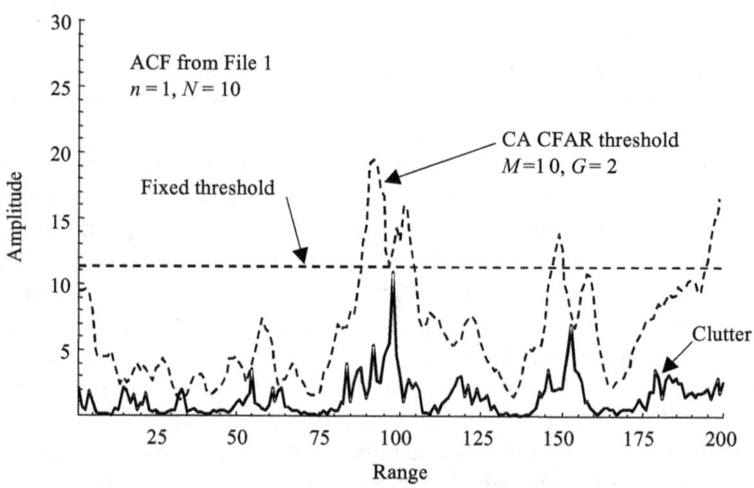

Figure 9.18 *Comparison of fixed and CA CFAR thresholds for the same average P_{FA}; clutter from File 1 (see Figure 9.15), $\nu = 1$, integrated over N = 10 pulses (independent speckle); double-sided CA CFAR, M = 10 and G = 2 (see Figure 9.2).*

across the cell-averager, the simulations above are repeated for each of the Files 1, 2 and 3, with clutter-to-noise ratios, CNR, of 10, 0 and −10 dB.

Figure 9.19 shows the results for File 1, for $N = 10$ pulses integrated. This figure also shows the results for CNR $= \infty$, reproduced from Figure 9.17. This shows the interesting result that the addition of a small amount of noise, with CNR $= 10$ dB, actually increases the CFAR gain. As the noise level increases further, the gain reduces

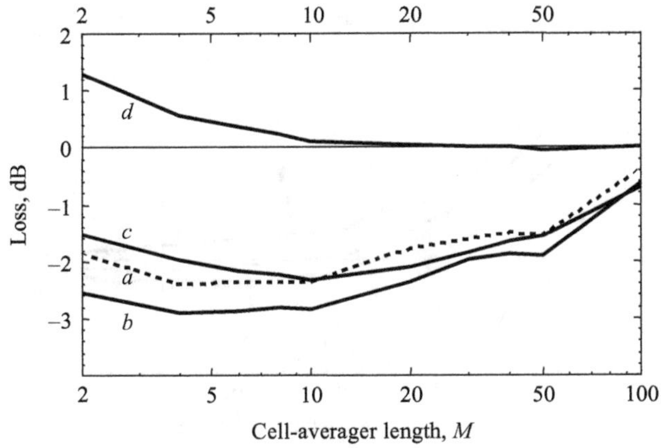

Figure 9.19 *CFAR loss v. M for 10 pulses integrated, N = 10 (assuming independent speckle), and G = 2, P_{FA} = 0.02; File 1 (ν = 1.0). a: CNR = ∞; b: CNR = 10 dB; c: CNR = 0 dB; d: CNR = −10 dB*

with the expected loss observed when the noise dominates (CNR = −10 dB). The effect of added noise in this case is quite subtle. The increased noise level reduces the value of threshold multiplier, α, needed for a fixed threshold and for the CA CFAR threshold. However, for CNR = 10 dB the reduction in α for the cell-averager is slightly greater than for the fixed threshold case, leading to an increase in CFAR gain. Clearly, the highly correlated nature of the higher clutter levels is still dominating the performance in the presence of a small amount of noise.

Figure 9.20 shows the equivalent results for the data in File 1, but with only single pulse returns being processed ($N = 1$). In this case, the added noise has a more expected effect and the CFAR gain is reduced as the CNR reduces.

Figure 9.21 shows the result for File 2 with $N = 10$. Here the addition of noise also has a very interesting effect. In the absence of noise, the CFAR loss is less than that for noise alone and a small CFAR gain is observed for longer cell-averager lengths. The addition of a small amount of noise (CNR = 10 dB) has the effect of reducing the CFAR gain slightly but the overall performance is similar for the short cell-averager lengths. For CNR = 0 dB, it is observed that the CFAR loss decreases for small cell-averager lengths. Finally, as the noise level increases to CNR = −10 dB, the CFAR loss rises to that expected for noise alone. The results for $N = 1$ are not shown, since the CFAR loss in clutter alone is very close to that for noise and the addition of further noise had little effect on the CFAR loss.

Finally, Figure 9.22 shows the effect of adding noise to File 3 and $N = 10$. Here the clutter exhibits little spatial correlation across the cell-averager and the CFAR loss in the absence of noise is very high. As now expected, the addition of thermal noise significantly reduces the CFAR loss for short cell-averager lengths.

In summary, it is observed here that if the clutter exhibits significant spatial correlation across the range returns used in a cell-averager, then the use of short

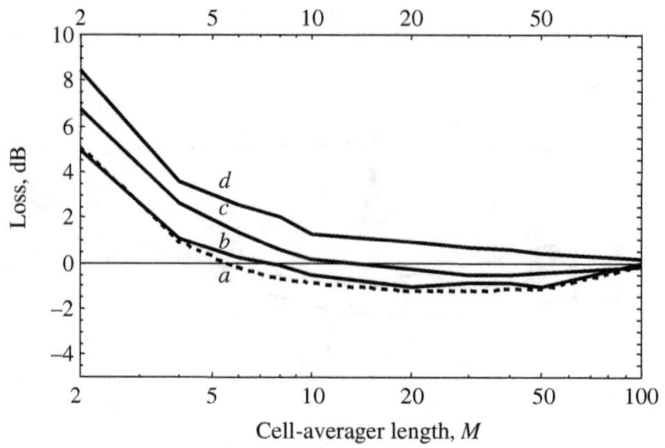

Figure 9.20 *CFAR loss v. M for a single pulse, N = 1, and G = 2, P_{FA} = 0.02; File 1 (ν = 1.0)*
a: CNR = ∞; b: CNR = 10 dB; c: CNR = 0 dB; d: CNR = −10 dB

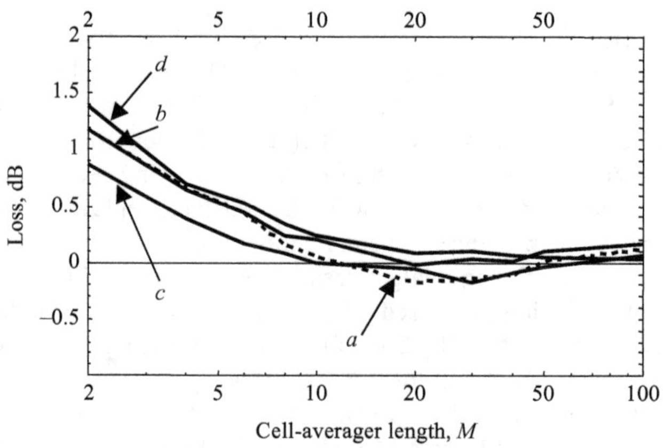

Figure 9.21 *CFAR loss v. M for 10 pusles integrated, N = 10 (assuming independent speckle), and G = 2, P_{FA} = 0.02; File 2 (ν = 5.0)*
a: CNR = ∞; b: CNR = 10 dB; c: CNR = 0 dB; d: CNR = −10 dB

cell-averager lengths can give performance improvements (CFAR gain) compared to performance with a fixed threshold. However, it must be remembered that in uncorrelated clutter the use of too short a cell-averager length can also introduce very large CFAR losses. Thermal noise is always present to some extent and it is shown here that this can significantly reduce the high CFAR loss that is observed in clutter alone, as discussed in Section 9.2.3.6. Interestingly, added thermal noise can in some circumstances result in small increases in CFAR gain where this is observed in highly correlated clutter.

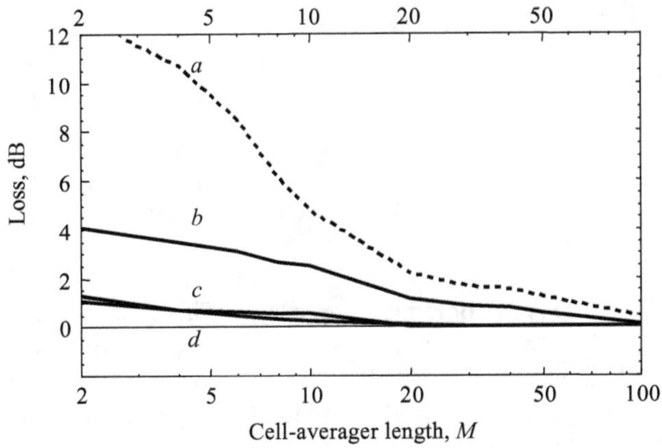

Figure 9.22 *CFAR loss v. M for 10 pulses integrated, $N = 10$ (assuming independent speckle), and $G = 2$, $P_{FA} = 0.02$; File 3 ($\nu = 0.5$) a: $CNR = \infty$; b: $CNR = 10$ dB; c: $CNR = 0$ dB; (d) $CNR = -10$ dB*

9.2.4 *Linear prediction techniques*

The modelling of the spatial correlation of sea clutter using an autocorrelation function (ACF), such as in Section 9.2.3.8, suggests that some form of linear prediction technique could be used to exploit this additional information. Darwish and Cooper [11] proposed the use of a log-lattice filter to predict the clutter level based on the amplitude of surrounding range cells. This technique was tested on some recorded I-band radar data with a range resolution of 5 m. A reduction in CFAR loss of 2.4 dB compared to a double-sided cell-averager was claimed for a cell-averager length of 8 range cells.

A related approach was proposed by Watts [12], based on an autoregressive analysis of clutter spatial correlation. In this technique, it is assumed that the mean level of the cell-under test can be predicted as a weighted sum of succeeding and/or preceding range samples. So for a double-sided cell-averager, of total length M and with gap G either side of the cell under test, as in Figure 9.2, we now have:

$$t = \frac{\alpha}{M}\left(\sum_{i=1}^{M/2} a_i E_{i+G} + \sum_{i=-M/2}^{-1} a_i E_{i-G}\right) \tag{9.37}$$

where a_i, $i = 1, \ldots, M/2$, are the weights applied symmetrically to the cell-averagers up and down range of the cell under test. It is assumed that the weights are normalised so that $2\sum_{i=1}^{M/2} a_i = M$. Equation (9.37) is equivalent to (9.7) when all a_i are unity. Using Burg's maximum entropy method, it can readily be shown that the weights that give the minimum error in the estimation of the mean level of the cell under test are given by:

$$\begin{pmatrix} R(0) & \cdot & \cdot & \cdot & R(M/2-1) \\ R(1) & R(0) & \cdot & \cdot & R(M/2-2) \\ \cdot & \cdot & \cdot & \cdot & \cdot \\ \cdot & \cdot & \cdot & \cdot & \cdot \\ R(M/2-1) & \cdot & \cdot & R(1) & R(0) \end{pmatrix} \begin{pmatrix} a_1 \\ a_2 \\ \cdot \\ \cdot \\ a_{M/2} \end{pmatrix} = \begin{pmatrix} R(1+G) \\ R(2+G) \\ \cdot \\ \cdot \\ R(M/2+G) \end{pmatrix} \tag{9.38}$$

Here $R(j)$ is the range ACF for range sample lag j. This equation must be inverted to solve for a_i. Clearly an estimate of the ACF is also required for this process. Assessment of this technique showed a small advantage over a conventional CA CFAR in some circumstances, but with considerable additional processing complexity.

It might be noted that the transient coherent nature of sea clutter spatial structure described in Chapter 2 may also contribute to the limited performance of this type of scheme, requiring frequent updates to the estimate of the ACF.

9.2.5 *Non-linear predictors*

Considerable interest was aroused for a few years by the work of Haykin *et al.* [13,14], who analysed sea clutter using non-linear techniques and concluded that sea clutter is generated by an underlying chaotic process. If sea clutter were indeed a non-linear process, it should be possible to find a non-linear predictor to improve detection performance. Cowper *et al.* [15], working with several different radar data sets, compared the use of non-linear and linear predictors for target detection in maritime radar. They found that non-linear predictors performed no better than linear predictors, using a training set length of 4000 samples. Subsequent work by Unsworth *et al.* [16] investigated the techniques used to assess the chaotic behaviour of sea clutter. They showed that the techniques that had previously been used to assess the chaotic nature of sea clutter were unreliable and incorrectly classified stochastic processes as chaotic. They concluded that the sea clutter that they had analysed is stochastic in nature.

9.3 Adaptation to changing clutter pdf

The cell-averaging CFAR is essentially designed to follow variations in the clutter mean level, either on a large scale as illustrated in Figure 9.1, or locally as illustrated in Figure 9.18. In the discussion so far it has been assumed that the threshold multiplier, α, and by implication, the clutter statistics are known *a priori* or can be estimated exactly in some way. Of course, this will often not be a valid assumption and errors in estimating the correct value of threshold multiplier can be a further source of error, resulting in increased false alarms or reduced detection sensitivity.

In general, the detection threshold required to maintain a constant false alarm rate depends on the amplitude statistics of non-Gaussian (non-Rayleigh envelope)

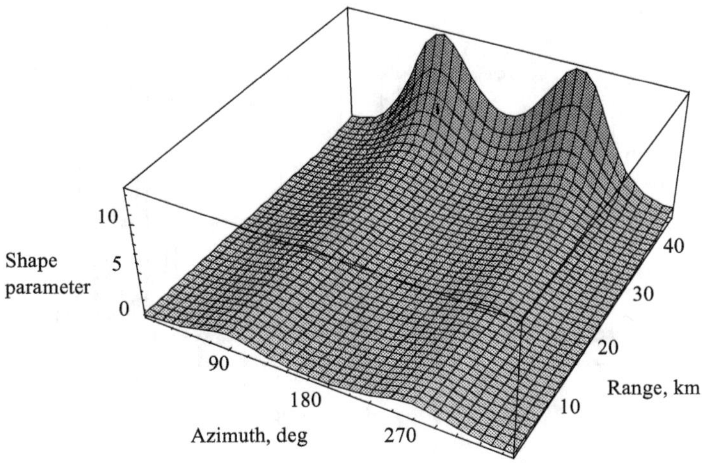

Figure 9.23 Variation of K distribution shape parameter with range and azimuth (see Figure 9.1)

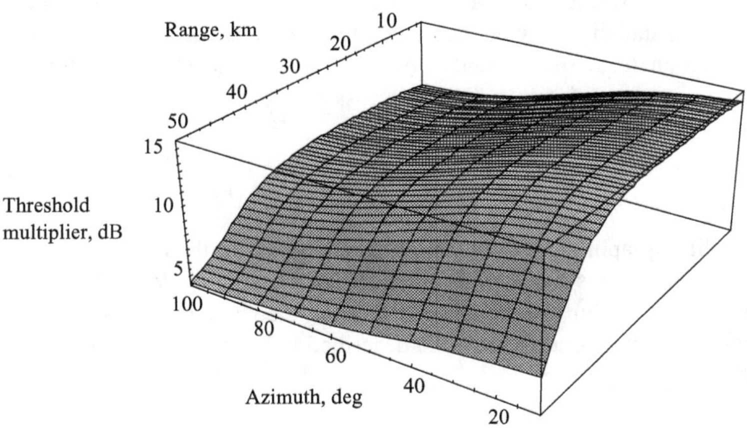

Figure 9.24 Variation of threshold multiplier, α, with range and azimuth (see Figures 9.1 and 9.23)

distributed clutter. For a CA CFAR system, it may also be dependent on any spatial correlation in the clutter, as discussed above. Figure 9.23 shows how the clutter shape parameter may vary as a function of range and azimuth for an airborne radar in the same scenario as Figure 9.1. The associated variation in threshold required over range and bearing may be many dB. This is shown in Figure 9.24, for a 90° sector of the radar angular coverage. The calculation of the appropriate value requires the local clutter statistics to be estimated, directly or indirectly. Any error in estimating the statistics leads to variations in false alarm rate, dependent on the

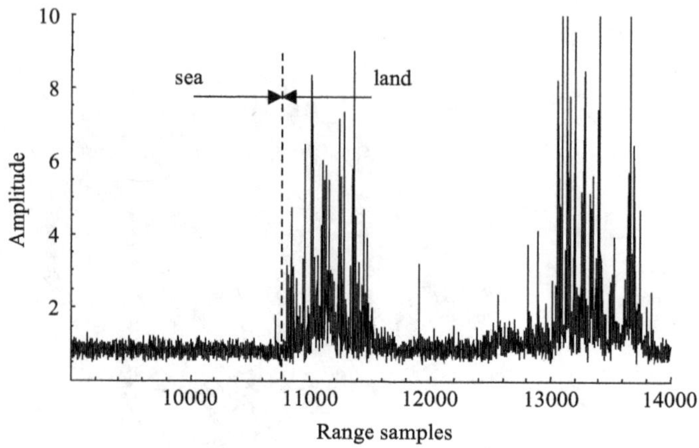

Figure 9.25 Range profile of the interface between land and sea, recorded from an airborne radar

estimation methods used and the spatial and temporal variation of the characteristics. These simulated results show a large variation but also one that occurs smoothly. Often the real statistics may change much more rapidly, as illustrated in Figure 9.25 for a transition from sea to land. The associated rapid change in threshold multiplier required to maintain a CFAR across this boundary would be many dB in this case.

9.3.1 Fitting to a family of distributions

One simplifying approach to the problem of estimating the clutter distribution is to assume that it belongs to a family of distributions, such as Weibull or Log-Normal. In some cases, distribution-robust detectors can then be devised. An example of this approach for Weibull clutter has been described in Reference 17. For Weibull clutter the PDF is given by

$$P(E) = \frac{\beta}{a^{\beta}} E^{\beta-1} \exp\left(-(E/a)^{\beta}\right); \qquad 0 \le E \le \infty \tag{9.39}$$

where β is a shape parameter and a is a scale parameter.

A test statistic, T, is generated:

$$T = \frac{\xi_0 - (1/M)\sum_{i=1}^{M} \xi_i}{\sqrt{(1/M)\sum_{i=1}^{M}\left(\xi_i - (1/M)\sum_{j=1}^{M} \xi_j\right)^2}} \tag{9.40}$$

where $\xi_i = \log_e E_i$, ξ_0 is the cell under test and M independent samples of clutter are used surrounding the cell under test.

It can be seen that the test statistic, T, is invariant under changes of shape or scale of $P(z)$ so that it has the necessary CFAR properties. However, it requires a logarithmic transformation and several numerical operations (sum, difference, square, square

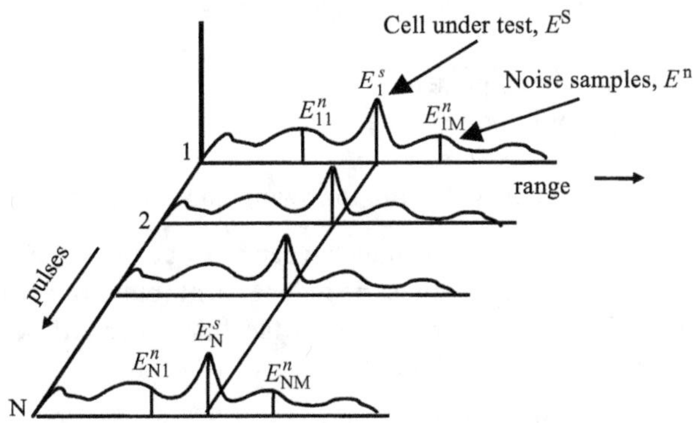

Figure 9.26 *N successive pulse returns, with M noise samples surrounding the cell under test*

root etc.) which may be difficult to achieve in real time in some implementations. In addition it incurs losses compared with the ideal detector. For $\beta = 2$ (Rayleigh clutter) and $M = 50$, the CFAR loss for a design $P_{FA} = 10^{-4}$ is 2.5 dB. In more spiky clutter, with say $\beta = 0.5$, the loss would be about 10 dB. However, such losses may be quite acceptable if it is likely that the shape parameter would be poorly estimated, giving the potential for much higher losses or large increases in P_{FA}.

This test also works in log-normal clutter, although the threshold setting for a given value of P_{FA} is different. In general, it is not CFAR for other distributions or with added thermal noise. Also, real clutter rarely conforms to models to this level of detail and deviations from the assumptions made can lead to significant variations in false alarm rate and detection sensitivity.

9.3.2 Distribution-free detection

In some circumstances completely distribution-free detectors can be devised, such as those based on rank-ordered statistics [18]. An example of such a distribution-free detector is given here.

Figure 9.26 shows the generalised case of N pulse returns and M samples surrounding the cell under test on each pulse.

Instead of operating on the amplitude of the signals, we calculate the rank of the cell under test within the surrounding clutter cells. The rank for each pulse return is then integrated over M pulses and tested against a threshold.

Using the notation of Figure 9.26, the rank of the sample under test, E^s_j, in M noise samples E^n_{jk} is defined as r_j, where

$$r_j = \sum_{k=1}^{M} u\left(E^s_j - E^n_{jk}\right) \tag{9.41}$$

and $u(x)$ is a unit step function. The test statistic T is given by

$$T = \sum_{j=1}^{N} r_j \begin{array}{ll} \geq T_0 & \text{signal present} \\ < T_0 & \text{only noise} \end{array} \tag{9.42}$$

This test is non-parametric provided the following conditions are met:

- The pdf of all observations on a single sweep must be the same.
- All the statistics $r_1, r_2, \ldots, r_M$ must be independent (but observations on a single sweep need not be independent nor need the first-order pdf be the same from sweep-to-sweep.

For $N = 16$, $M = 16$, $P_{FA} = 10^{-6}$, $P_D = 0.5$ and a Swerling Case 2 target, this detector gives loss of 4 dB in thermal noise relative to the optimum detection threshold.

A simplified version of this test has been described in Reference 19, where it is now arranged that $r'_j = 1$ if $r_j = M$ and 0 otherwise. Thus, when there is no target present, $P(r'_j = 1) = 1/(1 + M)$.

This may be derived by inspection or as follows:

$$\text{If } P(y > Y) = F(Y), \qquad \text{then } P(r'_j = 1|Y) = [F(Y)]^M,$$

$$\text{so that } P(r'_j = 1) = \int [F(Y)]^M \, dF(Y) = \frac{1}{1 + M}$$

Given this result,

$$P(T = k) = \binom{N}{k} \left(\frac{1}{1 + M}\right)^k \left(1 - \frac{1}{1 + M}\right)^{N-k}$$

$$P_{FA} = \left(\frac{M}{M + 1}\right)^N \sum_{k=T_0}^{N} \binom{N}{k} M^{-k} \tag{9.43}$$

These methods can be truly CFAR but they also incur losses compared to ideal performance. They require a number of assumptions to be met, including independence of the clutter samples from pulse to pulse. Thus the tests do not normally work in sea clutter since the returns on successive pulses are not independent (the speckle component may be independent but the underlying mean level will be highly correlated from one pulse to the next).

9.3.3 Estimation of the K distribution shape parameter

The best approach in real clutter is usually to attempt to estimate local statistics directly and set a threshold accordingly. Clearly, the accuracy that can be achieved will be dependent on the size of the clutter patch used to estimate the statistics and the rate of response required in changing conditions, such as the transition from land to sea. The earlier section addressed the methods used to estimate the clutter mean level. Here, we discuss direct methods for estimating the shape parameter of K-distributed clutter.

9.3.3.1 Matching moments

One approach is to match the moments of the data sample to those of a K distribution. The moments of clutter-plus-noise are discussed in Appendix 1, Section A1.12. In the usual situation where the noise level is known, the clutter parameters may be estimated by applying equation (A1.182). Thus

$$\begin{aligned}\langle z\rangle_c &= \langle z\rangle_{C+N} - p_n \\ \langle z^2\rangle_c &= \langle z^2\rangle_{C+N} - 4\langle z\rangle_{C+N}\, p_n + 2p_n^2\end{aligned} \tag{9.44}$$

Using the noise free result for the K distribution

$$\frac{\langle z^2\rangle_C}{\langle z\rangle_C^2} = 2\left(1+\frac{1}{\nu}\right) \tag{9.45}$$

we obtain

$$\tilde{\nu} = \frac{2(\langle z\rangle_{C+N} - p_n)^2}{\langle z^2\rangle_{C+N} - 2\langle z\rangle_{C+N}^2} \tag{9.46}$$

where $\langle z^n\rangle_{C+N}$ is the nth moment of the intensity of the data and $\tilde{\nu}$ is the estimate of ν, the K distribution shape parameter.

If the noise power, p_n, is unknown we need the third moment of clutter from (A.182),

$$\langle z^3\rangle_C = \langle z^3\rangle_{C+N} - 9\langle z^2\rangle_{C+N}\, p_n + 18\langle z\rangle_{C+N}\, p_n^2 - 6p_n^3 \tag{9.47}$$

and the third normalised moment of the K distribution,

$$\frac{\langle z^3\rangle_C}{\langle z\rangle_C^3} = 6\left(1+\frac{1}{\nu}\right)\left(1+\frac{2}{\nu}\right) \tag{9.48}$$

which together give the result

$$\begin{aligned}\tilde{\nu} &= \frac{18\left(\langle z^2\rangle_{C+N} - 2\langle z\rangle_{C+N}^2\right)^3}{\left(12\langle z\rangle_{C+N}^3 - 9\langle z\rangle_{C+N}\langle z^2\rangle_{C+N} + \langle z^3\rangle_{C+N}\right)^2} \\ \tilde{p}_n &= \langle z\rangle_{C+N} - \left(\frac{\tilde{\nu}}{2}\left(\langle z^2\rangle_{C+N} - 2\langle z\rangle_{C+N}^2\right)\right)^{1/2}\end{aligned} \tag{9.49}$$

Accurate estimation of the higher moments requires very large data samples and is often impractical, especially for dynamically changing conditions. Better methods of estimating ν in the absence of added thermal noise are discussed in Section 6.9.2 and Reference 20. These are based on the estimates of the mean of the intensity of the data, $\langle z\rangle$ and the mean of the logarithm of the data, $\langle \log z\rangle = (1/M)\sum_{i=1}^{M} \log z_i$. A particularly useful method is

$$\frac{\langle z \log z\rangle}{\langle z\rangle} - \langle \log z\rangle = \frac{1}{N} + \frac{1}{\tilde{\nu}} \tag{9.50}$$

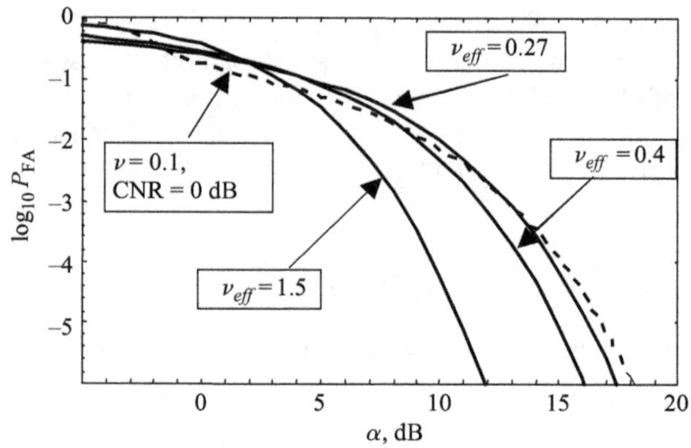

Figure 9.27 Attempted fits of the K distribution ($CNR = \infty$) to data with $\nu = 0.1$ and $CNR = 0\,dB$; see text for details

where $\tilde{\nu}$ is the estimate of ν and N is the number of pulses integrated, assuming independent samples of the speckle component from pulse to pulse.

However, in practice, the clutter is likely to contain added noise and under these circumstances the estimator of equation (9.50) performs poorly. As discussed above, in the presence of added noise, the clutter-to-noise ratio, CNR, must also be determined. This could be achieved by matching first, second and third intensity moments, as in (9.49), but the accuracy of the higher order estimates is likely to be poor unless the sample sizes are very large. However, in a practical radar it is usually possible to estimate the CNR from a knowledge of signal level and the associated receiver noise figure and gain settings.

Figure 9.27 shows various estimates of the shape parameter for clutter with added noise ($\nu = 0.1$, CNR $= 0\,$dB) but where no knowledge of CNR is assumed. The dashed line shows the true distribution. Matching intensity moments on the assumption that the data are noise-free leads to a proposed effective value of ν,

$$\nu_{\text{eff}} = \nu(1 + 1/\text{CNR})^2 \tag{9.51}$$

with $\nu_{\text{eff}} = 0.4$ in this example. It can be seen that this is not a very good fit to the tail.

The use of the mean of the log of the data, as described above, gave an estimate $\nu_{\text{eff}} = 1.5$ using 10,000 independent data samples, but clearly this is also not a good estimate.

9.3.3.2 Matching to the tail of the distribution

If there is added thermal noise, but with the CNR unknown, another pragmatic approach is to match to the tail of the distribution with an effective value of shape parameter, ν_{eff}. One method of matching to the tail of the distribution of clutter plus noise is by direct comparison of cumulative distributions. An example of real data fitted in this way is shown in Figure 9.28. An indication of the accuracy that can be

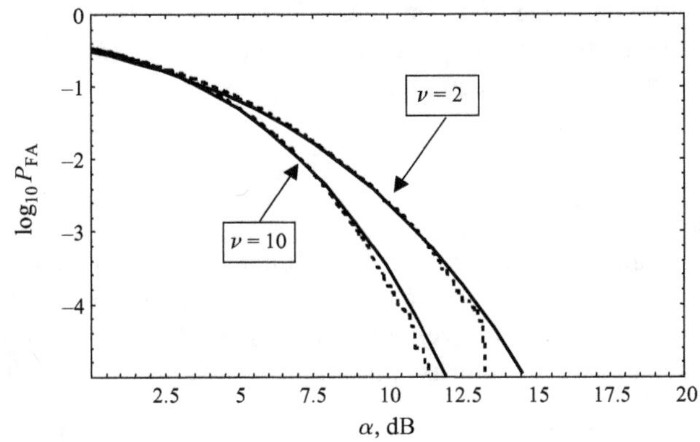

Figure 9.28 Cumulative distribution of intensity of real data fitted to the K distribution; α = threshold/mean intensity.
——— : fitted distribution
---------: real data

Table 9.1 Mean and s.d. of estimates of ν obtained by matching to the tails of the distributions over 1000 independent samples

ν	Mean ν_{eff}	s.d. ν_{eff}	s.d. log estimator
10	10.11	2.22	0.09
1	1.02	0.10	0.04
0.5	0.52	0.06	0.04

achieved in the absence of noise is summarised in Table 9.1 for $N = 10$ pulses integrated and 1000 independent samples used to estimate the distribution. Also shown is the accuracy of the log estimator (9.50) based on moments for clutter alone. A variation of ± 1 s.d. in the magnitudes shown is equivalent to an error of about ± 0.5 dB in α.

This method was also used to fit the data in Figure 9.27, which has a low CNR in very spiky clutter. Direct matching to the tail of the distribution at $P_{FA} = 10^{-3}$ gives a value of $\nu_{eff} = 0.27$ but the match does not hold as the P_{FA} reduces, as can be seen in Figure 9.27. Clearly an estimate of CNR is required if possible.

A more detailed analysis of the accuracy of fitting data to the tail of the distribution, in the absence of noise, has been undertaken. As the fit is being achieved by estimating the value of P_{FA} at a given value of α, the accuracy of the fit can be directly related to the accuracy of the estimation of P_{FA}.

Assume a total data size of M samples. If k threshold crossings are observed for a given threshold, α, then the best estimate of P_{FA} is $\hat{p} = k/M$. This estimate has a variance (as can be seen from equation A1.10) given by

$$\mathrm{var}(\hat{p}) = \frac{p(1-p)}{M} \tag{9.52}$$

where p is the true value of P_{FA}.

The lowest value of P_{FA} that can be estimated by this method is $\hat{p} = 1/M$ (i.e. $k = 1$ threshold crossing) but clearly this estimate is poor, with a variance of $p\hat{p}(1-p)$. The accuracy of a given estimate is dependent on k and for the analysis here the variance of the estimator is approximated as

$$\mathrm{var}(\hat{p}) = \frac{\hat{p}p(1-p)}{k} \cong \frac{p^2}{k} \quad (p \ll 1) \tag{9.53}$$

For a square law detector, using the same notation as equations (9.32) to (9.35), the fit to a K distribution is achieved by solving the following equation for ν:

$$\hat{p} = \int_0^\infty P(x)\frac{1}{\Gamma(N)}\Gamma\left(N, \frac{\alpha N(\nu/b + p_n)}{x + p_n}\right) dx \tag{9.54}$$

where

$$P(x) = \frac{x^{\nu-1}}{\Gamma(\nu)} b^\nu \exp(-b\nu) \tag{9.55}$$

and x is the local clutter power, ν/b is the overall mean clutter power, α is the threshold multiplier and p_n is the noise power. Errors in $\hat{p}$ clearly result in errors in $\hat{\nu}$. The accuracy of $\hat{\nu}$ is then a function of the value of p, the number of pulses integrated, N, and the number of threshold crossings, k, used to estimate $\hat{p}$.

Figure 9.29 illustrates the resulting errors in $\hat{\nu}$ assuming errors in $\hat{p}$ of ± 1 s.d. about p (where s.d. is the standard deviation of $\hat{p}$), for $N = 1$ pulse integrated and $p = 10^{-2}$. The error in $\hat{\nu}$ is expressed as $\log_{10}[\nu/\hat{\nu}]$ where ν is the true value and $\hat{\nu}$ is the value estimated. This is plotted in Figure 9.29 as a function of $\log_{10}\nu$, over the range $0.3 \leq \nu \leq 10$ and for $k = 2, 5, 10, 100$. In each case, the errors are shown separately for ($p + 1$ s.d.) and ($p - 1$ s.d.).

It is clear that the fractional error in $\hat{\nu}$ is not a strong function of ν itself. However, for a given value of k, the accuracy of estimation of $\hat{\nu}$ improves as p reduces. It is also found that the accuracy also improves with pulse-to-pulse integration (N). In each case, the improved accuracy is associated with a larger quantity of base data being used, for a given value of k.

It is also found that using, say, $k = 100$ threshold crossings with raw data (see Figure 9.29) is comparable with using $k = 10$ threshold crossings with integrated data ($N = 10$). However, in practice with a scanning radar, there may be a benefit from using integrated data. Assume that over a radar beamwidth S spokes of data are collected. If pulse-to-pulse integration is used, the effective sample size, M, to estimate p is apparently reduced by a factor S. However, the raw data over one

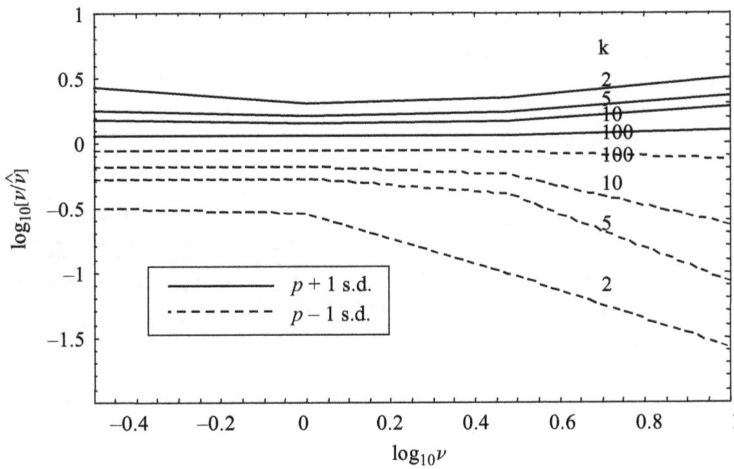

Figure 9.29 Accuracy of estimating ν, for N = 1 pulse integrated and p = 10^{-2}

beamwidth do not provide S independent samples of a K distribution, as the underlying mean level does not change significantly. The reduction in effective sample size by integration is less than S, resulting in some benefit to the accuracy of the fit.

Errors in estimating the shape parameter, ν, can also be interpreted as errors in estimating the resulting radar sensitivity. Figure 9.30 shows the equivalent conditions to Figure 9.29 but now $(\alpha - \hat{\alpha})$ is plotted as a function of ν. Here α is the value of normalised threshold required to set p for a shape parameter ν, whilst $\hat{\alpha}$ is the value required to set the same value of p with a shape parameter $\hat{\nu}$. The parameter $(\alpha - \hat{\alpha})$ can be interpreted as the error in the estimation of system sensitivity.

9.3.4 *Estimation of a Weibull shape parameter*

The Weibull distribution is also sometimes used as a model for designing CFAR detection systems in sea clutter. This may be satisfactory if the overall raw data amplitude distribution alone is of interest (i.e. neither pulse-to-pulse integration nor added thermal noise are present). The Weibull distribution for an intensity z is given by

$$P(z) = \beta \frac{z^{\beta-1}}{a^{\beta}} \exp\left(-\left(\frac{z}{a}\right)^{\beta}\right); \quad 0 \leq z \leq \infty \tag{9.56}$$

where $\langle z \rangle = a\Gamma(1 + 1/\beta)$. The shape parameter β can be estimated using the U parameter [21]:

$$U = \langle \log z \rangle - \log\langle z \rangle \tag{9.57}$$

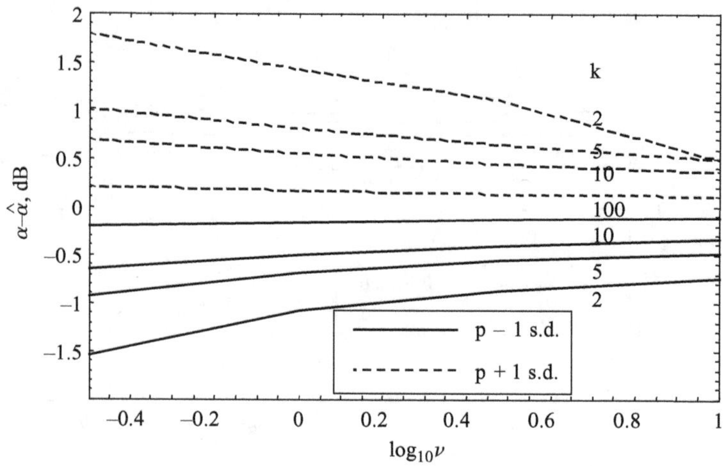

Figure 9.30 Sensitivity accuracy, for $N = 1$ pulse integrated and $p = 10^{-2}$

U relates to the value of β by

$$U(\beta) = -\frac{\gamma}{\beta} - \log(\Gamma(1 + 1/\beta)) \tag{9.58}$$

where γ is Euler's constant. Equation (9.58) can be inverted to find β, and then the threshold for a given value of P_{FA} is given by

$$Y = a(-\log P_{FA})^{1/\beta} \tag{9.59}$$

In a similar way to the estimator for the K distribution shape parameter given in equation (9.50), this estimator for β approaches the accuracy of the maximum likelihood estimator approach. However, in an equally similar manner, this estimator fails in the presence of added thermal noise.

The design of a CFAR system using the U estimator is described in Reference 21, where the problems of added noise are discussed and an alternative approach called Graphical Estimation is also proposed. This involves a direct fit to the tail of the distribution of the radar data, taking into account any added noise. This is a very similar approach to that described in Section 9.3.3.2. As expected, the Graphical Estimator provides better performance in the presence of added noise.

9.4 Other CFAR detection techniques

9.4.1 Site specific CFAR

In Reference 22 a site specific technique is described for maintaining CFAR performance in littoral waters. In essence, this relies on the use of terrain elevation data to model the expected land clutter returns. For example, the accurate knowledge of

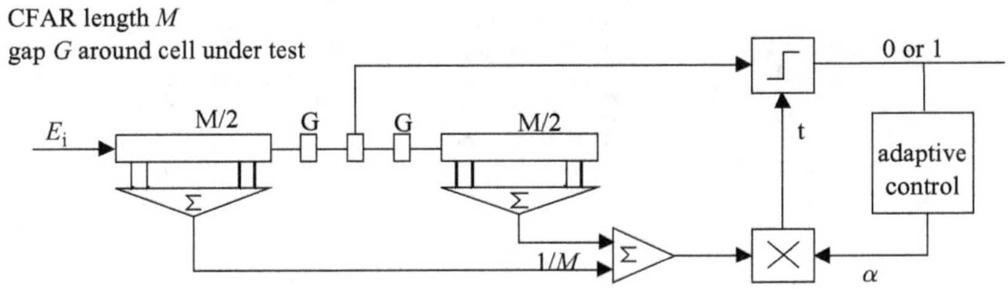

Figure 9.31 Cell-averaging CFAR with closed loop feedback control

the land-sea interface (e.g. see illustration in Figure 9.25) may be used to ensure that the threshold can change rapidly across the interface and that statistical estimates are not contaminated by returns from different terrains. Such modelling can also be used to predict areas that are expected to be screened from the radar, such as areas of sea behind islands or rocks. Knowledge of sea state and wind conditions can be used to predict the sea clutter conditions. The combined model outputs are then used to predict the required threshold multipliers. This is a good example of knowledge-based signal processing.

9.4.2 Closed loop systems

Figure 9.31 shows a version of the CA CFAR detector that uses a closed loop estimation system to adapt the value of threshold multiplier, α. Here the P_{FA} at the threshold is measured and used in a closed control loop to adjust the threshold multiplier. The accuracy can be analysed using methods similar to those in Section 9.3.3.2. This approach is mentioned in Reference 21 where it is noted that this type of scheme is typically implemented over large areas and over many consecutive scans in order to obtain sufficient statistics that wanted targets are not discounted. This can be regarded as a clutter mapping process. It may be difficult to implement from a moving platform when the clutter statistics may be varying very rapidly, preventing the false alarm statistics from being accurately measured.

9.4.3 Exploitation of transient coherence

Contemporary CFAR systems use data from the current scan (i.e. data in one set of range cells and perhaps a limited set of adjacent azimuth cells) to estimate the clutter level in a test cell. Attempts to predict the level in the test cell through some process adapting to the local clutter correlation properties may well be frustrated by the apparent non-stationarity manifest in the data, as discussed in Chapter 2 and mentioned in Section 9.2.4.

In Chapter 2, it is shown how analysis of the power spectrum of a range-time intensity plot of clutter (an ω–k plot) can provide information on the spatial and temporal correlations in the sea clutter and the non-linear effects that appear to be responsible for the non-stationary behaviour of the clutter. Tough *et al.* [23] have suggested how

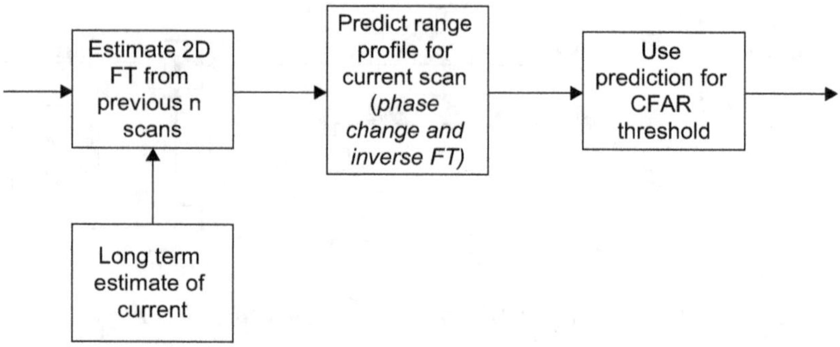

Figure 9.32 CFAR processing scheme based on Fourier transform (FT) of range-time intensity plot

such an analysis might be exploited in a CFAR detection system. They proposed that a history of previous scans could be used to build up an estimate of the sea wave dispersion relation, which may then be used to predict the clutter level in the current scan. The previous n scans are used to generate the 2D Fourier transform of the most recent range-time intensity plot. The estimate of the current range profile can then be generated by the introduction of suitable phase shifts, to effect the propagation over a time step, and subsequent Fourier inversion. The effects of wave modulation and spike features resulting from breaking waves should add coherently within the continuously updated Fourier transform, and so be predicted with some accuracy. Target-like features arising from debris etc. should average out and so not impact unduly on the estimate of the current clutter profile. In this way it may be possible to achieve improved CFAR performance, at the price of an increased, but not unrealistic, computational burden. A CFAR detector architecture illustrating this approach is shown in Figure 9.32.

9.4.4 Scan-to-scan integration

Pulse-to-pulse integration prior to thresholding is often used to increase radar sensitivity. Benefit may also be obtained from integrating from scan to scan (see Chapter 8). In particular, integration over periods of several seconds can help to discriminate against discrete clutter spikes (see Chapters 2 and 4). These spikes may appear target-like and persist over lifetimes of typically between 0.1 and 2 s. Their discrete nature in range means that they do not decorrelate with frequency agility. Real targets can be distinguished from clutter spikes if they persist for significantly longer.

In order to achieve scan-to-scan integration of targets, they need to be tracked, especially if the radar had a high spatial resolution. All possible target motions must be tested and there is usually a maximum target velocity specified for the integrator. Any motion of the radar platform must also be compensated. Such scan-to-scan integration is an example of a track-before-detect process [24].

In principle, scan-to-scan integration may be applied to analogue signals, following pulse-to-pulse integration within a beam dwell. In practice, a detection threshold is usually applied after pulse-to-pulse integration. Then, binary detections may be integrated from scan to scan. Another approach is to extract tracker plots from the binary detections (e.g. an estimate of the coordinates of the target centroid) and pass these to the scan-to-scan integrator. Clearly, there are many variations of processing possible. For example, there may be a choice of scan rate, allowing a trade-off between pulse-to-pulse integration, scan update rate and scan-to-scan integration, within a fixed total dwell time. These strategies are discussed in Chapter 8 and in References 25 and 26. Quantifying performance for the suppression of discrete clutter spikes requires the development of improved models for the lifetime of spikes.

9.5 Practical CFAR detectors

The techniques that have been described and analysed in this chapter can be exploited in various ways to develop practical CFAR detection systems. If a CA CFAR system is to be used to estimate the local mean level, choices must be made amongst the many possible configurations, such as the length (M) of the cell-averagers, the gap (G) surrounding the cell under test, and the various schemes described in Section 9.2.3. It may be appropriate to have these parameters or choice of scheme selectable by the radar operator or automatically adapting, according to the prevailing conditions.

The method for setting the threshold multiplier must also be determined. The CA CFAR illustrated in Figure 9.2 suggests that the surrounding clutter may be used directly to estimate the distribution shape and hence the threshold multiplier, α. The various methods for estimating the clutter shape parameter in Section 9.3 could be assessed for this purpose. As discussed in Section 9.1, the clutter conditions may vary very widely with range and azimuth and a problem is likely to be encountered in achieving a sufficient number of independent clutter samples within an area of constant statistics. One approach may be to gather statistics over a number of scans. Other estimates of clutter statistics, such as the U estimator [21] described in Section 9.3, may also be used. A closed-loop system (Section 9.4.2) that directly estimates the P_{FA} may also be a practical solution, although there are similar considerations on the number of samples required to give a satisfactory estimate, especially if a low value of P_{FA} is required.

Scan-to-scan integration (Section 9.4.4) may also be used to improve detection sensitivity and, in particular, to discriminate against short persistence clutter spikes.

Much of the discussion in this chapter has assumed radar processing of the envelope or intensity of the returns, often with pulse-to-pulse frequency agility. These techniques can also be applied to pulsed Doppler radars, following detection of the output from a Doppler filter bank or a moving target indication (MTI) canceller. In the case of a Doppler filter bank, a separate CFAR threshold control is usually needed for each filter output, since the clutter residue magnitude and statistics vary, depending on the filter frequency response.

References

1 W.R. Wallace, 'The use of track-before-detect in pulse-Doppler radar', Radar 2002, IEE Conference Publication 490, October 2002, pp. 315–319.

2 M.W. Long, '*Radar reflectivity of land and sea*', Artech House, New York, 1983.

3 M.M. Finn and R.S. Johnson, 'Adaptive detection mode with threshold control as a function of spatially sampled clutter-level estimates', *RCA Review*, **30**, 414–465, 1968.

4 V.Y.F. Li and K.M. Miller, 'Target detection in radar: Current status and future possibilities', *J. Navigat.*, **50**(2), 303–313, 1997.

5 I.S. Gradshteyn and I.W. Ryzhik, '*Table of integrals, series, and products*', Academic Press, 4th edition, 1965 (Equation 8.350 p. 940).

6 D. Shnidman, 'Radar detection probabilities and their calculation', *IEEE* **AES-31**(3), 928–950, 1995.

7 V.G. Hansen and J.H. Sawyers, 'Detectability loss due to 'Greatest Of' selection in a cell-averaging CFAR', *IEEE Trans. AES*, **16**(1), 115–118, 1980.

8 H. Rohling, 'Radar CFAR thresholding in clutter and multiple target situations', *IEEE Trans. AES*, **AES-19**(4), 608–621, 1983.

9 S. Watts, 'Radar detection prediction in sea clutter using the compound K-distribution model', *Proc. IEE*, **132** (Part F, No. 7), 613–620, 1985.

10 S. Watts, 'Cell-averaging CFAR gain in spatially correlated K distributed sea clutter', *IEE Proc. Radar, Sonar, Navig.*, **143**(5), 321–327, 1996.

11 H. Darwish and D.C. Cooper, 'The log-lattice detector for radar targets in sea clutter', IEE Colloquium on 'Discrimination and identification methods in radar and sonar systems', Digest No. 1986/69, paper 9, 6 May 1986.

12 S. Watts, 'Improved cell-averaging CFAR for maritime surveillance radar', UK Patent 2191052, 1981.

13 S. Haykin and X.B. Li, 'Detection of signals in chaos', *Proc. IEEE*, **83**, 95–122, 1995.

14 S. Haykin and S. Puthusserypady, 'Chaotic dynamics of sea clutter: an experimental study', Proceedings of the IEE Conference on Radar, 1997, pp. 75–79.

15 M.R. Cowper, B. Mulgrew and C.P. Unsworth, 'Investigation into the use on non-linear predictor networks to improve performance of maritime surveillance radar detectors', *IEE Proc. – Radar, Sonar Navig.*, **148**(3), 103–111, 2001.

16 C.P. Unsworth, M.R. Cowper, S. McLaughlin and B. Mulgrew, 'Re-examining the nature of radar sea clutter', *IEE Proc. – Radar, Sonar Navig.*, **149**(3), 105–114, 2002.

17 G.B. Goldstein, 'False alarm regulation in log-normal and Weibull clutter', *IEEE Trans.*, **AES-9**(1), 84–92, 1973.

18 V.G. Hansen and B.A. Olsen, 'Non-parametric radar extraction using a generalised sign test', *IEEE Trans.*, **AES-7**(5), 942–950, 1971.

19 G.M. Dillard and C.E. Antoniak, 'A practical distribution-free detection procedure for multiple range-bin radar', *IEEE Trans.*, **AES-6**(5), 629–635, 1970.

20 D. Blacknell and R.J.A. Tough, '*Parameter estimation for the K distribution based on [zlog(z)]*', *IEE Proc. Radar, Sonar, Navig.*, **148**, 309–312, 2001.

21 W. Dawber and J. Branson, 'Comparison of CFAR algorithms in littoral clutter data', 4th TTCP Workshop on Defence Applications of Signal Processing, DASP 05, Utah, 17 March–1 April 2005.

22 W. Dawber and J. Branson, 'Use of site specific radar modelling to improve CFAR performance in the littoral', IEEE International Radar Conference Radar 2005, Washington, DC, 9–12 May 2005, pp. 161–166.

23 R.J.A. Tough, K.D. Ward and P.W. Shepherd, 'The modelling and exploitation of spatial correlation in spiky sea clutter', EMRS DTC 2nd Annual Conference, 16–17 June 2005, paper A1.

24 E.J. Hughes and M. Lewis, 'An intelligent agent-based track-before-detect system applied to a range and velocity ambiguous radar', EMRS DTC 1st Annual Conference, May 2004, paper A4.

25 M. McDonald and S. Lycett, 'A comparison of fast scan versus slow scan radars', International Radar Conference Radar 2004, Toulouse 19–21 October 2004, paper 10P-RCMT-255.

26 S. Watts and A.G. Stove, 'Non-coherent integration strategies in scanning radar for the detection of targets in sea clutter and noise', International Radar Conference Radar 2004, Toulouse 19–21 October 2004, paper 10P-SP-229.

Chapter 10
The specification and measurement of radar performance

10.1 Introduction

One of the ultimate aims of modelling sea clutter is to inform the development of improved radar systems that meet the operational needs of their users. However, the translation from a user's requirements to the entry into service of an equipment that meets these needs in an acceptable way is a very complex process. From the viewpoint of a radar designer, an important part of the process is the methodology for specifying and measuring radar detection performance. This methodology is the subject of this chapter, with emphasis being placed on the issues relating to maritime radars and the detection of targets in sea clutter.

Future high performance radars are likely to be increasingly complex with multiple modes and extensive data and signal processing. In particular, adaptive processing modes are likely to be employed, ranging from adaptive beamforming with electronically steered arrays to automatic detection schemes for small targets in clutter. The adaptive nature of these systems will mean that performance and 'fitness for purpose' will be determined by their dynamic characteristics. Traditional static measures, such as detection range against a given target, will no longer adequately define the capabilities of the radar system. However, dynamic performance is very difficult to specify and even more difficult to measure quantitatively under practical trials conditions [1].

It is becoming increasingly important to ensure that the methodology for specifying the requirements for adaptive radar systems is well understood and agreed by both procurers and suppliers of radars [2,3]. Any specification used for procurement must be measurable; the detection performance of a modern adaptive radar by practical trials is likely to be very difficult, as well as time consuming and expensive. Therefore it is expected that computer modelling and simulation will become increasingly important in this aspect of procurement, in addition to their traditional role in design and performance prediction.

The overall process of procuring a radar system involves very many engineering disciplines and user requirements such as reliability, maintainability and through life support. This chapter is concerned only with the specification and measurement of detection performance, and pays particular attention to the problems posed by adaptive radars. Section 10.2 addresses the issues concerned with specifying radar performance, including examples of how specific performance indicators can be specified. Performance prediction using computer and mathematical modelling plays an important part in the initial design process and is described in Section 10.3. Once a radar has been developed to meet a specification, its actual performance must be measured to assess whether it has indeed achieved the predicted performance; this measurement of radar performance is described in Section 10.4. Although mathematical modelling is likely to be increasingly important, the use of radar trials will nearly always form an essential part in the measurement of performance. Section 10.5 discusses the issues associated with taking quantitative measurements from trials or from radar data recorded during trials.

10.2 Performance specification issues

10.2.1 Discussion

The performance of a radar should ideally be specified by the customer, in a manner that unambiguously describes the requirements. Equally importantly, any such specification needs to be understood by the supplier in order to generate a technical design. For this process to work well, there must be a close working relationship between the customer and supplier.

A simplified representation of the customer-supplier relationship is shown in Figure 10.1. The cycle starts with the customer's needs. These may be defined

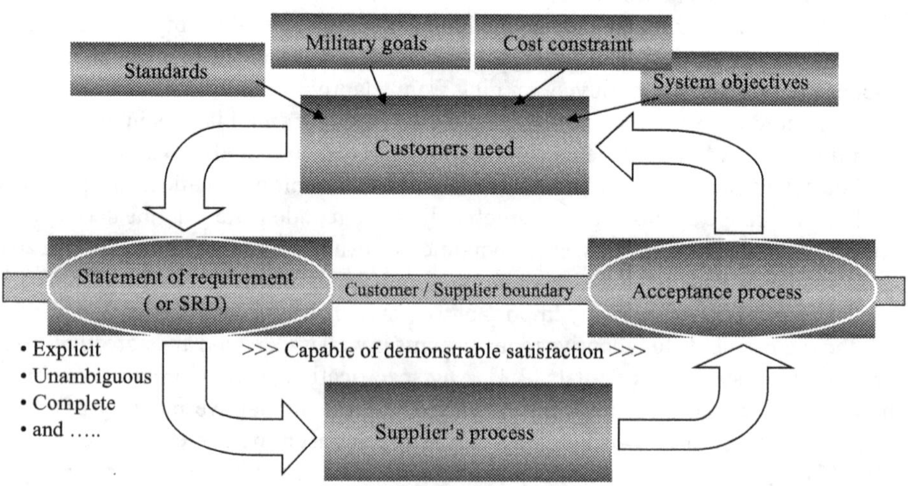

Figure 10.1 Customer-supplier relationship [2,3]

by many interacting issues, including operational needs, cost, logistics and so on. They may then be formulated as a statement of requirement (often as a System Requirement Document, SRD) that is offered to potential contractors. The contractor has to interpret this statement of requirement to produce a technical specification for the equipment. The solution finally delivered to the customer should be demonstrably capable of meeting the agreed specification, with performance determined by an agreed acceptance process.

The relationship between the customer and supplier is governed by two main elements. These are the statement of the requirement and the method of acceptance.

A clear distinction must be made between capability-based operational specifications, which essentially describe the end-user's needs in the context of military or other operations, and metric specifications, against which the radar manufacturer can demonstrate compliance of the radar system. It may be tempting for the customer to believe that involvement in the overall process can be limited to the operational requirement, leaving other details to contractors. However, past experience has shown that this does not usually result in the best possible equipment for the user, since the detailed customer requirements cannot usually be encapsulated in an operational specification. A close customer-supplier relationship is very desirable when procuring a modern adaptive radar; the operational specification can then be explained to the supplier, while the customer develops an understanding of the full implication of the metric specification.

Any radar specification represents a balance between risk, cost and value for both customer and supplier. Inevitably, this results in compromises that should be properly understood before a fixed contract is agreed. The important points are summarised below.

First, the customer may not be able to specify every aspect of required performance comprehensively at the start of the process. Modern radars are very complex and, while the customer may have a good idea of his or her high level requirements, gaps and inconsistencies in the specification are likely to exist. In addition, the supplier's interpretation of a specification may not match the customer's expectations. Because of the system complexity, many of these problems need not become apparent until late into the programme.

Second, unambiguous specifications may not be possible in the time available during contract negotiations. For fixed price contracts, both sides require that the specifications, work breakdown and so on are fully detailed and unambiguously defined prior to contract. However, the nature of competitive bidding is often such that there is insufficient time to explore fully the detailed specification points and the compromises between cost, risk and value. Once the contract is awarded, any changes to the specification require revision of the contract itself. This is costly, time-consuming and often not ultimately beneficial to either supplier or customer. Some customers use a 'Product Definition Document' to combine the interpretation of the specification and the supplier's proposal into a common document against which acceptance and delivery can take place.

Finally, prior to final contract award, a more detailed specification must be agreed with the supplier, to determine exactly what performance is required to be delivered. At this stage the following points should be noted:

- performance specifications should ideally be not be too detailed and should be restricted to essential performance parameters;
- suppliers should, if possible, be afforded the flexibility within their contract to trade-off detailed sub-system parameters and performance whilst still meeting the essential performance parameters;
- any specified performance should be measurable;
- the methodology for final acceptance should be agreed before contract award.

As discussed above, the nature of radar development often requires that changes to the specification are made after the award of contract. This may be due to unforeseen development problems or, equally, be due to a change in operational requirement by the customer. It is likely to be in the best interests of both customer and supplier if changes to the specification can be managed without recourse to formal changes to the contract. However, if capability-based operational specifications are included, it is important for the customer that these are still met, despite any trade-off in the detailed sub-system parameters.

10.2.2 Adaptive radars

As mentioned earlier, a particular challenge for the specification of radar performance is presented by the adaptivity characteristic of modern radars. Most, if not all, modern radars include a number of adaptive features, which are becoming more and more extensive and sophisticated. There are a several reasons for this. Some are concerned with overcoming or ameliorating practical limitations in the implementation of the radar; others are concerned with optimising the operation of the radar in a particular environment. Adaptivity is often introduced to allow a single multi-function radar to carry out efficiently a set of tasks that formerly required multiple separate radars. It is important to realise that many forms of adaptation may improve performance in one respect at the expense of reducing it in others. Furthermore, adaptive processes are by their nature dynamic, and the transient behaviour of an adaptive system must be given careful consideration. It is these features of adaptive systems in particular that present significant challenges for the procurement process.

Several common examples of radar adaptive operation are listed below. Some of these examples are commonplace and well established, but the majority are relatively new and some have yet to be incorporated in mainstream designs:

- Automatic gain control (AGC)
- Automatic calibration
- Constant-False-Alarm-Rate (CFAR)
- Adaptive beamforming (ABF)
- Adaptive mode selection
- Automatic radar management

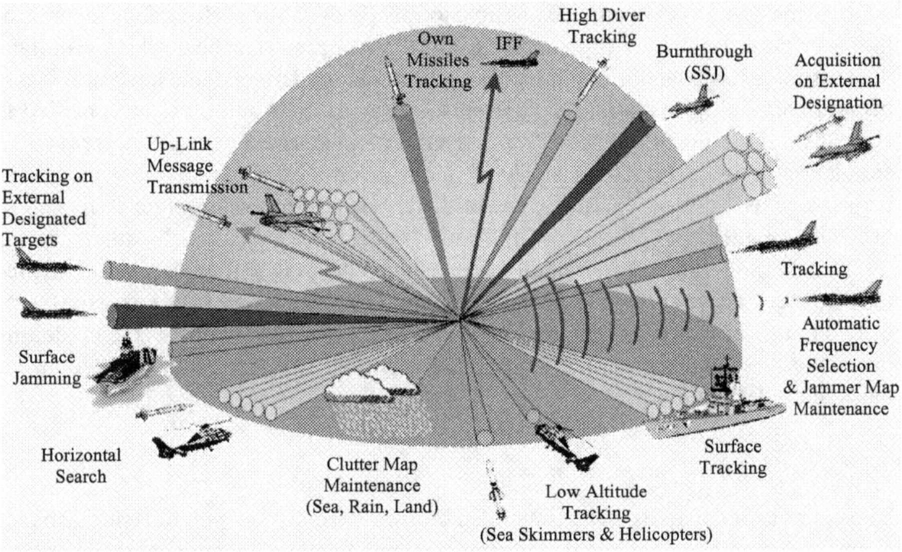

Figure 10.2 Multifunction radar tasks [reproduced with the kind permission of BAESystems Integrated Systems Technology]

- Adaptive tracking
- Adaptive reconfiguration
- Adaptive Moving Target Indication (MTI)

An illustration of multifunction radar tasks is shown in Figure 10.2. Operation of this type of radar requires many of the adaptive features listed above.

Overall, the purpose of adaptive systems in radar is to allow the system to react to its operating environment (in the broadest sense, including both internal and external factors) in such a way as to maximise its performance against certain specific requirements.

10.2.3 Specification of adaptive systems

The specification of adaptive performance involves an understanding of the dynamic performance of the radar, which is probably scenario-dependent. Indeed, it may not be obvious how to quantify 'performance' effectively. Certainly, traditional 'static' performance measures, such as detection range or track initiation range, may no longer be very meaningful in terms of parameters that define performance that makes the system 'fit for purpose'. Measures such as the susceptibility to processor overload or the time to adapt in changing conditions may be more important to the user but are also very difficult to quantify and measure from practical trials. However, a feature of any useful specification point is that it should be verifiable or measurable. One solution to this is to specify and measure aspects of performance by using modelling and simulation.

In some cases, it may only be possible to specify dynamic performance in terms of the algorithms used in the signal processing. This approach can be used if a prototype demonstrator has been built and tested or if the required algorithms have previously been proved on other operational systems. Performance would now be specified in terms of a deterministic response to prescribed situations that could be proved by simulation or special static tests. The dynamic performance in real scenarios is then taken to be an inherent characteristic of the specified processing algorithms and so need not be measured as part of the radar's formal acceptance. In other cases a mixture of modelling and simulation, together with practical trials measurements, may be required. In these circumstances, it may be satisfactory to measure only one or two points in the radar performance envelope in order to 'validate' the modelling predictions. All other specification points could then be accepted or rejected on the basis of modelling alone.

10.2.4 Practical performance specification

The issues raised above can be illustrated by considering some practical performance specification issues. A typical performance specification for an airborne maritime surveillance radar might be identified as the ability to detect, say, a 5 m^2 target in sea state 3 at a range of at least 25 nmi in clear air, with a detection probability of 0.9 and with one false alarm per minute. This is a basic statement of a user's needs but it is very inadequate as a specification of the radar performance that is actually required. The performance of an operational radar is characterised by many other features, which are not simply summarised in terms of a detection range and false alarm rate in a generalised environment.

In particular, concepts such as probability of detection and probability of false alarm are conventionally steady state performance measures, for operation in thermal noise or uniformly distributed clutter. In practice, the radar adjusts its thresholds dynamically in response to a spatially and temporally varying environment (e.g. CA CFAR detector, as described in Chapter 9). This leads to a dynamic variation in detection and false alarm rate that cannot be described simply by traditional methods.

10.2.4.1 Probability of false alarm, P_{FA}

A typical specification may call for (up to) one false alarm per minute or one false alarm over a given number of scans of the radar. This may appear to adequately define the needs of the radar operator but it will be seen that such a limited statement is often not sufficient to specify the ability of the radar to control false alarms. First, it must be clear what constitutes a false alarm. For some radars, this may be a fleeting false paint on the display that causes the operator to act. For others, it may be the initiation of a false track, with obviously different implications for the radar performance. As another example, returns from land seen by a maritime surveillance radar may not be regarded as false alarms, but as useful information for the operator. More importantly, the false alarm rate in an adaptive radar is likely to vary over the display due to failure of the adaptive constant false alarm rate (CFAR) circuits to respond sufficiently fast to changes. This is discussed in the following section.

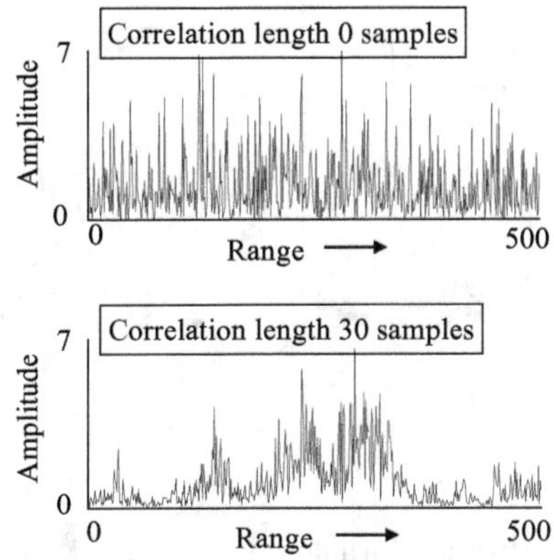

Figure 10.3 Correlated clutter

10.2.4.2 Spatial Variation of P_{FA}

In a practical radar the detection threshold always has to adapt to maintain a CFAR. This adaptation must follow receiver gain variation (e.g. due to STC or AGC) or changes in clutter level in range and look-direction (with clutter cross-section dependent on grazing angle, wind direction, spatial resolution, etc.).

An important measure of adaptive threshold performance is the variation of the false alarm rate. Even if the average false alarm rate is as specified, it may be that the false alarms are not uniformly distributed over the display, but concentrated in specific areas of 'difficult' clutter. One reason for a non-uniform spatial distribution of alarms may be that the clutter has some discernible spatial structure, which interacts with the adaptive threshold. Although sea clutter is not always very well described by a spatial spectrum (i.e. it is not reliably a stationary process), a simple application of spatial correlation gives characteristics that are reasonably representative of practical conditions.[1] Figure 10.3 shows examples of simulated clutter with different correlation lengths in the Gamma component. The clutter amplitude has a K distribution pdf with shape parameter $\nu = 0.5$.

Figure 10.4 illustrates the different local variations of P_{FA} that can be encountered, dependent on the type of adaptive threshold being used. The identical left and right-hand side upper plots show the local mean clutter level (i.e. with the speckle component removed) for simulated K-distributed clutter with a shape parameter $\nu = 0.5$ and a correlation length of 30 samples. Also shown is the fixed threshold that would be set using a knowledge only of the global statistics of the clutter (i.e. a K distribution with shape parameter $\nu = 0.5$), to give an average P_{FA} of 10^{-4}. An adaptive threshold might be able to follow the local mean fluctuations, and also shown on the two upper plots is the adaptive threshold set by a cell-averaging (CA) CFAR, with an averager length of 5 cells either side of the cell under test with an additional gap or guard cell of

[1] Further details of methods for simulating K-distributed variates with different autocorrelation functions can be found in Chapter 5.

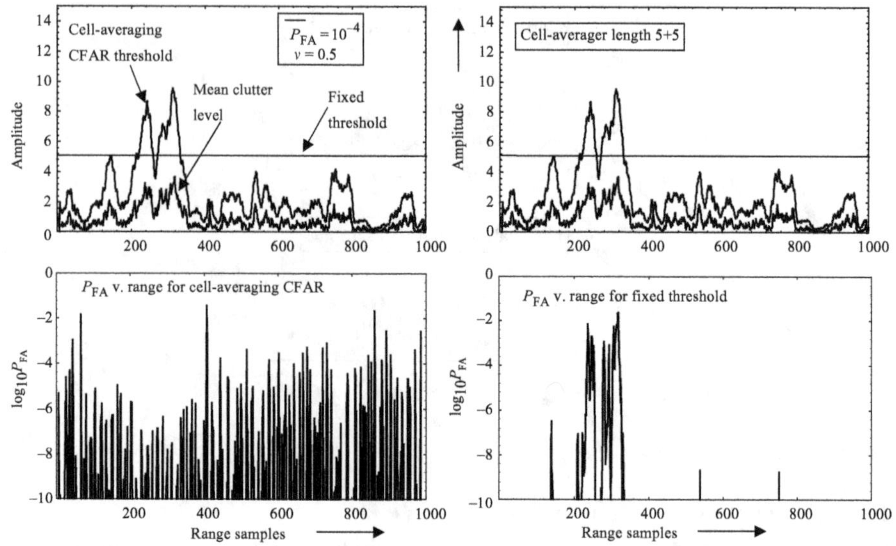

Figure 10.4 P_{FA} vs. range for CA CFAR and fixed threshold

1 cell either side (see Chapter 9 for further details of CA CFAR systems). The lower plots of Figure 10.4 show the variation in P_{FA} with range. For the conventional fixed threshold almost all the alarms would be concentrated around the area of locally high clutter. The average P_{FA} is 10^{-4} but the peak P_{FA} is more than 10^{-2}. The threshold set by the CA CFAR is calculated from the running estimate of local mean, scaled by a threshold multiplier, α; in this example, $\alpha = 9.6$ dB to achieve $P_{FA} = 10^{-4}$. The fixed threshold is set relative to the global mean, with $\alpha = 15.6$ dB to achieve the same value of P_{FA}. Clearly, the average level of threshold set by the CA CFAR is much lower, contributing to a 'CFAR gain' (see Section 9.2.3.7). The CA CFAR also gives a much more uniform distribution of alarms. There are locally high values, but these are associated with edge-like features in the mean level variation which may be much more short-lived than the large scale high amplitude features more likely to be associated with the long waves and sea swell. Of course an ideal adaptive threshold would follow the mean level exactly and give a P_{FA} that is constant with range. This is termed the 'ideal CFAR' threshold (see Section 9.2.3.7). In thermal noise or Gaussian clutter, the fixed threshold and 'ideal CFAR' solutions are identical and P_{FA} does not vary with range.

P_{FA} is likely to vary widely across the whole radar display in difficult conditions. Figure 10.5 illustrates the false alarm density for a radar operating in littoral waters with an adaptive threshold, generated from recording of raw data from an airborne radar. The false alarms have been accumulated over several scans for illustrative purposes. Apart from the returns from the land, which are not perceived as false alarms in the context of maritime surveillance radar, it can be seen that the alarm density varies widely.

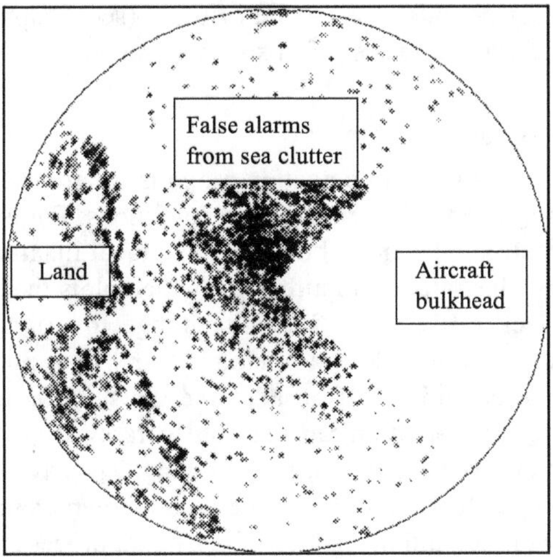

Figure 10.5 False alarm map on radar PPI

Near the centre of the display, the alarms are clearly higher towards the top of the display, due to the clutter conditions. Also, the algorithm used for this example produced a constant false alarm rate along a given radar spoke, rather than a constant area alarm rate which would require a lower false alarm probability near the centre of the display. Another point to note is the region at maximum range just after the land for a clockwise scanning radar. Here the adaptive threshold has been desensitised by the land, resulting in too low a false alarm rate and therefore an associated loss in detection sensitivity. This plot illustrates a number of problems which may be encountered but which are not covered by standard performance specifications.

The temporal correlation of false alarms is also of interest, in addition to their spatial distribution. Localised areas of persistently high clutter returns may tend to give false alarms on successive scans from the same area on the radar display. This may prevent any expected reduction in false alarms from scan-to-scan integration. Some benefit should be achieved by maintaining as high a spatial resolution as possible during scan-to-scan integration, prior to final collapsing onto the display. This in turn may require the scan-to-scan integration algorithms to cater for any expected target motion, to ensure targets are not lost due to range-walk problems. However, the spatial resolution should match the expected target extent to avoid partitioning losses. The exact locations of false alarms and their correlation from scan-to-scan depend to some extent on the form of detection processing used. A simple CA CFAR threshold control tends to give alarms at clutter edges (between land and sea or sea and precipitation, etc.). Procedures such as the Greater-Of CFAR (see Section 9.2.3.1) can be used to suppress alarms at edges whilst a Smaller-Of CFAR maintains sensitivity up to the clutter edge. In very spiky sea clutter with little spatial correlation (e.g. the top plot in Figure 10.3), cell-averagers cannot follow any local fluctuations. With the threshold set to give the desired P_{FA}, alarms occur from the occasional very large spike. These spikes may persist for periods of a second or more (see Chapter 2), which may be comparable with the scan period of a fast scanning radar.

10.2.4.3 Probability of detection, P_D

The simple concept of maximum detection range is often inadequate in even quite simple environments, due to changing clutter conditions. For example, P_D may not decrease monotonically with range, but may have local maxima and minima. This type of behaviour is illustrated in Figure 8.19, which plots the minimum detectable target Radar Cross Section (RCS) as a function of range, for a particular radar scenario. A single measure of maximum detection range clearly does not properly define the performance in these conditions. If an adaptive threshold is used to control false alarms, the nature of this adaptation can also considerably affect the local variation of target detectability and so the apparent target fading. P_D may also vary widely over the radar display, dependent not only on the target's own characteristics but on its interaction with the clutter and the nature of the signal processing. This is discussed further, below.

10.2.4.4 Spatial variation of P_D

An example of the spatial variation of P_D with different thresholds is shown in Figure 10.6, which plots the variation in P_D with range in spatially correlated clutter, for a fixed threshold and for a CA CFAR. The CA CFAR has an average length of 5 cells either side of the cell under test, with an additional gap or guard cell of 1 cell either side. In both cases the average P_D is 0.5 and the average P_{FA} is 10^{-4} (as illustrated in Figure 10.4). For the fixed threshold, P_D does not vary significantly with range and the detected target would not appear to fluctuate significantly as it changed range (detected on successive scans, say). For the CA CFAR, P_D varies significantly with range, dependent on the local clutter conditions. For this detection threshold, the target would appear to fade significantly as it changed range. At first sight, the fixed threshold performance might seem preferable. However, it should be noted that in the example illustrated, the average signal-to-clutter ratio, *SCR*, required to achieve an average P_D of 0.5 is 15 dB for the fixed threshold detector and 6.7 dB for the CA CFAR (see the discussion on CFAR gain in Section 9.2.3.8). Also shown in Figure 10.6 is the variation of P_D with range for ideal CFAR performance (see Section 9.2.3.7). This is very similar to the variation observed for the CA CFAR but it is found that the average signal-to-clutter ratio, *SCR*, required to achieve an average P_D of 0.5 is now only 0.2 dB.

Similar spatial variations of P_D are observed with binary integrators, dependent on the setting of the second threshold [4].

The example in Figure 10.6 shows how estimation of the clutter mean level may introduce losses. In addition, if the clutter statistical distribution is varying dynamically (e.g. as a function of range and radar look direction), then a CFAR system must also attempt to adapt the threshold multiplier, α, to maintain a constant P_{FA}, given an estimate of the local mean level from a cell-averager. The ability of the threshold multiplier to respond to changing conditions will also clearly affect P_{FA}

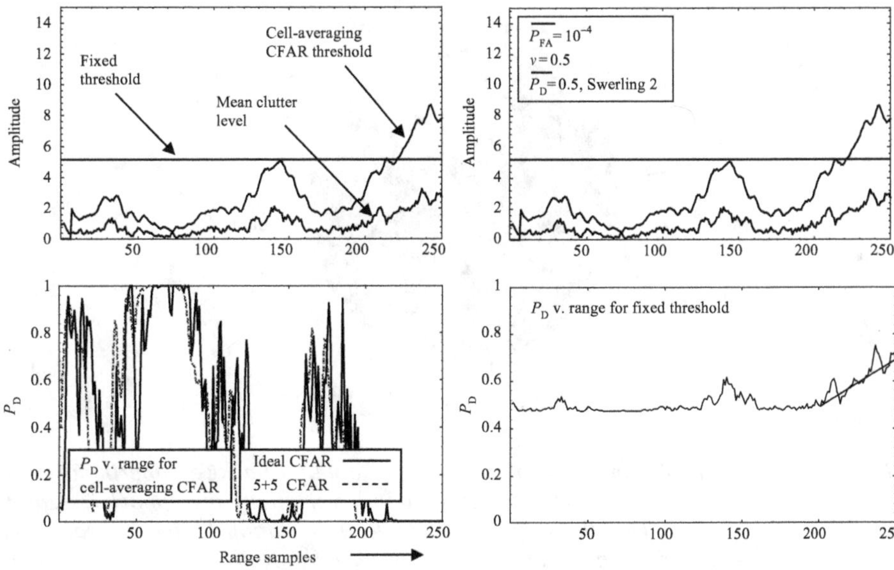

Figure 10.6 P_D *vs. range for CA CFAR and fixed threshold;* $P_{FA} = 10^{-4}$, $P_D = 0.5$; $SCR = 0.2$ *dB for ideal CFAR, 6.7 dB for CA CFAR, 15 dB for fixed threshold*

and P_D. A detection range display [5,6] shows the variation of P_D for a given target size in the presence of an adaptive threshold system. This display uses the local value of threshold multiplier, α, being used by the radar (combined with knowledge of the receiver gain settings, the thermal noise level and noise-limited performance, antenna tilt and viewing geometry), to estimate the smallest target RCS that can be detected at a given point on the sea surface. These estimates are presented to the operator as contours superimposed on the normal PPI display. An example of a poorly adapting system is shown in Figure 10.7. These results were achieved with real airborne radar data and a particular instance of an experimental CFAR system. Here it can be seen that the presence of land has caused the threshold to be raised, reducing sensitivity. As the radar scans off the land (clockwise scan as in Figure 10.5) the adaptive circuits are taking many degrees of scan to recover, resulting in a large loss of sensitivity.

Figure 10.8 shows an example of a well-adapted performance, for the same radar data, but with an improved CFAR detection system. Here it can be seen that the detection threshold has reacted very rapidly to the change from land to sea.

10.3 Performance prediction

Having considered the basic specification of performance, we now turn to performance assessment. In Chapter 8 this subject was addressed for the purposes of system design and optimisation; here we consider it from the perspective of the assessment

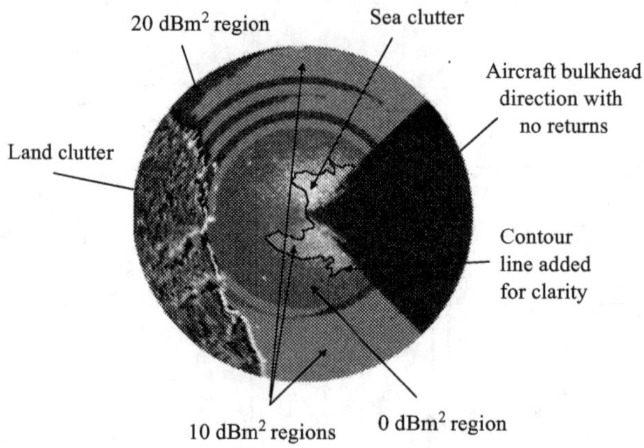

Figure 10.7 Contours showing minimum detectable target RCS superimposed on raw radar returns, showing spatial variations in a poorly adapting system. The black sector is due to obscuration by the aircraft bulkhead with a nose mounted airborne radar. Note the region of reduced sensitivity near the centre, due to high sea clutter returns

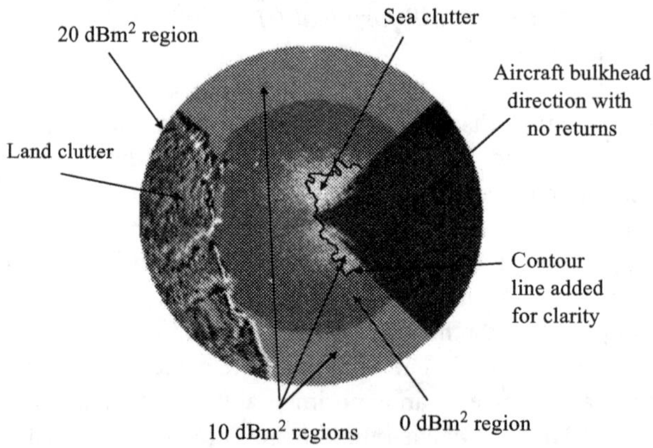

Figure 10.8 Contours showing minimum detectable target RCS superimposed on raw radar returns, showing spatial variations in a well adapting system. The black sector is due to obscuration by the aircraft bulkhead with a nose mounted airborne radar. Note the area of reduced sensitivity at the centre of the display is smaller than in Figure 10.7

of competitive bids (based on a specification) and the demonstration of compliance of a system to a specification.[2] It turns out that in order to undertake the assessment

[2] As a consequence many of the aspects discussed in this section are also covered in Chapter 8, and there is inevitably some overlap.

in an unambiguous manner, many of the details of the modelling methods have to be included in the original specification.

One difficulty of calculating performance in sea clutter derives from the inability to describe accurately the environment in which the radar must perform. As an example, consider a requirement to operate in a given sea state. Whilst this may define the user's needs quite well, the performance of a radar can vary very widely in different sea conditions that may be described by the same sea state. The radar supplier has to interpret such a specification in terms of the characteristics of the radar returns and the response of the radar's detection processing. For detection in sea clutter, the following characteristics will be needed, as a minimum:

- reflectivity
- amplitude statistics
- temporal correlation or spectrum
- spatial characteristics and correlations
- discrete clutter spike characteristics

There will be a wide spread of these values for apparently similar sea conditions. These characteristics are expected to depend on a wide range of radar parameters such as:

- radar frequency
- radar polarisation
- radar spatial resolution
- use of pulse-to-pulse frequency agility, etc.

In addition, the clutter characteristics will be dependent on the environment and viewing geometry, such as:

- grazing angle
- wind speed and direction
- sea swell and direction

As described before, the dynamic performance of the radar may be dependent on the characteristics of particular operating scenarios, such as littoral or open-ocean environments, the presence of land clutter, high target densities, target fading, target-clutter interaction, etc. It is also dependent on the detailed signal processing algorithms used. Detection range alone is clearly unlikely to be a sufficient measure of performance [1,7].

If the performances of competing radars are to be compared meaningfully by the analysis of predicted performance, all these target and clutter characteristics must be specified. If they are not, each radar supplier will make his own assumptions. It follows that a prerequisite for comparing performance is the adoption of a common modelling approach. However, any predicted performance must ultimately be related to that actually observed in practice and it is essential that the models used represent the real world as closely as possible.

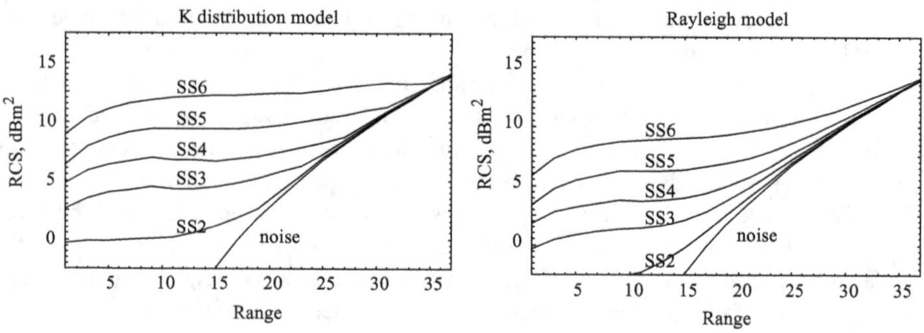

Figure 10.9 Detection performance in different sea states (SS) modelled by K distribution and Rayleigh models

10.3.1 Clutter amplitude statistics

An important feature when specifying performance in clutter is the model to be used for amplitude statistics. A simple model might assume Rayleigh statistics. An additional assumption might specify independent samples from pulse to pulse if frequency agility is used. However, real clutter often deviates significantly from this simple model, as described in earlier chapters.

Figure 10.9 compares the performance predicted by a simple model using Rayleigh statistics and that predicted by a compound K distribution model. The plots show the relative values of the minimum radar cross-section (RCS) as a function of range and sea state for a specific viewing geometry and radar characteristics. The marked difference between the predicted performances is clear. Another feature to notice on these plots is the relatively flat performance variation with range. This is a consequence of the variation with range (and hence grazing angle) of clutter reflectivity, the size of the illuminated clutter patch and the clutter distribution shape parameter. As will be discussed later, this means that the detection range for a real target with a fluctuating RCS is very difficult to measure with any precision.

10.3.2 Clutter speckle component

It is often assumed that independent samples of the speckle component of the compound distribution can be achieved from pulse to pulse with frequency agility. Conversely, it is usually assumed that, for a fixed frequency radar, the speckle component will remain correlated over the relatively short integration period of a beam dwell for a scanning radar.

In practice, for fixed frequency operation the speckle component will decorrelate over a period of 5–10 ms due to the inherent motion of the clutter scatterers. Furthermore, antenna scanning will introduce some decorrelation of the clutter speckle component, due to the change of scatterers that are illuminated from pulse to pulse [8]. Over a typical beam dwell of a scanning radar, the degree of decorrelation is small. One measure of decorrelation is the effective number of independent pulses, N_{eff}, that contribute to integration gain over the beam dwell. For non-coherent integration in Gaussian clutter, for 10 pulses transmitted across the 3 dB antenna beamwidth dwell, $N_{\text{eff}} \approx 1.4$, compared to $N_{\text{eff}} \approx 1$ for a stationary beam or $N_{\text{eff}} \approx 10$ for complete pulse-to-pulse decorrelation of the pulse returns. Similar results are found for K-distributed clutter.

For coherent operation, the temporal decorrelation of the clutter is observed as a spectral broadening of the returns. In the same way, the internal motion of the clutter and the antenna scanning contribute to the clutter spectrum.

Decorrelation of the speckle component with frequency agility is discussed in Chapter 2 and Chapter 8. In Reference 9 it is shown that frequency steps equal to the radar pulse bandwidth cannot fully decorrelate the clutter speckle component from pulse to pulse. However, in a practical situation, the combination of pulse-to-pulse frequency agility, temporal decorrelation due to the clutter's inherent motion and antenna scanning can often produce effective pulse-to-pulse decorrelation of the clutter speckle component [8].

It should also be noted that scanning can introduce some decorrelation of the underlying mean level, which will affect the shape of the clutter distribution for higher values of P_{FA} (say, $P_{\text{FA}} > 10^{-3}$) after pulse-to-pulse integration [8].

10.3.3 False alarms

False alarms are another cause of difficulty when specifying performance. These may variously be specified in terms of a probability of false alarm (P_{FA}) or a false alarm rate (in alarms per unit time). The P_{FA} has most meaning to the radar designer at the point where the first detection threshold is set. After this point the detections may be collapsed in range and/or azimuth, integrated from scan-to-scan and so on, before finally being displayed (see Chapter 8). The P_{FA} at the first threshold is a function of the clutter and noise statistics and the threshold value. It is not dependent on the range scale or search area. The false alarm rate, usually measured at the display, is determined by the P_{FA} and the area (i.e. range and bearing extent) of the radar coverage on the display. The exact relationship between P_{FA} and alarm rate is not straightforward. The P_{FA} is related to the mean time between alarms or the false alarm time, t_a, by $P_{\text{FA}} = t_d/t_a$, where t_d is the mean alarm duration. A simple analysis suggests that $t_d \approx 1/B$, where B is the receiver noise bandwidth. However a more detailed analysis [10] for thermal noise reveals that in general $t_d < 1/B$ and is a function of P_{FA}. The lower the value of P_{FA}, the lower the value of t_d. Typically for $P_{\text{FA}} \sim 10^{-6}$, $t_d = 0.5/B$. For thermal noise, the relationship is a function of the filter characteristics of the radar receiver. For clutter the situation may be more complex, with the false alarm duration being determined by the radar receiver filter characteristics and the spatial extent of the clutter peaks. If the clutter peaks are effectively derived from point-source scattering centres, then the duration of alarms is similar to that for noise.

Analysis of clutter recorded from various airborne radars shows that clutter peaks tend to be slightly wider than noise peaks, but are still apparently dominated by the frequency response of the radar receiver. Following pulse-to-pulse integration, some

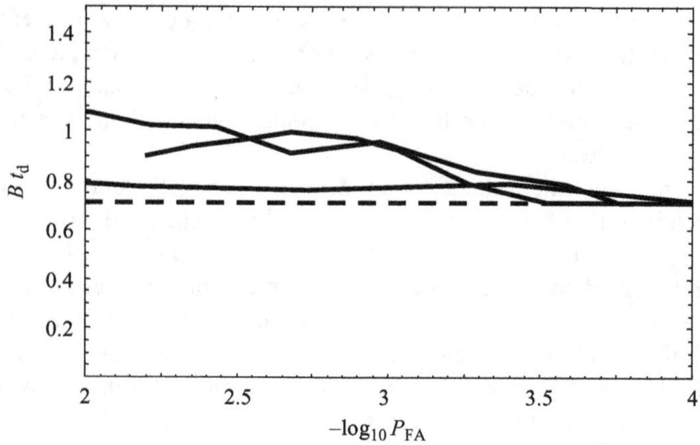

Figure 10.10 *Average alarm duration t_d as a function of P_{FA} for a receiver sampling rate of 1.4B; dotted line: thermal noise; solid lines: examples of clutter*

large spikes are observed to have a greater range extent, giving a slightly larger value of t_d than noise for a given P_{FA}. Figure 10.10 shows the average duration of alarms measured in clutter after pulse-to-pulse integration with a frequency agile radar. The radar pulse bandwidth is of the order of 70 MHz. The range sampling frequency is about $1.4B$. This means that the shortest alarm duration that can be measured is $t_d = 0.7/B$. For P_{FA} in the range 10^{-2}–10^{-4}, the average alarm duration in noise is therefore expected to be $0.7/B$ and this is confirmed in Figure 10.10. Also shown are equivalent measurements in clutter and it can be seen that the limiting value of t_d is not reached until about $P_{FA} = 10^{-4}$, indicating a more range extensive response than noise.

10.4 Measuring performance

Currently it is not clear what measures should be used to quantify the performance of an adaptive radar. Traditional measures, such as the range by which an approaching target has been detected or a track initiated, may still be important for some radar modes but are not sufficient for a full description of performance. These traditional performance measures may be described as steady state or 'static' measures. It is usually assumed that the radar is properly set up for the environment and that the appropriate performance can be measured by observing the radar display or data outputs. The accuracy with which such measurements can be made is discussed later.

The adaptive nature of future radars implies that 'dynamic' performance measures are required. The adaptive performance of a radar is determined by its behaviour in rapidly changing scenarios. This behaviour may be quantified in terms of, say, speed of response to a given change or susceptibility to overload by multiple targets or interference, as well as in terms of detection ranges, track accuracies and so on. This dynamic behaviour is very scenario-dependent and difficult to measure or specify in terms of simple pass or fail criteria in a trial.

The determination of the essential performance characteristics of an adaptive radar and agreement on how these are to be specified and measured are therefore very important aspects of future radar developments. To 'measure' the performance of a radar there is a need to quantify the significance of measurements, and a need to understand what is a 'failed' or a 'successful' trial by the customer. The measurement of radar performance is an area in procurement open to great debate because of the variable levels of performance implicit in a radar operating in a real environment.

Sub-system measurements are important when deciding whether a radar is operating as envisaged, particularly when trying to understand real life effects in the total system. As systems become more complex, a simulation-based procurement process, with specification flow down to sub-system level encourages good sub-system acceptance with modelling providing the system level acceptance. This process depends critically on the system level appropriateness of the design. Different parts of the system have their own measurement requirements and separate consideration may have to be given to areas such as RF processing, signal and data processing, adaptive processes etc. Aspects of performance whose measurement is particularly difficult include the provision of controlled target and clutter environments, propagation uncertainty and the definition of false alarms.

10.4.1 Trials

Acceptance trials are an integral part of any radar development programme. These are intended to show that the radar achieves the functionality required and that it is 'fit for purpose'. Where possible, such trials are also required to assess the radar's performance against its specification quantitatively. However, in many cases the measurement of performance can be very difficult and may require many hours of trials.

Even in the relatively straightforward case of a target with known RCS in noise-limited detection, performance can be difficult to measure accurately with limited number of trials (see Section 10.5). The quantitative assessment of clutter-limited detection, such as small targets on the sea surface, is much more difficult. The radar performance is predicted on the basis of a given target radar cross-section and given clutter characteristics (reflectivity, amplitude statistics, etc.). On a particular trial it is generally impossible to reproduce any of the specified conditions. Targets on the sea surface have an apparent RCS that fluctuates widely due to influences such as multipath reflections, wave shadowing, wave splashing and the presence of target wakes. Fluctuations of ± 10 dB are quite possible for a small target. The clutter reflectivity is notoriously variable, with spreads of ± 5 dB for apparently similar conditions (e.g. wind speed or sea state). The average values of clutter reflectivity and target RCS used to predict performance are likely to have been derived from many hundreds of hours of measurements to obtain average values. If the radar's average performance is to be reproduced it is clear that similar lengths of trials are required. This is usually quite impractical and the results are often subject to disagreement on the conditions actually prevailing (e.g. was it really sea state 3 at the range where the target was measured?). In these circumstances, it is necessary to resort to alternative means of proving performance in a timely and cost effective way.

For an adaptive radar, assessment of performance is even more problematic since the state of the radar is constantly varying. As discussed above, performance may vary widely over the radar display and change rapidly with range and scan angle; its dependence on additional environmental factors, such as spatial correlation in the clutter, adds further uncertainty to the assessment of trials results.

10.4.2 Factory measurements

Noise-limited detection performance may be amenable to measurement by practical trial, as discussed above. However, such performance can be derived with a high degree of confidence from factory measurements. The noise-limited performance can also be determined from measurements of the radar parameters (transmitter power, receiver noise-figure, microwave losses, etc.) and a detailed analysis of the signal processing characteristics (pulse compression gain, pulse-to-pulse integration, sampling errors, etc.). In this way detection performance can be assessed accurately, and in a well defined fashion, at the cost of neglecting the variations in target RCS and propagation conditions encountered in a real trial.

There may be some radar characteristics that can only be measured with the radar installed on its platform. These could include system coherence (local oscillator, transmitter, etc.) under vibration, installed antenna sidelobes, etc. However, it should be possible to make independent measurements of these features and to assess their effect on noise-limited performance by analysis and modelling, as discussed in the following section.

If factory measurements are to be used to predict average performance over a production run, it is important to estimate the spread of parameter values that are expected. Measurement of a specific trials installation may be quite precise, but estimates of the spread in performance of production systems require a statistical analysis of the spread of parameter values and the resultant radar performance.

10.4.3 Modelling and simulation

It must now be clear that the measurement of performance of adaptive radars, especially in clutter-limited conditions, is generally either impossible or too time-consuming if trials alone are used. However, the radar performance can still be specified and predicted in terms of detailed modelling and simulation. For this purpose, the models of the environment may well be based on extensive independent trials and be quite representative of realistic conditions. The basic radar parameters can be measured very accurately in the factory. The radar signal processing, including adaptive processes, is likely to be deterministic for a given input and its performance can be very accurately modelled or simulated for known data inputs. It should therefore be possible to predict with some confidence how the radar would perform under the conditions specified in the models. Provided that the models and simulations are sufficiently detailed, and agreed between the buyer and seller, they should be sufficient to provide a quantitative measurement of radar performance.

Of course, the buyer requires a radar that not only meets the required quantifiable performance, but is also 'fit for purpose'. The radar should perform satisfactorily over a wide range of real conditions, as observed in a trial. Such trials are not, however, used to provide comprehensive performance measurement, but to establish that the modelled results are representative of real results. The main unknowns on a trial are the target, the environment, and the operation of adaptive processing in the radar. This can be assessed by instrumenting the trial to obtain estimates of parameters such as signal-to-clutter and clutter-to-noise ratios in the local region of the target. For adaptive processing in clutter-limited conditions, estimates may also be made of the local clutter statistics, spatial correlation, etc. These measurements may then be analysed to assess how the conditions differed from those assumed in the model. The model may then be modified appropriately to predict detection performance and these predictions compared with actual measured results in the trial.

Clearly the methods for measurement of performance must be assessed independently for each new radar, taking into account the operating environment and operational requirements. The link between performance predicted by modelling and actual performance requires a considerable amount of preparatory work. In particular, models and simulations must be developed which are sufficiently representative of real performance to the satisfaction of buyer and seller. Whilst this may be very difficult, there is really no alternative due to the prohibitive expense of measurement of performance by trials alone.

If performance predicted by modelling is to be validated by spot trials (as suggested here), it is important to understand the accuracy with which the relevant environmental parameters and detection performance can be measured. The statistical significance of any results must also be understood. These issues are addressed in the following sections.

10.5 Measurement methods and accuracies

As has been discussed in earlier sections, assessment of performance by practical trials requires the measurement of selected features. Obvious features to be measured include detection range and probabilities of detection and false alarm. In some circumstances these features do not reliably define the useful performance of the radar and other measures, such as the minimum detectable target RCS, may need to be measured. Often, trials measurements are used to validate modelling and so the trials need to measure and record the prevailing environmental conditions, such as clutter reflectivity and amplitude statistics, and intermediate processing parameters, such as the signal-to-clutter and signal-to-noise ratios.

It is important to understand both how selected features are to be measured and also the accuracy to which measurements can be made. Finally, statistical criteria must be developed for establishing success or failure of the trials, based on the measurements made.

The following sections assess measurement of probability of detection and detection range, in both noise and clutter; probability of false alarm; signal-to-noise and signal-to-clutter ratios; and clutter amplitude statistics. The final section looks at statistical criteria for assessing success or failure of trials.

10.5.1 Probability of detection

There is very little material in the radar literature on the measurement of detection performance from radar trials, although a good analysis of the problem has been given by Stove and Hurd [11]. For noise-limited detection, probability of detection as a function of detection range can be assessed directly by counting detections over a number of scans (the blip-to-scan ratio). In sea clutter, the notion of detection range is less useful and in this case, it is often better to assess the minimum detectable target at a given range. These ideas are assessed in more detail below.

10.5.1.1 Blip-to-scan ratio

The detection range against a target in a trial can be estimated by observing detections on successive radar scans. The 'blip-to-scan' ratio (i.e. the number of detections, or blips, of a given target, counted over several scans) provides a direct estimate of the probability of detection.

The most straightforward case is for a target at a fixed range from the radar. If k detections are observed over n scans, the best estimate of the probability of detection, P_D, is $\hat{p} = k/n$. If the true value of P_D is p, this estimate has a variance $\text{var}(\hat{p}) = p(1 - p)/n$. For a stationary target with $p = 0.5$ and $n = 50$, the standard deviation of $\hat{p}$ is 0.07. This can be interpreted as an equivalent error in estimating signal-to-noise ratio (SNR) or detection range. Take the example of a non-fading target, with $P_{FA} = 10^{-6}$, following the non-coherent integration of 10 pulses. A variation of P_D of ± 0.07 about a mean of 0.5 is equivalent to a spread of SNR of 0.5 dB or a 3% variation in range.

For a fading target the problem is more severe. If the target has Swerling Case 1 fading characteristics, an uncertainty in P_D of ± 0.07 corresponds to a spread of SNR of 1.9 dB, or an 11% spread in range. To achieve an accuracy in range prediction equivalent to that obtained for the non-fading target would require a P_D with a spread of about ± 0.02, implying about 625 independent detections (scans of the radar in this example).

If the radar platform or target is moving, the estimation of P_D is much more difficult as it is continuously changing. Given a sufficient number of trials, the P_D at each range can be estimated. However, the number of trials required may be large, especially for fast moving targets when few observations are available. Also, averaging over several scans in a single trial introduces errors in estimating detection range, including the possibility of introducing a bias.

Figure 10.11 [11] shows a theoretical detection curve for a noise-limited fast moving target with 1, 7 and 17 point moving-averages of the data. It may be noted that the 50% detection range is almost unchanged by averaging but the other ranges

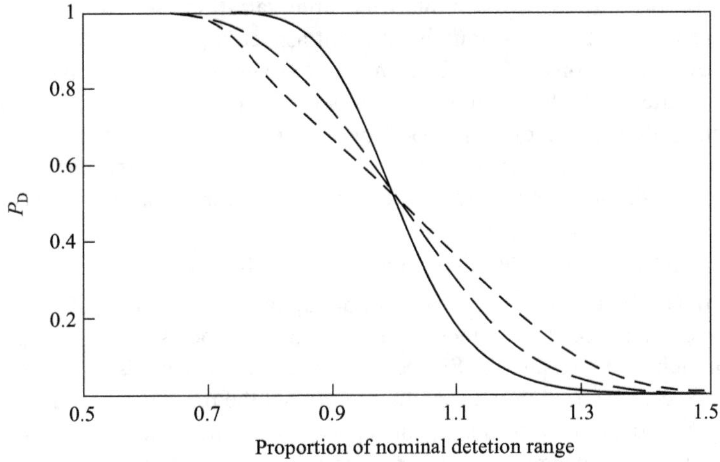

Figure 10.11 *Effect of different levels of smoothing on a detection curve; solid line: raw data; long dashes: 7 point moving average; short dashes: 17 point moving average*

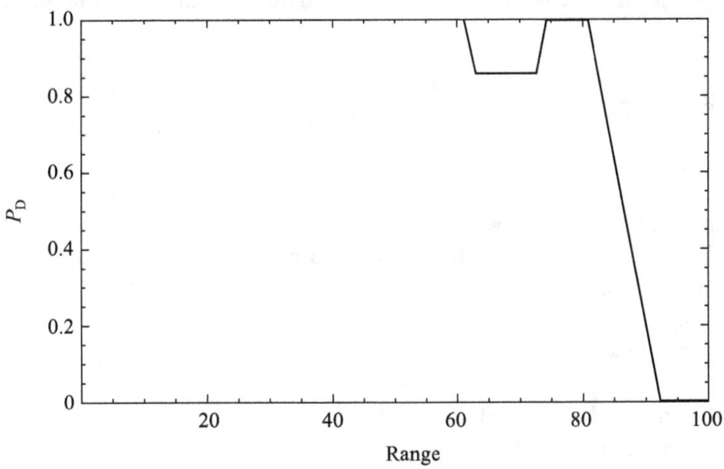

Figure 10.12 *Blip-to-scan ratio curve for noise-limited performance, averaged over 7 scans*

are moved. This suggests that estimation of detection range for $P_D \approx 0.5$ is preferable for moving targets.

Figure 10.12 [11] shows the blip-to-scan ratio for noise limited detection of an air target, taken from a radar trial. Detections are averaged over 7 scans. The sharp fall-off in range allows the noise-limited detection range to be estimated quite accurately.

The main issues with this type of trial relate to the uncertainties of predicting the target RCS and the environmental conditions. The RCS of a complex target, such as a ship or aircraft, usually exhibits very large variations as a function of viewing angle. Even a target with notionally constant aspect angle with respect to the radar might show fluctuations of 2–3 dB from scan to scan, due to small changes in aspect angle associated with the motion of the target and radar platform. Such fluctuations lead to equivalent variation in probability of detection from scan to scan and hence to variations in the estimated detection range. Further uncertainty may be introduced due to multipath interference, especially for targets near or on the sea surface.

10.5.1.2 Estimation of signal-to-noise ratio, SNR

Some of the problems associated with estimating the detection range from the blip-to-scan ratio can be avoided if detection performance can be estimated in a single scan. One approach is to measure the SNR of the test target on successive scans. Estimates of SNR can then be used with modelling to predict the probability of detection at the output of the radar (i.e. at the display or at the input to a track-while scan system, etc.). This assumes that the signal and data processing chain has previously been validated by laboratory or factory measurements (which should be quite straightforward for noise-limited detection). Alternatively, the SNR can be measured at any convenient range and the value compared with predicted performance for the trials conditions.

As shown at the beginning of Chapter 8, the single pulse SNR, referred to the receiver just prior to detection, can be predicted quite accurately from the radar range equation:

$$\mathrm{SNR} = \frac{p_t G(\theta,\phi)}{4\pi R^2}\,\frac{\sigma}{4\pi R^2}\,\frac{G(\theta,\phi)\lambda^2}{4\pi}\,\frac{\tau}{kT_0 F_\mathrm{n} L_\mathrm{a} L_\mu} \tag{10.1}$$

where

p_t is the peak transmit power
$G(\theta, \phi)$ is the one-way antenna gain, as a function of the look direction in azimuth, θ, and elevation, ϕ
σ is the target RCS
R is the range to the target
k is Boltzman's constant
T_0 is the standard noise temperature (290°K)
F_n is the receiver noise figure
τ is the pulse length
λ is the radar wavelength
L_a is the atmospheric loss
L_μ is the microwave loss in the radar

In principle, all these parameters, except for σ and L_a, can be measured in the factory or at the radar installation. The atmospheric loss, L_a, can be a major uncertainty in a practical trial, especially over long path lengths where local conditions along the path may be unknown. The target RCS, σ, may also be quite variable, as discussed above. In a practical trial, care must also be taken to accurately determine the position of the target in the antenna beam, so that the correct value of $G(\theta, \phi)$ may be used.

Measurement of SNR at a convenient point in the radar receiver requires the radar to be instrumented so that the appropriate radar signals can be recorded for subsequent analysis. The noise level can be measured in the absence of a target or other radar returns, preferably when the radar is not transmitting. The target can be measured separately, even if seen against a background of clutter. Provided that the linearity of the receiver and any changes of gain are accurately known, the SNR can be calculated. Factory measurements of the radar systems can then be used to predict the SNR, given the target RCS, viewing geometry, range and so on.

A number of different measurements can be used to develop confidence in this process:

- laboratory measurements of the radar components to predict the signal to noise ratio
- field measurements of the SNR
- comparison of the radar display with the results of simulating the signal processor, to determine that the signal processing is behaving as expected
- comparison of observed and predicted performance in partially clutter-limited conditions, which also require the noise-limited performance to be correct

Variations of target RCS with viewing angle and the effects of multipath interference can be mitigated by the use of special test targets in multipath-free environment. For air targets, towed spheres have been used. On the sea surface, Luneberg lenses or spheres on buoys or towed platforms have been used, although multipath interference is always likely to be a problem for targets on or near the sea surface.

In some cases, the complexity of the target and the unknown multipath interference may make it impossible to accurately predict the target RCS. In such cases, the measured values of SNR on each scan can be used to predict the probability of detection, P_D, for the prevailing conditions and the results compared with the estimates of P_D observed on the trial. This approach requires an independent assessment of the receiver characteristics to ensure the performance being measured is acceptable and in line with earlier predictions of performance used to sell the radar to the customer. A similar approach may often be needed for measurement of detection performance in sea clutter, as discussed below.

10.5.1.3 Detection in sea clutter

The blip-to-scan ratio can also be used to estimate the probability of detection for targets in sea clutter. However, detection range is not usually a useful concept, as discussed in Section 10.3.1 (see Figure 10.9). As an example, Figure 10.13 [11] shows the estimate of P_D, based on the estimate of the blip-to-scan ratio over 7 scans, for a trial of an airborne radar detecting a target on the sea surface. Figure 10.14 [11] shows the same results averaged over 17 scans. Also shown in Figure 10.14 is the predicted performance for the prevailing conditions, which shows that the overall expected performance was reflected in the trial.

One approach to estimating performance in sea clutter is to measure the P_D at an appropriate range and then to use modelling to estimate the target RCS that would

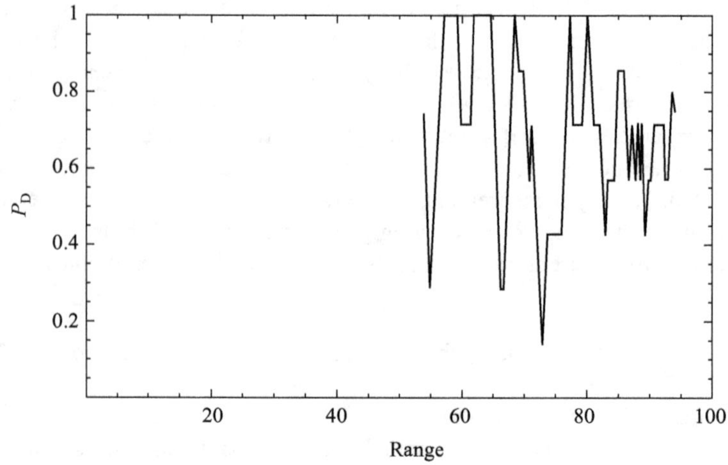

Figure 10.13 *Blip-to-scan ratio curve for clutter-limited performance – averaged over 7 scans; Solid line: blip-to-scan ratio estimate of P_D*

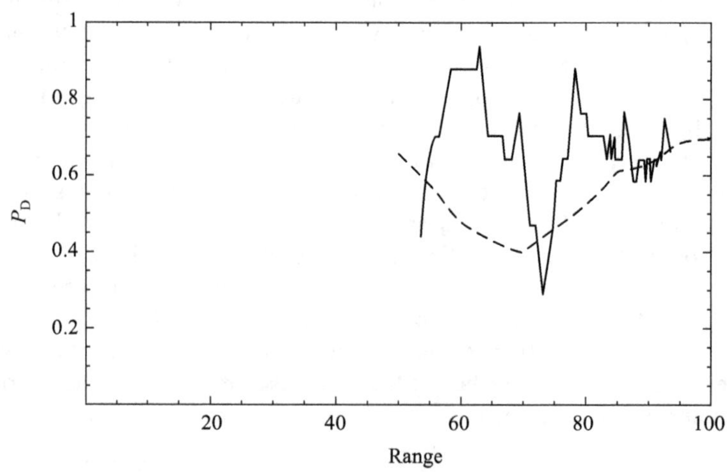

Figure 10.14 *Blip-to-scan ratio curve for clutter-limited performance – averaged over 17 scans; Solid line: blip-to-scan ratio estimate of P_D; Dotted line: modelled performance*

have given that performance in the prevailing conditions. This can then be compared with the known RCS of the target to assess whether the radar sensitivity is as it should be. Clearly, the problems of knowing the target RCS accurately may limit the utility of this approach.

In addition, prediction of target RCS to give a particular P_D requires a detailed knowledge of the prevailing clutter conditions, so that the techniques described in

Chapter 6 can be applied. Methods of estimating the K distribution shape parameter from radar data and their associated accuracies are described in Chapter 6 (Section 6.9) and Chapter 9 (Section 9.3). Measurement of the clutter characteristics requires instrumentation of the radar, so that the actual conditions in the vicinity of the target can be measured. It is also possible to measure the signal-to-clutter-plus-noise ratio from the trials data and to use this to predict the value of P_D, to be compared with that achieved in the trial. This method has the disadvantage of requiring the additional assumption of the satisfactory performance of the radar receiver, but this can be overcome to a large extent by testing in noise as described in the last section.

10.5.2 Probability of false alarm P_{FA}

Any measure of probability of detection must be associated with a probability of false alarm, if it is to have any value as a measure of performance. Section 10.2.4.2 discusses how the P_{FA} may be expected to vary spatially and temporally across the radar display. One measure of performance may be the variability of P_{FA} across the display. The relationship between P_{FA} and false alarm time must also be understood (see Section 10.3.3). If the P_{FA} is varying across the display, the local P_{FA} associated with an estimate of P_D must be measured.

In practice, it is often very difficult to measure low values of P_{FA}. A typical requirement for a surveillance radar might ideally be for a very small number of false alarms per scan or over a period of time. In thermal noise, the P_{FA} can be measured by counting alarms over a long period of time. In addition, the P_{FA} may be quite accurately predicted by calculation, knowing the threshold value that is set. In clutter, the problem is much more complicated. In the open ocean, large areas of clutter may be observed, but the conditions (reflectivity, amplitude statistics, etc.) will vary dynamically as a function of space and time. Assuming some form of CFAR control of the detection threshold, it might be possible to estimate low values of P_{FA} by observing over long periods of time. However, in practice, it will not be possible to observe clutter alone. There may be large numbers of targets, ranging from ships through to flotsam and jetsam, all of which may give alarms.

The false alarm rate at earlier stages of signal processing, say prior to pulse-to-pulse or scan-to-scan integration will be higher and will be more amenable to measurement. However, such measurements are only useful for performance assessment if the relationship between P_{FA} at different stages of the processing is understood.

10.5.3 Statistical analysis of trials

It should be clear from the previous discussions that the measurement of radar performance from trials can be very difficult. Sometimes, the lack of knowledge of the prevailing conditions and the complexity of the adaptive nature of the radar performance mean that the measurement of performance by trials alone is not possible. More often, it may be possible to measure performance but only by averaging results over a very large number of trials, which are very costly to conduct, especially if they require the use of expensive platforms such as military aircraft or large ships. It is usually desirable to limit the number of trials to the minimum possible.

In all cases, it is necessary to understand the statistics of the measurements made on trials, in order to estimate the number of trials required and the confidence that can be placed in the results. An introduction to the statistical analysis of trials is given in Reference 11 and more detailed mathematical techniques are described in Reference 12.

When planning a trial, the first stage is to assess its aims and to design the experiment. If the aim is to establish the actual performance of the radar, as might be required for proving trials, it will be necessary to place upper and lower limits on the uncertainty of the measured performance. In other words, values must be assigned to the probabilities that the measured performance will be better or worse, by some agreed margins, than the true performance. However, if the aim is only to determine whether the radar achieves or exceeds its specified performance, it is usually only necessary to assign a value to the probability that the test will succeed for a system that does not meet the requirement (see later discussion on Buyer's Risk and Seller's Risk).

The second stage of planning is to estimate how many independent measurements (e.g. estimates of detection range etc.) are needed within the trial to obtain the required degree of confidence in the results. This requires an understanding of the statistics of the measurements to be made. These may be derived by analysis or Monte Carlo simulation, although this latter approach can introduce further sources of error. However, these sources of error should not have a significant effect on the trials results themselves, but may introduce errors into the degree of confidence that the customer or supplier place on the results.

10.5.3.1 Sequential testing

One approach to using the lowest possible number of measurements in a trial is sequential testing as described in Appendix 1, Section A1.12 and Reference 12. In this approach, results are assessed after each measurement. The cumulative results (estimates of detection range etc.) over all runs to date are then compared to two thresholds. If performance falls below the lower threshold, the radar is deemed to have failed to achieve the desired performance and the trial is terminated. Conversely, if the performance exceeds the upper threshold, the radar is deemed to have met its desired performance and the trial is terminated. For intermediate values of performance the trial is continued for another measurement. The values of threshold are continuously adjusted as a function of the number of measurements that have been made. This process is illustrated in Figure 10.15.

As an example of sequential testing, suppose that it is desired to estimate a parameter, m, from measurements of a random variable x (e.g. m could be the rms error of a sequence of measurements of the quantity x), using the minimum number of measurements (samples). If the specified value of m is m_0, it may be judged that it would be unwise to accept the system if the value exceeded m_1. In a case like this, where m is some measure of absolute error (i.e., its ideal value is zero and it is always

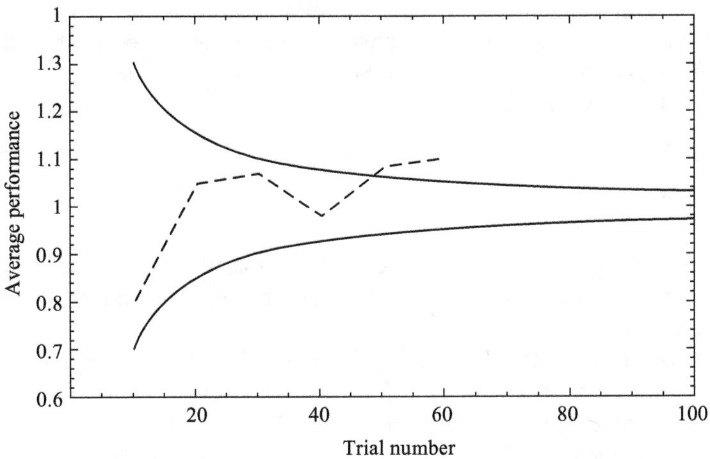

Figure 10.15 *Illustration of a sequential test, solid lines: acceptance and rejection thresholds, dotted line: performance results*

positive), the ratio m_1/m_0 is defined as λ, the Discrimination Ratio. λ is also known as the performance acceptance margin.

Now the trial is arranged to test the following hypotheses:

$$H_0 : m = m_0$$

$$H_1 : m = m_1 = \lambda m_0$$

H_0 is the Null Hypothesis and the system is accepted if it is met. H_1 is the Alternative Hypothesis and the system is rejected if it is met. There are potential errors associated with the acceptance or rejection of a system by using these hypotheses. In particular, a Type 1 error (known as the Seller's Risk, α) is said to be made if the Null Hypothesis is rejected when the hypothesis is true. A Type 2 error (known as the Buyer's Risk, β) is said to have been made if the Null Hypothesis is accepted when the hypothesis is false. It is therefore assumed that the systems will be accepted if the Null Hypothesis is met with a probability of at least $(1 - \alpha)$. Similarly the system will be rejected if the Alternative Hypothesis is met, with a probability of $(1 - \beta)$.

The values of α, β, m_0 and λ must be negotiated between the buyer and seller prior to any testing taking place. If the value of λ is chosen below some small value, say about 1.06, the number of tests required to make a decision increases rapidly. As λ increases, the number of tests required to make a decision reduces but the confidence in the accuracy of the outcome also reduces. As a rule of thumb, it is desirable for $1.06 < \lambda < 1.2$; although this range depends upon the pdf of the measurements, $P(x|m)$. The upper and lower thresholds depend on both $P(x|m)$ and the number of measurements made; further details can be found in Reference 12.

A potential difficulty with sequential testing is that the total number of measurements required is unknown prior to any trial. However, it is possible to estimate the average number of measurements required and the actual value is unlikely to exceed 3 times this. In addition, it is possible to place an upper limit on the number of measurements (known as Truncated Sequential Testing), although this affects the choice of values for α and β.

References

1 S. Watts, ‘Performance measures for airborne maritime surveillance radars’, IEE Colloquium on specifying and measuring performance of modern radar systems, 6 March 1998.
2 S. Watts, H.D. Griffiths, J.R. Holloway, A.M. Kinghorn, D.G. Money, D.J. Price, A.M. Whitehead, A.R. Moore, M.A. Wood and D.J. Bannister, The specification and measurement of radar performance, IEE International Conference Radar 2002, 14–17 October 2002, IEE Publication No. 490, pp. 542–546.
3 S. Watts, H.D. Griffiths, J.R. Holloway, A.M. Kinghorn, D.G. Money, D.J. Price, A.M. Whitehead, A.R. Moore, M.A. Wood and D.J. Bannister, The specification and measurement of radar performance – future research challenges, *J. Defence Sci.*, **8**(2), 2003, 83–91.
4 S. Watts, ‘Radar detection prediction in sea clutter using the compound K-distribution model’, *Proc. IEE*, **132**(F, 7), 613–620, 1985.
5 S. Watts and A.J. Gordon, The Searchwater family of radars, Proceedings of the IEE International Conference Radar 97, 14–16 October 1997, IEE Publication No. 449, pp 334–338.
6 S. Watts, ‘Method and apparatus for estimating the detection range of a radar’, UK Patent 2297213 (17 Jan 1996); USA Patent 5703592; France Patent 9600418.
7 S. Watts, Specification and measurement of performance for airborne maritime surveillance radars, International Conference, Radar 99, Brest, 17–21 May, 1999, Session 2.6.
8 S. Watts and G. Knight, Performance prediction for scanning radars with correlated returns from K-distributed sea clutter, International Radar Conference Radar 2004, Toulouse 19–21 October 2004.
9 K.D. Ward, R.J.A. Tough and P.W. Shepherd, Modelling sea clutter: correlation, resolution and non-Gaussian statistics, IEE Conference Publication 449, Radar 97, pp. 95–99.
10 S. Watts, Duration of radar false alarms in band-limited Gaussian noise, *IEE Proc. Radar, Sonar, Navig.*, **146**(6), 273–277, 1999.
11 A.G. Stove and D.L. Hurd, Performance evaluation for modern radars, Proceedings of the IEEE International Radar Conference, Adelaide, September 2003, pp. 547–553 (paper 44)
12 A. Wald, *Sequential Analysis*, Dover Books, 1973

Appendix 1
Elements of probability theory

A1.1 Introduction

In this appendix we give an informal review of probability theory and other matters relevant to the modelling of clutter and its impact on radar systems. Our selection of topics is rather eclectic and our treatment pragmatic; the principal aim is to remove those obstacles to the evaluation of radar performance presented by any unfamiliarity with the concepts and practice involved in calculating probabilities. Much fuller accounts can be found in standard texts such as References 1 and 2. By collecting together the relevant definitions and didactic developments, we also ensure a greater continuity and brevity in the main text.

Historically, probability theory grew out of a desire, on the part of the rich and indolent, to gain some advantage in the games of chance with which they filled their leisure hours. Even today this scenario provides us with a useful introduction to the subject. Simple die and coin tossing games introduce the concept of a random variable; common sense rules that allow us to characterise these games' outcomes provide the basis for manipulating probabilities and random variables. These ideas can be generalised to cover other situations, in particular those where the outcome (e.g. a received signal) takes a continuum of values and develops in time. Combining and processing these outcomes gives us new random variables; how do we characterise their properties? Once again we will move from simple discrete examples to the more complicated situations we encounter in practice. Particular attention will be paid to the distinction between correlated and independent events and the concepts of joint and conditional probabilities. This leads us to Bayes' theorem, which underpins much of the estimation of parameters that characterise observed processes in probabilistic terms. Some commonly used models, including the binomial, Poisson, gamma and normal processes, will be described along the way. This is followed by a brief review of the description of the dynamics of a Gaussian and non-Gaussian processes, in terms of Langevin equations and power spectra. We conclude by collecting together some more specialised, but interesting and useful results, which find application in several parts of the book. Most of the material in this appendix is relatively straightforward and underpins much of the discussion in Chapters 4–9; other parts, in particular some of Sections A1.6 (multivariate Gaussian processes) and A1.11 (stochastic differential equations) are more advanced and of more specialised application. All told, we hope that, along with the account of some useful special functions given in the second appendix, this review is sufficient to stand our readers in good stead when they apply the material in this book to the solution of real radar problems.

A1.2 Finite numbers of discrete events

Perhaps the simplest game of chance is that in which a single unbiased coin is tossed and the uppermost face (H or T) is recorded. In any one game one cannot predict the outcome, obviously. If we were to repeat the game many times we would be surprised if there was a great preponderance of either heads or tails. Experience and intuition would lead us to expect approximately equal numbers of heads and tails and that ultimately, in a very large number of trials, the numbers of heads and tails would be effectively indistinguishable. We 'feel' that heads and tails are equally likely to occur in any one trial and that this quantifiable indeterminism in one event is nonetheless predictably manifest in the outcome of a large number of such events. In the coin tossing game we recognise two outcomes, H and T. Either outcome is one of these two options; we identify the frequency of occurrence of either outcome with the ratio of number of ways it can be achieved to the total number of outcomes of a single trial. Throwing a single die provides another example: the die has six faces corresponding to six outcomes. The probability of any one of these turning up in a single throw is taken to be 1/6. This intuitive approach, identifying the outcome of an exercise in enumeration with a frequency or 'probability' provides us with a blueprint from which we can build up an extensive calculational framework. There are many ways of introducing the basic ideas ('axioms') and interpretation of probabilistic calculus, whose conflicting claims of rigour and appeal to common sense do much to make the subject seem magical. Our approach will be rather lacking in rigour, and has us extrapolate from these simple examples until we make contact with the problems in which we are interested. Straight away we see that we have already highlighted two salient features of a probability: that it is positive (or zero) and that the sum of the probabilities of all possible outcomes of a trial is one.

A slightly more complicated problem of enumeration does have significant practical consequences: in how many ways might we pick n objects from N, without replacing them? There are N ways of picking the first; its removal limits one's choice when picking the second; there are $N-1$ ways of doing this. Continuing in this way we find that there are

$$N(N-1)(N-2)\cdots(N-n+1) = N!/(N-n)! \tag{A1.1}$$

ways of picking n objects without replacement. This calculation distinguishes between sets of objects that differ in the order in which they are drawn out; if we do not wish to distinguish between these orderings we must remove this multiple counting by dividing by the number of ways in which n objects can be ordered to give

$$\frac{N!}{n!(N-n)!} \tag{A1.2}$$

Calculations of this kind have applications even more significant than calculating the odds of winning the National Lottery. Rather than considering the outcome of a single coin tossing game, we could consider how many heads might be thrown in a sequence of N games. So far we have convinced ourselves that it will be about $N/2$. Can we do better than this and calculate the probability of there being n heads? Again we seek to enumerate the possible outcomes. All told there are 2^N outcomes to the sequence of N games. Of these $N!/(N-n)!n!$ will consist of n heads. Thus the probability of there being n heads thrown in N games is given by

$$P(n \mid N) = \frac{N!}{2^N (N-n)!n!} \tag{A1.3}$$

So far we have just worked out probabilities; now we are going to use these to work out average (mean, expectation) values. Were we to carry out a large number of N fold coin tossing experiments, what would be the average number of heads we would obtain? We work this out as

$$\langle n \rangle_N = \sum_{n=0}^{N} nP(n \mid N) = \frac{N!}{2^N} \sum_{n=0}^{N} \frac{n}{n!(N-n)!} = \frac{N}{2} \tag{A1.4}$$

an answer we might have anticipated. The mean square number of heads is given by

$$\langle n^2 \rangle_N = \sum_{n=0}^{N} n^2 P(n \mid N) = \frac{N!}{2^N} \sum_{n=0}^{N} \frac{n^2}{n!(N-n)!} = \frac{N(N+1)}{4} \tag{A1.5}$$

Other expectation values can be worked out similarly. Equations (A1.4) and (A1.5) provide us with an insight of the spread of n about its mean value; a commonly used measure of this is the normalised variance

$$\frac{\langle n^2 \rangle_N - \langle n \rangle_N^2}{\langle n \rangle_N^2} = \frac{1}{N} \tag{A1.6}$$

We see that this gets smaller as the number of games per trial gets larger, again in accord with the intuitive picture we are adopting here.

This analysis can be extended to other binary experiments, where the outcomes are not the same (e.g. is there a target there or not). Rather than construct a set of outcomes that incorporate the bias in the tossed coins, we seek to combine probabilities of individual events more efficiently. If two events E_1, E_2 are mutually exclusive outcomes of a trial (e.g. a head and a tail) then the probability that one or the other will occur is given by

$$P(E_1 \cup E_2) = P(E_1) + P(E_2) \tag{A1.7}$$

If two independent events E_1, E_2 have probabilities $P(E_1), P(E_2)$ then the probability of their both occurring is given by

$$P(E_1, E_2) = P(E_1)P(E_2) \tag{A1.8}$$

Suppose that, when we toss a biased coin, the probability of a head occurring is p. (As head and tail exhaust all possible outcomes of the trial the probability of a tail is $1-p$.) So, if we toss N coins and obtain a configuration of n heads the probability of that given configuration arising is $p^n(1-p)^{N-n}$. However there are $N!/(N-n)!n!$ such configurations, each with n heads. So, to get the probability of any of these, equally probable and mutually exclusive, combinations occurring in the trial, we form

$$P(n \mid N) = \frac{p^n(1-p)^{N-n}N!}{(N-n)!n!} \tag{A1.9}$$

Using this we find that

$$\begin{aligned}
&\sum_{n=0}^{N} P(n \mid N) = 1 \\
&\sum_{n=0}^{N} nP(n \mid N) = Np \\
&\sum_{n=0}^{N} nP(n \mid N) = N^2p^2 + Np(1-p)
\end{aligned} \tag{A1.10}$$

which respectively confirm that the sum of probabilities of all possible outcomes is unity, and generalise (A1.4) and (A1.5) to the biased case.

A1.3 An infinite number of discrete events

We now consider what happens to this distribution in the limit of a very large N. Two quite different limiting forms of the binomial distribution, with different physical interpretations, can be derived. In the first of these we retain discrete values of n and allow p to scale with N as

$$p = \frac{\bar{N}}{N} \tag{A1.11}$$

If N is allowed to increase without limit then

$$\begin{aligned}
P(n) &= \lim_{N\to\infty} P(n \mid N) \\
&= \lim_{N\to\infty} \frac{N^n}{n!}\left(\frac{\bar{N}}{N}\right)^n \left(1 - \frac{\bar{N}}{N}\right)^{N-n} = \exp(-\bar{N})\frac{\bar{N}^n}{n!}
\end{aligned} \tag{A1.12}$$

This Poisson distribution, is characterised by the single parameter $\bar{N}$ that can be identified with the mean value of the population

$$\bar{N} = \sum_{n=0}^{\infty} nP(n) \tag{A1.13}$$

The mean square can be calculated similarly as

$$\sum_{n=0}^{\infty} n^2 P(n) = \bar{N}(\bar{N} + 1) \tag{A1.14}$$

A distribution, closely related to the Poisson model, has a $P(n)$ of the form

$$P(n) = \frac{1}{(1 + \bar{N}/\alpha)^{\alpha}} \frac{\Gamma(n + \alpha)}{n!\Gamma(\alpha)} \left(\frac{\bar{N}/\alpha}{1 + \bar{N}/\alpha} \right)^n \tag{A1.15}$$

This so-called negative binomial distribution tends to the Poisson form as the parameter α increases without limit. We note that we now have

$$\sum_{n=0}^{\infty} nP\,(n) = \bar{N}, \quad \sum_{n=0}^{\infty} n^2 P\,(n) = \bar{N} \left(\bar{N} \left(1 + \frac{1}{\alpha} \right) + 1 \right) \tag{A1.16}$$

The Poisson and negative binomial models are particularly useful in our modelling of clutter as radiation returned from a fluctuating population of scatterers (see Chapter 4).

A different limiting form of the binomial distribution emerges when both n and N become large, so that their ratio can be thought of as a continuous variable. If we set $n = xN$ and invoke Stirling's approximation (A2.17) we find that

$$\begin{aligned} \log(P(n \mid N)) \sim{} & Nx \log p + N(1 - x) \log(1 - p) + (N + 1/2) \\ & \times \log N - (N - (1 - x) + 1/2) \log(N(1 - x)) \\ & - (Nx + 1/2) \log(xN) - 1/2 \log(2\pi) \end{aligned} \tag{A1.17}$$

From this we see that the maximum in (the log of) $P(n \mid N)$ is located at

$$x_m = p + O(1/N) \tag{A1.18}$$

Expanding the exponent around this point we find that

$$\log(P(n \mid N)) \approx -\frac{1}{2} \log(2\pi) - \frac{1}{2} \log(Np(1 - p)) - \frac{(n - Np)^2}{2Np(1 - p)} \tag{A1.19}$$

or

$$P(n \mid N) \approx \frac{1}{\sqrt{2\pi Np(1 - p)}} \exp \left(-\frac{(n - Np)^2}{2Np(1 - p)} \right) \tag{A1.20}$$

This has the expected Gaussian form. A comparison with plots of the binomial distribution shows that this approximation is not bad for $N = 5$ and acceptably good for $N = 10$ (Figure A1.1).

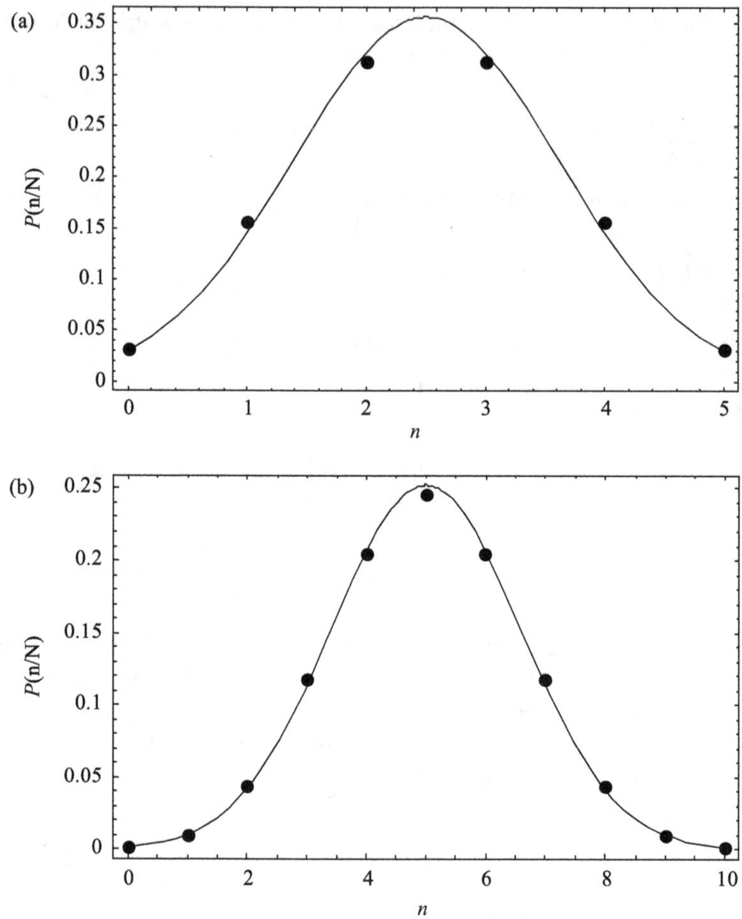

Figure A1.1 *A comparison of the exact form (A1.3) (points) and Gaussian approximation (A1.20) (full line) to $P(n \mid N)$, (a) $N = 5$ and (b) $N = 10$*

A1.4 Continuous random variables

So far we have focused our attention on outcomes with discrete values and provided an illustration of the basic properties that we might expect of a probability (positivity, normalisability, use in the calculation of expectation values). It is relatively straightforward to extend these formal concepts to an outcome that can take a continuum of values x. We do this by defining a probability distribution F and its associated probability density P. Thus we define:

$F(x) =$ Probability that the random variable takes a value less than or equal to x

Note that $F(-\infty) = 0$; $F(\infty) = 1$. The probability density is defined by

$$P(x) = \frac{dF(x)}{dx} \tag{A1.21}$$

which has the rather loose interpretation that the probability that the random variable takes values between x and $x+dx$ is given by $P(x)dx$. Sometimes, to avoid confusion F and P are labelled with both the random variable's symbol (usually **bold**) and the value that it takes. Thus we might have $P_{\mathbf{x}}(x)$. The probability that the random variable takes a value exceeding some threshold is given by

$$1 - F(x_T) = \int_{x_T}^{\infty} P(x)dx; \tag{A1.22}$$

this provides us with a template for the calculation of a probability of detection or false alarm. Expectation values are given in terms of integrals over the probability density function

$$\langle f \rangle = \int_{-\infty}^{\infty} f(x)P(x)dx \tag{A1.23}$$

Among these expectation values are the moments

$$\langle x^n \rangle = \int_{-\infty}^{\infty} x^n P(x)dx \tag{A1.24}$$

and the characteristic function

$$C(k) = \langle \exp(ikx) \rangle = \int_{-\infty}^{\infty} \exp(ikx)P(x)dx \tag{A1.25}$$

which are related through

$$C(k) = \sum_{n=0}^{\infty} \frac{(ik)^n}{n!} \langle x^n \rangle. \tag{A1.26}$$

It is possible for F to have all the properties of a probability, and yet for not all of the moments of the corresponding density to exist; in this case the characteristic function cannot be expanded in a Taylor series as we have done here. However, for the clutter models we are interested in, all the moments exist and the characteristic function is well behaved. In some cases, where the random variable takes only positive values, it is convenient to define the characteristic function as the Laplace transform of the pdf:

$$C(s) = \langle \exp(-sx) \rangle. \tag{A1.27}$$

Some examples of probability density functions or continuous random variables are:

the Gaussian distribution

$$P(x) = \frac{1}{\sqrt{2\pi\sigma^2}} \exp(-(x-m)^2/2\sigma^2) \qquad -\infty < x < \infty, \tag{A1.28}$$

the gamma distribution

$$P(x) = \frac{b^\nu}{\Gamma(\nu)} x^{\nu-1} \exp(-bx) \qquad 0 \le x < \infty \tag{A1.29}$$

and the Cauchy distribution

$$P(x) = \frac{a}{\pi} \frac{1}{a^2 + x^2} \qquad -\infty < x < \infty. \tag{A1.30}$$

The Gaussian and gamma distributions arise in the solution of many practical problems. Their probability distributions can be expressed in terms of the error (erf(t)) and incomplete gamma ($\Gamma(\nu, t)$) functions, defined by

$$\operatorname{erf}(t) = \frac{2}{\sqrt{\pi}} \int_0^t \exp(-p^2) dp$$

$$\Gamma(\nu, t) = \int_t^\infty p^{\nu-1} \exp(-p) dp$$

$$\Gamma(N, t) = (N-1)! \exp(-t) \sum_{k=0}^{N-1} \frac{t^k}{k!}, \quad \text{for integral } N \tag{A1.31}$$

These have been studied in some detail (see Reference 3, and Chapters 6 and 7) and their values are readily computable.

If we have two random variables $\mathbf{x}, \mathbf{y}$ taking values x, y then, much as in the single variable case, we can define a joint probability distribution $F_{\mathbf{xy}}(x, y)$, the probability that $\mathbf{x}, \mathbf{y}$ take values less than or equal to x, y. (Note that $F(\infty, \infty.) = 1$.) Using this joint probability distribution we can evaluate the corresponding probability density function

$$P_{\mathbf{xy}}(x, y) = \frac{\partial^2}{\partial x \partial y} F(x, y) \tag{A1.32}$$

from which we can determine expectation values

$$\langle f \rangle = \int_{-\infty}^{\infty} dx \int_{-\infty}^{\infty} dy f(x, y) P_{\mathbf{xy}}(x, y) \tag{A1.33}$$

Given $F_{\mathbf{xy}}(x, y)$ we can find the corresponding probability distribution of values taken by $\mathbf{x}$ as $F_{\mathbf{xy}}(x, \infty)$; the corresponding probability density function is found by integrating over all possible values of y to give the marginal density:

$$P_{\mathbf{x}}(x) = \int_{-\infty}^{\infty} dy\, P_{\mathbf{xy}}(x, y) \tag{A1.34}$$

In general $P_{\mathbf{xy}}(x, y)$ is not merely the product of the marginal densities of x and y, i.e.

$$P_{\mathbf{xy}}(x, y) \neq P_{\mathbf{x}}(x) P_{\mathbf{y}}(y) \tag{A1.35}$$

if the equality does hold then the variables $\mathbf{x}, \mathbf{y}$ are said to be independent.

The correlation function of $\mathbf{x}$ and $\mathbf{y}$ is given by

$$\langle xy \rangle = \int_{-\infty}^{\infty} dx \int_{-\infty}^{\infty} dy\, xy P_{\mathbf{xy}}(x, y) \tag{A1.36}$$

For correlated variables $\langle xy \rangle \neq \langle x \rangle \langle y \rangle$; for independent variables the correlation function factorises: $\langle xy \rangle = \langle x \rangle \langle y \rangle$.

When $\mathbf{x}, \mathbf{y}$ are independent we would expect the value taken by one to be unaffected by the value taken by the other; strongly correlated random variables would be expected to show a significant coupling between the values they take. This idea can be formalised in terms of a conditional probability density. The probability of two events occurring (e.g. the outcome HH of a game in which two coins are tossed) is $P(E_1, E_2)$. The probability that the event E_1 occurs, given that the event E_2 has occurred is referred to as the conditional probability and is given by

$$P(E_1 \mid E_2) = \frac{P(E_1, E_2)}{P(E_2)} \tag{A1.37}$$

In the un-biased two coin game, for example, we have

$$P(H) = P(T) = 1/2;$$

$$P(H, H) = P(H, T) = P(T, H) = P(T, T) = 1/4$$

$$P(H \mid T) = \frac{P(H, T)}{P(T)} = 1/2 \tag{A1.38}$$

This idea can be generalised to the case where the 'events' are continuous random variables. Thus the probability that $\mathbf{x}$ takes a value less than or equal to x, given that $\mathbf{y}$ takes a value less than or equal to y is given by the conditional distribution

$$F_{\mathbf{xy}}(x \mid y) = \frac{F_{\mathbf{xy}}(x, y)}{F_{\mathbf{y}}(y)} \tag{A1.39}$$

The corresponding conditional density, giving the probability that $\mathbf{x}$ takes a value between x and $x + dx$, given that $\mathbf{y}$ takes a value between y and $y + dy$ is given by $P_{\mathbf{xy}}(x \mid y)dx$ where $P_{\mathbf{xy}}(x \mid y) = P_{\mathbf{xy}}(x, y)/P_{\mathbf{y}}(y)$. Two random variables $\mathbf{x}, \mathbf{y}$ are said to be independent if

$$F_{\mathbf{xy}}(x,y) = F_{\mathbf{y}}(y)F_{\mathbf{x}}(x); \qquad F_{\mathbf{xy}}(x\mid y) = F_{\mathbf{x}}(x) \tag{A1.40}$$

or, equivalently,

$$P_{\mathbf{xy}}(x,y) = P_{\mathbf{y}}(y)P_{\mathbf{x}}(x); \qquad P_{\mathbf{xy}}(x\mid y) = P_{\mathbf{x}}(x) \tag{A1.41}$$

As $P_{\mathbf{xy}}(x,y) = P_{\mathbf{xy}}(x\mid y)P_{\mathbf{y}}(y) = P_{\mathbf{xy}}(y\mid x)P_{\mathbf{x}}(x)$ the conditional density specifying the 'probability of x given y' can be related to the 'probability of y given x' through Bayes' theorem, which states that

$$P_{\mathbf{xy}}(x\mid y) = \frac{P_{\mathbf{xy}}(y\mid x)P_{\mathbf{x}}(x)}{P_{\mathbf{y}}(y)} = \frac{P_{\mathbf{xy}}(y\mid x)P_{\mathbf{x}}(x)}{\int P_{\mathbf{xy}}(y\mid x)dx} \tag{A1.42}$$

In some circles this result is dismissed as mere algebra, in others it is identified as the fundamental point of contact between probability theory and the real world. In Chapters 6 and 7 we see that it plays a central role in the detection of and discrimination between signals.

These ideas are readily extended to three and more random variables, as is discussed in Section 8.1 of Reference 2.

A1.5 Functions of random variables

If you have a function that maps values of **x** into those of some other random variable **y**, what is the pdf of its value

$$y = y(x)? \tag{A1.43}$$

In essence this is a simple problem in the integral calculus, as long as the function maps one to one. Merely by changing variables we have

$$P_{\mathbf{y}}(y) = P_{\mathbf{x}}(x(y))\frac{dx}{dy} \tag{A1.44}$$

where $x = x(y)$ is the inverse of the function y. As an illustration, consider the pdf of the logarithm of a gamma distributed variable:

$$y = \log(x) \qquad x = \exp(y) \qquad \frac{dx}{dy} = \exp(y)$$

$$P_{\mathbf{x}}(x) = \frac{b^\nu}{\Gamma(\nu)}x^{\nu-1}\exp(-bx) \qquad 0 \le x < \infty$$

$$P_{\mathbf{y}}(y) = \frac{b^\nu}{\Gamma(\nu)}\exp(\nu y - b\exp(y)) \qquad -\infty < y < \infty \tag{A1.45}$$

The multi-variable case proceeds similarly. Consider two random variables **x**, **y** from which we form two new random variables **u**, **v**, with values given by

$$u = u(x,y), \qquad v = v(x,y) \tag{A1.46}$$

we assume that this mapping can be inverted through

$$x = x(u, v), \qquad y = y(u, v) \tag{A1.47}$$

The joint density of u,v is given by

$$P_{\mathbf{uv}}(u, v) = P_{\mathbf{xy}}(x(u, v), y(u, v)) \left| \frac{\partial(x, y)}{\partial(u, v)} \right| = \frac{P_{\mathbf{xy}}(x(u, v), y(u, v))}{|\partial(u, v)/\partial(x, y)|} \tag{A1.48}$$

where the Jacobian of the transformation is given by

$$\frac{\partial(u, v)}{\partial(x, y)} = \begin{vmatrix} \dfrac{\partial u}{\partial x} & \dfrac{\partial u}{\partial y} \\ \dfrac{\partial v}{\partial x} & \dfrac{\partial v}{\partial y} \end{vmatrix} = \frac{1}{\partial(x, y)/\partial(u, v)} \tag{A1.49}$$

If $\mathbf{x}, \mathbf{y}$ are combined to form a single new random variable $\mathbf{z}$ we can find the density of its values as follows. We consider the transformation to the new variables $\mathbf{z}, \mathbf{y}$, whose values have the joint density

$$P_{\mathbf{zy}}(z, y) = \frac{P_{\mathbf{xy}}(x(z, y), y)}{|\partial(z, y)/\partial(x, y)|} = \frac{P_{\mathbf{xy}}(x(z, y), y)}{|\partial z/\partial x|_{x=x(z,y)}} \tag{A1.50}$$

The density of the values of $\mathbf{z}$ is then obtained by integrating over y:

$$P_{\mathbf{z}}(z) = \int_{Y} P_{\mathbf{zy}}(z, y)\, dy = \int_{Y} \frac{P_{\mathbf{xy}}(x(z, y), y)}{|\partial z/\partial x|_{x=x(z,y)}} dy \tag{A1.51}$$

Some care is needed in carrying out this integration, to ensure that Y, the range of integration, is consistent with the value taken by z.

These formal developments can be illustrated by a few practically important examples. First, we consider the sum $\mathbf{z}$ of the independent random variables $\mathbf{x}$ and $\mathbf{y}$. Using the methods described above we see that the pdf of its values z is given by the convolution of the pdfs of the constituent parts:

$$P_{\mathbf{z}}(z) = \int_{-\infty}^{\infty} P_{\mathbf{x}}(z - y) P_{\mathbf{y}}(y) dy \qquad \mathbf{z} = \mathbf{x} + \mathbf{y} \tag{A1.52}$$

From the properties of the Fourier and Laplace transforms we see that the characteristic function of z is given by the product of the characteristic functions of the x and y. This result is very useful. Thus, if we add two independent random variables from identical gamma distributions, the pdf of their sum is given by:

$$P_{\mathbf{z}}(z) = \frac{b^{2\nu}}{\Gamma(\nu)^2} \exp(-bz) \int_0^z (z - x)^{\nu-1} x^{\nu-1} dx = \frac{b^{2\nu}}{\Gamma(2\nu)} z^{2\nu-1} \exp(-bz) \tag{A1.53}$$

If the random variables are drawn from gamma distributions with identical scale parameters b, but different shape parameters ν_1, ν_2, essentially the same argument goes through to give us

$$P_y(y) = \frac{b^{\nu_1+\nu_2} y^{\nu_1+\nu_2-1} \exp(-by)}{\Gamma(\nu_1+\nu_2)} \tag{A1.54}$$

In the most general case, in which both the scale and shape parameters of the two distributions differ, we have

$$P_y(y) = \frac{(b_1 b_2)^{\nu}\sqrt{\pi}}{\Gamma(\nu)}((b_1-b_2)y)^{1/2-\nu} \times \exp\left(-\frac{(b_1+b_2)y}{2}\right) I_{\nu-1/2}\left(\frac{(b_1-b_2)y}{2}\right) \tag{A1.55}$$

Here I is a modified Bessel function of the first kind; the symmetry of this expression in the two different scale parameters is obvious. The derivation of this result is a useful exercise in the application of the material in Appendix 2.

A similar analysis yields the pdf of the product of two random variables. Thus, if $\mathbf{z} = \mathbf{xy}$, then

$$P_{\mathbf{z}}(z) = \int P_{\mathbf{x}}(z/y)\frac{1}{|y|}P_{\mathbf{y}}(y)dy \tag{A1.56}$$

and is the Mellin convolution [4] of the pdfs of the constituent random variables. For example, the product of two gamma variables, drawn from distributions characterised by shape and scale parameters ν_1, b_1 and ν_2, b_2 has a pdf given by

$$\begin{aligned} P(z) &= \frac{b_1^{\nu_1} b_2^{\nu_2}}{\Gamma(\nu_1)\Gamma(\nu_2)} z^{\nu_1-1} \int_0^\infty y^{1-\nu_1} \exp(-b_1 z/y) y^{\nu_2-1} \exp(-b_2 y) dy \\ &= 2\frac{(b_1 b_2)^{(\nu_1+\nu_2)/2}}{\Gamma(\nu_1)\Gamma(\nu_2)} z^{(\nu_1+\nu_2)/2-1} K_{\nu_1-\nu_2}\left(2\sqrt{b_1 b_2 z}\right) \end{aligned} \tag{A1.57}$$

A1.6 The normal process

As the normal or Gaussian process, which we saw emerging from the binomial distribution in a continuous limit, finds widespread application as a statistical model, we will review its properties in a little more detail. First, we consider a single random variable (univariate process) $\mathbf{x}$ whose pdf is

$$P_{\mathbf{x}}(x) = \frac{1}{\sqrt{2\pi\sigma^2}} \exp(-(x-m)^2/2\sigma^2) \tag{A1.58}$$

It is easy to show that m is the mean value taken by $\mathbf{x}$; σ^2 is similarly identified with the variance of the process

$$\langle x\rangle = \int_{-\infty}^{\infty} x P_{\mathbf{x}}(x)dx = m$$

$$\langle (x-\langle x\rangle)^2\rangle = \langle x^2\rangle - \langle x\rangle^2 = \int_{-\infty}^{\infty} x^2 P_{\mathbf{x}}(x)dx - m^2 = \sigma^2 \tag{A1.59}$$

The characteristic function of the distribution is given by

$$\begin{aligned}\langle \exp(ikx)\rangle &= \frac{1}{\sqrt{2\pi\sigma^2}}\int_{-\infty}^{\infty} dx \exp(ikx)\exp(-(x-m)^2/2\sigma^2)\\ &= \exp(ikm)\exp(-k^2\sigma^2/2)\end{aligned} \tag{A1.60}$$

We have seen (in equation A1.52) that the pdf of the sum of two independent random variables is the convolution of their separate pdfs; consequently the characteristic function of the distribution of sums is the product of those of the individual random variables. As the product of two exponential functions is the exponential of the sum of their exponents, it follows that the sum of two Gaussian variables is itself a Gaussian random variable. Its mean value is the sum of the means of the two added processes; its variance is the sum of their variances. Consequently the sum of any number of Gaussian random variables is itself Gaussian. The moments of the Gaussian distribution can be read off directly from the characteristic function. To tidy things up a little we will let the mean value m be zero. We then find that

$$\langle \exp(ikx)\rangle = \sum_{n=0}^{\infty}\frac{(ik)^n}{n!}\langle x^n\rangle = \sum_{p=0}^{\infty}\frac{(-k^2/2)^p}{p!}\langle x^2\rangle^p \tag{A1.61}$$

so that, on equating coefficients of powers of k,

$$\langle x^{2n}\rangle = \frac{(2n)!}{n!2^n}\langle x^2\rangle^n \tag{A1.62}$$

This result provides an example of the factorisation property of expectation values of products of Gaussian variables. Assuming for simplicity that their means are all zero, this is equal to the product of the expectation values of pairs of the variables, summed over all possible distinct decompositions of the original product into pairs. Thus we might have

$$\langle x_1x_2x_3x_4\rangle = \langle x_1x_2\rangle\langle x_3x_4\rangle + \langle x_1x_3\rangle\langle x_2x_4\rangle + \langle x_1x_4\rangle\langle x_2x_3\rangle\langle x^4\rangle = 3\langle x^2\rangle^2 \tag{A1.63}$$

(It is an interesting exercise to develop much of the theory of Gaussian processes from this property alone.)

The bivariate normal process consists of two random variables **x**, **y** whose joint probability density function is

$$P_{\mathbf{xy}}(x,y) = \frac{1}{2\pi\sigma^2\sqrt{(1-r^2)}} \exp\left(-\frac{(x^2+y^2-2rxy)}{2\sigma^2(1-r^2)}\right) \qquad -\infty < x, y < \infty \tag{A1.64}$$

The marginal distributions of x and y are themselves Gaussian, with zero means, and equal variances:

$$\langle x^2 \rangle = \int_{-\infty}^{\infty} x^2 P_{\mathbf{x}}(x)dx = \sigma^2 = \langle y^2 \rangle \tag{A1.65}$$

x and **y** are not independent; we can see this straight away from

$$P_{\mathbf{xy}}(x,y) \neq P_{\mathbf{x}}(x)P_{\mathbf{y}}(y) \tag{A1.66}$$

The correlation function of x and y is

$$\langle xy \rangle = \int_{-\infty}^{\infty} dx \int_{-\infty}^{\infty} dy xy P_{\mathbf{xy}}(x,y) = r\sigma^2 \tag{A1.67}$$

These calculations allow us to interpret the parameters occurring in the original pdf (A1.64). Should x and y be un-correlated then $P_{\mathbf{xy}}(x,y) = P_{\mathbf{x}}(x)P_{\mathbf{y}}(y)$ holds. A related set of random variables, which can be interpreted as an amplitude and phase, have values defined by

$$\begin{aligned} q &= \sqrt{x^2+y^2} \\ \phi &= \tan^{-1}(y/x) \end{aligned} \tag{A1.68}$$

Their joint pdf is

$$P_{\mathbf{q}\boldsymbol{\phi}}(q,\phi) = \frac{q}{2\pi\sigma^2\sqrt{1-r^2}} \exp\left(-\frac{q^2}{2\sigma^2(1-r^2)}(1-2r\cos\phi\sin\phi)\right)$$
$$0 \le q < \infty, \quad 0 \le \phi < 2\pi \tag{A1.69}$$

from which we can derive the marginal distributions

$$\begin{aligned} P_{\mathbf{q}}(q) &= \frac{q}{\sigma^2\sqrt{1-r^2}} \exp\left(-\frac{q^2}{2\sigma^2(1-r^2)}\right) I_0\left(\frac{rq^2}{2\sigma^2(1-r^2)}\right) \\ P_{\boldsymbol{\phi}}(\phi) &= \frac{\sqrt{1-r^2}}{2\pi}\frac{1}{1-r\sin 2\phi} \end{aligned} \tag{A1.70}$$

The distribution of the product of the two correlated Gaussian quantities x, y can also be calculated, using (A1.56).

$$\mathbf{u} = \mathbf{xy}$$

$$P_{\mathbf{u}}(u) = \int \frac{dx}{|x|} P_{\mathbf{xy}}(x, u/x) = \frac{\exp(ur/\sigma^2(1-r^2))}{\pi\sigma^2\sqrt{1-r^2}} K_0\left(\frac{u}{\sigma^2(1-r^2)}\right) \tag{A1.71}$$

Here K_0 is a modified Bessel function of the second kind, whose properties are discussed in some detail in the second appendix.

These ideas can be extended to the multivariate case.[1] The n random variables $x_1, x_2, \ldots, x_n$, represented by the vector $\mathbf{x}$, have mean values $m_1, m_2, \ldots, m_n \equiv \mathbf{m}$; their covariance matrix $\mathbf{K}$ has elements

$$K_{ij} = \langle (x_i - m_i)(x_j - m_j) \rangle = K_{ji} \tag{A1.72}$$

The joint pdf of $x_1, x_2, \ldots, x_n$ can then be written as

$$P(\{x_1, \ldots, x_n\}) = \frac{\exp(-\frac{1}{2}(\mathbf{x}^{\mathrm{T}} - \mathbf{m}^{\mathrm{T}}) \cdot \mathbf{K}^{-1} \cdot (\mathbf{x} - \mathbf{m}))}{(2\pi)^{n/2}\sqrt{\det \mathbf{K}}} \tag{A1.73}$$

while the associated characteristic function takes the form

$$\langle \exp(i\mathbf{u} \cdot \mathbf{x}) \rangle = \exp(i\mathbf{u} \cdot \mathbf{m}) \exp(-\tfrac{1}{2}\mathbf{u}^{\mathrm{T}} \cdot \mathbf{K} \cdot \mathbf{u}) \tag{A1.74}$$

We note that (A1.64), the pdf of two correlated Gaussian variables, is a special case of (A1.73).

The covariance matrix $\mathbf{K}$ is necessarily square, symmetric and positive definite; thus it has some interesting and useful properties. For an n by n matrix, the eigen problem

$$\mathbf{K}^{-1} \cdot \mathbf{u}_n = \lambda_n \mathbf{u}_n \tag{A1.75}$$

leads to

$$\det(\mathbf{K}^{-1} - \lambda \mathbf{1}) = 0 \tag{A1.76}$$

which has n, not necessarily distinct, positive real roots. The eigenvectors corresponding to different eigenvalues must be orthogonal; those corresponding to degenerate eigenvalues can also have this property imposed upon them, by construction. To identify the eigenvalues we must locate the zeroes of an nth order polynomial in λ. Analytically, this can be done straightforwardly for $n = 2$, but rapidly becomes first prolix ($n = 3, 4$) then, in general, impossible ($n \geq 5$) with increasing n. Numerically however, the location of the roots of (A1.76) is relatively straightforward. Once these eigenvalues are known, the corresponding eigen vectors can be determined, and a

[1] The remainder of this section presents material that is more advanced than that in the rest of the appendix. This is quite germane to the analysis of correlated signals discussed in Chapter 7; readers to whom these matters are of little or no interest can happily move on to Section A1.7.

similarity transform that simultaneously diagonalises $\mathbf{K}^{-1}$ and $\mathbf{K}$ constructed. Thus we can set

$$\mathbf{K}^{-1} \cdot \mathbf{u}_n = \lambda_n \mathbf{u}_n$$
$$\mathbf{u}_n^{\mathrm{T}} \cdot \mathbf{u}_m = \delta_{n,m}$$
$$\mathbf{S} = \{\mathbf{u}_1, \mathbf{u}_2, \ldots, \mathbf{u}_n\}, \qquad \mathbf{S}^{\mathrm{T}} \cdot \mathbf{S} = \mathbf{1} \tag{A1.77}$$

so that

$$\mathbf{S}^{\mathrm{T}} \cdot \mathbf{K}^{-1} \cdot \mathbf{S} = \mathbf{L}; \qquad \mathbf{K}^{-1} = \mathbf{S} \cdot \mathbf{L} \cdot \mathbf{S}^{\mathrm{T}} \tag{A1.78}$$

Here $\mathbf{L}$ is a diagonal matrix with elements

$$L_{mn} = \delta_{m,n}\lambda_m \tag{A1.79}$$

with no summation on the repeated index. The transformation (A1.78) can be quite useful, for example it facilitates the evaluation of normalisation integral

$$\mathrm{Norm}(\mathbf{K}) = \int d^n x \exp(-\tfrac{1}{2}\mathbf{x}^{\mathrm{T}} \cdot \mathbf{K}^{-1} \cdot \mathbf{x}) \tag{A1.80}$$

First we make the linear change in variables

$$\mathbf{x} = \mathbf{S} \cdot \mathbf{u} \tag{A1.81}$$

as $\mathbf{S}^{\mathrm{T}} \cdot \mathbf{S} = \mathbf{1}$ this transformation has a unit Jacobian. Equation (A1.80) now simplifies dramatically as the exponent reduces to a sum of squares

$$\begin{aligned}\mathrm{Norm}(\mathbf{K}) &= \int d^n u \exp\left(-\frac{1}{2}\mathbf{u}^{\mathrm{T}} \cdot \mathbf{S} \cdot \mathbf{K}^{-1} \cdot \mathbf{S} \cdot \mathbf{u}\right) \\ &= \int d^n u \exp\left(-\frac{1}{2}\mathbf{u}^{\mathrm{T}} \cdot \mathbf{L} \cdot \mathbf{u}\right) = \int d^n u \exp\left(-\frac{1}{2}\sum_{k=1}^{n} \lambda_k u_k^2\right) \\ &= \frac{(2\pi)^{n/2}}{\prod_{k=1}^{n} \lambda_k^{1/2}} = \frac{(2\pi)^{n/2}}{\sqrt{\det(\mathbf{L})}} = (2\pi)^{n/2}\sqrt{\det \mathbf{K}}\end{aligned} \tag{A1.82}$$

The characteristic function (A1.74) can be evaluated in much the same way; this fundamental result can also be derived by recognising $\mathbf{q} \cdot (\mathbf{x} - \mathbf{m})$ as a single zero mean Gaussian random variable and exploiting the factorisation properties of its mean powers through (A1.62).

In addition to providing the normalisation constant in (A1.73) Norm($\mathbf{K}$) is also a generating function from which we can derive mean values. Thus we have

$$\begin{aligned}\langle x_k x_l \rangle &= -\frac{2}{\mathrm{Norm}(\mathbf{K})} \frac{\partial \mathrm{Norm}(\mathbf{K})}{\partial (\mathbf{K}^{-1})_{kl}} = \frac{\partial \log(\det(\mathbf{K}^{-1}))}{\partial (\mathbf{K}^{-1})_{kl}} \\ &= \frac{1}{\det(\mathbf{K}^{-1})} (\mathrm{Adj}\mathbf{K}^{-1})_{kl} = K_{kl}\end{aligned} \tag{A1.83}$$

Manipulations of covariance matrices, that exploit the eigen-system (A1.77) and identities such as (A1.83), underpin a great deal of discussion of signal processing, some of which we make contact with in Chapter 7.

A rather different insight into the structure of the covariance matrix can be gained from the construction of a set of statistically independent random variables **y**, from the correlated quantities **x**. Thus we consider the set of quantities

$$\begin{aligned}\hat{y}_1(1) &= x_1\\ &\vdots\\ \hat{y}_n(1) &= x_n\end{aligned} \tag{A1.84}$$

and construct $\hat{y}_k(2) = \hat{y}_k(1) - \hat{y}_1(1)\langle\hat{y}_k(1)\hat{y}_1(1)\rangle; k \geq 2$, each of which is, necessarily, un-correlated with $\hat{y}_1(1) = x_1$. The correlation function $\langle\hat{y}_k(1)\hat{y}_1(1)\rangle$ is identified as the element K_{1k} of the covariance matrix. This process is then iterated through

$$\hat{y}_k(m) = \hat{y}_k(m-1) - \hat{y}_{m-1}(m-1)\frac{\langle\hat{y}_k(m-1)\hat{y}_{m-1}(m-1)\rangle}{\langle\hat{y}^2_{m-1}(m-1)\rangle} k \geq m \tag{A1.85}$$

whose output has the following correlation properties

$$\begin{aligned}\langle\hat{y}_j(m)\hat{y}_k(m)\rangle &= \langle\hat{y}_j(m-1)\hat{y}_k(m-1)\rangle\\ &\quad - \frac{\langle\hat{y}_j(m-1)\hat{y}_{m-1}(m-1)\rangle\langle\hat{y}_k(m-1)\hat{y}_{m-1}(m-1)\rangle}{\langle\hat{y}^2_{m-1}(m-1)\rangle},\\ &\quad j,k \geq m\\ \langle\hat{y}_j(1)\hat{y}_k(1)\rangle &= K_{jk}\end{aligned} \tag{A1.86}$$

This yields a set of n un-correlated random variables

$$y_k = \hat{y}_k(k), \quad 1 \leq k \leq n \tag{A1.87}$$

Finally we can normalise each of these to have a unit mean square value

$$v_k = \frac{y_k}{\sqrt{\langle y_k^2\rangle}}, \quad 1 \leq k \leq n \tag{A1.88}$$

(This procedure can be thought of as an application of the Gram Schmidt orthogonalisation [5] to $\{x_k\}$, the inner product now being identified with the correlation function.)

The transformation between **x** and **v** is linear and can be represented in matrix form as

$$\mathbf{v} = \mathbf{Q}\cdot\mathbf{x} \tag{A1.89}$$

where **Q** has a lower triangular form, i.e.

$$Q_{jk} = 0, \quad k > j \tag{A1.90}$$

The covariance of the **v** variables takes a particularly simple form

$$\langle \mathbf{v}\mathbf{v}^{\mathrm{T}} \rangle = \mathbf{1} \tag{A1.91}$$

Consequently we can express the covariance matrix **K** in terms of **Q** as follows:

$$\begin{aligned}
&\mathbf{x} = \mathbf{Q}^{-1} \cdot \mathbf{v}, \\
&\langle \mathbf{x}\mathbf{x}^{\mathrm{T}} \rangle = \langle \mathbf{Q}^{-1} \cdot \mathbf{v}(\mathbf{Q}^{-1} \cdot \mathbf{v})^{\mathrm{T}} \rangle = \mathbf{Q}^{-1} \cdot \langle \mathbf{v}\mathbf{v}^{\mathrm{T}} \rangle \cdot (\mathbf{Q}^{-1})^{\mathrm{T}} \\
&\mathbf{K} = \mathbf{Q}^{-1} \cdot (\mathbf{Q}^{-1})^{\mathrm{T}}
\end{aligned} \tag{A1.92}$$

This representation of a covariance matrix as the product of lower triangular matrix and its transpose is known as its Cholesky decomposition [6]; the inverse of the covariance matrix can be written in much the same way as

$$\mathbf{K}^{-1} = \mathbf{Q}^{\mathrm{T}} \cdot \mathbf{Q} \tag{A1.93}$$

This allows us to evaluate Norm(**K**) particularly simply. We note that the quadratic form in the exponent in the integrand can be written as

$$\mathbf{x} \cdot \mathbf{K} \cdot \mathbf{x} = \mathbf{x}^{\mathrm{T}} \cdot \mathbf{Q}^{\mathrm{T}} \cdot \mathbf{Q} \cdot \mathbf{x} = \mathbf{v}^{\mathrm{T}} \cdot \mathbf{v} \tag{A1.94}$$

A change in variables from **x** to **v** introduces a Jacobian scaling; thus we have

$$\mathrm{Norm}(\mathbf{K}) = \frac{1}{\det(\mathbf{Q})} \int d^n v \exp\left(-\frac{1}{2}\mathbf{v}^{\mathrm{T}} \cdot \mathbf{v}\right) = (2\pi)^{n/n22}\sqrt{\det(\mathbf{K})} \tag{A1.95}$$

where we have identified the determinant of **Q** in terms of that of **K**, using (A1.93). As **Q** has a triangular form (cf. (A1.90)) its determinant is equal to the product of its diagonal elements, i.e

$$\det(\mathbf{Q}) = \prod_{k=1}^{n} \frac{1}{\sqrt{\hat{y}_k^2(k)}} \tag{A1.96}$$

It is interesting to compare this discussion of the Cholesky decomposition with that of the integration of the pdf (A1.73) (we set **m** to zero by making an appropriate choice of origin in **x** space) over the variables $x_1, x_2, \ldots, x_m, m < n$ to yield a Gaussian pdf for the remaining $n - m$ variables $x_{m+1}, \ldots, x_n$. This can be effected through a succession of completions of the square, in the exponent of the integrand.

Thus we have

$$\sum_{i,j=1}^{n} x_i(\mathbf{K}^{-1})_{ij}x_j$$

$$= x_1^2(\mathbf{K}^{-1})_{11} + 2x_1\sum_{j=2}^{n}(\mathbf{K}^{-1})_{1j}x_j + \sum_{i,j=2}^{n} x_i(\mathbf{K}^{-1})_{ij}x_j$$

$$= (\mathbf{K}^{-1})_{11}\left(\left(x_1 + \frac{\sum_{j=2}^{n}(\mathbf{K}^{-1})_{1j}x_j}{(\mathbf{K}^{-1})_{11}}\right)^2 - \frac{\sum_{j,k=2}^{n}(\mathbf{K}^{-1})_{1j}(\mathbf{K}^{-1})_{1k}x_jx_k}{(\mathbf{K}^{-1})_{11}^2}\right)$$

$$+ \sum_{i,j=2}^{n} x_i(\mathbf{K}^{-1})_{ij}x_j \qquad \text{(A1.97)}$$

An integration over x_1 introduces a factor of $\sqrt{(2\pi/(\mathbf{K}^{-1})_{11})}$; the quadratic form in the exponent can be expressed as

$$\frac{1}{2}\sum_{j,k=2}^{n} x_j((\mathbf{K}^{-1})_{1k} - (\mathbf{K}^{-1})_{1j}(\mathbf{K}^{-1})_{1k}/(\mathbf{K}^{-1})_{11})x_k \qquad \text{(A1.98)}$$

The subsequent steps in this process can be summarised by the recurrence relations

$$\Omega_{ij}(1) = (\mathbf{K}^{-1})_{ij}$$

$$\Omega_{ij}(k) = \left(\Omega_{ij}(k-1) - \frac{\Omega_{k-1,i}(k-1)\Omega_{k-1,j}(k-1)}{\Omega_{k-1,k-1}(k-1)}\right) \qquad i,j \geq k \qquad \text{(A1.99)}$$

Successive integrations on x_k generate factors of $\sqrt{2\pi\,/\,\Omega_{kk}(k)}$; at the same time the exponential quadratic form is reduced to

$$\frac{1}{2}\sum_{j,k=m+1}^{n} x_j\Omega_{jk}(m)\,x_k \qquad \text{(A1.100)}$$

The recurrence relations (A1.86) and (A1.99) are identical in form. However, (A1.86) takes the elements of the covariance matrix **K** as its starting point, while (A1.99) is initiated with those of its inverse $\mathbf{K}^{-1}$. Thus we see that (cf. (A1.96))

$$\prod_{k=1}^{n}\frac{1}{\sqrt{\Omega_{kk}(k)}} = \sqrt{\det(\mathbf{K})} \qquad \text{(A1.101)}$$

while the product of the first m terms is given in terms of the determinant of the $m \times m$ sub-matrix of $\mathbf{K}^{-1}$

$$\prod_{k=1}^{m} \frac{1}{\sqrt{\Omega_{kk}(k)}} = \frac{1}{\sqrt{\det\left({}^{(m)}\mathbf{K}^{-1}\right)}}$$

$${}^{(m)}K_{ij}^{-1} = K_{ij}^{-1}, \qquad 1 \le i, j \le m \tag{A1.102}$$

Combining these results, we find that the explicit form of the multivariate Gaussian obtained by integrating out m of the n variables in the original pdf is

$$\begin{aligned} P(x_{m+1}, \ldots, x_n) &= \int dx_1 \ldots dx_m P(x_1, \ldots, x_n) \\ &= \frac{\exp\left(-\frac{1}{2}\sum_{j,k=m+1}^{n} x_j \Omega_{jk}(m) x_k\right)}{(2\pi)^{(n-m)/2}\sqrt{\det({}^{(m)}\mathbf{K}^{-1})\det(\mathbf{K})}} \end{aligned} \tag{A1.103}$$

Another, simpler, form for $P(x_{m+1}, \ldots, x_n)$ can be written down, more or less by inspection

$$P(x_{m+1}, \ldots, x_n) = \frac{\exp\left(-\frac{1}{2}\sum_{j,k=m+1}^{n} x_j ((\mathbf{K}_{(n-m)})^{-1})_{jk} x_k\right)}{(2\pi)^{(n-m)/2}\sqrt{\det((\mathbf{K}_{(n-m)}))}} \tag{A1.104}$$

Here

$$(\mathbf{K}_{(n-m)})_{ij} = K_{ij}, \quad m < i, j \le n \tag{A1.105}$$

is obtained from the original covariance matrix by erasing its first m rows and columns. By comparing the two answers we obtain the rather striking results

$$\det(\mathbf{K}) = \frac{\det({}^{(m)}\mathbf{K})}{\det((\mathbf{K}^{-1})_{(n-m)})} = \frac{\det(\mathbf{K}_{(n-m)})}{\det({}^{(m)}(\mathbf{K}^{-1}))} \tag{A1.106}$$

which can be recognised as a special case of a theorem due to Jacobi [7].

Middleton [8] and Rice [9], give general accounts of the simpler properties of multivariate normal processes, while more detail can be found in the statistical literature [10].

A1.7 The time evolution of random processes

So far we have considered probability models that are independent of time. We will now consider how this constraint might be relaxed. As our first example we consider a population subject to processes of immigration, whose rate is independent of the population number N, and birth and death, whose rates are proportional to N. (Processes of this type find wide application in biometrics [11] and clutter modelling, as we will see with the emergence of the Negative Binomial distribution and its application in Chapter 4). So, if $P(N,t)$ is the probability that the population at time t is N, the condition that the total probability is conserved, i.e.

$$\sum_{N=0}^{\infty} P(N,t) = 1 \tag{A1.107}$$

leads to the equations

$$\frac{dP(N,t)}{dt} = B((N-1)P(N-1,t) - NP(N,t)) + M(P(N-1,t) - P(N,t)) + D((N+1)P(N+1,t) - NP(N,t)) \tag{A1.108}$$

where B, D and M are constants specifying the rates of birth, death and migration. The analysis of coupled ordinary differential equations of this type is simplified by the introduction of the generating function

$$C(s,t) = \sum_{N=0}^{\infty} s^N P(N,t) \tag{A1.109}$$

From (A1.108) and the requirement that negative population numbers have zero probability we find that this generating function satisfies the first order partial differential equation

$$\frac{\partial C(s,t)}{\partial t} = (s-1)\left[(Bs-D)\frac{\partial C(s,t)}{\partial s} + MC(s,t)\right] \tag{A1.110}$$

A full solution to (A1.110) can be derived and used, in conjunction with the definition (A1.109), to generate explicit expressions for the $P(N,t)$ [12]. Here we focus our attention on the stationary solution that determines the long-time, equilibrium, population probabilities. Setting the right hand side of (A1.110) to zero gives us two conditions:

$$s = 1 \tag{A1.111}$$

and

$$(Bs-D)\frac{dC(s,\infty)}{ds} + MC(s,\infty) = 0 \tag{A1.112}$$

The first of these embodies the conservation of probability condition (A1.107) (which holds for all times) while the second can be solved to yield

$$C(s,\infty) = \left(\frac{D-B}{D-Bs}\right)^{M/B} \tag{A1.113}$$

Expanding this using the binomial theorem we find that

$$P(N,\infty) = \left(\frac{D-B}{D}\right)^{M/B} \frac{\Gamma(M/B+N)}{\Gamma(M/B)N!}\left(\frac{B}{D}\right)^{N} \tag{A1.114}$$

which is identical with the negative binomial distribution (A1.15) if

$$\alpha = \frac{M}{B}, \qquad \bar{N} = \frac{M}{D - B} \tag{A1.115}$$

In the case where the birth rate B is set to zero the rate equation (A1.108) describes a Poisson process with a mean population given by the M/D.

A1.8 Power spectra and correlation functions

We will now consider two complementary ways in which the time evolution of a continuous random process might be analysed. The first approach, which makes the more direct contact with signal processing, is to regard a realisation of the process as a time series, observed over a period T. This record can be used to construct a correlation function as

$$\langle x(t)x(t+\tau)\rangle = \lim_{T\to\infty} \frac{1}{T} \int_{-T/2}^{T/2} x(t)x(t+\tau)dt \tag{A1.116}$$

The Fourier transform of the random variable, again observed over the interval T, is

$$\tilde{x}_T(\omega) = \int_{-T/2}^{T/2} \exp(i\omega t)x(t)dt \tag{A1.117}$$

(We recall that this is straightforward to do in practice, in a discrete form using an FFT.) This can be used to construct the correlation function (A1.116) as

$$\begin{aligned} &\lim_{T\to\infty} \frac{1}{(2\pi)^2} \frac{1}{T} \int_{-\infty}^{\infty} dt \int_{-\infty}^{\infty} d\omega_1 \int_{-\infty}^{\infty} d\omega_2 \tilde{x}_T(\omega_1)\tilde{x}_T(\omega_2) \\ &\quad \times \exp(-it(\omega_1+\omega_2) - i\omega_1\tau) \\ &= \frac{1}{2\pi} \int_{-\infty}^{\infty} d\omega_1 \exp(-i\omega_1\tau) \lim_{T\to\infty} \frac{|\tilde{x}_T(\omega_1)|^2}{T} \\ &= \int_{-\infty}^{\infty} d\omega_1 \exp(-i\omega_1\tau) S(\omega_1); \quad S(\omega_1) = \lim_{T\to\infty} \frac{|\tilde{x}_T(\omega_1)|^2}{2\pi T} \end{aligned} \tag{A1.118}$$

Here we have identified the correlation function as the Fourier transform of the correlation of the power spectrum of the process. This fundamental result is known as the Wiener Khintchine theorem. The factorisation property (cf. (A1.63)) determines all higher order moments of a Gaussian process in terms of its auto-correlation function; we now see how these are related to the power spectrum. In Chapter 5 we see how Gaussian (and non-Gaussian) processes with specified temporal and spatial correlation properties can be simulated exploiting this result.

A1.9 The complex Gaussian process

These ideas are easily extended to the complex Gaussian process, where we consider a complex signal[2]

$$\begin{aligned} Z(t) &= E_I(t) + iE_Q(t); \quad -T/2 < t < 2 < t < T/2 \\ &= 0; \quad \text{otherwise} \end{aligned} \tag{A1.119}$$

whose Fourier transform is given by

$$\hat{Z}(\omega) = \int_{-\infty}^{\infty} dt \exp(i\omega t)\, Z(t) \tag{A1.120}$$

From this we form the power spectrum as

$$S(\omega) = \frac{|\hat{Z}(\omega)|^2}{2\pi T} \tag{A1.121}$$

interpreted as the power in a given frequency component. This can be written in terms of the I and Q components of the signal as follows:

$$\begin{aligned} S(\omega) &= \lim_{T\to\infty} \frac{1}{2\pi T} \int_{-T/2}^{T2} dt_1 \int_{-T/2}^{T/2} dt_2 \exp(i\omega(t_1 - t_2)) \\ &\quad \times (E_I(t_1) + iE_Q(t_1))(E_I(t_2) - iE_Q(t_2)) \\ &= \frac{1}{2\pi} \int_{-\infty}^{\infty} dt \exp(i\omega t)[\langle E_I(0)E_I(t)\rangle \\ &\quad + \langle E_Q(0)E_Q(t)\rangle + i(\langle E_I(0)E_Q(t)\rangle - \langle E_Q(0)E_I(t)\rangle)] \end{aligned} \tag{A1.122}$$

[2] In a radar system the real and imaginary parts of the complex signal are derived by mixing the received signal with in-phase and quadrature local oscillators; hence the labelling with I and Q. Frequency analysis of the complex signal reveals the Doppler spectrum of the received radar signal.

The statistical equivalence and independence of the I and Q components of the signal result in

$$\begin{aligned}
&\langle E_I(t)E_I(t+\tau)\rangle = \langle E_Q(t)E_Q(t+\tau)\rangle = \psi\rho(\tau)\\
&\langle E_I(t)E_Q(t+\tau)\rangle = -\langle E_Q(t)E_I(t+\tau)\rangle = \psi\lambda(\tau)\\
&\langle E_I^2\rangle = \langle E_Q^2\rangle = \psi = \langle I\rangle/2\\
&\rho(\tau) = \frac{\int d\omega\cos(\omega\tau)S(\omega)}{\int d\omega S(\omega)}; \quad \lambda(\tau) = \frac{\int d\omega\sin(\omega\tau)S(\omega)}{\int d\omega S(\omega)}
\end{aligned} \tag{A1.123}$$

while the integral of the power spectrum over all frequencies yields the total mean power or intensity in the signal

$$\int_{-\infty}^{\infty} d\omega\, S(\omega) = \langle E_Q^2\rangle + \langle E_I^2\rangle = \langle E^2\rangle = \langle I\rangle \tag{A1.124}$$

The values of the I and Q components at two different times constitute a four dimensional Gaussian process that can be characterised by their joint pdf (cf. (A1.73)), which is in turn specified, through the covariance matrix (A1.72), by the power spectrum. Thus

$$P(E_I(t), E_Q(t), E_I(t+\tau), E_Q(t+\tau)) = \frac{\exp\left(-\frac{1}{2}\tilde{\mathbf{x}}\cdot\mathbf{K}^{-1}\cdot\mathbf{x}\right)}{(2\pi)^2\sqrt{\det\mathbf{K}}} \tag{A1.125}$$

where we identify $\mathbf{x}$ as the column vector

$$\mathbf{x} = \begin{pmatrix} E_I(t) \\ E_Q(t) \\ E_I(t+\tau) \\ E_Q(t+\tau) \end{pmatrix} \tag{A1.126}$$

$\tilde{\mathbf{x}}$ is its transpose and $\mathbf{K}$ is the covariance matrix

$$\mathbf{K} = \psi\begin{pmatrix} 1 & 0 & \rho & \lambda \\ 0 & 1 & -\lambda & \rho \\ \rho & -\lambda & 1 & 0 \\ \lambda & \rho & 0 & 1 \end{pmatrix} \tag{A1.127}$$

It is easy to show that

$$\det\mathbf{K} = \psi^4(1-\lambda^2-\rho^2)^2 = \psi^4(1-k^2)^2 \tag{A1.128}$$

and that

$$\mathbf{K}^{-1} = [\psi(1-k^2)]^{-1} \begin{pmatrix} 1 & 0 & -\rho & -\lambda \\ 0 & 1 & \lambda & -\rho \\ -\rho & \lambda & 1 & 0 \\ -\lambda & -\rho & 0 & 1 \end{pmatrix} \tag{A1.129}$$

The statistical properties of the amplitude and phase of the complex Gaussian signal can be derived through a straightforward change in variables (cf. (A1.50))

$$E(t) = \sqrt{E_I^2(t) + E_Q^2(t)}; \quad \theta(t) = \tan^{-1}\left(\frac{E_Q(t)}{E_I(t)}\right) \tag{A1.130}$$

Thus the single point statistics are governed by a Rayleigh distribution of amplitude and a uniform distribution of phase

$$P(E,\theta) = \frac{E}{2\pi\psi}\exp\left(-\frac{E^2}{2\psi}\right); \quad 0 \le E < \infty,\ 0 \le \theta < 2\pi \tag{A1.131}$$

while the joint distribution of the amplitudes and phases at two different times takes the slightly more complicated form

$$\begin{aligned} &P(E(t), E(t+\tau), \theta(t), \theta(t+\tau)) \\ &\quad = \frac{E(t)E(t+\tau)}{(2\pi\psi)^2(1-k^2)} \exp\left\{-\frac{1}{2\psi(1-k^2)}\right. \\ &\quad \left. \times \left[E(t)^2 + E(t+\tau)^2 - 2kE(t)E(t+\tau)\cos(\theta(t+\tau) - \theta(t) - \phi_0)\right]\right\} \end{aligned} \tag{A1.132}$$

where $\phi_0 = \tan^{-1}(\lambda/\rho)$.

A1.10 Spatially correlated processes

So far we have illustrated our discussion of continuous random variables in terms of time series and their temporal correlations. Random fields correlated at spatially separated points are also of considerable practical importance, as indeed are the temporal correlations between its values at different points. The power spectrum based approach can be generalised quite straightforwardly to this case. As an example we consider a random field $h(x, y, t)$ representing the height of the (sea) surface[3] above a plane $z = 0$, described by the Cartesian co-ordinates x, y, i.e.

$$z = h(x, y, t) \tag{A1.133}$$

[3] In a similar manner to Section 3.2 in Chapter 3.

A sample of this random field is measured over the spatial area $-L/2 \leq x, y \leq L/2$, for the period of time $-T/2 \leq t \leq T/2$ and a Fourier analysis performed (cf. (A1.117))

$$\hat{h}_{L,T}(k_x, k_y, \omega) = \int_{-L/2}^{L/2} dx \int_{-L/2}^{L/2} dy \int_{-T/2}^{T/2} dt \exp(i(\omega t - k_x x - k_y y))h(x, y, t) \tag{A1.134}$$

Arguing much as in the derivation of (A1.118) a spatio-temporal correlation function can be formed as

$$\begin{aligned} &\langle h(x + x', y + y', t + t')h(x', y', t'0) \\ &= \lim_{T,L\to\infty} \frac{1}{L^2 T} \int_{-L/2}^{L/2} dx' \int_{-L/2}^{L/2} dy' \int_{-T/2}^{T/2} dt' h(x + x', y + y', t + t')h(x', y', t') \\ &= \int dk_x \int dk_y \int d\omega \exp(i(k_x x + k_y y - \omega t))S(k_x, k_y, \omega) \end{aligned} \tag{A1.135}$$

where the power spectrum is now defined as

$$S(k_x, k_y, \omega) = \lim_{T,L\to\infty} \frac{|\hat{h}_{L,T}(k_x, k_y, \omega)|^2}{(2\pi)^3 L^2 T} \tag{A1.136}$$

In some situations the frequency and wave-vector are related through a physically imposed dispersion relation

$$\omega = \omega(k_x, k_y) \tag{A1.137}$$

In the case of gravity waves [13] this takes the form

$$\omega(k_x,k_y)^2 = g\sqrt{k_x^2 + k_y^2} \tag{A1.138}$$

The spatio-temporal correlation function can now be written as

$$\begin{aligned} &\langle h(x + x', y + y', t + t')h(x', y', t')\rangle \\ &= \frac{1}{2\pi} \int dk_x \int dk_y \int d\omega \exp(i(k_x x + k_y y - \omega(k_x, k_y)t))S(k_x, k_y) \end{aligned} \tag{A1.139}$$

with

$$S(k_x, k_y) = \lim_{L\to\infty} \frac{|\hat{h}_L(k_x, k_y)|^2}{(2\pi)^2 L^2}$$

$$\hat{h}_L(k_x, k_y) = \int_{-L/2}^{L/2} dx \int_{-L/2}^{L/2} dy \exp(-i(k_x x + k_y y))h(x, y) \tag{A1.140}$$

(Setting t to zero in (A1.139) yields the spatial correlation function.) As we see in Chapter 5 these Fourier techniques are particularly useful in the simulation of spatially correlated clutter, whose structure can interact with and degrade the performance of simple radar signal processing schemes; examples of this interaction are discussed in Chapters 2 and 9.

A1.11 Stochastic differential equations and noise processes

In some circumstances the dynamics of a random process can also be described in terms of an equation of motion, explicitly supplemented by a random driving term. This is referred to as a Langevin or stochastic differential equation [14]. The driving process is commonly modelled as white noise, which decorrelates instantaneously on the time scale of variations in the driven random variable. Gaussian factorisation properties are also assumed. Perhaps the simplest Langevin equation would be that describing a freely diffusing particle, whose position is x. Thus we might write

$$\frac{dx}{dt} = f(t); \quad \langle f(t_1) f(t_2) \rangle = 2\gamma\delta(t_1 - t_2) \tag{A1.141}$$

where γ will be identified shortly with a diffusion constant. We can solve (A1.141) formally as

$$\Delta x(t) = x(t) - x(0) = \int_0^t f(t_1) dt_1 \tag{A1.142}$$

The correlation function of this displacement is obtained, using the delta function correlation property (A1.141) of the driving noise,

$$\langle \Delta x(t_1) \Delta x(t_2) \rangle = 2\gamma \min(t_1, t_2) \tag{A1.143}$$

The special case of this result where $t_1 = t_2$ confirms our identification of γ with the diffusion constant. A Gaussian process evolving in time with a zero mean and the correlation property (A1.143) is called a Wiener process, and can be used to construct a rigorous formulation of the rather cavalier integration over white noise presented in (A1.142). This framework is frequently referred to as the Ito calculus, and finds wide application in financial and other modelling studies [15]. The assumed Gaussian properties of the white noise are manifest in those of the diffusive displacement; for example its mean fourth power is

$$\langle \Delta x(t)^4 \rangle = 3\langle \Delta x(t)^2 \rangle^2 \tag{A1.144}$$

The simple Langevin equation (A1.141) can be modified

$$\frac{dx(t)}{dt} = -\alpha x(t) + f(t) \tag{A1.145}$$

through the introduction of a decay term, proportional to α. This damping limits the effects of the random driving term; in contrast to the un-damped case (A1.141) the mean square value for x tends to a finite value at long times.

To see how this comes about we integrate (A1.145) formally to give

$$x(t) = \exp(-\alpha t)x(0) + \int_0^t \exp(-\alpha(t - t_1))f(t_1)dt_1 \tag{A1.146}$$

We now form the correlation function

$$\begin{aligned}\langle x(T)x(T+t)\rangle &= x(0)^2 \exp(-\alpha(2T+t)) \\ &\quad + \int_0^T dt_1 \int_0^{T+t} dt_2 \exp(-\alpha(2T+t-t_1-t_2))\langle f(t_1)f(t_2)\rangle \\ &= x(0)^2 \exp(-\alpha(2T+t)) + \frac{\gamma}{\alpha}\exp(-\alpha t)(1-\exp(-2\alpha T))\end{aligned} \tag{A1.147}$$

If we take the long time limit, and set t to zero, we obtain the expectation value

$$\langle x^2\rangle = \int_0^\infty dt_1 \int_0^\infty dt_2 \exp(-\alpha(2t - t_1 - t_2))\langle f(t_1)f(t_2)\rangle = \frac{\gamma}{\alpha} \tag{A1.148}$$

Equation (A1.148) embodies a simple 'fluctuation-dissipation' result, which requires the input white noise (fluctuation) and the linear resistance (dissipation) terms to determine an equilibrium mean square value. The power spectrum of the process can be derived from either the correlation function (A1.147) or the underlying SDE (A1.145). In the latter case we find, on Fourier transformation, that

$$\tilde{x}(\omega)(\alpha - i\omega) = \tilde{f}(\omega) \tag{A1.149}$$

so that

$$\frac{|\tilde{x}(\omega)|^2}{T} = \frac{1}{\alpha^2+\omega^2}\frac{|\tilde{f}(\omega)|^2}{T} \tag{A1.150}$$

As the power spectrum of white noise is 2γ at all frequencies the corresponding result for the process x is

$$S(\omega) = \frac{2\gamma}{\alpha^2+\omega^2} \tag{A1.151}$$

To make direct contact with the simple feedback algorithm for the generation of a correlated Gaussian process discussed in Chapter 5 we can make the identifications

$$\eta = \exp(-\alpha(t_n - t_{n-1})t)$$

$$\beta g_n = \int_{t_{n-1}}^{t_n} \exp(-\alpha(t_n - t)) f(t) dt$$

$$\langle g_n \rangle = 0$$

$$\beta^2 = \int_{t_{n-1}}^{t_n} \int_{t_{n-1}}^{t_n} dt_1\, dt_2 \exp(-\alpha(2t_n - t_1 - t_2)) \langle f(t_1) f(t_2) \rangle = (1 - \eta^2) \qquad \text{(A1.152)}$$

A simple physical realisation of this stochastic process is the position of an overdamped harmonic oscillator undergoing Brownian motion. In this case γ is again identified as a diffusion constant (conventionally denoted by D) while α is the ratio of the Hooke and friction coefficients. The result (A1.148) now takes the form of the celebrated Einstein relation between friction and diffusion constants.

Another way to describe these processes is in terms of the evolution of probability. Here we state the results and justify them by an example. The conditional probability $P(x, t \mid x_0)$ that a freely diffusing particle (satisfying A1.141), initially at x_0, is found at x at time t, satisfies the diffusion equation

$$\frac{\partial P(x, t \mid x_0)}{\partial t} = \gamma \frac{\partial^2 P(x, t \mid x_0)}{\partial x^2} \qquad \text{(A1.153)}$$

The corresponding quantity for the damped case (equation A1.145) satisfies

$$\frac{\partial P(x, t \mid x_0)}{\partial t} = \frac{\partial}{\partial x}(\alpha x P(x, t \mid x_0)) + \gamma \frac{\partial^2 P(x, t \mid x_0)}{\partial x^2} \qquad \text{(A1.154)}$$

the so-called Fokker Planck (FP) equation. $P(x, t \mid x_0)$ can be constructed quite straightforwardly for a Gaussian process. Referring to (A1.42) we see that

$$P_{\mathbf{y} \mid \mathbf{x}}(y \mid x) = \frac{P_{\mathbf{x},\mathbf{y}}(x, y)}{P_{\mathbf{x}}(x)}$$

so that

$$P(x, t \mid x_0) = \frac{\exp(-(x - x_0 r(t))^2 / 2\langle x^2 \rangle (1 - r(t)^2))}{\sqrt{2\pi \langle x^2 \rangle (1 - r(t)^2)}} \qquad \text{(A1.155)}$$

When we set

$$\langle x^2 \rangle = \frac{\gamma}{\alpha}; \quad r(t) = \exp(-\alpha t) \qquad \text{(A1.156)}$$

we see that $P(x, t \mid x_0)$ satisfies (A1.154), along with the initial condition

$$\lim_{t \to 0} P(x, t \mid x_0) = \delta(x - x_0) \qquad \text{(A1.157)}$$

In Appendix 2 we show how the fundamental solution (A1.155) of the Fokker Planck equation (A1.154) can be expanded in terms of Hermite polynomials which, in effect, encode the factorisation properties of a Gaussian process (A1.62), (A1.63). This expansion plays a central role in the simulation of non-Gaussian processes with specified correlation properties, described in Chapter 5.

The generalisation of this analysis to the multivariate case is fairly straightforward. To do this we adopt a vector notation and set up the Langevin equation as

$$\frac{d\mathbf{x}(t)}{dt} = \mathbf{A} \cdot \mathbf{x}(t) + \mathbf{f}(t) \tag{A1.158}$$

The matrix $\mathbf{A}$ is constant; to ensure that the system is stable all its eigen-values must have negative real parts. These coupled first order differential equations can be integrated formally to give

$$\mathbf{x}(t) = \exp(\mathbf{A}t)\mathbf{x}(0) + \int_0^t \exp(\mathbf{A}(t - t_1)) \cdot \mathbf{f}(t_1)dt_1 \tag{A1.159}$$

The white noise vector $\mathbf{f}$ has a correlation matrix of the form (which is yet to be determined)

$$\langle \mathbf{f}(t_1)\mathbf{f}(t_2)^{\mathrm{T}} \rangle = \delta(t_1 - t_2)\mathbf{G} \tag{A1.160}$$

In the long time limit, where all the transients have died down, we can construct the covariance matrix $\mathbf{B}$ of $\mathbf{x}$

$$\begin{aligned} \langle \mathbf{x}\mathbf{x}^{\mathrm{T}} \rangle &= \lim_{t\to\infty} \int_0^t dt_1 \int_0^t dt_2 \exp(\mathbf{A}(t - t_1))\mathbf{G}\exp(\mathbf{A}^{\mathrm{T}}(t - t_2))\delta(t_1 - t_2) \\ &= \int_0^\infty dt \exp(\mathbf{A}t)\mathbf{G}\exp(\mathbf{A}^T t) = \mathbf{B} \end{aligned} \tag{A1.161}$$

Integration by parts tell us that

$$\mathbf{AB} + \mathbf{BA}^{\mathrm{T}} = \int_0^\infty dt \frac{d(\exp(\mathbf{A}t)\mathbf{G}\exp(\mathbf{A}^{\mathrm{T}}t))}{dt} = -\mathbf{G} \tag{A1.162}$$

We see that, once we have prescribed our equilibrium statistics $\mathbf{B}$ and the relaxation processes $\mathbf{A}$, the covariance of the driving noise vector $\mathbf{f}$ can be identified. Thus given the dissipation, we can specify the fluctuations consistent with the prescribed covariance; (A1.162) is in effect the generalisation of the simple result (A1.148). Should the matrix $\mathbf{G}$ have any negative eigenvalues, $\mathbf{A}$ and $\mathbf{B}$ characterise a system that cannot be realised physically as its power spectrum necessarily contains components that have negative values.

To implement (A1.159) numerically we discretise the evolution equation as

$$\mathbf{x}_n = \mathbf{H} \cdot \mathbf{x}_{n-1} + \mathbf{V}_n$$
$$\mathbf{H} = \exp(\mathbf{A}(t_n - t_{n-1}))$$
$$\langle \mathbf{V}_n \rangle = \mathbf{0}$$
$$\langle \mathbf{V}_n \mathbf{V}_n^{\mathrm{T}} \rangle = \int_0^{t_n - t_{n-1}} \exp(\mathbf{A}t) \cdot \mathbf{G} \cdot \exp(\mathbf{A}^{\mathrm{T}} t) dt \tag{A1.163}$$

We now substitute our expression for **G**, and obtain the analogue of the last of (A1.152):

$$\begin{aligned}\left\langle \mathbf{V}_n \mathbf{V}_n^{\mathrm{T}} \right\rangle &= - \int_0^{t_n - t_{n-1}} \exp(\mathbf{A}t)(\mathbf{A} \cdot \mathbf{B} + \mathbf{B} \cdot \mathbf{A}^{\mathrm{T}}) \exp(\mathbf{A}^{\mathrm{T}} t) dt \\ &= - \int_0^{t_n - t_{n-1}} \frac{d(\exp(\mathbf{A}t) \cdot \mathbf{B} \cdot \exp(\mathbf{A}^T t))}{dt} \\ &= \mathbf{B} - \mathbf{H} \cdot \mathbf{B} \cdot \mathbf{H}^{\mathrm{T}}\end{aligned} \tag{A1.164}$$

The description of a process in terms of a conditional probability satisfying a FP equation assumes that it has no memory, i.e. is unaffected by its previous history. Written formally in terms of a conditional probability this condition becomes

$$P(x_n, t_n \,|\, x_{n-1}, t_{n-1}; x_{n-2}, t_{n-2}; \ldots x_1, t_1) = P(x_n, t_n \,|\, x_{n-1}, t_{n-1}) \tag{A1.165}$$

we also have

$$P\,(x, t \,|\, x_0, t_0) = \int dx' P\left(x, t \,|\, x', t'\right) P\left(x', t' \,|\, x_0, t_0\right); \quad t > t' > t_0 \tag{A1.166}$$

These properties are characteristic of a Markov process, and are frequently assumed in practice. This idea can be extended to a vector process, any given component of which may itself not be Markov. Modelling memory effects through extra 'hidden' processes evolving in the background is a widely used strategy in many applications. The requirement of Markov behaviour also imposes restrictions on the dynamics of a stochastic variable: a one dimensional Gaussian process that is Markov necessarily has an exponentially decaying correlation function [16].

It is interesting to note that contact can be made between the FP description of a continuous process and that of a discrete process in terms of rate equations. In the limit of a large mean population a negative binomial distribution (equations A1.15, A1.114 and A1.115) can be characterised by an effectively continuous variable x

$$x = \frac{N}{\bar{N}} \tag{A1.167}$$

whose pdf is

$$P(x) = \bar{N}P(\bar{N}x, \infty) \tag{A1.168}$$

Using Stirling's approximation to the gamma function we find that

$$\frac{\Gamma(\alpha + \bar{N}x)}{\Gamma(\bar{N}x + 1)} \approx (\bar{N}x)^{\alpha - 1} \tag{A1.169}$$

while

$$\lim_{\bar{N}\to\infty} \left(\frac{\bar{N}/\bar{N}\alpha}{1 + \bar{N}/\alpha}\right)^{\bar{N}x} = \exp(-\alpha x) \tag{A1.170}$$

Thus we see that x is indeed gamma distributed, i.e.

$$P(x) = \frac{\alpha^{\alpha} x^{\alpha - 1}}{\Gamma(\alpha)} \exp(-\alpha x) \tag{A1.171}$$

If we introduce $N = \bar{N}x$into the rate equation (A1.108), expand $P(\bar{N}x \pm 1, t)$ in Taylor series, and allow $\bar{N}$ to tend to infinity by letting the birth and death rates become equal, we can show (further details are given in Reference 14) that $P(x, t)$ satisfies the partial differential equation

$$\frac{\partial P(x,t)}{\partial t} = B\frac{\partial^2}{\partial x^2}(xP(x,t)) + M\frac{\partial}{\partial x}((x-1)P(x,t)) \tag{A1.172}$$

We note that (A1.171) is indeed the stationary solution of this FP type equation (with B and M related to α as in A1.115).

It is possible to extend the SDE description of time varying random processes to include non-linear drifts and state dependent noise powers, such as that just derived in our analysis of the continuous limit of the negative binomial distribution. Coupled SDEs describing correlated multi-variate processes and an equivalent description in terms of a FP equation can also be constructed. This formalism provides a starting point for the analysis of a wide variety of non-Gaussian processes. Particular attention is paid to the SDE/FP description of the K-distributed process in Reference 17 and, more recently, in References 18 and 19. Van Kampen [20] discusses the use of statistical concepts in the modelling of physical processes; this careful discussion provides a valuable complement to texts that focus attention primarily on the formal details of the manipulation of stochastic processes.

The extension to the spatial domain of the Langevin and rate equation descriptions of stochastic processes, which proved to be so useful in the analysis of temporal evolution, presents very significant difficulties. In essence, these derive from there being any number of non-overlapping paths between two points in a plane, for example, while there is only one between two points (times) on a line. Some progress has been made in the analysis of spatially distributed populations, subject to processes of birth, death and immigration, and coupled by inter-site migration; much of this work is described in the biological literature [21,22].

A1.12 Miscellaneous results

The concluding section of this appendix brings together several results that are used elsewhere in the book and provide interesting examples of the application of probability in the analysis of radar signal processing, and yet cannot be readily accommodated elsewhere. The analysis of data in terms of estimated moments of their pdf is of considerable practical importance; here we discuss the correction of these measured moments for the effects of additional thermal noise of known power, and analyse the statistics of estimators of normalised moments of the gamma distribution. We also briefly consider order statistics, which find application in analysis of CFAR and related detection processing (Chapters 6 and 9) and the matched filter detection of extended ocean surface features (Chapter 7) and the elements of sequential analysis employed in Chapter 10.

Correcting moments for the effect of noise: Let us first consider the effects of thermal noise on measurements of moments of the clutter distribution. In this case the complex signal $\mathbf{E}$ now consists of the clutter contribution $\mathbf{E}_c$ and a thermal noise component $\mathbf{n}$ (for convenience we represent the I and Q components in vector notation):

$$\mathbf{E} = \mathbf{E}_c + \mathbf{n} \tag{A1.173}$$

If in the absence of thermal noise the characteristic function of the distribution of the clutter signal takes the form

$$C_C(\mathbf{k}) = \langle \exp(i\mathbf{k} \cdot \mathbf{E}_c) \rangle_C \tag{A1.174}$$

As the phase of the signal $\mathbf{E}_c$ is uniformly distributed the characteristic function can be written as

$$C_C(\mathbf{k}) = \langle J_0(kE_c) \rangle_C = \sum_{r=0}^{\infty} \frac{(-k^2/4)^r}{r!r!} \langle E_c^{2r} \rangle_C = \sum_{r=0}^{\infty} \frac{(-k^2/4)^r}{r!r!} \langle z_c^r \rangle_C \tag{A1.175}$$

This can be thought of as a generating function for the intensity moments of the clutter. (The subscript C denotes an average over the clutter pdf.) If we form the characteristic function when noise is present we obtain

$$C_{C+N}(\mathbf{k}) = \langle \exp(i\mathbf{k} \cdot \mathbf{E}) \rangle_{C+N} = \langle \exp(i\mathbf{k} \cdot (\mathbf{E}_c + \mathbf{n})) \rangle_{C+N} \tag{A1.176}$$

As the clutter and noise processes are independent we can rewrite this as

$$\langle \exp(i\mathbf{k} \cdot \mathbf{E}) \rangle_{C+N} = \langle \exp(i\mathbf{k} \cdot \mathbf{E}_c) \rangle_C \langle \exp(i\mathbf{k} \cdot \mathbf{n}) \rangle_N \tag{A1.177}$$

Here the subscripts N and $C+N$ denote averages over the pdfs for the noise and noise plus clutter signals. If we once again assume that all phases are uniformly distributed we find that

$$\sum_{r=0}^{\infty} \frac{(-k^2/4)^r}{r!r!} \langle E^{2r} \rangle_{C+N} = \langle \exp(i\mathbf{k} \cdot \mathbf{n}) \rangle_N \sum_{p=0}^{\infty} \frac{(-k^2/4)^p}{p!p!} \langle z_c^r \rangle_C \tag{A1.178}$$

For thermal noise we have

$$P(\mathbf{n}) = \frac{1}{2\pi \langle n^2 \rangle} \exp(-n^2/\langle n^2 \rangle) \tag{A1.179}$$

and

$$\langle \exp(i\mathbf{k} \cdot \mathbf{n}) \rangle_N = \exp(-k^2 \langle n^2 \rangle /4) \tag{A1.180}$$

Therefore we can write

$$\sum_{q=0}^{\infty} \frac{(k^2/4)^q}{q!} \langle n^2 \rangle^q \sum_{r=0}^{\infty} \frac{(-k^2/4)^r}{r!r!} \langle E^{2r} \rangle_{C+N} = \sum_{p=0}^{\infty} \frac{(-k^2/4)^p}{p!p!} \langle z_c^p \rangle_C \tag{A1.181}$$

By equating coefficients of powers of $k^2/4$ we can now express the moments of the clutter intensity in terms of the measured intensity moments of clutter plus noise and the (assumed known) corrupting noise power through the relatively simple formula

$$\langle z_c^s \rangle_C = \sum_{q=0}^{s} (-1)^q \frac{\langle n^2 \rangle^q \langle z^{(s-q)} \rangle_{C+N} (s!)^2}{q!((s-q)!)^2} \tag{A1.182}$$

Correcting the moments of a limited number of samples: The analysis of local clutter power is frequently carried out in terms of estimators of normalised moments of its (assumed) gamma pdf. Given a set of N independent samples $\{x_j\}$ the estimate of un-normalised moment is

$$\hat{x}^n = \frac{1}{N} \sum_{j=1}^{N} x_j^n \tag{A1.183}$$

this estimator is unbiased, i.e.

$$\langle \hat{x}^n \rangle = \langle x^n \rangle = \frac{\Gamma(\nu + n)}{\Gamma(\nu) b^n} \tag{A1.184}$$

The estimators of the normalised moments constructed from these $\hat{x}^n$ are, however, biased, i.e.

$$\left\langle \frac{\hat{x}^n}{(\hat{x})^n} \right\rangle \neq \frac{\langle x^n \rangle}{\langle x \rangle^n} = \frac{\Gamma(n + \nu)}{\Gamma(\nu) \nu^n} \tag{A1.185}$$

This discrepancy must be allowed for in the data analysis; the variance of the estimator

$$\left\langle \frac{(\hat{x}^n)^2}{(\hat{x})^{2n}} \right\rangle - \left\langle \frac{\hat{x}^n}{(\hat{x})^n} \right\rangle^2 \tag{A1.186}$$

gives a quantitative indication of the spread expected in the estimator values. Early analyses of this problem [23] proceeded perturbatively, expressing the bias and variance to first order in N^{-1}; when these results are applied to very spiky data ($\nu \cong 0.01$ or less) they fail completely, yielding negative values for the variance. Attempts to extend the perturbation treatment to higher orders rapidly get bogged down in algebra. For gamma distributed x, however, it is possible to obtain exact, closed form expressions for the bias and variance, valid for any N. The key step in the analysis [24] is the introduction of the following representation of inverse powers of $\hat{x}$:

$$\frac{N^n}{\left(\sum_{j=1}^{N} x_j\right)^n} = \frac{N^n}{(n-1)!} \int_0^\infty ds\, s^{n-1} \exp\left(-s \sum_{j=1}^{N} x_j\right) \tag{A1.187}$$

As the joint pdf of the $\{x_j\}$ is given by

$$P\left(\{x_j\}\right) = \frac{b^{N\nu}}{\Gamma(\nu)^N} \left(\prod_{j=1}^{N} x_j^{\nu-1}\right) \exp\left(-b \sum_{j=1}^{N} x_j\right) \tag{A1.188}$$

the evaluation of (A1.185) is now much more straightforward:

$$\begin{aligned}
\left\langle \frac{\hat{x}^n}{(\hat{x})^n} \right\rangle &= \int dx_1 \cdots \int dx_N P(\{x_j\}) \sum_{j=1}^{N} x_j^n \int_0^\infty ds\, s^{n-1} \exp\left(-s \sum_{k=1}^{N} x_k\right) \\
&= \frac{N^n b^{\nu N}}{(n-1)!} \frac{\Gamma(N+\nu)}{\Gamma(\nu)} \int_0^\infty ds \frac{s^{n-1}}{(s+b)^{N\nu+n}} \\
&= \frac{\nu(\nu+1)\cdots(\nu+n-1)}{\nu^n(1+1/N\nu)\cdots(1+(n-1)/N\nu)} \\
&= \frac{\langle x^n \rangle}{\langle x \rangle^n} \prod_{j=1}^{n-1} (1 + j/N\nu)^{-1}
\end{aligned} \tag{A1.189}$$

(Note this depends explicitly on the combination $N\nu$, rather than just N. For very spiky clutter the 'effective sample number' $N\nu$ can be small even when N is of the order of 100.) This can be expanded to first order in N^{-1} as

$$\left\langle \frac{\hat{x}^n}{(\hat{x})^n} \right\rangle = \frac{\langle x^n \rangle}{\langle x \rangle^n} \left(1 - \frac{n(n-1)}{2N\nu}\right) + O(N^{-2}) \tag{A1.190}$$

which agrees with the perturbation result obtained in Reference 23. The variance (A1.86) can be evaluated in much the same way; details can be found in Reference 24. This reference also contains an analysis of very spiky data, derived from ship range profiles, along with further discussion of the statistics of the estimators of normalised moments.

Order statistics: Given a set of N independent values y of a random variable, these can be arranged and labelled as $y_1 < y_2 < \cdots < y_{N-1} < y_N$. So, for example, y_N

is the maximum value in the set, while y_1 is the minimum. Given the pdf $P(y)$ of the individual, independent, values of y, what is the pdf of, for example, the maximum or minimum value or, more generally, of y_k? Addressing these questions involves us in the study of so-called order statistics. The pdfs of y_N and y_1 can be derived quite straightforwardly. The probability that all N samples are less in value than $y_{\max}$ is given by $F(y_{\max})^N$, recalling that the probability distribution F is defined by

$$F(y) = \int_{-\infty}^{y} P(y')dy' \tag{A1.191}$$

This probability can be identified with the probability distribution $F_N(y_N)$ of y_N, i.e.

$$F_N(y_N) = F(y_N)^N = \int_{-\infty}^{y_N} P_N(y'_N)dy'_N \tag{A1.192}$$

The pdf $P_N(y_N)$ can now be determined by simple differentiation

$$P_N(y_N) = \frac{dF_N(y_N)}{dy_N} = NP(y_N)F(y_N)^{N-1} \tag{A1.193}$$

In much the same way we can show that the pdf of the least of the N sample values takes the form

$$P_1(y_1) = N(1 - F(y_1))^{N-1}P(y_1) \tag{A1.194}$$

To determine the pdf of the general y_k we first note that the joint pdf of the ordered values of the samples $y_1 < y_2 < \cdots < y_{N-1} < y_N$ is given by the following expression

$$\begin{aligned} P_{\text{ord}}(y_1, y_2, \ldots, y_N) &= N!P(y_1)P(y_2)\cdots P(y_N), \quad y_1 < y_2 < \cdots y_{N-1} < y_N \\ &= 0, \quad \text{otherwise} \end{aligned} \tag{A1.195}$$

Given this we can now obtain the pdf of y_k by integrating over all the other variables:

$$\begin{aligned} P_k(y_k) &= \int_{-\infty}^{\infty} dy_1 \cdots \int_{-\infty}^{\infty} dy_{k-1} \int_{-\infty}^{\infty} dy_{k+1} \cdots \int_{-\infty}^{\infty} dy_N \, P_{\text{ord}}(y_1, y_2, \cdots y_N) \\ &= N!P(y_k) \int_{-\infty}^{y_k} dy_{k-1}P(y_{k-1}) \int_{-\infty}^{y_{k-1}} dy_{k-2}P(y_{k-2}) \cdots \int_{-\infty}^{y_2} dy_1 P(y_1) \\ &\quad \times \int_{y_k}^{\infty} dy_{k+1}P(y_{k+1}) \int_{y_{k+1}}^{\infty} dy_{k+2}P(y_{k+2}) \cdots \int_{y_{N-1}}^{\infty} dy_N P(y_N) \\ &= \frac{N!}{(N-k)!(k-1)!} P(y_k)F(y_k)^{k-1}(1 - F(y_k))^{N-k} \end{aligned} \tag{A1.196}$$

We note that this reduces to (A1.193) and (A1.194) in the appropriate limits; we have made repeated use of (A1.191) and integration by parts in arriving at the final expression in (A1.196).

Sequential testing: In Chapter 6 we discuss how two hypotheses might be distinguished on the basis of measurements of a random quantity x. This can be drawn from one of two distributions, corresponding to the hypotheses distinguished by the values $\theta = \theta', \theta''$

$$\begin{aligned} H_0 &: \ P(x \mid \theta') \\ H_1 &: \ P(x \mid \theta'') \end{aligned} \tag{A1.197}$$

In the case where we make a fixed number N of independent tests we can define a likelihood for the N measurements as

$$L(N,\theta) = \prod_{j=1}^{N} P(x_j!\theta) \tag{A1.198}$$

and base our decision on the test reject H_0, $\theta = \theta'$ and accept $H_1, \theta = \theta''$ if $L(N,\theta')/L(N,\theta'') \leq k$.

Here we consider briefly the case where we do not fix N, but let it be a random variable taking integer values, starting at 1. So, given a stream of data $\{x_1, x_2, x_3, \ldots\}$ we calculate a sequence of random variables

$$\left\{ \frac{L(1,\theta')}{L(1,\theta'')}, \frac{L(2,\theta')}{L(2,\theta'')}, \frac{L(3,\theta')}{L(3,\theta'')}, \ldots \right\} \tag{A1.199}$$

This series of values can be analysed effectively using the method of sequential probability ratio testing. In this, the hypothesis $H_0, \theta = \theta'$ is rejected (and $H_1, \theta = \theta''$ is accepted) if there exists a positive integer n so that $\{x_1, x_2, \ldots, x_n\}$ belongs to the set C_n satisfying

$$k_0 < \frac{L(j,\theta')}{L(j,\theta'')} < k_1; \quad j = 1, 2 \ldots, n-1 \tag{A1.200}$$

and

$$\frac{L(n,\theta')}{L(n,\theta'')} \leq k_0 \tag{A1.201}$$

If, however, $\{x_1, x_2, \ldots x_n\}$ belongs to the set B_n satisfying

$$k_0 < \frac{L(j,\theta')}{L(j,\theta'')} < k_1; \quad j = 1, 2 \ldots, n-1 \tag{A1.202}$$

and

$$\frac{L(n,\theta')}{L(n,\theta'')} \geq k_1 \tag{A1.203}$$

$H_0, \theta = \theta'$ is accepted (and $H_1, \theta = \theta''$ is rejected).

In practice we continue to observe sample data as long as

$$k_0 < \frac{L(n,\theta')}{L(n,\theta'')} < k_1 \tag{A1.204}$$

then terminate the sampling in one of two ways:

$$H_0,\ \theta = \theta' \text{ is rejected as soon as } \frac{L(n,\theta')}{L(n,\theta'')} \le k_0 \tag{A1.205}$$

$$H_0,\ \theta = \theta' \text{ is accepted as soon as } \frac{L(n,\theta')}{L(n,\theta'')} \ge k_1 \tag{A1.206}$$

It can be shown that this procedure must terminate for finite n, although this value is itself a random variable. With a suitably defined cost function one can show that the expectation value of n is significantly smaller than the fixed value N required to give the same cost.

Now we look at how the quantities k_0, k_1 are related to various probabilities of incorrect assignment (sometimes referred to as the buyer's and seller's risks). Let α be the probability of rejecting H_0 when this hypothesis is true, while β is the probability of accepting H_0 when this hypothesis is in fact false. With the sets of data values B_n, C_n defined above we can write

$$\alpha = \sum_{n=1}^{\infty} \int_{C_n} L(n,\theta'), \quad 1-\beta = \sum_{n=1}^{\infty} \int_{C_n} L(n,\theta'') \tag{A1.207}$$

Here we have introduced a shorthand notation for various integrals

$$\int_{R_n} L(n,\theta) \equiv \int_{R_n} \prod_{j=1}^{n} P(x_j \mid \theta) dx_1 dx_2 \cdots dx_n \tag{A1.208}$$

As the procedure can be shown to terminate with probability 1, we also have

$$1-\alpha = \sum_{n=1}^{\infty} \int_{B_n} L(n,\theta'), \quad \beta = \sum_{n=1}^{\infty} \int_{B_n} L(n,\theta'') \tag{A1.209}$$

If $\{x_1, x_2, \ldots x_n\}$ belongs to the set C_n then $L(n,\theta') \le k_0 L(n,\theta'')$ and we see that

$$\alpha \le k_0(1-\beta) \tag{A1.210}$$

If $\{x_1, x_2, \cdots x_n\}$ belongs to the set B_n then $L(n,\theta') \ge k_1 L(n,\theta'')$ and

$$1-\alpha \ge k_1\beta \tag{A1.211}$$

From this we see that the two risks and the upper and lower thresholds are subject to the inequalities

$$\frac{\alpha}{1-\beta} \le k_0; \quad k_1 \le \frac{1-\alpha}{\beta} \tag{A1.212}$$

If we now put forth guesses for the risks, which we denote by α_g, β_g, and form the test bounds as

$$k_0 = \frac{\alpha_g}{1-\beta_g}, \quad k_1 = \frac{1-\alpha_g}{\beta_g} \tag{A1.213}$$

it can then be shown that, when using these bounds, the sum of the true risks of missing a target and accepting a false alarm satisfy

$$\alpha + \beta \leq \alpha_g + \beta_g \tag{A1.214}$$

A discussion of the implications of this analysis or the radar procurement process is presented in Section 10.5.3.1; a full account of sequential analyses can be found in Wald's text [25]

References

1 W. Feller,'*An introduction to probability theory and its applications*', Vols. I, II, John Wiley, New York, 1971.
2 A. Papoulis: '*Probability, random variables and stochastic processes*', McGraw Hill, 1965.
3 M. Abramowitz and I.A. Stegun (Eds.), '*A handbook of mathematical functions*', Dover, New York, 1972.
4 I.N. Sneddon, '*The use of integral transforms*', section 4.4, Tata-McGraw Hill, New Delhi, 1974.
5 P.M. Morse and H. Feshbach, '*Methods of theoretical physics*' Vol. I, p. 781, McGraw Hill, New York, 1953.
6 G.H. Golub and C.F. Van Loan, '*Matrix computations*', Johns Hopkins University Press, 1989.
7 H. and B.S. Jeffreys, '*Mathematical physics*', chapter 4, University Press, Cambridge, 1966.
8 D. Middleton, '*An introduction to statistical communication theory*', McGraw Hill, New York, 1960.
9 S.O. Rice, 'Mathematical analysis of random noise', *Bell Syst. Tech. J.*, **23**, 282, 1944; **25**, 46, 1945.
10 M.G. Kendall, '*The advanced theory of statistics*', Vol. 1, Griffin, London, 1943.
11 E. Renshaw, '*Modelling biological populations in space and time*', University Press, Cambridge, 1991.
12 N.T.J. Bailey, '*The elements of stochastic processes*', section 8.7, John Wiley & Sons, New York, 1964.
13 J. Lighthill, '*Waves in fluids*', Chapter 3, University Press, Cambridge, 1978.
14 N. Wax (Ed.), '*Selected papers on noise and stochastic processes*', Dover, New York, 1954. This valuable collection of reprints contains refs 9 and 16.
15 B. Oksendal, '*Stochastic differential equations*', Springer Universitext, Berlin, 2000.

16 J.L. Doob, 'The Brownian movement and stochastic equations', *Ann. Math.*, **43**, 351, 1942.
17 R.J.A. Tough, 'A Fokker Planck description of K distributed noise', *J. Phys. A*, **20**, 551, 1987.
18 T.R. Field and R.J.A. Tough, ' Diffusion processes in electromagnetic scattering generating K distributed noise', *Proc. Roy. Soc.* **A459**, 2169, 2003.
19 T.R. Field and R.J.A. Tough, ' Stochastic dynamics of the scattering amplitude generating K distributed noise', *J. Math. Phys.*, **44**, 5212, 2003.
20 N.G. van Kampen, '*Stochastic processes in physics and chemistry*', North Holland, Amsterdam, 1984.
21 T.J. Bailey, 'Stochastic birth, death and migration processes for spatially distributed populations', *Biometrica*, **55**, 237–257, 1968.
22 E. Renshaw, 'Stepping stone models for population growth', *J. Appl. Probability*, **11**, 16–31, 1974.
23 E. Jakeman, '*Photon correlation and light beating spectroscopy.*', H.Z. Cummins and E.R. Pike, Eds., Plenum Press, London, 1974.
24 R.J.A. Tough, 'The estimators of normalised moments of the gamma distribution', *J. Phys.* **A 27**, 7883 (1994).
25 A. Wald, '*Sequential analysis*', John Wiley, New York, 1962.

Appendix 2

Some useful special functions

A2.1 Introduction

Much of the material in this book is presented in mathematical terms; these provide us with a compact and precise language in which to discuss engineering problems. Obviously this is not the place to develop a coherent and comprehensive account of 'engineering mathematics', such as that found in Reference 1, from which we then might pick the few topics we need. Nonetheless, when approaching the literature of sea clutter (or this book) for the first time, the reader can encounter a significant barrier, built from unfamiliar and slightly fantastical objects. Gamma, poly-gamma and hypergeometric functions, Bessel functions, both modified and un-modified, of every order and kind, all conspire to intimidate the beginner. In the days before the widespread availability of computers, the study of these functions formed part of the education of every scientist and engineer; in the final quarter of the 20th century this was supplanted, to a greater or lesser extent, by simulation and other numerical techniques. At the same time, computer based systems, such as Mathematica [2], were developed, that are able to carry out sophisticated mathematical manipulations and evaluate these unfamiliar functions. (The Mathematica web-site also supports an exhaustive on-line compilation of results from classical analysis.) So, paradoxically, this background material has become more useful and accessible at the same time that it has become less generally familiar. In this second appendix we will give a short review of some special functions that are useful in the study of clutter and its impact on radar performance – all you ever wanted to know about fancy sums, but were scared to ask.

A2.2 The gamma function and related topics

The gamma distribution (4.24) plays a central role in our clutter modelling, capturing the modulation of the local power of the clutter by the large-scale structure of the sea surface, as revealed by high-resolution radar. The gamma function itself appears as the normalising factor in the pdf

$$\Gamma(z) = \int_0^\infty \exp(-t)t^{z-1}dt \tag{A2.1}$$

This function occurs in many other contexts; some familiarity with its properties can be very helpful in the analysis of practical problems. Perhaps its most useful property follows from (A2.1) by integration by parts

$$\Gamma(z+1) = \int_0^\infty \exp(-t)t^z dt = -t^z \exp(-t)|_0^\infty + z\int_0^\infty \exp(-t)t^{z-1}dt = z\Gamma(z) \tag{A2.2}$$

The representation (A2.1) defines the gamma function for $\mathrm{Re}(z) > 0$, the region in which the integral converges; (A2.2), however, suggests that some sense can also be ascribed to the gamma function for $\mathrm{Re}(z) < 0$. To see how this might be done we split the range of integration in (A2.1) at t=1

$$\Gamma(z) = \int_0^1 \exp(-t)t^{z-1}dt + \int_1^\infty \exp(-t)t^{z-1}dt \tag{A2.3}$$

The second of these integrals converges for all values of z; if, for the moment, we assume that $\mathrm{Re}(z) > 0$, we can expand the exponential in the first and integrate term by term:

$$\Gamma(z) = \sum_{n=0}^\infty \frac{(-1)^n}{n!(n+z)} + \int_1^\infty \exp(-t)t^{z-1}dt \tag{A2.4}$$

We see that this series converges for all z, save for negative integer values; (A2.4) also satisfies (A2.2) and so provides a continuation of the gamma function into the region $\mathrm{Re}(z) < 0$. The singular behaviour of the gamma function close to a negative integer argument is captured by

$$\Gamma(-n+\delta) \sim \frac{(-1)^n}{n!\delta} \tag{A2.5}$$

When z takes positive integer values the gamma function reduces to the factorial function

$$\Gamma(n+1) = n! \tag{A2.6}$$

A great many integrals can be expressed in terms of gamma functions. The reader who feels the need to tone up his or her integral calculus skills might like to check that

$$\int_0^{\infty} \exp(-t^2)dt = \frac{\sqrt{\pi}}{2} = \frac{1}{2}\Gamma(1/2) \tag{A2.7a}$$

$$\int_0^1 t^{\nu-1}(1-t)^{\mu-1}dt = \int_0^{\infty} \frac{t^{\nu-1}}{(1+t)^{\mu+\nu}}dt = \frac{\Gamma(\nu)\Gamma(\mu)}{\Gamma(\nu+\mu)} \tag{A2.7b}$$

$$\int_0^{\pi/2} \cos^{2\nu-1}\theta \sin^{2\mu-1}\theta d\theta = \frac{1}{2}\frac{\Gamma(\nu)\Gamma(\mu)}{\Gamma(\nu+\mu)} \tag{A2.7c}$$

Innumerable examples of this kind can be found in older calculus texts, such as that of Gibson [3].

A particularly useful quantity is formed as a quotient of Gamma functions; again n is an integer

$$(\nu)_n = \frac{\Gamma(\nu+n)}{\Gamma(\nu)} = \nu(\nu+1)\cdots(\nu+n-1) \tag{A2.8}$$

This $(\nu)_n$, which is sometimes referred to as a Pochhammer symbol, was introduced in the discussion of the negative binomial distribution; (A1.15) can be re-written as

$$P(n) = \frac{1}{(1+\bar{N}/\alpha)^{\alpha}} \frac{(\alpha)_n}{n!} \left(\frac{\bar{N}\alpha}{1+\bar{N}/\alpha}\right)^n \tag{A2.9}$$

so that the normalisation condition

$$\sum_{n=0}^{\infty} P(n) = \frac{1}{(1+\bar{N}/\alpha)^{\alpha}} \sum_{n=0}^{\infty} \frac{(\alpha)_n}{n!} \left(\frac{\bar{N}/\alpha}{1+\bar{N}/\alpha}\right)^n$$

$$= \frac{1}{(1+\bar{N}/\alpha)^{\alpha}} \frac{1}{(1-(\bar{N}/\alpha)/(1+\bar{N}/\alpha))^{\alpha}} = 1 \tag{A2.10}$$

is satisfied as a consequence of the binomial theorem

$$\sum_{n=0}^{\infty} \frac{(\nu)_n}{n!} z^n = \frac{1}{(1-z)^{\nu}} \tag{A2.11}$$

We note that, when ν is a negative integer,

$$(-p)_n = (-)^n \frac{p!}{(p-n)!}, \quad n \le p$$

$$= 0 \quad n > p \tag{A2.12}$$

so that, in this case, the binomial expansion terminates as a polynomial.

The Pochhammer symbol (A2.8) can be used to construct series more general than the binomial expansion (A2.11); for example the hypergeometric and confluent hypergeometric series are defined by

$$ {}_2F_1(a,b;c;z) = \sum_{n=0}^{\infty} \frac{(a)_n(b)_n}{n!(c)_n} z^n \tag{A2.13} $$

$$ {}_1F_1(a;c;z) = \sum_{n=0}^{\infty} \frac{(a)_n}{n!(c)_n} z^n \tag{A2.14} $$

We note that, like its special case the binomial series (the series (A2.11) is none other than ${}_2F_1(\nu, a; a; z)$), the hypergeometric series (A2.13) has a radius of convergence of 1. However, the confluent hypergeometric series (A2.11) converges for all values of z, again like its special case the exponential series

$$ \exp(z) = {}_1F_1(a;a;z) \tag{A2.15} $$

The binomial theorem special case presents another feature of hypergeometric and related series. If any of a,b in (A2.13) or a in (A2.14) takes a negative integer value the series terminates as a polynomial; the orthogonal polynomials associated with the names of Legendre, Laguerre, Hermite and others can be represented in this way. These functions have been studied intensively for many years; their properties are understood and tabulated in detail. Some part of their value derives from their subsuming a great many individual functions, both elementary and more arcane, as special cases; the binomial and exponential functions just mentioned are particularly simple examples of this. Many standard texts [4,5], and the Mathematica web-site [6], provide exhaustive compilations of results of this kind. Middleton [7] highlights special forms of (A2.13) and (A2.14) that are particularly relevant to radar related issues.

Our discussions of the Poisson and Gaussian limiting forms of the binomial distribution in Section A1.3, and the gamma limit of the negative binomial distribution in Section A1.11, depend on an understanding of the behaviour of the gamma or factorial function with a large argument. The integral representation (A2.1) provides a convenient route to a useful result. We note that

$$ \Gamma(z) = \frac{1}{z} \int_0^{\infty} t^z \exp(-t)dt = z^z \int_0^{\infty} \exp(z(\log(t) - t))dt \tag{A2.16} $$

The exponent $z(\log(t) - t)$ has a single maximum at $t = 1$; so, for large z, the second integrand in (A2.16) is sharply peaked around this value. To capture the large z limiting behaviour we need only expand the exponent to quadratic order about $t = 1$, extend the limits of integration to $\pm\infty$ and carry out a single Gaussian integration to give us

$$\Gamma(z) = z^z \int_0^\infty \exp(z(\log(t) - t))\, dt \approx z^z \exp(-z) \int_{-\infty}^{\infty} \exp(-zp^2/2)\, dp$$

$$= z^z \exp(-z)\sqrt{\frac{2\pi}{z}} \tag{A2.17}$$

This result is known as Stirling's approximation, and is remarkably effective; substitution of $z = 11$ into (A2.17) captures over 99% of the value of $10! = 3628800$.

The logarithm of the gamma function and its derivatives arise in the discussion (Section 6.9) of parametric estimation. The analysis of $\log(\Gamma(z))$ is made easier if we can express the gamma function itself as a product. To do this we recall the elementary limiting result

$$\lim_{n\to\infty} \left(1 - \frac{t}{n}\right)^n = \exp(-t) \tag{A2.18}$$

and construct

$$G(z,n) = \int_0^n t^{z-1}\left(1 - \frac{t}{n}\right)^n dt$$

$$\Gamma(z) = \lim_{n\to\infty} G(z,n) \tag{A2.19}$$

For the moment, we fix the value of n so that

$$G(z,n) = n^z \int_0^1 p^{z-1}(1-p)^n dp = \frac{n^z\Gamma(z)n!}{\Gamma(z+n+1)}$$

$$= \frac{n^z}{z(1+z)\cdots(1+z/n)} \tag{A2.20}$$

This representation again highlights the singular behaviour of the gamma function as z tends to zero; we note that the integral (A2.1) no longer converges in this limit. The product (A2.20) can be re-arranged as

$$G(z,n) = \frac{\exp\left(z\log n - z\sum_{z=1}^n 1/r\right)}{z\prod_{r=1}^n (1+z/r)\exp(-z/r)} \tag{A2.21}$$

We now let n tend to infinity, and identify Euler's constant γ as

$$\gamma = \lim_{n\to\infty} \sum_{r=1}^n \frac{1}{r} - \log n \approx 0.5772167 \tag{A2.22}$$

Thus we can express the gamma function as an infinite product

$$\Gamma(z) = \frac{\exp(-\gamma z)}{z\prod_{r=1}^\infty (1+z/r)\exp(-z/r)} \tag{A2.23}$$

From this we see that

$$\Gamma(z) \sim \frac{1}{z} - \gamma + O(z), \quad z \to 0 \tag{A2.24}$$

The digamma function is defined by

$$\begin{aligned}\psi(z) &= \frac{1}{\Gamma(z)} \frac{d\Gamma(z)}{dz} = \frac{1}{\Gamma(z)} \int_0^\infty \log(t) t^{z-1} \exp(-t) \\ &= \frac{d \log \Gamma(z)}{dz}\end{aligned} \tag{A2.25}$$

and can be expanded as

$$\psi(z) = -\frac{1}{z} - \gamma + \sum_{n=1}^{\infty} \left[\frac{1}{n} - \frac{1}{n+z} \right] \tag{A2.26}$$

Higher order derivatives of this function have simple series representations

$$\psi^{(m)}(z) = \frac{d^m}{dz^m} \psi(z) = (-1)^{m+1} m! \sum_{n=0}^{\infty} \frac{1}{(z+n)^{m+1}} \tag{A2.27}$$

As an example of the application of these results we evaluate the integral

$$\begin{aligned}\int_0^\infty (\log t)^2 \exp(-t) dt &= \frac{d^2\Gamma(z)}{dz^2} |_{z=1} \\ &= \psi^{(1)}(1) + \psi(1)^2 = \frac{\pi^2}{6} + \gamma^2\end{aligned}$$

A2.3 Some properties of the K distribution pdf

Much of the discussion in Chapter 4 is based on the compound representation of the clutter intensity pdf

$$P(z) = \frac{b^\nu}{\Gamma(\nu)} \int_0^\infty x^{\nu-2} \exp(-bx) \exp(-z/x) \tag{A2.28}$$

We have already mentioned that this can be expressed in terms of the modified Bessel function of the second kind (K for short); extensive accounts of the behaviour of this function can be found in several places [4,8]. Nonetheless, everything we need for our purposes can be derived from (A2.28); the few properties of the gamma function we have just discussed provide us with the tools to do this. Perhaps the simplest quantities characterising the K pdf are its moments. These can be evaluated as follows:

$$\begin{aligned}
\langle z^n \rangle &= \frac{b^\nu}{\Gamma(\nu)} \int_0^\infty dx x^{\nu-2} \exp(-bx) \int_0^\infty \exp(-z/x) z^n dz \\
&= n! \frac{b^\nu}{\Gamma(\nu)} \int_0^\infty dx x^{\nu+n-1} \exp(-bx) \\
&= n! \frac{\Gamma(\nu+n)}{b^n \Gamma(\nu)}
\end{aligned} \tag{A2.29}$$

The characteristic function is found, in much the same way, to be

$$\begin{aligned}
\langle \exp(-uz) \rangle &= \frac{b^\nu}{\Gamma(\nu)} \int_0^\infty dx x^{\nu-2} \exp(-bx) \int_0^\infty \exp(-uz - z/x)\, dz \\
&= \frac{b^\nu}{\Gamma(\nu)} \int_0^\infty \frac{x^{\nu-1} \exp(-bx)}{1+ux} dx \\
&= \frac{b}{u} \int_0^\infty \frac{\exp(-bs/u)}{(1+s)^\nu} ds
\end{aligned} \tag{A2.30}$$

This can also be expressed as an incomplete gamma function

$$\langle \exp(-uz) \rangle = \left(\frac{b}{u}\right)^\nu \exp(b/u) \int_{b/u}^\infty \frac{\exp(-t)}{t^\nu} dt \tag{A2.31}$$

An appropriate small u expansion of this result recovers the moments (A2.29).

The behaviour of the 'tail' of the intensity pdf (A2.28) can be deduced in the same way that we derived Stirling's approximation (A2.17)

$$P(z) = \frac{b^\nu}{\Gamma(\nu)} \int_0^\infty x^{\nu-2} \exp(-bx - z/x)\, dx$$

$$= \frac{b^\nu}{\Gamma(\nu)} z^{(\nu-1)/2} \int_0^\infty s^{\nu-2} \exp(-\sqrt{z}(bs + 1/s))\, ds$$

$$\approx \frac{b^{\nu/2+1}}{\Gamma(\nu)} z^{(\nu-1)/2} \exp(-2\sqrt{zb}) \int_{-\infty}^{\infty} \exp(-b\sqrt{bz}\, p^2)\, dp$$

$$\approx \frac{b^{(2\nu+1)/4}}{\Gamma(\nu)} z^{(2\nu-3)/4} \exp(-2\sqrt{zb})\sqrt{\pi} \qquad \text{(A2.32)}$$

Here we see an exponential decay in the square root of the intensity; this behaviour differs from the exponential decay in the intensity characteristic of the Rayleigh distribution.

Our final application of the compound representation of the K-distributed intensity pdf is in the derivation of its series expansion that reveals its behaviour for small values of the intensity. While this analysis is a little tricky, it is reasonably straightforward and is included to complete our specification of the pdf directly from the compound representation (A2.28), rather than from an unfamiliar, and equally tricky, body of work such as Sections 3.5 and 3.7 of Reference 8. At first sight, one might attempt to evaluate (A2.28) by expanding either the $\exp(-z/x)$ or $\exp(-bx)$ in the integrand, and integrating term by term. The first of these strategies is frustrated by the emergence of terms such as $\int_0^\infty \exp(-bx)x^{\nu-2-n}\, dx$; while the integrand is well behaved at large x, a non-integrable singularity is found at the origin when $n > \nu - 1$. The second approach runs into a different problem: the integrand in $\int_0^\infty x^{\nu-2+n} \exp(-z/x)\, dx$ is well behaved at the origin, but exhibits non-integrable behaviour for large x if $n + \nu > 1$. This suggests that we split the range of integration into two, making use of the $\exp(-z/x)$ expansion in the upper range of x out to infinity, and the $\exp(-bx)$ expansion for the range of x values down to zero. By making the change of variables $x \to 1/x$ in the second case we cast all the integrals we encounter into the form

$$\zeta_n(\alpha, p) = \int_1^\infty \exp(-pt)t^{\alpha-n} \qquad \text{(A2.33)}$$

For the moment we assume that α takes a non-integer value between 0 and 1; we will return to the integer value case shortly. Integrating by parts, we find that

$$\zeta_n(\alpha, p) = \frac{\exp(-p)}{n-1-\alpha} - \frac{p}{n-1-\alpha}\zeta_{n-1}(\alpha, p) \qquad \text{(A2.34)}$$

When we set n to unity we have

$$\zeta_1(\alpha, p) = \frac{\Gamma(\alpha)}{p^\alpha} - \int_0^1 \exp(-pt)t^{\alpha-1}dt = \frac{\Gamma(\alpha)}{p^\alpha} - \sum_{r=0}^{\infty} \frac{(-1)^r p^r}{r!(r+\alpha)} \tag{A2.35}$$

From these we see that

$$\zeta_n(\alpha, p) = (-p)^{n-1} p^{-\alpha} \frac{\Gamma(\alpha)\Gamma(1-\alpha)}{\Gamma(n-a)} + \sum_{r=0}^{\infty} \frac{(-p)^r}{r!(r+\alpha+1-n)} \tag{A2.36}$$

Now we can write the intensity pdf as

$$P(z) = \frac{b^\nu}{\Gamma(\nu)} \int_0^\infty x^{\nu-2} \exp(-bx) \exp(-z/x)\, dx$$

$$= \frac{b}{\Gamma(\alpha)} \int_0^\infty t^{\nu-2} \exp(-t) \exp(-y/t); \quad y = bz \tag{A2.37}$$

As suggested above, we split up the range of integration at $t = 1$ to give us

$$\begin{aligned} \int_0^\infty t^{\nu-2} \exp(-t) \exp(-y/t)\, dt &= \int_0^1 t^{\nu-2} \exp(-t) \exp(-y/t)\, dt \\ &\quad + \int_1^\infty t^{\nu-2} \exp(-t) \exp(-y/t)\, dt \\ &= \int_1^\infty \exp(-ty) \exp(-1/t) \frac{dt}{t^\nu} \\ &\quad + \int_1^\infty t^{\nu-2} \exp(-t) \exp(-y/t)\, dt \end{aligned} \tag{A2.38}$$

On expanding $\exp(-1/t)$ and $\exp(-y/t)$ in the first and second of these integrals respectively we find that

$$\begin{aligned} \int_0^\infty t^{\nu-2} \exp(-t) \exp(-y/t)\, dt &= \sum_{m=0}^{\infty} \frac{(-1)^m}{m!} (\zeta_{m+2}(2-\nu, y) \\ &\quad + y^m \zeta_{m+1}(\nu-1, 1)) \end{aligned} \tag{A2.39}$$

Introducing (A2.36) we find that

$$\int_0^\infty t^{\nu-2} \exp(-t) \exp(-y/t)\, dt$$

$$= \Gamma(2-\nu)\Gamma(\nu-1) \left\{ \sum_{m=0}^{\infty} \frac{y^m}{m!\Gamma(m+2-\nu)} - y^{\nu-1} \sum_{m=0}^{\infty} \frac{y^m}{m!\Gamma(\nu+m)} \right\}$$

$$- \sum_{r=0}^{\infty} \sum_{m=0}^{\infty} \frac{(-1)^{m+r}}{m!r!} \left\{ \frac{y^r}{r-m-\nu+1} + \frac{y^m}{r+\nu-1-m} \right\} \qquad \text{(A2.40)}$$

The doubly summed contribution vanishes in this case, as can be seen by interchanging the roles of m, r in the second term. This provides us with our expansion of the intensity pdf, for non-integer shape parameter

$$P(z) = b\Gamma(1-\nu) \sum_{m=0}^{\infty} \frac{(bz)^m}{m!} \left[\frac{(bz)^{\nu-1}}{\Gamma(\nu+m)} - \frac{1}{\Gamma(m+2-\nu)} \right] \qquad \text{(A2.41)}$$

This expression reveals an integrable singularity at the origin, should $0 < \nu < 1$. When the shape parameter takes integer values the analysis has to be modified slightly. Setting $\alpha = 0$ in (A2.34) presents no problems; deriving the analogue of (A2.36) requires a little more cunning. We define the function

$$f(p) = \int_0^1 \frac{(1-\exp(-pt))}{t} dt - \int_1^\infty \frac{\exp(-pt)}{t} dt \qquad \text{(A2.42)}$$

for which

$$\frac{df(p)}{dp} = \int_0^\infty \exp(-pt) dt = \frac{1}{p} \qquad \text{(A2.43)}$$

As $f(1) = \gamma$ (see (A2.22)) we now have

$$f(p) = \gamma + \log p \qquad \text{(A2.44)}$$

so that

$$\zeta_1(0,p) = \int_1^\infty \frac{\exp(-pt)}{t} dt = -\log z - \gamma - \sum_{m=1}^{\infty} \frac{(-p)^m}{m!m} \qquad \text{(A2.45)}$$

The method of induction leads us to the general result

$$\zeta_n(0,p) = \frac{(-p)^{n-1}}{(n-1)!}(-\log p + \psi(n)) - \sum_{m=0, m\neq n-1}^{\infty} \frac{(-p)^m}{m!(m-n+1)} \qquad \text{(A2.46)}$$

Having established these preliminary results we note from (A2.28) that, to expose the small intensity form of the K distribution pdf for integral values of the shape parameter, we must evaluate

$$\int_0^\infty t^s \exp(-t - y/t)dt \tag{A2.47}$$

where s is an integer greater than or equal to -1. To do this we proceed much as before

$$\int_0^\infty t^s \exp(-t - y/t)dt = \int_1^\infty \exp(-yt - 1/t)\frac{dt}{t^{s+2}} + \int_1^\infty t^s \exp(-t - y/t)dt$$
$$= \sum_{n=0}^\infty \frac{(-1)^n}{n!}(\zeta_{n+s+2}(0, y) + y^n \zeta_{n-s}(0, 1)) \tag{A2.48}$$

If we now substitute the result (A2.46) we find that, as in the derivation of (A2.41), the double summations that arise cancel out, leaving us with

$$\int_0^\infty t^s \exp(-t - y/t)dt = (-1)^{s+1} \sum_{n=0}^\infty \frac{y^{n+s+1}}{n!(n+s+1)!}$$
$$\times (-\log y + \psi(n+s+2) + \psi(n+1))$$
$$+ (-1)^{s+1} \sum_{n=0}^{s} \frac{y^n}{n!} \frac{\psi(n-s)}{\Gamma(n-s)} \tag{A2.49}$$

The last summation consists of a set of terms that is, at first sight, not defined. Using the results (A2.5) and (A2.26) we see that

$$\lim_{\delta \to 0} \frac{\psi(n-s+\delta)}{\Gamma(n-s+\delta)} = (s-n)!(-1)^{n+s+1}, \quad s \geq n \tag{A2.50}$$

which gives us

$$\int_0^\infty t^s \exp(-t - y/t)dt = (-1)^{s+1} \sum_{n=0}^\infty \frac{y^{n+s+1}}{n!(n+p+1)!}$$
$$\times (-\log y + \psi(n+s+2) + \psi(n+1))$$
$$+ \sum_{n=0}^{s} \frac{(-y)^n (s-n)!}{n!} \tag{A2.51}$$

This provides us with an expression for the pdf with an integer shape parameter, and reveals the logarithmic behaviour characteristic of this case.

A2.4 The Bessel functions I_n, J_n

In addition to the K Bessel function occurring in (4.25) we have also encountered the J and I functions; these have arisen in contexts where an exponential of a scalar product of two vectors in a plane is averaged over orientations of one of the vectors. Thus we have encountered J_0 and I_0 as special cases of

$$J_n(x) = \frac{1}{2\pi i^n} \int_{-\pi}^{\pi} \exp(ix \cos\theta) \cos(n\theta) d\theta$$

$$I_n(x) = \frac{1}{2\pi} \int_{-\pi}^{\pi} \exp(x \cos\theta) \cos(n\theta) d\theta \qquad \text{(A2.52)}$$

Here n must take integer values. This restriction can be relaxed; the Bessel functions of non-integer order do not figure prominently in the analysis of radar performance. So, just as the compound representation (A2.28) provided us with a starting point from which the properties of the K distribution pdf can be derived, we can take the integral representations (A2.52) as the basis of our discussion of the J and I Bessel functions.

The integral representations (A2.52) lead us directly to the small argument behaviour of the Bessel functions; we merely expand the exponential function and integrate term by term. Powers of cos θ can be expressed in terms of cosines of multiples of θ

$$\cos^{2m}\theta = \frac{(2m)!}{(m!)^2 2^{2m}} + \frac{1}{2^{2m-1}} \sum_{r=0}^{m-1} \frac{(2m)!}{r!(2m-r)!} \cos(2(m-r)\theta)$$

$$\cos^{2m+1}\theta = \frac{1}{2^{2m}} \sum_{r=0}^{m} \frac{(2m+1)!}{r!(2m+1-r)!} \cos((2(m-r)+1)\theta) \qquad \text{(A2.53)}$$

and the orthogonality property of the cosine functions

$$\frac{1}{2\pi} \int_0^{2\pi} \cos m\theta \cos n\theta d\theta = \delta_{m,n} \qquad \text{(A2.54)}$$

exploited to show that

$$J_n(x) = \left(\frac{x}{2}\right)^n \sum_{r=0}^{\infty} \left(-\frac{x^2}{4}\right)^r \frac{1}{r!(n+r)!}$$

$$I_n(x) = \left(\frac{x}{2}\right)^n \sum_{r=0}^{\infty} \left(\frac{x^2}{4}\right)^r \frac{1}{r!(n+r)!} \qquad \text{(A2.55)}$$

The series expansion of the Bessel function J_0 provides a simple route to the integral required in the inversion of the characteristic function (4.5)

$$\int_0^\infty dU U J_0(UE) \exp(-U^2x/4) = \sum_{n=0}^\infty \frac{(-E^2/4)^n}{n!n!} \int_0^\infty dU U^{2n+1} \exp(-U^2x/4)$$
$$= \frac{2}{x}\sum_{n=0}^\infty \frac{(-E^2/x)^n}{n!} = \frac{2}{x}\exp(-E^2/x) \qquad \text{(A2.56)}$$

A great many other integrals of Bessel functions can be evaluated by adopting this strategy of series expansion, term by term integration and re-summation (see Chapters 12 and 13 of Reference 8). The behaviour of Bessel functions with large arguments can also be derived from their integral representations. In the case of the I Bessel function the integrand in (A2.52) decays rapidly to zero either side of $\theta = 0$; it is sufficient to make a quadratic expansion of the exponent about this point, just as we did in the derivations of (A2.17) and (A2.32). Thus we have

$$I_n(x) \approx \frac{1}{2\pi}\int_{-\infty}^{\infty} \exp\left(x - \frac{x\theta^2}{2}\right) d\theta = \frac{\exp(x)}{\sqrt{2\pi x}} \qquad \text{(A2.57)}$$

The next highest order term in this asymptotic expansion is needed in Section 6.7; we will not derive this here, but merely note that

$$I_n(x) \sim \frac{\exp(x)}{\sqrt{2\pi x}}\left\{1 - \frac{(4n^2-1)}{8x} + \frac{(4n^2-1)(4n^2-9)}{2!(8x)^2}\cdots\right\} \qquad \text{(A2.58)}$$

In the case of the J Bessel function the integrand in (A2.53) is rapidly oscillating when x is large; the only significant contributions to the integral are derived from points where the phase of the integrand is stationary i.e. $\theta = 0, \pi$. Once again quadratic expansions of the exponent about these points enable us to capture the dominant behaviour of the integral

$$J_n(x) \approx \frac{1}{2\pi i^n}\left[\int_{-\infty}^{\infty} \exp(ix(1-\theta^2/2))d\theta + (-1)^n \int_{-\infty}^{\infty} \exp(-ix(1-\theta^2/2))d\theta\right]$$
$$= \frac{1}{2\pi i^n}\sqrt{\frac{2\pi}{x}}[\exp(ix - i\pi/4) + (-1)^n \exp(-ix + i\pi/4)]$$
$$= \sqrt{\frac{2}{\pi x}}\cos\left(x - \frac{n\pi}{2} - \frac{\pi}{4}\right) \qquad \text{(A2.59)}$$

Finally we note that the recurrence relations and differential equations satisfied by the Bessel functions can be deduced from their integral representations. Thus, for example, we have

$$J_{n-1}(x) + J_{n+1}(x) = \frac{1}{2\pi i^{n-1}} \int_0^{2\pi} \exp(ix \cos \theta) \times [\cos((n-1)\theta) - \cos((n+1)\theta)] d\theta \tag{A2.60}$$

The elementary trigonometric identity

$$\cos((n-1)\theta) - \cos((n+1)\theta) = 2 \sin(n\theta) \sin \theta$$

is substituted and the resulting expression integrated by parts

$$\begin{aligned} J_{n-1}(x) + J_{n+1}(x) &= \frac{1}{2\pi i^{n-1}} 2 \int_0^{2\pi} \exp(ix \cos \theta) \sin \theta \sin(n\theta) d\theta \\ &= \frac{1}{2\pi i^n} \frac{2}{x} \int_0^{2\pi} \exp(ix \cos \theta) \cos(n\theta) d\theta \\ &= \frac{2 J_n(x)}{x} \end{aligned} \tag{A2.61}$$

In much the same way we see that

$$\begin{aligned} J_{n-1}(x) - J_{n+1}(x) &= \frac{1}{2\pi i^{n-1}} \int_0^{2\pi} \exp(ix \cos \theta) [\cos((n-1)\theta) + \cos((n+1)\theta)] d\theta \\ &= \frac{1}{2\pi i^{n-1}} 2 \int_0^{2\pi} \exp(ix \cos \theta) \cos(n\theta) \cos \theta d\theta \\ &= 2 \frac{d}{dx} \frac{1}{2\pi i^n} \int_0^{2\pi} \exp(ix \cos \theta) \cos(n\theta) d\theta \\ &= 2 \frac{d J_n(x)}{dx} \end{aligned} \tag{A2.62}$$

These in turn imply that

$$\frac{n}{x} J_n(x) + \frac{d J_n(x)}{dx} = J_{n-1}(x), \qquad \frac{n}{x} J_n(x) - \frac{d J_n(x)}{dx} = J_{n+1}(x) \tag{A2.63}$$

and that the Bessel function satisfies the differential equation

$$\frac{d^2 J_n(x)}{dx^2} + \frac{1}{x} \frac{d J_n(x)}{dx} + \left(1 - \frac{n^2}{x^2}\right) J_n(x) = 0 \tag{A2.64}$$

These few properties of the Bessel J and I functions, which we have derived directly from (A2.52), provide an adequate background for their applications to the modelling and analysis of clutter.

A2.5 Expansions in Hermite and Laguerre polynomials

In our discussion of the simulation of correlated non-Gaussian noise processes in Chapter 5 we express the joint pdfs of correlated Gaussian and chi-squared variables as expansions in terms of Hermite and associated Laguerre polynomials. These functions are perhaps familiar to readers from their studies of quantum mechanics [9]. Their few properties we need can be derived directly and straightforwardly from first principles.

The pdf (A1.64) of two correlated Gaussian random variables, each with a unit variance, can be derived quite neatly from the appropriate characteristic function

$$\begin{aligned} P(x,y) &= \frac{1}{(2\pi)^2}\int dk_1 \int dk_2 \exp(-i(k_1x+k_2y))\langle \exp(i(k_1x+k_2y))\rangle \\ &= \frac{1}{(2\pi)^2}\int dk_1 \int dk_2 \exp(-i(k_1x+k_2y)) \exp\left(-\frac{\langle (k_1x+k_2y)^2\rangle}{2}\right) \\ &= \frac{1}{(2\pi)^2}\int dk_1 \int dk_2 \exp(-i(k_1x+k_2y)) \exp\left(-\frac{(k_1^2+k_2^2+2k_1k_2\rho)}{2}\right) \end{aligned} \qquad \text{(A2.65)}$$

Rather than evaluate this integral directly, we can decompose it into a sum of products of separate functions of x and y. Thus we have

$$P(x,y) = \frac{1}{(2\pi)^2}\sum_{n=0}^{\infty}\frac{\rho^n}{n!}\phi_n(x)\phi_n(y)$$

$$\begin{aligned} \phi_n(x) &= \int dk \exp(-ikx)(-ik)^n \exp(-k^2/2) \\ &= \left(\frac{d}{dx}\right)^n \int dk \exp(-ikx)\exp(-k^2/2) \\ &= \sqrt{2\pi}\left(\frac{d}{dx}\right)^n \exp(-x^2/2) \end{aligned} \qquad \text{(A2.66)}$$

The ϕ_n introduced here are very closely related to the standard Hermite polynomials

$$\phi_n(x) = (-1)^n\sqrt{2\pi}\exp(-x^2/2)\frac{1}{2^{n/2}}H_n(x/\sqrt{2}) \qquad \text{(A2.67)}$$

where

$$H_n(u) = (-1)^n \exp(u^2) \left(\frac{d}{du}\right)^n \exp(-u^2) \tag{A2.68}$$

and can be expressed in terms of the confluent hypergeometric function (A2.14) as

$$\begin{aligned} H_{2n}(x) &= (-1)^n 2^{2n} (1/2)_n {}_1F_1(-n, 1/2, x^2) \\ H_{2n+1}(x) &= (-1)^n 2^{2n+1} (3/2)_n x {}_1F_1(-n, 3/2, x^2) \end{aligned} \tag{A2.69}$$

Thus we can write the joint pdf of x and y as

$$P(x, y) = \frac{\exp(-(x^2 + y^2)/2)}{2\pi} \sum_{n=0}^{\infty} \frac{\rho^n}{2^n n!} H_n(x/\sqrt{2}) H_n(y/\sqrt{2}) \tag{A2.70}$$

The Hermite polynomials can be accessed through Mathematica, and are described in detail in many texts [4,10] in which a result equivalent to (A2.70) is identified as the Hardy Hille expansion.

Let us now consider how we might expand the joint pdf of two correlated intensities, adopting much the same approach. From (A1.132) we see, on setting $\psi = 1$ for simplicity, that this joint pdf can be written in terms of a modified Bessel function of the first kind

$$P(z_1, z_2) = \frac{1}{4(1 - k_0^2)} \exp\left(-\frac{z_1 + z_2}{2(1 - k_0^2)}\right) I_0\left(\frac{k_0\sqrt{z_1 z_2}}{(1 - k_0^2)}\right) \tag{A2.71}$$

The required characteristic function is given by

$$\begin{aligned} \langle \exp(-s_1 z_1 - s_2 z_2) \rangle &= \int_0^\infty dz_1 \int_0^\infty dz_2 \exp(-s_1 z_1 - s_2 z_2) P(z_1, z_2) \\ &= \frac{1}{((1 + 2s_1)(1 + 2s_2) - 4k_0^2 s_1 s_2)} \end{aligned} \tag{A2.72}$$

This result can be obtained by expanding the Bessel function, integrating each term, and re-summing the resulting series. If N such intensities z_1, z_2 are summed separately to give the correlated resultants Z_1, Z_2, their characteristic function is given by

$$\langle \exp(-s_1 Z_1 - s_2 Z_2) \rangle = \frac{1}{((1 + 2s_1)(1 + 2s_2) - 4k_0^2 s_1 s_2)^N} \tag{A2.73}$$

This characteristic function can be inverted directly to yield the result

$$\begin{aligned} P(Z_1, Z_2) = {} & \frac{1}{4(1 - k_0^2)(N - 1)!} \left(\frac{Z_1 Z_2}{4k_0^2}\right)^{(N-1)/2} \\ & \times \exp\left(-\frac{Z_1 + Z_2}{2(1 - k_0^2)}\right) I_{N-1}\left(\frac{k_0\sqrt{Z_1 Z_2}}{(1 - k_0^2)}\right) \end{aligned} \tag{A2.74}$$

A representation of this joint pdf of Z_1, Z_2 analogous to (A2.70) can be constructed; in doing this we will make the acquaintance of the associated Laguerre polynomials, and provide a basis for some of the discussion in Sections 5.8. Much as before, we 'chop up' the characteristic function, making use of the binomial theorem

$$\frac{1}{((1+2s_1)(1+2s_2)-4k_0^2 s_1 s_2)^N} = \frac{1}{((1+2s_1)(1+2s_2))^N}\sum_{n=0}^{\infty}\frac{(N)_n}{n!}\left(\frac{4k_0^2 s_1 s_2}{(1+2s_1)(1+2s_2)}\right)^n \quad \text{(A2.75)}$$

The required representation of the pdf can now be generated by Laplace inversion

$$P(Z_1, Z_2) = \sum_{n=0}^{\infty}\frac{(N)_n(4k_0^2)^n}{n!}\varphi_{N,n}(Z_1)\varphi_{N,n}(Z_2)$$

$$\varphi_{N,n}(Z) = \frac{1}{2\pi i}\int_{\gamma-i\infty}^{\gamma+i\infty}\exp(sZ)\frac{s^n}{(1+2s)^{n+N}}ds \quad \text{(A2.76)}$$

The contour integral defining $\varphi_{N,n}$ can be evaluated to give

$$\varphi_{N,n}(Z) = \frac{Z^{N-1}\exp(-Z/2)}{2^{n+N}}L_n^{(N-1)}(Z/2) \quad \text{(A2.77)}$$

Here we have identified the associated Laguerre polynomial

$$L_n^{(\alpha)}(u) = \frac{\exp(u)}{u^\alpha n!}\left(\frac{d}{du}\right)^n(\exp(-u)u^{n+\alpha}) \quad \text{(A2.78)}$$

which can be expressed as a confluent hypergeometric function through

$$L_n^{(\alpha)}(u) = \frac{(\alpha+1)_n}{n!}{}_1F_1(-n, \alpha+1, u) \quad \text{(A2.79)}$$

When (A2.78) is introduced into (A2.76) we find that

$$P(Z_1, Z_2) = \frac{\exp(-(Z_1+Z_2)/2)}{4(N-1)!}\left(\frac{Z_1 Z_2}{4}\right)^{N-1} \times \sum_{n=0}^{\infty}\frac{n!}{(N+n-1)!}k_0^{2n}L_n^{(N-1)}(Z_1/2)L_n^{(N-1)}(Z_2/2) \quad \text{(A2.80)}$$

Like (A2.71), this result has a counterpart, the so-called Mehler expansion, in the literature of special functions [10].

References

1 K.A. Stroud and D.J. Booth, *'Advanced engineering mathematics'*, Palgrave McMillan, Basingstoke, 2003.

2 S. Wolfram, *' The Mathematica book'*, University Press, Cambridge, 1996.
3 G.A. Gibson, *'Advanced calculus'*, Macmillan, London, 1944.
4 M. Abramowitz and I.A. Stegun, *'Handbook of mathematical functions'*, Dover, New York, 1965.
5 I.S. Gradshteyn and I.M Ryzhik, *'Table of integrals, series and products'*, Academic Press, New York, 1965.
6 http://functions.wolfram.com/1
7 D. Middleton, *'Introduction to statistical communication theory'*, Appendix 1, McGraw Hill, New York, 1960.
8 G.N. Watson, *'A treatise on the theory of Bessel functions'*, University Press, Cambridge, 1966.
9 L.D. Landau and E.M. Lifshitz, *'Quantum mechanics – non-relativistic theory'*, Pergamon Press, Oxford, 1958.
10 N.N. Lebedev, *'Special functions and their applications'*, Dover, New York, 1972.

Appendix 3

Scattering from a corrugated surface

Virtually all the detailed numerical calculations of low grazing angle EM scattering described in this book, and the published literature, represent the sea surface as a corrugated structure. As we discussed in Chapter 3, in this special case the solution of the vector Maxwell's equations reduces to that of an appropriate scalar Helmholtz equation. In this appendix we will discuss this simpler problem in some detail. Using this formalism we are able to demonstrate theoretical techniques used in the solution of scattering problems and derive the equations implemented in the Forward/Back computational method with relatively little effort. By exploiting reciprocity and a simple Fourier representation of the Helmholtz equation Green's function we can derive several practically important results more concisely and clearly than is done in the literature. This development of perturbation and other approximate theories also presents their results in a form that makes direct comparison with the numerical results quite straightforward.

This appendix will be long and perhaps overburdened with technical detail. Possibly as a result of this, it should provide a reasonably thorough introduction to the theory of electromagnetic scattering in the low grazing angle regime that is particularly pertinent to the discussion of sea clutter. The reader who wishes to carry out controlled calculations of this kind and to interpret their output sensibly, has no alternative but to study their theoretical background in some depth. Much of the relevant material is spread throughout the physics and engineering literature and is expressed in a variety of notations. It is hoped that, by focusing our attention on the less general scalar formulation of the problem, we will be able to lead the reader through the finer points of the theory and its implementation, and elucidate details too often omitted in articles in the literature, without incurring an unacceptable overhead of formulaic obfuscation. The authors sincerely hope that they can succeed in achieving this aim, as they remember only too well the difficulties they experienced when encountering this material for the first time.

A3.1 The integral formulation of the scalar scattering problem

Typically, the sea surface is represented as an interface between regions of space in which an appropriately chosen scalar field (ψ_V or ψ_H from Section 3.6) satisfies one of two different Helmholtz equations. Thus, in region 1 above the sea surface, we have

$$\nabla^2\psi(\mathbf{r}) + k_1^2\psi(\mathbf{r}) = \xi(\mathbf{r}) \tag{A3.1}$$

while, in region 2 below the sea surface

$$\nabla^2\psi(\mathbf{r}) + k_2^2\psi(\mathbf{r}) = 0 \tag{A3.2}$$

Here $\xi(\mathbf{r})$ is an inhomogeneity, localised in space, which generates the field incident on the sea-surface; the wave-numbers occurring in (A3.1) and (A3.2) are defined by

$$k_1^2 = \omega^2\varepsilon_0\mu_0 \tag{A3.3}$$

and

$$k_2^2 = \varepsilon k_1^2 \tag{A3.4}$$

(When the wave number's dependence on the dielectric properties of medium is clear from its context, and is not particularly pertinent to the discussion, we will denote it simply by k.) At this point we recall Gauss' theorem from vector analysis

$$\int_S \mathbf{A}(\mathbf{r})\cdot d\mathbf{S} = \int_V \nabla\cdot\mathbf{A}(\mathbf{r})dV \tag{A3.5}$$

which equates the net 'outflow' of a vector field $\mathbf{A}$ through a closed surface S with the integral of the divergence $\nabla\cdot\mathbf{A}(\mathbf{r})$ of that vector field over the volume V enclosed within S. We apply this to a vector constructed from the gradients of two scalar functions $\phi(\mathbf{r}), \varphi(\mathbf{r})$

$$\mathbf{A}(\mathbf{r}) \equiv \phi(\mathbf{r})\nabla\varphi(\mathbf{r}) - \varphi(\mathbf{r})\nabla\phi(\mathbf{r})$$

Gauss's theorem now leads us to a result known as Green's theorem

$$\int_S d\mathbf{S}\cdot(\phi(\mathbf{r})\nabla\varphi(\mathbf{r}) - \varphi(\mathbf{r})\nabla\phi(\mathbf{r})) = \int_V dV(\phi(\mathbf{r})\nabla^2\varphi(\mathbf{r}) - \varphi(\mathbf{r})\nabla^2\phi(\mathbf{r})) \tag{A3.6}$$

The solution of Helmholtz equations such as (A3.1) and (A3.2) is facilitated by the introduction of Green's functions satisfying

$$(\nabla^2 + k_{1,2}^2)G_0^{(1,2)}(\mathbf{r},\mathbf{r}') = \delta(\mathbf{r}-\mathbf{r}') \tag{A3.7}$$

Explicit forms for these Green's functions will be discussed shortly; for the moment we note the presence of the delta-function driving term in (A3.7). Using this, and applying Green's theorem, we find that at a point $\mathbf{r}$ above the sea surface

$$\psi(\mathbf{r}) = \psi_{\text{in}}(\mathbf{r}) + \int (G_0^{(1)}(\mathbf{r}, \mathbf{x}')\zeta(\mathbf{x}') - \psi(\mathbf{x}')\mathbf{n}(\mathbf{x}') \cdot \nabla' G_0^{(1)}(\mathbf{r}, \mathbf{x}'))dS' \quad \text{(A3.8)}$$

while below the surface

$$\psi(\mathbf{r}) = -\int (\alpha G_0^{(2)}(\mathbf{r}, \mathbf{x}')\zeta(\mathbf{x}') - \psi(\mathbf{x}')\mathbf{n}(\mathbf{x}') \cdot \nabla' G_0^{(2)}(\mathbf{r}, \mathbf{x}'))dS' \quad \text{(A3.9)}$$

Thus the field is specified in terms of the field $\psi(\mathbf{x}')$ and the normal component of its gradient $\zeta(\mathbf{x}') = \mathbf{n}(\mathbf{x}') \cdot \nabla' \psi(\mathbf{x}')$ on the upper side of the interface (whose co-ordinates are denoted by $\mathbf{x}'$) and, in the region above the surface, an incident field is identified as

$$\psi_{\text{in}}(\mathbf{r}) = \int dV' G_0^{(1)}(\mathbf{r}, \mathbf{r}')\xi(\mathbf{r}') \quad \text{(A3.10)}$$

The values taken by the surface field and its normal gradient in the two media are related by the boundary conditions (as described in terms of the EM fields in Section 3.6)

$$\psi(\mathbf{x}^{(1)}) = \psi(\mathbf{x}^{(2)}), \qquad \zeta(\mathbf{x}^{(1)}) = \alpha\zeta(\mathbf{x}^{(2)}) \quad \text{(A3.11)}$$

Equations for the surface field and normal derivative can be derived by considering the limiting process in which $\mathbf{r}$ in (A3.8) and (A3.9) approach the interface from above and below respectively. In performing this limiting procedure one, speaking loosely, picks up only half of the contribution of the delta function derived from the Green's function. (Reference 1 gives a detailed account of this analysis.) Thus we obtain the coupled surface field integral equations

$$\psi(\mathbf{x}) = 2\psi_{\text{in}}(\mathbf{x}) + 2\int (G_0^{(1)}(\mathbf{x}, \mathbf{x}')\zeta(\mathbf{x}') - \psi(\mathbf{x}')\mathbf{n}(\mathbf{x}') \cdot \nabla' G_0^{(1)}(\mathbf{x}, \mathbf{x}'))dS'$$

$$\psi(\mathbf{x}) = -2\int (\alpha\, G_0^{(2)}(\mathbf{x}, \mathbf{x}')\zeta(\mathbf{x}') - \psi(\mathbf{x}')\mathbf{n}(\mathbf{x}') \cdot \nabla' G_0^{(2)}(\mathbf{x}, \mathbf{x}'))dS' \quad \text{(A3.12)}$$

The equations appropriate to H polarisation are retrieved when we set $\alpha = 1$, and to V polarisation when $\alpha = \varepsilon$. If it is assumed that the sea is a perfect conductor, the equations (A3.12) de-couple (since ψ_{H} and ζ_{V} are zero on the surface) and we find that

$$\psi(\mathbf{x}) = 2\psi_{\text{in}}(\mathbf{x}) - 2\int \psi(\mathbf{x}')\mathbf{n}(\mathbf{x}') \cdot \nabla' G_0^{(1)}(\mathbf{x}, \mathbf{x}')dS' \quad \text{(A3.13)}$$

for V polarisation, while

$$\zeta(\mathbf{x}) = 2\mathbf{n}(\mathbf{x}) \cdot \nabla\psi_{\text{in}}(\mathbf{x}) + 2\int \zeta(\mathbf{x}')\mathbf{n}(\mathbf{x}) \cdot \nabla G_0^{(1)}(\mathbf{x}, \mathbf{x}')dS' \quad \text{(A3.14)}$$

in the H polarised case.

Here we have presented the integral formulation in fairly general terms, without specifying the form of the scattering surface as either one (i.e. corrugated) or two-dimensional, or the explicit form taken by the Green's function. We do this, partly for the sake of clarity and brevity (equations burdened with Hankel functions and the explicit representation of the scattering interface are necessarily rather cumbersome), and to retain a greater level of generality. We will construct explicit results appropriate to the corrugated surface here; nonetheless their extension to two-dimensional scattering surfaces, if required, should be straightforward.[1]

The second of the integral equations (A3.12) leads to an implicit relationship between the field $\psi(\mathbf{x}')$ and its normal derivative $\zeta(\mathbf{x}')$ on the scattering surface, providing a so-called extended boundary condition. Once these equations (A3.12) have been solved, almost invariably by some numerical method, the scattered field is calculated from

$$\psi_{\mathrm{scat}}(\mathbf{R}) = \int (G_0^{(1)}(\mathbf{r},\mathbf{x}')\zeta(\mathbf{x}') - \psi(\mathbf{x}')\mathbf{n}(\mathbf{x}')\cdot\nabla' G_0^{(1)}(\mathbf{R},\mathbf{x}'))dS' \tag{A3.15}$$

The normalised scattering cross section can now be calculated, as

$$\sigma^0 = \lim_{R\to\infty} \frac{2\pi R}{L|\psi_{\mathrm{in}}|^2}\langle|\psi_{\mathrm{scat}}(\mathbf{R})|^2\rangle \tag{A3.16}$$

for a corrugated surface (where L is the illuminated length), and as

$$\sigma^0 = \lim_{R\to\infty} \frac{4\pi R^2}{A|\psi_{\mathrm{in}}|^2}\langle|\psi_{\mathrm{scat}}(\mathbf{R})|^2\rangle \tag{A3.17}$$

for a two-dimensional surface (where A is the illuminated area).

A3.2 Helmholtz equation Green's functions in two and three dimensions

The integral equations we have just derived establish a framework within which we can calculate the radar back-scatter from the interface. If we are to make further progress in this analysis we require a tractable representation of the Green's function. There are essentially two, complementary, approaches to this problem. The first, in which the Green's function is expressed explicitly as a function of $\mathbf{r}, \mathbf{r}'$, is particularly useful in numerical work; in the second we exploit Fourier and complex variable methods. A full specification of the Green's function includes the boundary conditions it satisfies; here we will adopt those describing outgoing radiation in free space. The Fourier representation of a Green's function useful in the solution of this problem can be derived directly from (A3.7):

$$(\nabla^2 + k^2)G(\mathbf{r},\mathbf{r}') = \delta(\mathbf{r}-\mathbf{r}')$$

$$(k^2 - q^2)\tilde{G}(\mathbf{q},\mathbf{r}') = \exp(i\mathbf{q}\cdot\mathbf{r}')$$

$$G(\mathbf{r},\mathbf{r}') = -\frac{1}{(2\pi)^n}\int d^n q \frac{\exp(i\mathbf{q}\cdot(\mathbf{r}'-\mathbf{r}))}{q^2-k^2} \tag{A3.18}$$

[1] However, we must stress that scalar scattering from two-dimensional surfaces cannot be applied in general to EM scattering, which is vector in nature and only becomes a scalar problem for a one-dimensional surface; this is shown in Section 3.6. Thus our use of the terms Horizontal (H) and Vertical (V) polarisation only has any meaning for EM scattering from a 1D surface. Some other scattering problems, such as acoustics, are scalar in all cases.

We now obtain explicit expressions for the Green's functions describing outgoing radiation in unbounded two and three-dimensional spaces. In the latter case we set n in (A3.18) to 3 and adopt a spherical polar co-ordinate system in which the polar axis is identified with the direction of $\mathbf{r}-\mathbf{r}'$. The integral over the polar angle ϕ is easy; in this way we obtain

$$
\begin{aligned}
G(\mathbf{r}-\mathbf{r}') &= -\frac{1}{(2\pi)^2}\int_0^\infty dq\frac{q^2}{q^2-k^2}\int_0^\pi d\theta\,\sin\theta\,\exp(iq|\mathbf{r}-\mathbf{r}'|\cos\theta)\\
&= -\frac{1}{(2\pi)^2 i}\int_0^\infty dq\frac{q}{q^2-k^2}(\exp(iq|\mathbf{r}-\mathbf{r}'|)-\exp(-iq|\mathbf{r}-\mathbf{r}'|))\\
&= -\frac{1}{(2\pi)^2 2i}\int_{-\infty}^\infty dq\,\exp(iq|\mathbf{r}-\mathbf{r}'|)\left(\frac{1}{q-k}+\frac{1}{q+k}\right) \qquad \text{(A3.19)}
\end{aligned}
$$

It remains to evaluate this integral which, as it stands, might seem to be divergent. We overcome this difficulty when we impose the boundary condition that G represents an outgoing wave. This we do by supplementing the wave number k with a small positive imaginary part. The integral (A3.19) can now be evaluated using Cauchy's theorem [2] as

$$
\begin{aligned}
G(\mathbf{r},\mathbf{r}') &= -\frac{1}{(2\pi)^2 i}\int_{-\infty}^\infty dq\,\exp(iq|\mathbf{r}-\mathbf{r}'|)\left(\frac{1}{q-k-i\delta}+\frac{1}{q+k+i\delta}\right)\\
&= -\frac{1}{4\pi|\mathbf{r}-\mathbf{r}'|}\exp(ik|\mathbf{r}-\mathbf{r}'|) \qquad \text{(A3.20)}
\end{aligned}
$$

Thus we see that the three-dimensional free space Green's function can be expressed in terms of simple functions and is readily interpreted as an outgoing spherical wave emanating from a point source.

The evaluation of the two-dimensional free space Green's function proceeds in much the same way; in this case we encounter the Bessel functions introduced in Section A2.4. Having set $n=2$ in (A3.18), we chose a Cartesian co-ordinate system, in which the x direction coincides with that of $\mathbf{r}-\mathbf{r}'$.

$$
\begin{aligned}
G(\mathbf{r},\mathbf{r}') &= -\frac{1}{(2\pi)^2}\int_{-\infty}^\infty dq_x\,\exp(iq_x|\mathbf{r}-\mathbf{r}'|)\int_{-\infty}^\infty \frac{dq_y}{q_y^2+q_x^2-k^2-i\delta}\\
&= -\frac{i}{4\pi}\int_{-\infty}^\infty dq_x\frac{\exp(iq_x|\mathbf{r}-\mathbf{r}'|)}{\sqrt{k^2-q_x^2}} = -\frac{i}{4\pi}\int_{-\infty}^\infty \frac{\exp(ikt|\mathbf{r}-\mathbf{r}'|)}{\sqrt{1-t^2}}dt
\end{aligned}
$$
(A3.21)

This integral can be broken up as follows

$$\int_{-\infty}^{\infty} \frac{\exp(ikt|\mathbf{r}-\mathbf{r}'|)}{\sqrt{1-t^2}}dt = \int_{-1}^{1} \frac{\exp(ikt|\mathbf{r}-\mathbf{r}'|)}{\sqrt{1-t^2}}dt - 2i\int_{1}^{\infty} \frac{\cos(kt|\mathbf{r}-\mathbf{r}'|)}{\sqrt{t^2-1}}dt \tag{A3.22}$$

Here we have identified the phase of the square root of a negative number through

$$\sqrt{1-t^2} = i\sqrt{t^2-1}; \qquad t > 1 \tag{A3.23}$$

This choice is consistent with the requirement that the Green's function represents an outgoing wave. The first of the terms in (A3.22) is recognised as the Bessel function J_0 defined in (A2.52); the second can be identified with another, linearly independent solution of (A2.64) (with $n = 0$)

$$Y_0(x) = -\frac{2}{\pi}\int_{1}^{\infty} \frac{\cos(xt)}{\sqrt{t^2-1}}dt \tag{A3.24}$$

It is easily verified, by substitution and an integration by parts, that this is a solution of (A2.64). The J and Y functions can be combined as a Hankel function

$$H_0^{(1)}(x) = J_0(x) + iY_0(x) \tag{A3.25}$$

A stationary phase argument similar to that leading to (A2.59) shows that the large argument behaviour of the Y function (A3.24) is

$$Y_0(x) \sim \sqrt{\frac{2}{\pi x}}\sin(x - \pi/4) \tag{A3.26}$$

Along with (A3.25), this establishes that the two-dimensional Green's function represents an outgoing cylindrical wave as

$$\begin{aligned} G(\mathbf{r},\mathbf{r}') &= -\frac{i}{4}(J_0(k|\mathbf{r}-\mathbf{r}'|) + iY_0(k|\mathbf{r}-\mathbf{r}'|)) = -\frac{i}{4}H_0(k|\mathbf{r}-\mathbf{r}'|) \\ &\sim -i\frac{\exp(ik|\mathbf{r}-\mathbf{r}'|)}{\sqrt{8\pi ik|\mathbf{r}-\mathbf{r}'|}}, \qquad k|\mathbf{r}-\mathbf{r}'| \to \infty \end{aligned} \tag{A3.27}$$

The J_0 and Y_0 functions are readily evaluated numerically; this makes the explicit representation (A3.27) of the Green's function particularly useful in computational work. The change in variable $t = \cosh\mu$ and a shift in the path of integration (justified by an appeal to Cauchy's theorem) transform (A3.21) into a contour integral representation of the Hankel function

$$H_0^{(1)}(kr) = \frac{1}{\pi i} \int_{-\infty - i\pi}^{\infty} \exp(ikr \cosh \mu) \, d\mu$$

$$= \frac{1}{\pi i} \int_{-\infty - i\pi/2}^{\infty + i\pi/2} \exp(ikr \cosh \mu) \, d\mu \qquad \text{(A3.28)}$$

This representation of the Hankel function is exploited in our discussion of the impedance boundary condition (Section A3.4) and in the calculation of the adjunct plane contributions in the F/B method.

A3.3 Derivation of the Fresnel formulae

To demonstrate the utility of the Green's functions and Fourier representations we have just introduced, we will use them to discuss the scattering of scalar waves by a planar interface and derive the familiar Fresnel formulae.[2]

In the case of a planar interface the surface normal **n** is orthogonal to the gradient of the Green's function for points within the scattering surface. Thus we can combine the equations (A3.12) to give us

$$\psi_{\text{in}}(\mathbf{x}) = -\int d^2x' \zeta(\mathbf{x}')(G^{(1)}(\mathbf{x}, \mathbf{x}') + \alpha G^{(2)}(\mathbf{x}, \mathbf{x}')) \qquad \text{(A3.29)}$$

This can be solved through the application of Fourier methods. In doing this we have to evaluate the transform of the Green's function in the $z = 0$ plane. Reference 3 gives an account of this calculation that draws on various properties of the Bessel functions; if we use (A3.18) we can proceed more straightforwardly. The required Fourier transform takes the form

$$\int d^2x \exp(i\mathbf{q}_{\text{H}} \cdot \mathbf{x}) G(\mathbf{x})$$

$$= -\frac{1}{(2\pi)^3} \int \frac{d^3p}{(p^2 - k^2 - i\delta)} \int d^2x \exp(i\mathbf{x} \cdot (\mathbf{q}_{\text{H}} + \mathbf{p}_{\text{H}})) \qquad \text{(A3.30)}$$

If we now identify the delta function from

$$\delta(\mathbf{q}_{\text{H}} + \mathbf{p}_{\text{H}}) = \frac{1}{(2\pi)^2} \int d^2x \exp(i\mathbf{x} \cdot (\mathbf{q}_{\text{H}} + \mathbf{p}_{\text{H}})) \qquad \text{(A3.31)}$$

[2] This calculation is done using scalars in 3D. The results may be applied to EM scattering because there is no variation in the dimension orthogonal to the plane of incidence.

we obtain

$$\int d^2x \exp(i\mathbf{q}_{\mathrm{H}} \cdot \mathbf{x})G(\mathbf{x}) = -\frac{1}{2\pi} \int_{-\infty}^{\infty} \frac{dp_z}{p_z^2 - (k^2 - q_{\mathrm{H}}^2) - i\delta} = -\frac{i}{2\sqrt{k^2 - q_{\mathrm{H}}^2}}$$

$$k = k_1 \text{ (in region 1);} \qquad k = k_2 \text{ (in region 2)} \tag{A3.32}$$

We can now use (A3.32) to solve (A3.29). Assuming the incident field to be a plane wave with wave vector $\mathbf{k}_0$, the surface field integral equation becomes, taking account of the presence of the delta function,

$$(2\pi)^2\delta(\mathbf{q}_{\mathrm{H}} + \mathbf{k}_{0\mathrm{H}}) = \tilde{\zeta}(\mathbf{q}_{\mathrm{H}}) \left\{ \frac{i}{2k_z^{(1)}} + \frac{i\alpha}{2k_z^{(2)}} \right\}$$

$$\mathbf{k}_{0\mathrm{H}} = (\mathbf{1} - \mathbf{nn}) \cdot \mathbf{k}_0$$

$$k_z^{(1)} = \sqrt{k_1^2 - k_{0\mathrm{H}}^2}, \qquad k_z^{(2)} = \sqrt{k_2^2 - k_{0\mathrm{H}}^2} \tag{A3.33}$$

so that

$$\tilde{\zeta}(\mathbf{q}_{\mathrm{H}}) = -2(2\pi)^2 i\delta(\mathbf{q}_{\mathrm{H}} + \mathbf{k}_{0\mathrm{H}}) \frac{k_z^{(1)} k_z^{(2)}}{k_z^{(2)} + \alpha k_z^{(1)}} \tag{A3.34}$$

The surface field can be deduced from this surface normal derivative

$$\tilde{\psi}(\mathbf{q}_{\mathrm{H}}) = 2\tilde{\psi}_{\mathrm{in}}(\mathbf{q}_{\mathrm{H}}) + 2\tilde{\zeta}(\mathbf{q}_{\mathrm{H}})\tilde{G}^{(1)}(\mathbf{q}_{\mathrm{H}}) = 2(2\pi)^2\delta(\mathbf{q}_{\mathrm{H}} + \mathbf{k}_{0\mathrm{H}}) \frac{\alpha k_z^{(1)}}{k_z^{(2)} + \alpha k_z^{(1)}} \tag{A3.35}$$

These results make contact with the standard Fresnel coefficients [4]. For V polarisation we have

$$\alpha = \varepsilon \qquad R = \frac{\varepsilon k_z^{(1)} - k_z^{(2)}}{\varepsilon k_z^{(1)} + k_z^{(2)}}; \qquad 1 + R = \frac{2\varepsilon k_z^{(1)}}{\varepsilon k_z^{(1)} + k_z^{(2)}} \tag{A3.36}$$

This identifies the surface magnetic field. In the horizontally polarised case

$$\alpha = 1 \qquad R = \frac{k_z^{(1)} - k_z^{(2)}}{k_z^{(1)} + k_z^{(2)}}; \qquad 1 + R = \frac{2k_z^{(1)}}{k_z^{(2)} + k_z^{(2)}} \tag{A3.37}$$

which identifies the surface electric field. These surface fields enable us to determine the scattered fields by Fourier inversion; we use Cauchy's theorem to pick out the reflected or transmitted wave, depending on whether we are above or below the scattering interface.

The real space surface fields corresponding to (A3.34) and (A3.35) take the forms

$$\zeta(\mathbf{x}) = -2i \exp(i\mathbf{k}_{0\mathrm{H}} \cdot \mathbf{x}) \frac{k_z^{(1)} k_z^{(2)}}{k_z^{(2)} + \alpha k_z^{(1)}}$$

$$\psi(\mathbf{x}) = 2\alpha \exp(i\mathbf{k}_{0\mathrm{H}} \cdot \mathbf{x}) \frac{k_z^{(1)}}{k_z^{(2)} + \alpha k_z^{(1)}} \tag{A3.38}$$

To calculate the scattered field we require that $z > 0$, and proceed to evaluate

$$\psi_{\mathrm{scat}}(\mathbf{r}) = \int (G_0^{(1)}(\mathbf{r}, \mathbf{x}')\mathbf{n}(\mathbf{x}') \cdot \nabla' \psi(\mathbf{x}') - \psi(\mathbf{x}')\mathbf{n}(\mathbf{x}') \cdot \nabla' G_0^{(1)}(\mathbf{r}, \mathbf{x}'))dS' \tag{A3.39}$$

Introducing the Fourier representations of the Green's function and its normal derivative we find that

$$\begin{aligned} \psi_{\mathrm{scat}}(\mathbf{r}) &= \frac{i}{\pi} \frac{k_z^{(1)}}{k_z^{(2)} + \alpha k_z^{(1)}} \int d^3 p \frac{\exp(-i\mathbf{p} \cdot \mathbf{r})\delta(\mathbf{k}_{0\mathrm{H}} + \mathbf{p}_{\mathrm{H}})(k_z^{(2)} + \alpha p_z)}{p^2 - k^2 - i\eta} \\ &= \frac{i}{2\pi} \frac{k_z^{(1)}}{k_z^{(2)} + \alpha k_z^{(1)}} \exp(i\mathbf{k}_{0\mathrm{H}} \cdot \mathbf{r}_{\mathrm{H}}) \int_{-\infty}^{\infty} dp_z \exp(-ip_z z) \\ &\quad \times \left[\frac{k_z^{(2)}}{k_z^{(1)}} \left(\frac{1}{p_z - k_z^{(1)}} - \frac{1}{p_z + k_z^{(1)}} \right) + \alpha \left(\frac{1}{p_z - k_z^{(1)}} + \frac{1}{p_z + k_z^{(1)}} \right) \right] \end{aligned} \tag{A3.40}$$

As z is greater than zero we close the contour of integration in the lower half of the complex plane and pick up the contribution

$$\psi_{\mathrm{scat}}(\mathbf{r}) = \exp(i\mathbf{k}_{0\mathrm{H}} \cdot \mathbf{r}_{\mathrm{H}}) \exp\left(i\sqrt{k^2 - k_{\mathrm{H}}^2} z\right) \frac{\alpha k_z^{(1)} - k_z^{(2)}}{k_z^{(2)} + \alpha k_z^{(1)}} \tag{A3.41}$$

where we take the positive square root. The transmitted wave is found similarly;

$$\begin{aligned} \psi_{\mathrm{trans}}(\mathbf{r}) &= -\frac{i\alpha}{\pi} \frac{k_z^{(1)} \exp(i\mathbf{k}_{0\mathrm{H}} \cdot \mathbf{r}_{\mathrm{H}})}{k_z^{(2)} + \alpha k_z^{(1)}} \int_{-\infty}^{\infty} dp_z \frac{\exp(-ip_z z)}{p_z - k_z^{(2)}} \\ &= \frac{2\alpha k_z^{(1)} \exp(i\mathbf{k}_{0\mathrm{H}} \cdot \mathbf{r}_{\mathrm{H}}) \exp(-izk_z^{(2)})}{k_z^{(2)} + \alpha k_z^{(1)}} \end{aligned} \tag{A3.42}$$

when $z < 0$.

This simple analysis demonstrates that the integral equation approach to scattering by a planar dielectric interface is able to reproduce the familiar textbook Fresnel results and provides an example of the simplicity and usefulness in application of Fourier integral representation (A3.18) of the Green's function.

A3.4 Approximate de-coupling of the integral equations – the impedance boundary condition

The so-called extended boundary condition, which is the underside integral equation of the pair (A3.12),

$$\psi(\mathbf{x}) = -2\int (\alpha G_0^{(2)}(\mathbf{x},\mathbf{x}')\zeta(\mathbf{x}') - \psi(\mathbf{x}')\mathbf{n}(\mathbf{x}')\cdot\nabla' G_0^{(2)}(\mathbf{x},\mathbf{x}'))dS' \qquad \text{(A3.43)}$$

provides us with a relationship between the surface field and its normal derivative that can be used to simplify the top-side equation. As it stands, however, this relationship is non-local and so is computationally and analytically rather intractable. If we specialise to the case of a planar dielectric interface, equation (A3.43) reduces to

$$\psi(\mathbf{x}) = -2\alpha\int dS'\zeta(\mathbf{x}')G_0^{(2)}(\mathbf{x},\mathbf{x}') \qquad \text{(A3.44)}$$

This relationship becomes progressively more localised as the 'range' of the Green's function becomes smaller. In a conducting medium, the Green's function does indeed decay away, within a 'skin depth' characteristic of its penetration by radiation. So, in a strongly conducting medium, whose refractive index has a large positive imaginary part, we would expect the following approximation to hold:

$$\psi(\mathbf{x}) \approx -2\alpha\zeta(\mathbf{x})\int dS' G^{(2)}(\mathbf{x},\mathbf{x}') = \frac{i\alpha\zeta(\mathbf{x})}{k_2} \qquad \text{(A3.45)}$$

where, in the second step, we have used (A3.32). This provides us with a simple impedance boundary condition (IBC). (A3.45) can be used to eliminate the normal derivative of the field from the first of (A3.12) to give us the single, approximate, integral equation satisfied by the surface field alone:

$$\psi(\mathbf{x}) = 2\psi_{\text{in}}(\mathbf{x}) - 2\int dS'(\mathbf{n}(\mathbf{x}')\cdot\nabla' G^{(1)}(\mathbf{x},\mathbf{x}') + \frac{ik_2}{\alpha}G^{(1)}(\mathbf{x},\mathbf{x}'))\psi(\mathbf{x}') \qquad \text{(A3.46)}$$

In the case of a planar interface this reduces to

$$\psi(\mathbf{x}) = 2\psi_{\text{in}}(\mathbf{x}) - 2\frac{ik_2}{\alpha}\int dS' G^{(1)}(\mathbf{x},\mathbf{x}')\psi(\mathbf{x}') \qquad \text{(A3.47)}$$

which can now be solved to give us the surface field

$$\psi(\mathbf{x}) = \frac{2\alpha k_z^{(1)}}{k_2 + \alpha k_z^{(1)}}\exp(i\mathbf{k}_{0\text{H}}\cdot\mathbf{x}) \qquad \text{(A3.48)}$$

This result can be compared with the exact result (A3.38) obtained for a planar interface; they differ only through the replacement of $k_z^{(2)}$ in the exact result by $k_0\sqrt{\varepsilon}$ in the IBC result.

So far we have restricted our discussion to the case where the dielectric interface is planar, and a full solution can be derived. It can be argued that (A3.45) might be invoked, even when the interface is not flat; this approximation is widely used in the computational scattering literature [5]. Corrections to the simple IBC (A3.45) resulting from surface curvature have been discussed in some detail [6, 7]; much of the greatest part of the effort involved in these calculations is expended in characterising the effects of curvature of a general surface in three-dimensional space. In the simpler case of a corrugated surface, such as that we are considering here, these effects can be accommodated within a fairly simple analysis. To proceed we must do two things: firstly we parameterise the co-ordinates of a point on the scattering surface in terms of the path-length l along that surface, i.e. as

$$y = y(l), \qquad z = z(l) \tag{A3.49}$$

The second preliminary is the identification of the real space form of the Green's function appropriate to this geometry. From (A3.28) we see that

$$G_0^{(2)}(\mathbf{x}', \mathbf{x}) = -\frac{1}{4\pi} \int_{-\infty - i\pi/2}^{\infty + i\pi/2} \exp(ik_2|\mathbf{x}' - \mathbf{x}| \cosh\mu)\, d\mu \tag{A3.50}$$

This Green's function describes evanescent propagation and so decays rapidly as a function of $|\mathbf{x}' - \mathbf{x}|$; thus we can write, much as in the planar interface case,

$$\psi(\mathbf{x}) = -2[\alpha\zeta(\mathbf{x}) \int G_0^{(2)}(\mathbf{x}, \mathbf{x}')dl' - \psi(\mathbf{x}) \int \mathbf{n}(\mathbf{x}') \cdot \nabla' G_0^{(2)}(\mathbf{x}, \mathbf{x}')dl'] \tag{A3.51}$$

We now make the expansions

$$\begin{aligned} \mathbf{n}(l') &= \mathbf{e}_z(y'(0) + y''(0)l') - \mathbf{e}_y(z'(0) + z''(0)l') \\ \mathbf{x}'(l') &= \mathbf{e}_z(z'(0)l' + z''(0)l'^2/2) + \mathbf{e}_y(y'(0)l' + y''(0)l'^2/2) \end{aligned} \tag{A3.52}$$

and

$$\mathbf{n}(l') \cdot \mathbf{x}'(l') \approx \frac{1}{2}(y''(0)z'(0) - z''(0)y'(0))l'^2; \qquad |\mathbf{x}'(l')| \approx |l'| \tag{A3.53}$$

where primes on y, z denote differentiation with respect to path length l. In conjunction with equation (A3.50), (A3.51) now leads to

$$\psi(\mathbf{x}(l)) = \frac{i}{k_2}(\alpha\zeta(\mathbf{x}(l)) + \frac{\psi(\mathbf{x}(l))}{2}(y''(l)z'(l) - z''(l)y'(l))) \tag{A3.54}$$

when use is made of the result

$$\int_0^{\infty + i\pi/2} \frac{d\mu}{\cosh\mu} = 2\int_0^1 \frac{dp}{1 + p^2} = \frac{\pi}{2} \tag{A3.55}$$

Finally (A3.54) can be recast as

$$\psi(l)\left(1 - \frac{i(y''(l)z'(l) - y'(l)z''(l))}{2k_2}\right) = \frac{i\alpha}{k_2}\zeta(l) \tag{A3.56}$$

This provides us with a curvature-corrected IBC, appropriate to a corrugated surface [8].

A3.5 Scattering by a perfectly conducting surface

As we have already seen the integral equations (A3.12) decouple immediately in the perfectly conducting limit, for both Vpol (A3.13) and Hpol (A3.14). These special cases provide us with a framework in which various approximate and numerical calculations of the scattered radiation can be carried out relatively simply. This lets us identify the principles involved; once this is done we can develop similar solutions of the coupled equations describing scattering by an imperfect conductor.

A3.5.1 The physical optics or Kirchoff approximation

Perhaps the simplest approximate solution to the surface field integral equations (A3.13) and (A3.14) is that provided by the Born approximations

$$\psi(\mathbf{x}) = 2\psi_{\text{in}}(\mathbf{x}) \tag{A3.57a}$$

to the former and

$$\zeta(\mathbf{x}) = 2\zeta_{\text{in}}(\mathbf{x}) \tag{A3.57b}$$

to the latter. We note that these provide exact solutions in the case where the scattering surface is planar; as a consequence (A3.57a) and (A3.57b) are frequently referred to as tangent plane approximations. The scattered field can now be calculated directly from

$$\psi_{\text{scatt}}(\mathbf{R}) = \int dS'(G_0^{(1)}(\mathbf{R},\mathbf{x}')\zeta(\mathbf{x}') - \psi(\mathbf{x}')\mathbf{n}(\mathbf{x}')\cdot\nabla' G_0^{(1)}(\mathbf{R},\mathbf{x}')) \tag{A3.58}$$

In practice we are interested in the far field limit of this expression. Thus we introduce the approximation (A3.27) to give

$$G_0^{(1)}(\mathbf{R},\mathbf{x}') = -\frac{i}{4}\mathrm{H}_0^{(1)}(ik|\mathbf{R}-\mathbf{x}'|) \sim -i\frac{\exp(ikR)}{\sqrt{8\pi ikR}}\exp(-i\mathbf{k}\cdot\mathbf{x}'), \qquad \mathbf{k} = \frac{\mathbf{R}}{R}$$

$$\nabla' G_0^{(1)}(\mathbf{R},\mathbf{x}') \sim \mathbf{k}\frac{\exp(ikR)}{\sqrt{8\pi ikR}}\exp(-i\mathbf{k}\cdot\mathbf{x}') \tag{A3.59}$$

In the Vpol case (where ζ is zero on the surface), we take a plane wave incident field $\psi_0(\mathbf{r}) = \exp(i\mathbf{k}_0\cdot\mathbf{r})$, and obtain the scattered field as

$$\psi_{scat}(\mathbf{R}) = \frac{\exp(ikR)}{\sqrt{2\pi ikR}}\int dS' \exp(i\mathbf{k}_0\cdot\mathbf{r}')\mathbf{n}(\mathbf{r}')\cdot\mathbf{k}\exp(-i\mathbf{k}\cdot\mathbf{r}') \tag{A3.60}$$

The integral over the scattering surface S can be simplified by noting that

$$dS'\mathbf{n}(\mathbf{r}') = dy'\left(\mathbf{e}_z - \mathbf{e}_z\frac{\partial\eta(y')}{\partial y'}\right) \tag{A3.61}$$

specialising to a back-scatter geometry, with $\mathbf{k} = -\mathbf{k}_0$, then leads to

$$\psi_{bs}(\mathbf{R}) = -\frac{\exp(ikR)}{\sqrt{2\pi ikR}} \int dy' \exp(i\mathbf{k}_0 \cdot \mathbf{r}') \left(k_{0z} - k_{0y} \frac{\partial \eta(y')}{\partial y'} \right) \exp(i\mathbf{k}_0 \cdot \mathbf{r}')$$

$$\mathbf{r}' = \mathbf{e}_y y' + \mathbf{e}_z \eta(y') \tag{A3.62}$$

while an integration by parts reduces this further:

$$\begin{aligned}\psi_{bs}(\mathbf{R})_V &= -\frac{\exp(ikR)}{\sqrt{2\pi ikR}} \frac{k^2}{k_{0z}} \int dy' \exp(2i(k_{0y}y' + k_{0z}\eta(yy))) \\ &= -\frac{\exp(ikR)}{\sqrt{2\pi ikR}} \frac{k^2}{k_{0z}} C(2k_{0y}, 2k_{0z})\end{aligned} \tag{A3.63}$$

Here we have introduced the random variable

$$C(2k_{0y}, 2k_{0z}) = \int dy' \exp(2ik_{0y}y') \exp(2ik_{0z}\eta(y')) \tag{A3.64}$$

The Hpol calculation (where ψ is set to zero on the surface) goes through in much the same way; the only difference is in the introduction of a phase change that does not affect the RCS derived from the model.

$$\psi_{bs}(\mathbf{R})_H = \frac{\exp(ikR)}{\sqrt{2\pi ikR}} \frac{k^2}{k_{0z}} C(2k_{0y}, 2k_{0z}) \tag{A3.65}$$

The mean value of the RCS of the rough corrugated surface can then be calculated as follows

$$\begin{aligned}\sigma^0 &= \lim_{R\to\infty} \frac{2\pi R}{L} \langle |\psi_{bs}(\mathbf{R})|^2 \rangle \\ &= \frac{1}{L} \frac{k^3}{k_{0z}^2} \int dy_1 \int dy_2 \exp(2ik_{0y}(y_1 - y_2)) \langle \exp(2ik_{0z}(\eta(y_1) - \eta(y_2))) \rangle\end{aligned} \tag{A3.66}$$

If we assume that the rough surface has a Gaussian height distribution, we may now use the result (cf. (A1.74))

$$\langle \exp(2ik_{0z}(\eta(y_1) - \eta(y_2))) \rangle = \exp(-4k_{0z}^2 \langle \eta^2 \rangle (1 - \rho(y_1 - y_2))) \tag{A3.67}$$

and change to co-ordinates y_1 and $y = y_1 - y_2$

$$\sigma^0 = \frac{1}{L} \frac{k^3}{k_{0z}^2} \int dy_1 \int dy \exp(2ik_{0y}y) \exp(-4k_{0z}^2 \langle \eta^2 \rangle (1 - \rho(y))) \tag{A3.68}$$

Integration on y_1 merely 'measures' the surface length L which then cancels out of the normalised RCS (A3.66). In this way we obtain

$$\sigma^0 = \frac{k^3}{k_{0z}^2} \int dy \exp(2ik_{0y}y) \exp(-4k_{0z}^2 \langle \eta^2 \rangle (1 - \rho(y))) \tag{A3.69}$$

We note that the result is independent of polarisation; its small height limiting form relates the RCS directly to the power spectrum of the surface height fluctuations:

$$\begin{aligned}\sigma^0 &\approx 4k^3\langle\eta^2\rangle \int dy \exp(2ik_{0y}y)\rho(y) \\ &= 4\pi k^3(S(-2k_{0y}) + S(2k_{0y}))\end{aligned} \tag{A3.70}$$

Here we have omitted the specular reflection term that does not contribute to LGA back-scatter.

A3.5.2 Small height perturbation theory – PC case

One of the most striking features of LGA back scattering (and the resulting sea clutter) is its dependence on polarisation; VV and HH sea clutter are often markedly different in their mean RCS and fluctuation characteristics. The physical optics or Born approximation model we have just described does not reproduce this behaviour. The analysis leading to (A3.63) and (A3.69) can be extended in various ways that capture a polarisation dependence in the RCS, essentially by including some of the effects of multiple scattering. The first of these is the systematic generation of the small height perturbation expansion; in this we ensure that all terms of order $\langle\eta^2\rangle$ are retained in the analysis.

In the V pol case we go beyond the Kirchoff approximation, by iterating the surface field integral equation (A3.13). In this way we find that

$$\begin{aligned}\psi(\mathbf{r}) &= 2\psi_0(\mathbf{r}) - 2\int dS'\psi(\mathbf{r}')\mathbf{n}(\mathbf{r}')\cdot\nabla' G_0^{(1)}(\mathbf{r},\mathbf{r}') \\ &\approx 2\psi_0(\mathbf{r}) - 4\int dS'\psi_0(\mathbf{r}')\mathbf{n}(\mathbf{r}')\cdot\nabla' G_0^{(1)}(\mathbf{r},\mathbf{r}')\end{aligned} \tag{A3.71}$$

The first term is that leading to the RCS (A3.50); the contribution to the scattered field derived from the second can be written as

$$\psi_{\text{scat}}^{(2)}(\mathbf{R})_V = 4\int dS \int dS' \exp(i\mathbf{k}_0\cdot\mathbf{r}')\mathbf{n}(\mathbf{r}')\cdot\nabla' G_0^{(1)}(\mathbf{r},\mathbf{r}')\mathbf{n}(\mathbf{r})\cdot\nabla G_0^{(1)}(\mathbf{R},\mathbf{r}) \tag{A3.72}$$

We now introduce the far field form (A3.59) of the gradient of the Green's function $G_0^{(1)}(\mathbf{R},\mathbf{r})$; the Green's function $G_0^{(1)}(\mathbf{r},\mathbf{r}')$ that models scattering between points on the rough surface is represented by the Fourier integral (A3.18), appropriate to all separations of $\mathbf{r}$ and $\mathbf{r}'$. (Working with the Fourier representation leads to a simpler calculation than does the use of the explicit real space form of the Green's function favoured by Holliday [3].) This leads us to

$$\psi^{(2)}_{\text{scat}}(\mathbf{R})_V = -2\frac{i\exp(ikR)}{\sqrt{2\pi ikR}}\frac{1}{(2\pi)^2}\int d^2q\frac{1}{q^2-k^2-i\alpha}$$
$$\times\int dS\exp(i(\mathbf{q}-\mathbf{k})\cdot\mathbf{r})\mathbf{n}(\mathbf{r})\cdot\mathbf{k}$$
$$\times\int dS'\exp(i(\mathbf{k}_0-\mathbf{q})\cdot\mathbf{r}')\mathbf{n}(\mathbf{r}')\cdot\mathbf{q} \tag{A3.73}$$

which can be cast into a more compact form

$$\psi^{(2)}_{\text{scat}}(\mathbf{R})_V = -2\frac{i\exp(ikR)}{\sqrt{2\pi ikR}}\frac{1}{(2\pi)^2}\int d^2q\frac{1}{q^2-k^2-i\alpha}\left(k_z+\frac{k_y(q_y-k_y)}{q_z-k_Z}\right)$$
$$\times\left(q_Z+\frac{q_y(k_{0y}-q_y)}{k_{0Z}-q_z}\right)C(k_{0y}-q_y,k_{0Z}-q_z)$$
$$\times C(q_y-k_y,q_z-k_Z) \tag{A3.74}$$

by integrating by parts. Now we specialise this to the back scattering case: $\mathbf{k}=-\mathbf{k}_0$

$$\psi^{(2)}_{\text{bs}}(\mathbf{R})_V = 2i\frac{\exp(ikR)}{\sqrt{2\pi ikR}}\frac{1}{(2\pi)^2}\int d^2q\frac{1}{q^2-k^2-i\alpha}\left(k_{0z}+\frac{k_{0y}(q_y+k_{0y})}{q_z+k_{0z}}\right)$$
$$\times\left(q_Z+\frac{q_y(k_{0y}-q_y)}{k_{0Z}-q_z}C(k_{0y}-q_y,k_{0Z}-q_z)\right)$$
$$\times C(q_y+k_{0y},q_z+k_{0Z}) \tag{A3.75}$$

The small height expansion of the product of C functions (cf. (A3.64))

$$[C(k_{0y}-q_y,k_{0Z}-q_z)C(k_{0y}+q_y,k_{SZ}+q_Z)$$
$$\sim(2\pi)i\tilde{\eta}(2k_{0y})[(k_{0Z}+q_Z)\delta(k_{0y}-q_y)+(k_{0Z}-q_Z)\delta(k_{0y}+q_y)]$$

leads to

$$\psi^{(2)}_{\text{bs}}(\mathbf{R})_V = \frac{-i2\exp(ikR)}{\sqrt{2\pi ikR}}\tilde{\eta}(2k_{0y})\frac{1}{\pi}\int\frac{dq_z}{q_z^2-K_Z^2-i\alpha}(k^2q_Z-k_{0y}^2k_Z) \tag{A3.76}$$

The first integral in (A3.76) vanishes, as its integrand is an odd function; the second can be evaluated using Cauchy's theorem. Thus we find that

$$\psi^{(2)}_{\text{bs}}(\mathbf{R})_{\text{V}} = 2i\frac{\exp(ikR)}{\sqrt{2\pi ikR}}\tilde{\eta}(2k_{0y})k_{0y}^2 \tag{A3.77}$$

If we make the small height expansion of the KA result and combine the two, the total Vpol scattered field, to first order in the surface height fluctuations, is given by

$$\psi^{\text{SPM}}_{\text{bs}}(\mathbf{R})_{\text{V}} = 2i\frac{\exp(ikR)}{\sqrt{2\pi ikR}}\tilde{\eta}(2k_{0y})(k_{0y}^2+k^2) \tag{A3.78}$$

A similar calculation can be carried through for the Hpol case, but in this case we calculate ζ on the surface; for back-scatter and the small height approximation the contribution to the scattered field is the same. As the KA field has the opposite sign in this polarisation we find that

$$\psi_{\mathrm{bs}}^{\mathrm{SPM}}(\mathbf{R})_{\mathrm{H}} = 2\frac{i\exp(ikR)}{\sqrt{2\pi ikR}}\tilde{\eta}(2k_{0y})(k_{0y}^2 - k^2) \tag{A3.79}$$

Equations (A3.78) and (A3.79) lead to the following, rather different, expressions for the VV and HH pol RCS values:

$$\begin{aligned} \sigma_{\mathrm{HH}}^0 &= 4\pi k^3 \cos^4\theta_i (S(-2k_{0y}) + S(2k_{0y})) \\ \sigma_{\mathrm{VV}}^0 &= 4\pi k^3 (1 + \sin^2\theta_i)^2 (S(-2k_{0y}) + S(2k_{0y})) \end{aligned} \tag{A3.80}$$

An equivalent analysis of the scattering by a two-dimensional rough surface can be carried out, incorporating the Green's function (A3.20) and its Fourier representation. This leads us straightforwardly to the results

$$\begin{aligned} \text{“}\sigma_{\mathrm{VV}}^0\text{”} &= 8\pi k^4 (1 + \sin^2\theta_i)^2 (S(-2\mathbf{k}_{0\mathrm{H}}) + S(2\mathbf{k}_{0\mathrm{H}})) \\ \text{“}\sigma_{\mathrm{HH}}^0\text{”} &= 8\pi k^4 \cos^4\theta_i (S(-2\mathbf{k}_{0\mathrm{H}}) + S(2\mathbf{k}_{0\mathrm{H}})) \end{aligned} \tag{A3.81}$$

Here the quotation marks remind us that a scalar scattering theory is not, in general, a suitable vehicle for the discussion of EM scattering by a two-dimensional surface. In the case of perturbation theory, however, the results (A3.81) are identical with those obtained from a real space analysis of the vector Stratton Chu equations by Holliday [3].

A3.5.3 The half-space and reciprocal field formalisms

The iterative derivation of this perturbation theory result is rather involved, even in the relatively simple perfectly conducting case. As we will want to extend this analysis to the more complex, imperfectly conducting, case, it is useful to consider how it might be abbreviated. One way in which this can be done is that discussed, in the context of a scalar analysis of the scattering of sound by a rough surface, by Berman and Perkins [9]; the corresponding analysis of EM scattering is presented in a paper by Shaw and Dougan [10]. These authors propose a scattering theory based, not on a free space Green's function such as (A3.20), but on a Green's function, constructed by the method of images, that captures the appropriate $\psi = 0$ or $\zeta = 0$ boundary condition on the $z = 0$ planar interface. The derivation of the surface field integral equation (SFIE) sketched out in Section A3.1 does not depend critically on the introduction of the free space Green's function; any other Green's function (satisfying different boundary conditions) can be employed provided it obeys the equation

$$(\nabla^2 + k^2)G(\mathbf{r}, \mathbf{r}') = \delta(\mathbf{r} - \mathbf{r}') \tag{A3.82}$$

If we specialise to the PC Vpol case, where $\zeta(\mathbf{x}) = 0$ on the scattering surface, we obtain the expression

$$\psi(\mathbf{r}) = \psi_0(\mathbf{r}) - \int \psi(\mathbf{x}')\mathbf{n}(\mathbf{x}') \cdot \nabla' G(\mathbf{r}, \mathbf{x}') dS' \tag{A3.83}$$

for the field above the scattering surface. Here the first term in (A3.83) takes the form

$$\psi_0(\mathbf{r}) = \int \xi(\mathbf{r}')G(\mathbf{r}, \mathbf{r}')dV' \tag{A3.84}$$

Shaw and Dougan argue that more effective computational and approximate schemes can be derived from an integral formulation whose Green's function captures a significant part of the modelled scattering process. (We recall that, if G satisfied the Vpol boundary condition of a vanishing normal derivative at the scattering surface, then the evaluation of $\psi_0(\mathbf{r})$ solves the scattering problem exactly, as the integral contribution in (A3.83) vanishes identically.) The half space Green's function

$$\begin{aligned} G_{\text{HS}}(\mathbf{r}, \mathbf{r}') &= (G_0^{(1)}(\mathbf{r}, \mathbf{r}') + G_0^{(1)}(\tilde{\mathbf{r}}, \mathbf{r}')) = (G_0^{(1)}(\mathbf{r}, \mathbf{r}') + G_0^{(1)}(\mathbf{r}, \tilde{\mathbf{r}}')) \\ \tilde{\mathbf{r}} &= (\mathbf{1} - 2\mathbf{e}_z\mathbf{e}_z) \cdot \mathbf{r} \end{aligned} \tag{A3.85}$$

mimics, to a certain extent, the forward scattering by an extended rough surface; the remainder of the scattered field (which includes the practically important back-scatter) is contained in (cf. (A3.58))

$$\psi_{\text{scat}}(\mathbf{r}) = -\int \psi(\mathbf{x}')\mathbf{n}(\mathbf{x}') \cdot \nabla' G_{\text{HS}}(\mathbf{r}, \mathbf{x}') dS' \tag{A3.86}$$

So, given the Green's function (A3.85), which, in the large R limit, takes the form

$$G_{\text{HS}}(\mathbf{R}, \mathbf{x}) = -i\frac{\exp(ikR)}{\sqrt{8\pi ikR}}(\exp(-i\mathbf{k} \cdot \mathbf{x}) + \exp(-i\tilde{\mathbf{k}} \cdot \mathbf{x})) \tag{A3.87}$$

we can employ a reasonable approximation to the surface field, with the expectation that an improved representation of the scattered field will be obtained (relative to that obtained using the free space Green's function). Thus use of the tangent plane approximation (cf. (A3.57a)) $\psi(\mathbf{x}) = 2\exp(i\mathbf{k}_0 \cdot \mathbf{x})$ in (A3.86) should give us an approximation that is better than the Kirchoff or physical optics approximation (A3.63) obtained with the free space Green's function. Straightforward calculation then leads to

$$\psi_{\text{scat}}(\mathbf{R}) = \frac{\exp(ikR)}{\sqrt{2\pi ikR}} \begin{pmatrix} \left(k_z + \dfrac{(k_{0y} - k_y)k_y}{k_{0z} - k_z}\right) \tilde{C}(k_{0y} - k_y, k_{0z} - k_z) \\ -\left(k_z - \dfrac{(k_{0y} - k_y)k_y}{k_{0z} + k_z}\right) \tilde{C}(k_{0y} - k_y, k_{0z} + k_z) \end{pmatrix} \tag{A3.88}$$

In the back scattering limit, where $\mathbf{k} \to -\mathbf{k}_0$ this reduces to

$$\psi_{\text{scat}}(\mathbf{R}) = -2\frac{\exp(ikR)}{\sqrt{2\pi ikR}} \left(\frac{k^2}{k_{0z}}\tilde{C}(2k_{0y}, 2k_{0z}) + 2ik_{0y}^2\tilde{\eta}(2k_{0y})\right) \tag{A3.89}$$

whose small η expansion leads to the SPM result (A3.78). This confirms that use of a more realistic Green's function enhances the result obtained from the tangent plane, enabling it to capture multiple scattering events accessible only to iteration of the SFIE based on the free space form. In the Hpol case, the corresponding result obtained from the half space approach is

$$\psi_{\text{scat}}(\mathbf{R}) = 2\frac{\exp(ikR)}{\sqrt{2\pi ikR}}\left(\frac{k^2}{k_{0z}}\tilde{C}(2k_{0y}, 2k_{0z}) - 2ik_{0y}^2\tilde{\eta}(2k_{0y})\right) \tag{A3.90}$$

An attractive feature of these results [9] is their incorporation of the correct small perturbation limit and the Kirchoff approximation, which is valid for larger height perturbations away from grazing incidence. Shaw and Dougan have extended this analysis to the vector EM case [10] and adapted it to take account of the imperfect conductivity of the scattering medium [11]). Their observation that the use of a more 'realistic' Green's function simplifies much of the analysis suggests that this approach might be taken further.

To see how this might be done, we focus our attention on the PC problem once more, considering (A3.83) constructed from a Green's function G that provides the full solution of a reference problem, such as, for example, scattering by a smoothly varying surface. We assume that the source of the incident radiation is localised in the vicinity of $\mathbf{R}$, and write

$$\psi_0(\mathbf{x}) = \int G(\mathbf{x}, \mathbf{R}+\mathbf{s})\xi(\mathbf{R}+\mathbf{s})dV_s$$
$$\psi_{\text{scat}}(\mathbf{R}) = \psi(\mathbf{R}) - \psi_0(\mathbf{R}) = \int (G(\mathbf{R}, \mathbf{x}')\zeta(\mathbf{x}') - \psi(\mathbf{x}')\mathbf{n}(\mathbf{x}')\cdot\nabla' G(\mathbf{R}, \mathbf{x}'))dS' \tag{A3.91}$$

The free space and more refined Green's functions are related through an integral equation, obtained by integrating by parts,

$$G(\mathbf{R}, \mathbf{r}) = G_0^{(1)}(\mathbf{R}, \mathbf{r}) + \int d\mathbf{S}' \cdot (G_0^{(1)}(\mathbf{R}, \mathbf{r}')\nabla' G(\mathbf{r}', \mathbf{r})) - G(\mathbf{r}', \mathbf{r})\nabla' G_0^{(1)}(\mathbf{R}, \mathbf{r}')) \tag{A3.92}$$

from which it follows that

$$G(\mathbf{x}, \mathbf{R}+\mathbf{s}) \approx G(\mathbf{x}, \mathbf{R})\exp(i\mathbf{q}\cdot\mathbf{s}); \qquad \mathbf{q} = k\frac{\mathbf{R}}{R} \tag{A3.93}$$

Consequently we can write the reference field as

$$\psi_0(\mathbf{x}) \approx G(\mathbf{x}, \mathbf{R})Q(\mathbf{q}) \tag{A3.94}$$

where

$$Q(\mathbf{q}) = \int \xi(\mathbf{s})\exp(i\mathbf{q}\cdot\mathbf{s})dV_s \tag{A3.95}$$

We can then write the scattered field as

$$\psi_{\text{scat}}(\mathbf{R}) = \int (\zeta(\mathbf{x}')G(\mathbf{R},\mathbf{x}') - \psi(\mathbf{x}')\mathbf{n}(\mathbf{x}')\cdot\nabla' G(\mathbf{R},\mathbf{x}'))dS'$$
$$\approx \frac{1}{Q(\mathbf{q})}\int (\zeta(\mathbf{x}')\psi_0(\mathbf{x}') - \psi(\mathbf{x}')\mathbf{n}(\mathbf{x}')\cdot\nabla'\psi_0(\mathbf{x}'))dS' \qquad \text{(A3.96)}$$

with the reference field playing the role of the Green's function specifying the scattered field back at the point of transmission – in radar terms, the antenna. In doing this we are exploiting the reciprocity of the Helmholtz equations Green's function

$$G(\mathbf{r},\mathbf{r}') = G(\mathbf{r}',\mathbf{r}) \qquad \text{(A3.97)}$$

This expression for the scattered field can be used to calculate a scattering cross section; $Q(\mathbf{q})$ is eliminated with respect to an incident intensity, calculated using the free space Green's function. The explicit results for two-dimensional and corrugated surfaces are

$$\text{“}\sigma^0\text{”} = \lim_{R\to\infty}\frac{4\pi R^2|\psi_{\text{scat}}|^2}{A}$$
$$= \frac{1}{4\pi A}\left|\int (\zeta(\mathbf{x}')\psi_0(\mathbf{x}') - \psi(\mathbf{x}')\mathbf{n}(\mathbf{x}')\cdot\nabla'\psi_0(\mathbf{x}'))dS'\right|^2$$
$$\sigma^0 = \lim_{R\to\infty}\frac{2\pi R|\psi_{\text{scat}}|^2}{L}$$
$$= \frac{1}{4kL}\left|\int (\zeta(\mathbf{x}')\psi_0(\mathbf{x}') - \psi(\mathbf{x}')\mathbf{n}(\mathbf{x}')\cdot\nabla'\psi_0(\mathbf{x}'))dS'\right|^2 \qquad \text{(A3.98)}$$

Here ψ_0, $\mathbf{n}\cdot\nabla\psi_0$ are the field and its normal gradient induced by a plane wave of unit amplitude, incident upon the reference surface whose scattering is described by the Green's function G. It is interesting to note that a result equivalent to this was derived many years ago by Kerr [12], who exploited the electromagnetic Lorentz reciprocity explicitly, and so avoided the need for the Green's function formalism that was under development at the same time [13]. Of course, (A3.98) still requires estimates of the surface field and its normal gradient if it is to be used to calculate the scattering effectively. In the case of a small perturbation about the reference surface we can develop these quantities in power series, and derive SPM and related results in an economical manner. While this provides us with nothing new in the slightly rough planar PC scatterer case, the analysis we present is very useful in describing scattering by a roughened surface of arbitrary shape and imperfect conductivity.

We now consider the surface integral

$$\int (\zeta(\mathbf{x}')\psi_0(\mathbf{x}') - \psi(\mathbf{x}')\mathbf{n}(\mathbf{x}')\cdot\nabla'\psi_0(\mathbf{x}'))dS' \qquad \text{(A3.99)}$$

in a little more detail. To simplify matters, we first take the reference surface to be a plane; ψ_0, $\mathbf{n}\cdot\nabla\psi_0$ are then given by (A3.57). The true surface field and its normal gradient can be expanded in Taylor series around these reference fields, so that the integral (A3.99) becomes

$$\int d^2x \Big(\zeta_0(\mathbf{x})(\psi_0(\mathbf{x}) + \eta(\mathbf{x})\zeta_0(\mathbf{x})) - \psi_0(\mathbf{x})(\mathbf{e}_z - \nabla_H \eta(\mathbf{x})) \\ \times (\nabla\psi_0(\mathbf{x}) + \eta(\mathbf{x})\frac{\partial}{\partial z}\nabla\psi_0(\mathbf{x})) \Big) \\ = \int d^2x \left(\eta(\mathbf{x})\zeta_0(\mathbf{x})^2 + \psi_0(\mathbf{x})\nabla_H\eta(\mathbf{x}) \cdot \nabla_H\psi_0(\mathbf{x}) - \psi_0(\mathbf{x})\eta(\mathbf{x})\frac{\partial^2}{\partial z^2}\psi_0(\mathbf{x}) \right) \tag{A3.100}$$

where we ultimately retain only terms of first order in the surface height perturbation. As ψ_0 satisfies the Helmholtz equation the second derivative with respect to z can be eliminated through

$$\frac{\partial^2}{\partial z^2}\psi_0 = -(k^2 + \nabla_H^2)\psi_0 \\ \nabla_H^2 \equiv \nabla \cdot (\mathbf{1} - \mathbf{e}_z\mathbf{e}_z) \cdot \nabla \tag{A3.101}$$

This is substituted into (A3.100); an integration by parts gives the lowest order approximation to the scattering integral (A3.99) as

$$\int d^2x \eta(\mathbf{x})(\zeta_0(\mathbf{x})^2 - k^2\psi_0(\mathbf{x})^2 + \nabla_H\psi_0(\mathbf{x}) \cdot \nabla_H\psi_0(\mathbf{x})) \tag{A3.102}$$

Here we have presented our analysis in terms of a planar reference surface and Cartesian co-ordinates. This restriction is not necessary; the general development, couched in terms of co-ordinates induced by the reference surface, is essentially an exercise in tensor analysis [14] that we will not pursue here. The resulting expression for the scattering cross section takes the form

$$\text{“}\sigma^0_{\mathrm{HH}}\text{”} = \lim_{R\to\infty} \frac{4\pi R^2|\psi_{\mathrm{scat}}|^2}{A} = \frac{1}{4\pi A}\left| \int \zeta_0(\mathbf{x})^2\eta(\mathbf{x})dS_0 \right|^2 \tag{A3.103}$$

in the Hpol case and

$$\text{“}\sigma^0_{\mathrm{VV}}\text{”} = \lim_{R\to\infty} \frac{4\pi R^2|\psi_{\mathrm{scat}}|^2}{A} \\ = \frac{1}{4\pi A}\left| \int \eta(\mathbf{x})(\nabla_{\mathrm{T}}\psi_0(\mathbf{x}) \cdot \nabla_{\mathrm{T}}\psi_0(\mathbf{x}) - k^2\psi_0(\mathbf{x})^2)dS_0 \right|^2 \tag{A3.104}$$

for Vpol. Here $\eta(\mathbf{x})$ is the perturbation measured normal to the reference surface S_0, which contains the point $\mathbf{x}$; ∇_{T} denotes the component of the gradient tangential to the reference surface.

The corresponding results for a general corrugated surface take the forms

$$\sigma^0_{\mathrm{HH}} = \lim_{R\to\infty} \frac{2\pi R|\psi_{\mathrm{scat}}|^2}{L} = \frac{1}{4kL}\left|\int \zeta_0(\mathbf{x})^2\eta(\mathbf{x})dS_0\right|^2$$

$$\sigma^0_{\mathrm{VV}} = \lim_{R\to\infty} \frac{2\pi R|\psi_{\mathrm{scat}}|^2}{L}$$

$$= \frac{1}{4kL}\left|\int \eta(\mathbf{x})(\nabla_{\mathrm{T}}\psi_0(\mathbf{x})\cdot\nabla_{\mathrm{T}}\psi_0(\mathbf{x}) - k^2\psi_0(\mathbf{x})^2)dS_0\right|^2 \tag{A3.105}$$

It is a relatively straightforward matter to verify that these formulae reproduce PC small perturbation results in the case of a planar perfectly conducting reference surface.

A3.6 Scattering by an imperfectly conducting surface: small height perturbation theory

We have just seen how the exploitation of reciprocity, which casts the field established on a perfectly conducting reference surface in the role of a Green's function describing scattering in the presence of that surface, provides an effective method for the analysis of scattering by a superimposed small scale roughness. The modelling of scattering by an imperfectly conducting surface is necessarily more complicated; nonetheless it is possible to extend our earlier analysis to this practically important case.

We now consider the surface field integral equations of equation (A3.12)

$$\psi(\mathbf{x}) = 2\psi_0(\mathbf{x}) + 2\int (G^{(1)}(\mathbf{x},\mathbf{x}')\zeta(\mathbf{x}') - \psi(\mathbf{x}')\mathbf{n}(\mathbf{x}')\cdot\nabla' G^{(1)}(\mathbf{x},\mathbf{x}'))dS'$$

$$\psi(\mathbf{x}) = -2\int (\alpha G^{(2)}(\mathbf{x},\mathbf{x}')\zeta(\mathbf{x}') - \psi(\mathbf{x}')\mathbf{n}(\mathbf{x}')\cdot\nabla' G^{(2)}(\mathbf{x},\mathbf{x}'))dS' \tag{A3.106}$$

where the reference field is established as in (A3.91). The scattered field is given by

$$\psi_{\mathrm{scat}}(\mathbf{R}) = \int (G^{(1)}(\mathbf{R},\mathbf{x}')\zeta(\mathbf{x}') - \psi(\mathbf{x}')\mathbf{n}(\mathbf{x}')\cdot\nabla' G^{(1)}(\mathbf{R},\mathbf{x}'))dS' \tag{A3.107}$$

As the point $\mathbf{R}$ is above the ocean we also have the extended boundary condition

$$0 = \int (\alpha G^{(2)}(\mathbf{R},\mathbf{x}')\zeta(\mathbf{x}') - \psi(\mathbf{x}')\mathbf{n}(\mathbf{x}')\cdot\nabla' G^{(2)}(\mathbf{R},\mathbf{x}'))dS' \tag{A3.108}$$

Together (A3.107) and (A3.108) give us an expression for the scattered field in terms of the discontinuity in the Green's function across the scattering interface

$$\psi_{\mathrm{scat}}(\mathbf{R}) = \int \Big(\zeta(\mathbf{x}')(G^{(1)}(\mathbf{R},\mathbf{x}') - G^{(2)}(\mathbf{R},\mathbf{x}'))$$

$$- \psi(\mathbf{x}')\mathbf{n}(\mathbf{x}')\cdot\nabla'(G^{(1)}(\mathbf{R},\mathbf{x}') - \frac{1}{\alpha}G^{(2)}(\mathbf{R},(\mathbf{x}'))\Big)dS' \tag{A3.109}$$

Just as in our discussion of the PC case the Green's functions can be replaced by the surface fields for the reference problem.

$$\psi_{\text{scat}}(\mathbf{R}) = \frac{1}{Q(\mathbf{k})} \int \Bigg(\zeta(\mathbf{x}')(\psi_0^{(1)}(\mathbf{x}') - \psi_0^{(2)}(\mathbf{x}')) \\ - \psi(\mathbf{x}')\mathbf{n}(\mathbf{x}') \cdot \nabla'(\psi_0^{(1)}(\mathbf{x}') - \frac{1}{\alpha}\psi_0^{(2)}(\mathbf{x}')) \Bigg) dS' \tag{A3.110}$$

We now develop the fields either side of the interface, much as before: the field itself is continuous across the interface, while its normal derivative experiences a discontinuity. Thus we have the Taylor expansions and boundary conditions

$$\begin{aligned}
\psi_0^{(1)}(\mathbf{x}) &= \psi_0^{(1)}(\mathbf{x}_0) + \eta(\mathbf{x}_0)\zeta_0^{(1)}(\mathbf{x}_0) \\
\psi_0^{(2)}(\mathbf{x}) &= \psi_0^{(2)}(\mathbf{x}_0) + \eta(\mathbf{x}_0)\zeta_0^{(2)}(\mathbf{x}_0) \\
\psi_0^{(1)}(\mathbf{x}_0) &= \psi_0^{(2)}(\mathbf{x}_0), \\
\zeta_0^{(2)}(\mathbf{x}_0) &= \alpha\zeta_0^{(1)}(\mathbf{x}_0)
\end{aligned} \tag{A3.111}$$

which tell us that

$$\begin{aligned}
&(\psi_0^{(1)}(\mathbf{x}') - \psi_0^{(2)}(\mathbf{x}')) \\
&\mathbf{n}(\mathbf{x}') \cdot \nabla' \left(\psi_0^{(1)}(\mathbf{x}') - \frac{1}{\alpha}\psi_0^{(2)}(\mathbf{x}') \right)
\end{aligned} \tag{A3.112}$$

are both linear in $\eta(\mathbf{x}_0)$, so that $\zeta(\mathbf{x}')$, $\psi(\mathbf{x}')$ can be replaced by $\psi_0^{(1)}(\mathbf{x}_0)$, $\zeta_0^{(1)}(\mathbf{x}_0)$ in a perturbation theory. The analysis now proceeds much as in the perfectly conducting case. Thus we have

$$\zeta(\mathbf{x})(\psi_0^{(1)}(\mathbf{x}) - \psi_0^{(2)}(\mathbf{x})) \approx \zeta_0^{(1)}(\mathbf{x}_0)^2\eta(\mathbf{x}_0)(1-\alpha) \tag{A3.113}$$

The second term in the integrand is developed in much the same way

$$\begin{aligned}
&\psi(\mathbf{x})\mathbf{n}(\mathbf{x}) \cdot \nabla \left(\psi_0^{(1)}(\mathbf{x}) - \frac{1}{\alpha}\psi_0^{(2)}(\mathbf{x}) \right) \\
&\quad \approx \hat{\psi}^{(1)}(\mathbf{x}_0) \Bigg[\nabla_H \eta(\mathbf{x}_0) \cdot \left(\nabla_H \psi_0^{(1)}(\mathbf{x}_0) - \frac{1}{\alpha}\nabla_H \psi_0^{(2)}(\mathbf{x}_0) \right) \\
&\qquad + \eta(\mathbf{x}_0) \frac{\partial^2}{\partial z^2}(\psi_0^{(1)}(\mathbf{x}_0) - \frac{1}{\alpha}\psi_0^{(2)}(\mathbf{x}_0)) \Bigg]
\end{aligned} \tag{A3.114}$$

The Helmholtz equation tells us that

$$\frac{\partial^2}{\partial z^2}\left(\psi_0^{(1)}(\mathbf{x}_0) - \frac{1}{\alpha}\psi_0^{(2)}(\mathbf{x}_0)\right)$$
$$= -\left(k^{(1)^2}\psi_0^{(1)}(\mathbf{x}_0) - \frac{1}{\alpha}k^{(2)^2}\psi_0^{(2)}(\mathbf{x}_0) + \nabla_{\mathrm{H}}^2\left(\psi_0^{(1)}(\mathbf{x}_0) - \frac{1}{\alpha}\psi_0^{(2)}(\mathbf{x}_0)\right)\right) \tag{A3.115}$$

which the boundary conditions at the surface reduce to

$$\frac{\partial^2}{\partial z^2}\left(\psi_0^{(1)}(\mathbf{x}_0) - \frac{1}{\alpha}\psi_0^{(2)}(\mathbf{x}_0)\right)$$
$$= -k^2\left(1 - \frac{\varepsilon}{\alpha}\right)\psi_0^{(1)}(\mathbf{x}_0) - \frac{\alpha - 1}{\alpha}\nabla_{\mathrm{H}}^2\psi_0^{(1)}(\mathbf{x}_0) \tag{A3.116}$$

Bringing these results together we find that

$$\zeta(\mathbf{x})(\psi_0^{(1)}(\mathbf{x}) - \psi_0^{(2)}(\mathbf{x}))$$
$$- \psi(\mathbf{x})\mathbf{n}(\mathbf{x}) \cdot \nabla\left(\psi_0^{(1)}(\mathbf{x}) - \frac{1}{\alpha}\psi_0^{(2)}(\mathbf{x})\right)$$
$$\approx \eta(\mathbf{x}_0)\left[(1-\alpha)\zeta_0^{(1)}(\mathbf{x}_0)^2 + k^2\frac{\alpha - \varepsilon}{\alpha}\psi_0^{(1)}(\mathbf{x}_0)^2\right.$$
$$\left. - \frac{\alpha - 1}{\alpha}\nabla_{\mathrm{H}}\psi_0^{(1)}(\mathbf{x}_0) \cdot \nabla_{\mathrm{H}}\psi_0^{(1)}(\mathbf{x}_0)\right] \tag{A3.117}$$

This allows us to write the scattering integral in the relatively compact form

$$\int (\zeta(\mathbf{x}')(\psi_0^{(1)}(\mathbf{x}') - \alpha\psi_0^{(2)}(\mathbf{x}'))$$
$$- \psi(\mathbf{x}')\mathbf{n}(\mathbf{x}') \cdot \nabla'(\psi_0^{(1)}(\mathbf{x}') - \psi_0^{(2)}(\mathbf{x}')))dS'$$
$$\approx -k^2(\varepsilon - 1)\int \eta(\mathbf{x}_0)\psi_0^{(1)}(\mathbf{x}_0)^2\,dS_0, \quad \alpha = 1$$
$$- (\varepsilon - 1)\int \eta(\mathbf{x}_0)\left[\zeta_0^{(1)}(\mathbf{x}_0)^2 + \frac{1}{\varepsilon}\nabla_{\mathrm{H}}\psi_0^{(1)}(\mathbf{x}_0) \cdot \nabla_{\mathrm{H}}\psi_0^{(1)}(\mathbf{x}_0)\right]dS_0 \quad \alpha = \varepsilon \tag{A3.118}$$

These expressions can be generalised to the arbitrary curved reference surface by inspection, and reference to the PC case.

The expression for the scattering cross section in this imperfect conductor case is

$$
\begin{aligned}
``\sigma^{0}" &= \lim_{R\to\infty} \frac{4\pi R^2 |\psi_{\text{scat}}|^2}{|\psi_{\text{in}}|^2 A} \\
&= \frac{1}{4\pi A}\left| \int (\zeta(\mathbf{x}')(\psi_0^{(1)}(\mathbf{x}') - \alpha\psi_0^{(2)}(\mathbf{x}')) \right. \\
&\quad \left. - \psi(\mathbf{x}')\mathbf{n}(\mathbf{x}')\cdot\nabla'(\psi_0^{(1)}(\mathbf{x}') - \psi_0^{(2)}(\mathbf{x}')))dS' \right|^2 \\
&\approx \frac{k^4(\varepsilon-1)^2}{4\pi A}\left| \int \eta(\mathbf{x}_0)\psi_0^{(1)}(\mathbf{x}_0)^2 dS_0 \right|^2; \qquad \alpha = 1 \\
&\approx \frac{(\varepsilon-1)^2}{4\pi A}\left| \int \eta(\mathbf{x}_0)\left[\zeta_0^{(1)}(\mathbf{x}_0)^2 \right.\right. \\
&\quad \left.\left. + \frac{1}{\varepsilon}\nabla_{\mathrm{T}}\psi_0^{(1)}(\mathbf{x}_0)\cdot\nabla_T\psi_0^{(1)}(\mathbf{x}_0)\right] dS_0 \right|^2; \qquad \alpha = \varepsilon
\end{aligned} \tag{A3.119}
$$

while in the case of a corrugated surface we have

$$
\begin{aligned}
\sigma^{0} &= \lim_{R\to\infty} \frac{2\pi R |\psi_{\text{scat}}|^2}{L} \\
&= \frac{1}{4kL}\left| \int \left(\zeta(\mathbf{x}')(\psi_0^{(1)}(\mathbf{x}') - \alpha\psi_0^{(2)}(\mathbf{x}')) \right.\right. \\
&\quad \left.\left. -\psi(\mathbf{x}')\mathbf{n}(\mathbf{x}')\cdot\nabla'(\psi_0^{(1)}(\mathbf{x}') - \psi_0^{(2)}(\mathbf{x}'))\right) dS' \right|^2 \\
&\approx \frac{k^4(\varepsilon-1)^2}{4kL}\left| \int \eta(\mathbf{x}_0)\psi_0^{(1)}(\mathbf{x}_0)^2 dS_0 \right|^2; \qquad \alpha = 1 \\
&\approx \frac{(\varepsilon-1)^2}{4kL}\left| \int \eta(\mathbf{x}_0)[\zeta_0^{(1)}(\mathbf{x}_0)^2 + \frac{1}{\varepsilon}\nabla_{\mathrm{T}}\psi_0^{(1)}(\mathbf{x}_0)\cdot\nabla_{\mathrm{T}}\psi_0^{(1)}(\mathbf{x}_0)] dS_0 \right|^2; \\
&\quad \alpha = \varepsilon
\end{aligned} \tag{A3.120}
$$

Here we have a simple and general expression for scattering by small-scale roughness, superimposed on a smoothly varying surface that supports known values of the field and its normal gradient. This analysis is used as the basis of numerical calculations in Chapter 3. If (A3.119) is specialised to a slightly roughened two-dimensional planar interface we obtain the 'RCS' as

$$
\begin{aligned}
``\sigma^0_{\mathrm{HH}}" &= 8\pi k^4 \cos^4\theta(\mathbf{k}_0)(S_s(-2\mathbf{k}_{0\mathrm{H}}) + S_s(2\mathbf{k}_{0\mathrm{H}}))|g_{\mathrm{HH}}|^2 \\
``\sigma^0_{\mathrm{VV}}" &= 8\pi k^4 \cos^4\theta(\mathbf{k}_0)(S_s(-2\mathbf{k}_{0\mathrm{H}}) + S_s(2\mathbf{k}_{0\mathrm{H}}))|g_{\mathrm{VV}}|^2 \\
g_{\mathrm{HH}} &= \frac{\varepsilon - 1}{(\cos\theta(\mathbf{k}_0) + \sqrt{\varepsilon - 1 + \cos^2\theta(\mathbf{k}_0)})^2}, \\
g_{\mathrm{VV}} &= \frac{(\varepsilon - 1)(\varepsilon(1 + \sin^2\theta(\mathbf{k}_0)) - \sin^2\theta(\mathbf{k}_0))}{(\varepsilon\cos\theta(\mathbf{k}_0) + \sqrt{\varepsilon - 1 + \cos^2\theta(\mathbf{k}_0)})^2} \\
\mathbf{k}_{0\mathrm{H}} &= \mathbf{k}_0 \cdot (\mathbf{1} - \mathbf{e}_z\mathbf{e}_z); \qquad \cos\theta(\mathbf{k}_0) = \frac{-\mathbf{k}_0 \cdot \mathbf{e}_z}{k}
\end{aligned}
\tag{A3.121}
$$

The factors g_{HH}, g_{VV} incorporate the Fresnel coefficients derived in Section A3.3. The corresponding results for a corrugated, imperfectly conducting surface are

$$
\begin{aligned}
\sigma^0_{\mathrm{HH}} &= 4\pi k^3 \cos^4\theta\,(S_s(-2k_{0y}) + S_s(2k_{0y}))|g_{\mathrm{HH}}|^2 \\
\sigma^0_{\mathrm{VV}} &= 4\pi k^3 \cos^4\theta(S_s(-2k_{0y}) + S_s(2k_{0y}))|g_{\mathrm{VV}}|^2
\end{aligned}
\tag{A3.122}
$$

Here S_s is the power spectrum of the small-scale surface height fluctuations η and is related to their auto correlation function through the Wiener Khintchine theorem (see Section A1.10).

These perturbation theory results were first obtained by Peake [15]. We note that, while the results (A3.121) have been derived here from a scalar scattering theory, they are, much as in the perfectly conducting case, identical with those obtained from the full vector analysis of EM scattering from a two-dimensional rough surface.

A recurring theme of our discussion of radar scattering by the ocean has been the wide range of spatial and temporal scales that characterise both the sea surface and the correlation properties of the clutter. The two-scale or composite model provides a simple caricature of this complex situation. In this, scattering by the small-scale structure is described by the perturbation theory results, based on a planar interface that we have just derived. In Section 3.5 we discuss how this resonant scattering is modulated by the large scale structure of the sea surface, captured by the long wavelength, small wave number contribution to the power spectrum of the sea surface height fluctuations. This model can be specialised to the case where the sea surface is corrugated; the only significant change is that induced by the replacement of $8\pi k^4 \cos^4\theta(\mathbf{k})[S_s(2(\mathbf{k}_{\mathrm{H}}(\mathbf{n}) + k_z\nabla\eta)) + S_s(-2(\mathbf{k}_{\mathrm{H}}(\mathbf{n}) + k_z\nabla\eta))]$, in the two-dimensional case, by $4\pi k^3 \cos^4\theta(\mathbf{k})[S_s(2(k_y(\mathbf{n}) + k_z\partial\eta(y)/\partial y)) + S_s(-2(k_y(\mathbf{n}) + k_z\partial\eta(y)/\partial y))]$, in the corrugated case. The underlying arguments justifying and qualifying the use of the composite model are of course unchanged.

A3.7 Numerical solutions of the scattering problem

The approximate analyses of scattering by rough surfaces described so far provide a useful vehicle for the qualitative discussion of the generation of radar sea clutter.

However, none of them fully captures the shadowing and multiple-scattering effects that we expect to be of considerable importance in the LGA regime. In particular, the analysis of the contributions made to the clutter by returns from specific ocean surface features will require an altogether more detailed approach, provided, for example, by the direct calculation of the EM scattering from an incipiently breaking wave. Numerical methods that allow us to do this will be described in this section.

A3.7.1 Scattering from a perfect conductor

The SFIE formulation of the scattering problem also allows us to derive and discuss numerical methods for the solution of the LGA scattering problem. In the PC special case the equations satisfied by the field and its normal gradient are automatically de-coupled, so that a single linear integral equation can be solved to determine the surface field, from which the scattered field is calculated by quadrature. The numerical implementation of the integrations implicit in (A3.13) and (A3.14) proceeds via discretisation, typically though the trapezium rule [16]; this converts the integral equation into a set of linear equations that can be represented in matrix form as

$$\mathbf{f} = \mathbf{f}_0 + \mathbf{M} \cdot \mathbf{f} \tag{A3.123}$$

(The detailed identification of $\mathbf{f}$, $\mathbf{f}_0$ and $\mathbf{M}$ will be discussed shortly.) Thus the solution of the SFIE is reduced to the inversion of the matrix $\mathbf{M}$. When the scatterer is of finite spatial extent this matrix will itself be of finite size, and standard techniques of numerical linear algebra can be brought to bear on the problem. This, in essence, is the method of moments [17] that has been applied widely in the calculation of RCS characteristics [18]. In our case, where the scatterer is the sea surface, and is illuminated at low grazing angles, significant problems arise in the application of this method. An effectively infinite scattering object cannot be accommodated realistically in a computer of finite capacity; something must be done to reduce the sample of scattering surface to a manageable size. If one were merely to 'chop out' a finite portion of surface, and calculate its RCS, artefacts corresponding to scattering from its sharp edges would dominate the results, particularly in cases where the back-scattering from the surface is relatively weak. In recent years significant progress has been made in the effective elimination of these edge effects, so that it is now possible to calculate the RCS of realistic sea surface features in LGA geometries.

Perhaps the most successful approach to this problem has been Holliday *et al.*'s development of equations of the general form (A3.123), that describe the scattering by a finite rough surface, supplemented by adjunct planes that extend to infinity. The contributions of these supplementary surfaces can be calculated, assuming that they support the surface fields established on an infinite planar interface. These are then incorporated in the driving term $\mathbf{f}_0$; the matrix $\mathbf{M}$ now need only describe the interactions of the finite portion of ocean surface with the incident radiation and that scattered from the adjunct planes. Other approaches to this problem have been suggested, including the incorporation of a weighted incident beam [19] and the elimination of edge effects by periodic replication, used in conjunction with Petit's Fourier based methods [20].

To motivate this discussion we will consider the solution of (A3.123) in a little more detail. In principle a solution can be obtained straightforwardly once we have constructed the inverse of the matrix $\mathbf{1} - \mathbf{M}$

$$\mathbf{f} = (\mathbf{1} - \mathbf{M})^{-1} \cdot \mathbf{f}_0 \tag{A3.124}$$

The direct generation of this inverse is nonetheless a non-trivial task and presents a heavy computational burden when $\mathbf{M}$ becomes reasonably large. An iterative solution of (A3.123), that represents $(\mathbf{1} - \mathbf{M})^{-1}$ as a geometric series

$$(\mathbf{1} - \mathbf{M})^{-1} = \mathbf{1} + \sum_{n=1}^{\infty} \mathbf{M}^n \tag{A3.125}$$

is computationally more convenient, if the iterative process converges reasonably quickly. This approach to the solution of (A3.123) is none other than Jacobi's method [21] and can be applied when the eigenvalues of $\mathbf{M}$ are sufficiently small. In general, however, rapid convergence cannot be assured; we might sensibly expect it to be effective only in the case of very weak scattering. There is, however, a situation in which the equation (A3.123) can be solved non-iteratively. Should $\mathbf{M}$ have either an upper

$$\mathbf{M} = \mathbf{U}; \qquad U_{ij} = 0, \qquad j < i \tag{A3.126}$$

or lower

$$\mathbf{M} = \mathbf{L}; \qquad L_{ij} = 0, \qquad N \geq j > i \tag{A3.127}$$

triangular form, it can be inverted in a finite number of steps. Thus

$$\mathbf{f} = \mathbf{f}_0 + \mathbf{U} \cdot \mathbf{f} \tag{A3.128}$$

is equivalent to the set of linear equations

$$f_i = f_{0i} + \sum_j U_{ij} f_j \tag{A3.129}$$

which, in turn, given the upper triangular structure of $\mathbf{U}$, reduce to

$$\begin{aligned} f_1 &= f_{01} \\ f_2 &= f_{02} + U_{21} f_1 \\ &\vdots \\ f_N &= f_{0N} + U_{N1} f_1 + U_{N2} f_2 \cdots U_{N,N-1} f_{N-1} \end{aligned} \tag{A3.130}$$

We can solve the first of (A3.130) immediately; the resulting f_1 can now give us f_2; $\mathbf{f}$ can thus be calculated by successive substitutions of this type. The structure of the inverse of $\mathbf{U}$ generated in this way is manifest in the termination of the binomial expansion (3.125), as a consequence of

$$(\mathbf{U}^m)_{ij} = 0, \qquad m \geq N \tag{A3.131}$$

Should $\mathbf{M} = \mathbf{L}$, (A3.123) can be handled in much the same way. Moving on, we now consider $\mathbf{M}$ to be a sum of upper and lower triangular parts, the latter being a small perturbation to the former. In this case we can write

$$\begin{aligned} \mathbf{f} &= \mathbf{f}_0 + (\mathbf{U} + \mathbf{L}) \cdot \mathbf{f} \\ (\mathbf{1} - \mathbf{U}) \cdot \mathbf{f} &= \mathbf{f}_0 + \mathbf{L} \cdot \mathbf{f} \end{aligned} \tag{A3.132}$$

Equation (A3.130) showed that $(\mathbf{1} - \mathbf{U})$ can be inverted directly; we can cast (A3.132) into a form suitable for solution by iteration

$$\mathbf{f} = (\mathbf{1} - \mathbf{U})^{-1} \cdot \mathbf{f}_0 + (\mathbf{1} - \mathbf{U})^{-1} \cdot \mathbf{L} \cdot \mathbf{f} \tag{A3.133}$$

Thus we expect

$$\mathbf{f} = \sum_n ((\mathbf{1} - \mathbf{U})^{-1} \cdot \mathbf{L})^n \cdot (\mathbf{1} - \mathbf{U})^{-1} \cdot \mathbf{f}_0 \tag{A3.134}$$

to converge rapidly. This method of matrix inversion is again quite standard, and is known as Gauss Seidel iteration [21]. Together (A3.132) and (A3.133) are equivalent to the operator identity

$$(\mathbf{1} - \mathbf{U} - \mathbf{L})^{-1} = (\mathbf{1} - \mathbf{U})^{-1} + (\mathbf{1} - \mathbf{U})^{-1} \cdot \mathbf{L} \cdot (\mathbf{1} - \mathbf{U} - \mathbf{L})^{-1} \tag{A3.135}$$

widely used in the generation of perturbation expansions [22].

As we shall see shortly, we can associate upper and lower diagonal matrices with forward and backward scattering processes. Thus we should expect the iterative procedure (3.134) to be particularly effective should forward scattering be the dominant process. If, however, both forward and backward scattering processes contribute significantly, a more rapidly convergent iterative solution can be constructed. The operator identity

$$\mathbf{1} - \mathbf{U} - \mathbf{L} = (\mathbf{1} - \mathbf{U}) \cdot (\mathbf{1} - \mathbf{L}) - \mathbf{U} \cdot \mathbf{L} \tag{A3.136}$$

gives us

$$(\mathbf{1} - \mathbf{U} - \mathbf{L}) \cdot \mathbf{f} = (\mathbf{1} - \mathbf{U}) \cdot (\mathbf{1} - \mathbf{L}) \cdot \mathbf{f} - \mathbf{U} \cdot \mathbf{L} \cdot \mathbf{f} = \mathbf{f}_0 \tag{A3.137}$$

Then (A3.135) can be recast as

$$\begin{aligned} &(\mathbf{1} - \mathbf{U} - \mathbf{L})^{-1} \\ &\quad = (\mathbf{1} - \mathbf{L})^{-1} \cdot (\mathbf{1} - \mathbf{U})^{-1} + (\mathbf{1} - \mathbf{L})^{-1} \cdot (\mathbf{1} - \mathbf{U})^{-1} \cdot \mathbf{U} \cdot \mathbf{L} \cdot (\mathbf{1} - \mathbf{U} - \mathbf{L})^{-1} \end{aligned} \tag{A3.138}$$

We note that $\mathbf{U}$ and $\mathbf{L}$ appear on an 'equal footing' here. The series generated by iteration

$$\mathbf{f} = \sum_{n=0}^{\infty} ((\mathbf{1} - \mathbf{L})^{-1} \cdot (\mathbf{1} - \mathbf{U})^{-1} \cdot \mathbf{U} \cdot \mathbf{L})^n \cdot (\mathbf{1} - \mathbf{L})^{-1} \cdot (\mathbf{1} - \mathbf{U})^{-1} \cdot \mathbf{f}_0 \tag{A3.139}$$

might reasonably be expected to converge more rapidly in this case, than either the Jacobi (A3.125) or Gauss Seidel (A3.134) solution.

We now consider how we might apply these iterative techniques to the solution of the Vpol SFIE (A3.13). The ocean surface is represented as $z = \eta(y)$, with y taking values $-\infty < y < \infty$. The regions $y < -L/2$, $y > L/2$ correspond to the adjunct planes; in these regions $\eta = 0$. Thus we can re-write the SFIE, incorporating the contributions in these regions into the inhomogeneous term in the integral equation:

$$\psi(y\mathbf{e}_y) = 2\psi_{\text{in}}(y\mathbf{e}_y); \qquad |y| > L/2$$

$$\begin{aligned}\psi(\mathbf{x}) = 2\psi_{\text{in}}(\mathbf{x}) &- 4\int_{-\infty}^{-L/2} \bar{G}_0^{(1)}(\mathbf{x}, \mathbf{e}_y y)\psi_{\text{in}}(\mathbf{e}_y y)dy \\ &- 4\int_{L/2}^{\infty} \bar{G}_0^{(1)}(\mathbf{x}, \mathbf{e}_y y)\psi_{\text{in}}(\mathbf{e}_y y)dy - 2\int_{|y|<L/2} \bar{G}_0^{(1)}(\mathbf{x}, \mathbf{x}')\psi(\mathbf{x}')dS'\end{aligned} \tag{A3.140}$$

where

$$\bar{G}_0^{(1)}(\mathbf{x}, \mathbf{x}') \equiv \mathbf{n}(\mathbf{x}') \cdot \nabla' G_0^{(1)}(\mathbf{x}, \mathbf{x}') \tag{A3.141}$$

To make further progress we must now identify the explicit forms of ψ_{in} and $\bar{G}_0^{(1)}$. As we will eliminate edge effects through our adjunct plane construction, we can quite reasonably represent the incident field as a plane wave:

$$\psi_{\text{in}}(\mathbf{x}) = \exp(i\mathbf{k}_0 \cdot \mathbf{x}), \qquad \mathbf{k}_0 = -k(\mathbf{e}_y \sin\theta_i + \mathbf{e}_z \cos\theta_i) \tag{A3.142}$$

This has the benefits of computational simplicity and, more importantly, providing a straightforward and unambiguous route to RCS values. At this point it is convenient to factor out the phase variation in the incident plane wave from the surface field

$$\psi(\mathbf{x}) = \exp(i\mathbf{k}_0 \cdot \mathbf{x})\varphi(\mathbf{x}) \tag{A3.143}$$

The SFIE now takes the form

$$\begin{aligned}\varphi(\mathbf{x}) = 2 &- 4\int_{-\infty}^{-L/2} \bar{G}_0^{(1)}(\mathbf{x}, \mathbf{e}_y y) \exp(i\mathbf{k}_0 \cdot (\mathbf{e}_y y - \mathbf{x}))dy \\ &- 4\int_{L/2}^{\infty} \bar{G}_0^{(1)}(\mathbf{x}, \mathbf{e}_y y) \exp(i\mathbf{k}_0 \cdot (\mathbf{e}_y y - \mathbf{x}))dy \\ &- 2\int_{|y|<L/2} \bar{G}_0^{(1)}(\mathbf{x}, \mathbf{x}')\varphi(\mathbf{x}') \exp(i\mathbf{k}_0 \cdot (\mathbf{x}' - \mathbf{x}))dS'\end{aligned} \tag{A3.144}$$

The explicit form of the free space Green's function (A3.27) leads to

$$\bar{G}_0^{(1)}(\mathbf{x}, \mathbf{x}') = \frac{\mathbf{n}(\mathbf{x}') \cdot (\mathbf{x}' - \mathbf{x})}{|\mathbf{x} - \mathbf{x}'|} \frac{ik}{4} H_1^{(1)}(k|\mathbf{x} - \mathbf{x}'|) \tag{A3.145}$$

the special case of which

$$\bar{G}_0^{(1)}(\mathbf{x}, y\mathbf{e}_y) = -\frac{\mathbf{e}_z \cdot \mathbf{x}}{|\mathbf{x} - y\mathbf{e}_y|}\frac{ik}{4}H_1^{(1)}(k|\mathbf{x} - y\mathbf{e}_y|) \tag{A3.146}$$

is required in the evaluation of the adjunct plane contributions. We note that the integrand derived from the $y < -L/2$ adjunct plane is rapidly oscillating, and so describes a back scattered process, while that for which $y > L/2$ is much less rapidly oscillating, and describes a forward scattering process. (We would expect this from the direction of the incident radiation implicit in (3.142).) We see from the explicit form of $\bar{G}_0^{(1)}(\mathbf{x}, \mathbf{x}')$ that these adjunct plane contributions vanish at those points on the scattering surface for which $\eta(y) = 0$; in practice this condition is imposed over a significant region $L_{\text{eff}}/2 < |y| \leq L/2$ over which numerical evaluation is not required. Nonetheless, the numerical evaluation of remaining adjunct plane contributions poses significant problems, borne of the oscillatory and very slowly decaying large argument behaviour of the associated integrands (cf. (A3.27)); we will postpone our discussion of these issues to Section A3.9.

Having identified the driving terms in (A3.140) we now convert the integral equation into a matrix equation of the form (A3.132). We focus our attention on the scattering region $|y| < L/2$; points on the scattering surface are conveniently parameterised by

$$\mathbf{x} = \mathbf{e}_z z(l) + \mathbf{e}_y y(l) \tag{A3.147}$$

where l is the distance measured along the curve $z = \eta(y)$; this allows very steep waves and waves that 'curl over' at the top to be modelled. (Other representations can be used; in some cases that in terms of y is more convenient.)

We assume that the scattering surface is sufficiently smooth that its first and second derivatives exist, and can be calculated conveniently; a spline-fitted representation is particularly useful in this context. Thus the normal to this scattering surface can be expressed as

$$\mathbf{n}(\mathbf{x}) = \mathbf{e}_z \frac{dy(l)}{dl} - \mathbf{e}_y \frac{dz(l)}{dl} \tag{A3.148}$$

Points along the surface are now identified through

$$\mathbf{x}_n = \mathbf{e}_z z(n\Delta l) + \mathbf{e}_y y(n\Delta l) \tag{A3.149}$$

We construct N dimensional vectors from the surface field and driving term values

$$
\begin{aligned}
\mathbf{f} &= \{f_n, n = 1, \cdots, N\}; \qquad f_n = \varphi(\mathbf{x}_{n-1}) \\
\hat{\mathbf{f}}_0 &= \{\hat{f}_{0n}; n = 1, \cdots, N\}; \\
\hat{f}_{0n} &= 2 - 4 \int_{-\infty}^{-L/2} \bar{G}_0^{(1)}(\mathbf{x}_n, \mathbf{e}_y y) \exp(i\mathbf{k}_0 \cdot (\mathbf{e}_y y - \mathbf{x}_n)) dy \\
&\quad - 4 \int_{L/2}^{\infty} \bar{G}_0^{(1)}(\mathbf{x}_n, \mathbf{e}_y y) \exp(i\mathbf{k}_0 \cdot (\mathbf{e}_y y - \mathbf{x}_n)) dy
\end{aligned} \tag{A3.150}
$$

The numerical quadrature over the scattering region is effected by the trapezium rule, implemented as a matrix multiplication. We intend to separate this matrix into its diagonal and upper and lower triangular parts and consider the diagonal elements first. This requires a little care, as the Hankel function displays a singular behaviour for small arguments

$$
H_1^{(1)}(z) \sim -\frac{2i}{\pi z} \tag{A3.151}
$$

However, reference to (A3.52) shows that

$$
\begin{aligned}
\mathbf{n}(\mathbf{x}') \cdot (\mathbf{x}' - \mathbf{x}) &\approx \frac{1}{2}(y''(l)z'(l) - y'(l)z''(l))(\delta l)^2 \\
|\mathbf{x}' - \mathbf{x}| &\approx \delta l
\end{aligned} \tag{A3.152}
$$

revealing that $\bar{G}_0^{(1)}(\mathbf{x}, \mathbf{x}')$ is in fact well behaved as $\mathbf{x} \to \mathbf{x}'$. Consequently we can construct the diagonal matrix elements as

$$
E_{mn} = -\frac{\Delta l}{2\pi} \delta_{m,n} a_m b_n (y''(n\Delta l)z'(n\Delta l) - z''(n\Delta l)y'(n\Delta l)) \tag{A3.153}
$$

The evaluation of the off diagonal elements presents no problems of principle, as Hankel functions of integer order are readily computed using standard methods [23]. Upper and lower triangular matrices are constructed in the following way

$$
\begin{aligned}
U_{mn} &= -2\Delta l (1 - E_{mm})^{-1} b_n \exp(i\mathbf{k}_0 \cdot (\mathbf{x}_n - \mathbf{x}_m)) \bar{G}_0^{(1)}(\mathbf{x}_m, \mathbf{x}_n); \qquad m < n \\
&= 0; \qquad m \geq n \\
L_{mn} &= 0; \qquad m \leq n \\
&= -2\Delta l (1 - E_{mm})^{-1} a_m \exp(i\mathbf{k}_0 \cdot (\mathbf{x}_n - \mathbf{x}_m)) \bar{G}_0^{(1)}(\mathbf{x}_m, \mathbf{x}_n); 0; \qquad m > n
\end{aligned} \tag{A3.154}
$$

and the driving vector is modified

$$
\mathbf{f}_0 = (\mathbf{1} - \mathbf{E})^{-1} \cdot \hat{\mathbf{f}}_0 \tag{A3.155}
$$

Equations (A3.153) and (A3.154) incorporate the factors

$$a_m = \tfrac{1}{2}, \quad m = 1; \qquad a_m = 1, \quad \text{otherwise}$$
$$b_n = \tfrac{1}{2}, \quad n = N \quad b_n = 1, \quad \text{otherwise} \tag{A3.156}$$

that capture the endpoint weightings implicit in the trapezium rule. Thus we have finally cast the discretised SFIE into the form $\mathbf{f} = \mathbf{f}_0 + (\mathbf{U} + \mathbf{L}) \cdot \mathbf{f}$. Holliday *et al.* [24] suggested that this equation be solved in the following, physically motivated, fashion. **f** is represented as a sum of 'forward' and backward' (scattering) contributions

$$\mathbf{f} = \mathbf{f}_\mathrm{F} + \mathbf{f}_\mathrm{B} \tag{A3.157}$$

which are assigned initial values incorporating the tangent plane approximation

$$\mathbf{f}_\mathrm{F}^{(0)} = \mathbf{f}_0, \quad \mathbf{f}_\mathrm{B}^{(0)} = 0 \tag{A3.158}$$

The equation is then solved through the iterative scheme

$$\mathbf{f}_\mathrm{F}^{(n)} = \mathbf{f}_0 + \mathbf{U} \cdot (\mathbf{f}_\mathrm{F}^{(n)} + \mathbf{f}_\mathrm{B}^{(n-1)})$$
$$\mathbf{f}_\mathrm{B}^{(n)} = \mathbf{L} \cdot (\mathbf{f}_\mathrm{F}^{(n)} + \mathbf{f}_\mathrm{B}^{(n)}) \tag{A3.159}$$

which is readily implemented, as each of these equations can be solved by back substitution, as in (A3.130). Surprisingly few iterations of this scheme are needed to establish an effectively invariant solution; some physical reasons for this will be discussed shortly. The first forward iteration, i.e.

$$\mathbf{f}_\mathrm{F}^{(1)} = (\mathbf{1} - \mathbf{U})^{-1} \cdot \mathbf{f}_0 \tag{A3.160}$$

captures all processes built up exclusively from forward scattering events. (It is interesting to note that a solution to the LGA scattering problem of this type was first suggested by Spivack [25], who worked within the parabolic wave equation approximation and so completely neglected back-scattering processes.) The solution of the second, back scattering, equation then yields

$$\mathbf{f}_\mathrm{B}^{(1)} = (\mathbf{1} - \mathbf{L})^{-1} \cdot \mathbf{L} \cdot \mathbf{f}_\mathrm{F}^{(1)} = \mathbf{L} \cdot (\mathbf{1} - \mathbf{L})^{-1} \cdot (\mathbf{1} - \mathbf{U})^{-1} \cdot \mathbf{f}_0 \tag{A3.161}$$

The approximation to **f** generated by the first forward/backward (F/B) iteration is therefore

$$\mathbf{f}^{(1)} = (\mathbf{1} - \mathbf{L} + \mathbf{L})(\mathbf{1} - \mathbf{L})^{-1} \cdot (\mathbf{1} - \mathbf{U})^{-1} \cdot \mathbf{f}_0$$
$$= (\mathbf{1} - \mathbf{L})^{-1} \cdot (\mathbf{1} - \mathbf{U})^{-1} \cdot \mathbf{f}_0 \tag{A3.162}$$

the first term in the solution (A3.139). Subsequent iterations of (A3.159) generate the higher order terms; the nth partial sum of (A3.139)

$$\mathbf{f}^{(\mathbf{n})} = \sum_{m=0}^{n-1} ((\mathbf{1} - \mathbf{L})^{-1} \cdot (\mathbf{1} - \mathbf{U})^{-1} \cdot \mathbf{U} \cdot \mathbf{L})^m \cdot (\mathbf{1} - \mathbf{L})^{-1} \cdot (\mathbf{1} - \mathbf{U})^{-1} \cdot \mathbf{f}_0 \tag{A3.163}$$

is related to the nth forward and backward iterates obtained by Holliday's method through

$$\mathbf{f}_{\mathrm{B}}^{(n)} = \mathbf{L} \cdot \mathbf{f}^{(n)} \qquad \mathbf{f}_{\mathrm{F}}^{(n)} = (\mathbf{1} - \mathbf{L}) \cdot \mathbf{f}^{(n)} \tag{A3.164}$$

In the case of a smoothly varying, perfectly conducting surface subject to LGA illumination, much the greatest part of the scattering is in the forward direction. The 'first pass' (A3.162) of Holliday's F/B iteration is therefore sufficient to capture the salient features (such as more pronounced shadowing effects) of the surface field. The finer details of the surface field, established by weaker back-scattering processes, can then be determined in relatively few iterations. One of the most attractive features of Holliday's analysis of the PC scattering is its exposure of the physical components of the calculation; as we shall see shortly, much of this is lost when we consider scattering by an imperfect conductor. Although the earlier formal discussion based on (A3.138) does not have quite the physical motivation of Holliday's approach, it is more compact, and places the F/B method in the wider context of Jacobi and Gauss Seidel iterative matrix inversion procedures.

Once the surface field has been determined one can evaluate the back-scattered field, and hence the normalised RCS. Thus we have (cf. (A3.16))

$$\sigma^0 = \lim_{R\to\infty} \frac{|\psi_{\mathrm{s}}(\mathbf{R})|^2 2\pi R}{|\psi_{\mathrm{in}}|^2 L} \tag{A3.165}$$

where L is the length of the scattering surface. The far field form of the Green's function

$$G_0^{(1)}(\mathbf{R}, \mathbf{x}') \sim \sqrt{\frac{1}{8\pi k R}} \exp(-3i\pi/4) \exp(ikR) \exp(-i\mathbf{k}_{\mathrm{s}} \cdot \mathbf{x}'); \qquad \mathbf{k}_{\mathrm{s}} = \frac{k\mathbf{R}}{R} \tag{A3.166}$$

then leads to

$$\sigma_{VV}^0 = \frac{1}{4kL} \left| \int \exp(2i\mathbf{k}_0 \cdot \mathbf{x}')\, \mathbf{n}(\mathbf{x}') \cdot \mathbf{k}_0 \varphi(\mathbf{x}') dS' \right|^2 \tag{A3.167}$$

This incorporates the contributions to the scattered field from the adjunct planes; these are calculated in terms of Fresnel integrals and eliminate spurious edge scattering effects that would otherwise emerge from a computation based on a finite scatterer. The analysis of the Hpol scattering by a perfectly conducting surface proceeds in much the same way. The starting equation is now (cf. (A3.14))

$$\zeta(\mathbf{x}) = 2\mathbf{n}(\mathbf{x}) \cdot \nabla\psi_{\mathrm{in}}(\mathbf{x}) + 2 \int \zeta(\mathbf{x}')\bar{G}_0^{(1)}(\mathbf{x}', \mathbf{x}) dS' \tag{A3.168}$$

Introduction of the plane wave incident field, whose phase is factored out of $\zeta(\mathbf{x})$, leads us to

$$\zeta(\mathbf{x}) = \exp(i\mathbf{k}_0 \cdot \mathbf{x})\xi(\mathbf{x})$$

$$\xi(\mathbf{x}) = 2i\mathbf{k}_0 \cdot \mathbf{n}(\mathbf{x}) + 2\int \xi(\mathbf{x}')\bar{G}_0^{(1)}(\mathbf{x}', \mathbf{x}) \exp(i\mathbf{k}_0 \cdot (\mathbf{x}' - \mathbf{x}))dS' \qquad \text{(A3.169)}$$

This is very similar to (A3.144); the only differences are in the form of the driving term and the change of sign and order of arguments of the $\bar{G}_0^{(1)}$ term. Consequently it can be cast in essentially the same form as (A3.144), and solved by the same F/B iteration method. Table A3.1 shows the explicit forms the quantities $\mathbf{f}, \mathbf{f}_0, \mathbf{U}$ and $\mathbf{L}$ take in this case. Once the SFIE has been solved the normalised RCS can now be calculated from the result

$$\sigma^0_{HH} = \frac{1}{4kL}\left|\int \exp(2i\mathbf{k}_0 \cdot \mathbf{x}')\xi(\mathbf{x}')dS'\right|^2 \qquad \text{(A3.170)}$$

which, once again, includes adjunct plane contributions.

A3.7.2 Scattering from an imperfect conductor; modification of the F/B method

In the previous section we saw how Vpol and Hpol scattering from a corrugated, perfectly conducting surface can be calculated accurately and efficiently. In particular, the F/B method emerged as a physically motivated and practically useful method for the inversion of the underlying surface field integral equations. We now consider to what extent this approach can be extended to the calculation of scattering by an imperfectly conducting surface; the treatment will be less detailed than that of perfect conductor scattering and attempts, instead, to highlight the principles of a rather

Table A3.1 Identification of $\mathbf{f}, \mathbf{f}_0, \mathbf{U}$ *and* $\mathbf{L}$ *for Hpol scattering by PC surface*

f_n	$\xi(\mathbf{x}_n)$ $\quad \zeta(\mathbf{x}) = \exp(i\mathbf{k}_0 \cdot \mathbf{x})\xi(\mathbf{x})$	
$\hat{f}_{0n}$	$2i\mathbf{n}(\mathbf{x}_n) \cdot \mathbf{k}_0 + 4ik_{0z} \int_{-\infty}^{-L/2} \bar{G}_0^{(1)}(\mathbf{e}_y y, \mathbf{x}_n) \exp(i\mathbf{k}_0 \cdot (\mathbf{e}_y y - \mathbf{x}_n))dy$ $+ 4ik_{0z} \int_{L/2}^{\infty} \bar{G}_0^{(1)}(\mathbf{e}_y y, \mathbf{x}_n) \exp(i\mathbf{k}_0 \cdot (\mathbf{e}_y y - \mathbf{x}_n))dy$	
U_{mn}	$= 2\Delta l(1 + E_{mm})^{-1} b_n \exp(i\mathbf{k}_0 \cdot (\mathbf{x}_n - \mathbf{x}_m))\bar{G}_0^{(1)}(\mathbf{x}_n, \mathbf{x}_m);$	$m < n$
	$= 0; \quad m \geq n$	
L_{mn}	$= 0; \quad m \leq n$	
	$= 2\Delta l(1 + E_{mm})^{-1} a_m \exp(i\mathbf{k}_0 \cdot (\mathbf{x}_n - \mathbf{x}_m))\bar{G}_0^{(1)}(\mathbf{x}_n, \mathbf{x}_m); 0;$	$m > n$

lengthy analysis. (Explicit details of the implementation of this method are contained in Reference 26.) The finite conductivity of sea water, which is manifest in observable Brewster angle and related effects in LGA sea scattering of microwaves, provides us with sufficient motivation for this study; PC scattering calculations are not able to account for RCS measurements made in the controlled environment of a wave tank, as discussed in Chapter 3.

Once again we adopt the scalar field formulation of the scattering problem. Our starting point is the coupled integral equations (A3.12) satisfied by the field and its normal gradient at the scattering surface

$$\psi(\mathbf{x}) = 2\psi_{\text{in}}(\mathbf{x}) + 2\int (G_0^{(1)}(\mathbf{x},\mathbf{x}')\zeta(\mathbf{x}') - \bar{G}_0^{(1)}(\mathbf{x},\mathbf{x}')\psi(\mathbf{x}'))dS'$$

$$\psi(\mathbf{x}) = -2\int (G_0^{(2)}(\mathbf{x},\mathbf{x}')\zeta(\mathbf{x}') - \bar{G}_0^{(2)}(\mathbf{x},\mathbf{x}')\psi(\mathbf{x}'))dS' \qquad \text{(A3.171)}$$

Even at this early stage, we are encountering problems that did not arise in the PC calculations; in that case we merely had to consider a single SFIE, that could be solved by direct inversion. There is of course, one case in which the equations (A3.171) can be de-coupled straightforwardly: that of the infinite planar dielectric interface, discussed in Section A3.3. There we see that, for an incident plane wave $\psi_{\text{in}}(\mathbf{r}) = \exp(i\mathbf{k}_0 \cdot \mathbf{r})$ the surface field is given by

$$\psi(\mathbf{x}) = \frac{2\alpha k_z^{(1)}}{\alpha k_z^{(1)} + k_z^{(2)}} \exp(i\mathbf{k}_{\text{H}0} \cdot \mathbf{x}) \qquad \text{(A3.172)}$$

(We recall that $\mathbf{x}$ denotes a point in the scattering surface; here the subscript H reminds us that the scattering surface is the horizontal $z = 0$ plane.) As this planar interface scattering provides a 'first cut' at the solution of the scattering problem (much as the tangent plane $\psi(\mathbf{x}) = 2\psi_{\text{in}}(\mathbf{x})$ result does in the Vpol PC case), Holliday *et al.* [26] suggest a trial solution of the form

$$\psi(\mathbf{x}) = \hat{\psi}(\mathbf{x}) + \delta\psi(\mathbf{x}) \qquad \text{(A3.173)}$$

where

$$\hat{\psi}(\mathbf{x}) = \frac{2\alpha k_z^{(1)}}{\alpha k_z^{(1)} + k_z^{(2)}} \exp(i\mathbf{k}_0 \cdot \mathbf{x}) = \beta\psi_{\text{in}}(\mathbf{x}) \qquad \text{(A3.174)}$$

and $\mathbf{x}$ need no longer be confined to the $z = 0$ plane. The $\hat{\psi}(\mathbf{x})$ field and the second of the integral equations (A3.166) then specify $\hat{\zeta}(\mathbf{x})$ as the solution of

$$\hat{\psi}(\mathbf{x}) = -2\int (G_0^{(2)}(\mathbf{x},\mathbf{x}')\hat{\zeta}(\mathbf{x}') - \bar{G}_0^{(2)}(\mathbf{x},\mathbf{x}')\hat{\psi}(\mathbf{x}'))dS' \qquad \text{(A3.175)}$$

In the scattering region a discretised version of this integral equation can be generated by iteration, taking (cf. (A3.38))

$$\hat{\zeta}^{(0)}(\mathbf{x}) = \frac{-2ik_z^{(1)}k_z^{(2)}}{\alpha k_z^{(1)} + k_z^{(2)}} \psi_{\text{in}}(\mathbf{x}) \qquad \text{(A3.176)}$$

as a starting point. (Obviously $\hat{\zeta}(\mathbf{x})$ takes this value in the adjunct plane regions.) This procedure is fairly robust, as a consequence of the rapid decay of the Green's function describing propagation in the imperfectly conducting medium. Once $\hat{\zeta}(\mathbf{x})$ is known, a second trial solution can be introduced

$$\zeta(\mathbf{x}) = \hat{\zeta}(\mathbf{x}) + \delta\zeta(\mathbf{x}) \tag{A3.177}$$

When (A3.173) and (A3.177) are substituted into (A3.171) we obtain

$$\begin{aligned} \delta\psi(\mathbf{x}) &= D(\mathbf{x}) + 2\int (G_0^{(1)}(\mathbf{x},\mathbf{x}')\delta\zeta(\mathbf{x}') - \bar{G}_0^{(1)}(\mathbf{x},\mathbf{x}')\delta\psi(\mathbf{x}'))dS' \\ \delta\psi(\mathbf{x}) &= -2\int (\alpha G_0^{(2)}(\mathbf{x},\mathbf{x}')\delta\zeta(\mathbf{x}') - \bar{G}_0^{(2)}(\mathbf{x},\mathbf{x}')\delta\psi(\mathbf{x}'))dS' \end{aligned} \tag{A3.178}$$

where

$$D(\mathbf{x}) = (2-\beta)\psi_{\text{in}}(\mathbf{x}) + 2\int (G_0^{(1)}(\mathbf{x},\mathbf{x}')\hat{\zeta}(\mathbf{x}') - \bar{G}_0^{(1)}(\mathbf{x},\mathbf{x}')\hat{\psi}(\mathbf{x}'))dS' \tag{A3.179}$$

Given the values of $\hat{\psi}(\mathbf{x}), \hat{\zeta}(\mathbf{x})$, we can construct this driving term. The integral in (A3.179) is taken over the entire imperfectly conducting interface, including the adjunct planes. In the regions $|y| > L/2$ we can set

$$\psi(\mathbf{x}) = \hat{\psi}(\mathbf{x}); \qquad \zeta(\mathbf{x}) = \hat{\zeta}^{(0)}(\mathbf{x}) \tag{A3.180}$$

and attempt to evaluate the integrals much as in the PC case. Terms derived from $\bar{G}_0^{(1)}(\mathbf{x},\mathbf{x}')\hat{\psi}(\mathbf{x}')$ can indeed be evaluated just as before; those that arise from $G_0^{(1)}(\mathbf{x},\mathbf{x}')\hat{\zeta}_0(\mathbf{x}')$ are slightly more problematic, principally because they take non-zero values over the entire region $|y| \le L/2$. Holliday *et al.* avoid the evaluation of these terms altogether by exploiting the transparency identity derived, in the context of a vector analysis, by Shaw and Dougan [11]. The scalar analysis is, as usual, more compact. The SFIE

$$\psi(\mathbf{x}) = 2\psi_{\text{in}}(\mathbf{x}) + 2\int (G_0^{(1)}(\mathbf{x},\mathbf{x}')\zeta(\mathbf{x}') - \bar{G}_0^{(1)}(\mathbf{x},\mathbf{x}')\psi(\mathbf{x}'))dS' \tag{A3.181}$$

is set up for the case of transmission of a plane wave through an interface established in the vacuum where it takes the form

$$\begin{aligned} \psi_{\text{in}}(\mathbf{x}) &= 2\psi_{\text{in}}(\mathbf{x}) + 2\int (G_0^{(1)}(\mathbf{x},\mathbf{x}')\zeta_{\text{in}}(\mathbf{x}') - \bar{G}_0^{(1)}(\mathbf{x},\mathbf{x}')\psi_{\text{in}}(\mathbf{x}'))dS' \\ \zeta_{\text{in}}(\mathbf{x}') &= \mathbf{n}(\mathbf{x}')\cdot\nabla'\psi_{\text{in}}(\mathbf{x}') \end{aligned} \tag{A3.182}$$

The second of equations (A3.171) reassuringly yields the same identity

$$\psi_{\text{in}}(\mathbf{x}) = -2\int (G_0^{(1)}(\mathbf{x},\mathbf{x}')\zeta_{\text{in}}(\mathbf{x}') - \bar{G}_0^{(1)}(\mathbf{x},\mathbf{x}')\psi_{\text{in}}(\mathbf{x}'))dS' \tag{A3.183}$$

In the case of a plane wave, propagating through the $z = 0$ interface, we have

$$\exp(i\mathbf{k}_0 \cdot \mathbf{x}) = 2i|k_{0z}| \int G_0^{(1)}(\mathbf{x}, \mathbf{x}') \exp(i\mathbf{k}_0 \cdot \mathbf{x}')dx' \tag{A3.184}$$

from which (A3.32) emerges as a special case when $\mathbf{x} = \mathbf{0}$. So, by taking (A3.182) and noting that, in the adjunct plane regions,

$$\hat{\zeta}_0(\mathbf{x}) = \frac{2k_z^{(1)}k_z^{(2)}}{\alpha k_z^{(1)} + k_z^{(2)}} \zeta_{\text{in}}(\mathbf{x}) = \chi \zeta_{\text{in}}(\mathbf{x}) \tag{A3.185}$$

we construct the identically vanishing quantity

$$\chi \psi_{\text{in}}(\mathbf{x}) + 2\chi \int (G_0^{(1)}(\mathbf{x}, \mathbf{x}')\zeta_{\text{in}}(\mathbf{x}') - \bar{G}_0^{(1)}(\mathbf{x}, \mathbf{x}')\psi_{\text{in}}(\mathbf{x}'))dS = 0 \tag{A3.186}$$

This is subtracted from the integral occurring in the driving term (A3.179); we find that

$$\begin{aligned} D(\mathbf{x}) = (2 - \beta - \chi)\psi_{\text{in}}(\mathbf{x}) + 2 \int (G_0^{(1)}(\mathbf{x}, \mathbf{x}')[\hat{\zeta}(\mathbf{x}') - \chi \zeta_{\text{in}}(\mathbf{x}')] \\ - \bar{G}_0^{(1)}(\mathbf{x}, \mathbf{x}')[\hat{\psi}(\mathbf{x}') - \chi \psi_{\text{in}}(\mathbf{x}')])dS' \end{aligned} \tag{A3.187}$$

As $\beta + \chi = 2$ the driving term can be expressed as

$$\begin{aligned} D(\mathbf{x}) = 2 \int\limits_{|y| \le L/2} (G_0^{(1)}(\mathbf{x}, \mathbf{x}')[\hat{\zeta}(\mathbf{x}') - \chi \zeta_{\text{in}}(\mathbf{x}')] - \bar{G}_0^{(1)}(\mathbf{x}, \mathbf{x}')[\hat{\psi}(\mathbf{x}') \\ - \chi \psi_{\text{in}}(\mathbf{x}')])dS' - 2 \int\limits_{|y| > L/2} \bar{G}_0^{(1)}(\mathbf{x}, \mathbf{x}')(\beta - \xi)\psi_{\text{in}}(\mathbf{x}')dS' \end{aligned} \tag{A3.188}$$

The adjunct plane contributions are now given in terms of $\bar{G}_0^{(1)}$, and can be evaluated in much the same way as those encountered in the PC calculation.

To complete the calculation we must de-couple the integral equations satisfied by $\delta\psi(\mathbf{x}), \delta\zeta(\mathbf{x})$; to do this we invoke a strategy very similar to that adopted in the derivation of the IBC in Section A3.4. Thus we set

$$\delta\psi(\mathbf{x}) \approx -2\alpha\delta\zeta(\mathbf{x}) \int G_0^{(2)}(\mathbf{x}, \mathbf{x}')dS' + 2\delta\psi(\mathbf{x}) \int \bar{G}_0^{(2)}(\mathbf{x}, \mathbf{x}')dS' \tag{A3.189}$$

The integrals over $G_0^{(2)}(\mathbf{x}, \mathbf{x}'), \bar{G}_0^{(2)}(\mathbf{x}, \mathbf{x}')$ are evaluated numerically; (A3.189) can now be used to eliminate $\delta\zeta(\mathbf{x})$ from

$$\delta\psi(\mathbf{x}) = D(\mathbf{x}) + 2 \int (G_0^{(1)}(\mathbf{x}, \mathbf{x}')\delta\zeta(\mathbf{x}') - \bar{G}_0^{(1)}(\mathbf{x}, \mathbf{x}')\delta\psi(\mathbf{x}'))dS' \tag{A3.190}$$

and the resulting equation is solved for $\delta\psi(\mathbf{x})$ by F/B iteration. The presence of the $G_0^{(1)}$ introduces additional terms in the $\mathbf{U}$, $\mathbf{L}$ and $\mathbf{E}$ matrices. Some care is needed in the evaluation of these last, diagonal terms, again as a result of the singular behaviour of the Hankel function. As

$$H_0^{(1)}(z) \sim 1 + \frac{2i}{\pi}(\log(z/2) + \gamma), \quad z \to 0 \tag{A3.191}$$

we see that

$$\int_{-\Delta/2}^{\Delta/2} H_0^{(1)}(k|l|)dl \approx \Delta\left(1 + \frac{2i}{\pi}\left(\log\left(\frac{k\Delta}{4}\right) + \gamma - 1\right)\right) \tag{A3.192}$$

from which the form of the extra contributions to **E** follow immediately. Here γ is Euler's constant and has the value 0.5772.

In the PC case the F/B iteration has an appealing and straightforward physical interpretation, much of which has been surrendered in its adaptation to the imperfectly conducting case, where it merely plays the role of a useful and efficient matrix inversion technique. Once $\delta\psi(\mathbf{x})$ has been found, $\delta\zeta(\mathbf{x})$ is recovered from (A3.190). The surface fields $\psi(\mathbf{x}), \zeta(\mathbf{x})$ can now be reconstructed and the scattered field evaluated from (A3.15).

A3.8 Incorporation of the impedance boundary condition in F/B calculations

Some of the simplicity and physical content of the F/B solution of the PC scattering problem can be retained in the calculation of scattering by an imperfect conductor if the impedance boundary condition is invoked at the outset. In the previous subsection it was introduced surreptitiously, through the approximation (A3.190), buried deep in Holliday *et al.*'s analysis. Thus if the normal field gradient $\zeta(\mathbf{x})$ is eliminated from (A3.191) by use of (A3.45) we obtain the single SFIE, satisfied by $\psi(\mathbf{x})$

$$\psi(\mathbf{x}) = 2\psi_{\text{in}}(\mathbf{x}) - 2\int (ik_2 G_0^{(1)}(\mathbf{x}, \mathbf{x}')/\alpha + \bar{G}_0^{(1)}(\mathbf{x}, \mathbf{x}'))\psi(\mathbf{x}')dS' \tag{A3.193}$$

which is very similar in structure to (A3.13), and so can be solved by much the same methods. Once again the **U**, **L** and **E** matrices describing the scattering processes taking place within the $|y| \le L/2$ region are modified by the presence of $G_0^{(1)}$ and can be evaluated much as they were in Holliday's imperfect conductor calculation. The adjunct plane driving terms are also modified; as the approximate SFIE incorporates the effects of the scattering surface we can no longer eliminate these $G_0^{(1)}$ terms by invoking a transparency identity. If we are to make further progress with the IBC implementation of the F/B method we need to compute these quantities effectively. We can postpone our discussion of the evaluation of the adjunct plane contributions no longer.

A3.9 Evaluation of adjunct plane contributions

The central problem encountered when determining the adjunct plane contributions is the evaluation of the integrals

$$\Xi_{0,\pm}(\eta(y),\Delta) = \int_{\Delta}^{\infty} H_0^{(1)}\left(k\sqrt{y'^2+\eta(y)^2}\right)\exp(\pm ik\sin\theta_i y')dy'$$

$$\Xi_{V,\pm}(\eta(y),\Delta) = \int_{\Delta}^{\infty} \mathbf{e}_z\cdot\nabla H_0^{(1)}\left(k\sqrt{y'^2+\eta(y)^2}\right)\exp(\pm ik\sin\theta_i y')dy'$$

$$\Xi_{H,\pm}(\eta(y),\Delta) = \int_{\Delta}^{\infty} \mathbf{e}_y\cdot\nabla H_0^{(1)}\left(k\sqrt{y'^2+\eta(y)^2}\right)\exp(\pm ik\sin\theta_i y')dy' \tag{A3.194}$$

These arise, respectively, in the evaluation of the $G_0^{(1)}$, Vpol $\bar{G}_0^{(1)}$ and Hpol $\bar{G}_0^{(1)}$ derived driving terms. Holliday *et al.* [27,28] identify much the same set of integrals, where they express the gradient of the $H_0^{(1)}$ Hankel function explicitly in terms of $H_1^{(1)}$ and distinguish notationally between the $\pm ik\sin\theta_i$ phase factors. Much of the difficulty encountered in effecting these numerical quadratures arises from the oscillating and very slowly decaying behaviour of the Hankel functions (see (A3.27)). Holliday *et al.* chose to confront these issues directly, carrying out their numerical integrations in real space. Integrals derived from $H_0^{(1)}(k\sqrt{y'^2+\eta(y)^2})\exp(ik\sin\theta_i y')$ have rapidly oscillating integrands; in [28] the V and H integrals are evaluated from complicated high frequency asymptotic expansions, generated by repeatedly integrating by parts. Their analogues, obtained from $H_0^{(1)}(k\sqrt{y'^2+\eta(y)^2})\exp(-ik\sin\theta_i y')$, have relatively slowly oscillating integrands. The associated integration is effected by subtracting the most slowly decaying contribution, derived from the asymptotic form of $H_1^{(1)}$, evaluating its integral analytically in terms of Fresnel integrals, and supplementing this value by a numerical quadrature of the more rapidly decaying difference [27]. Each of these methods can be applied, in principle, to the $G_0^{(1)}$ integrals; in practice, however, the high frequency approximation method is found to be quite inadequate (and the low frequency approximation invariably very time consuming) when evaluating the driving terms close to the edges of the adjunct planes. Another method of evaluating these integrals has to be devised in which we effectively suppress the oscillatory and slowly decaying behaviour of the integrands. An effective way of doing this is through the introduction of a contour integral representation of the Hankel function; exploiting the freedom afforded us by integration in the complex plane we can, loosely speaking, convert harmonic oscillation into exponential decay. To motivate our discussion we consider the evaluation of the Ξ_0 integral for $\eta(y)=0$. First, we recall (cf. (A3.28))

$$H_0^{(1)}(ky') = \frac{1}{i\pi} \int_{-\infty - i\pi/2}^{\infty + i\pi/2} \exp(iky' \cosh p) dp \tag{A3.195}$$

If we substitute this into the Ξ_0 integral we find that

$$\Xi_{0,\pm}(0, \Delta) = \frac{\exp(\pm ik\Delta \sin\theta_i)}{\pi k} \int_{-\infty - i\pi/2}^{\infty + i\pi/2} dp \frac{\exp(ik\Delta \cosh p)}{\cosh p \pm \sin\theta_i} \tag{A3.196}$$

At the very edge of the adjunct plane, where Holliday's methods encounter most difficulty, we set $\Delta = 0$ and obtain

$$\begin{aligned}\Xi_{0,\pm}(0, 0) &= \frac{1}{\pi k} \int_{-\infty - i\pi/2}^{\infty + i\pi/2} dp \frac{1}{\pm \sin\theta_i + \cosh p} \\ &= \frac{2}{\pi k} \int_0^\infty \frac{du}{u^2 \pm 2\sin\theta_i u + 1} = \frac{2}{\pi k \cos\theta_i}\left(\frac{\pi}{2} \pm \theta_i\right)\end{aligned} \tag{A3.197}$$

Note that, by forming the sum $\Xi_{0,+}(0,0) + \Xi_{0,-}(0,0)$, we provide yet another derivation of (A3.32).

To evaluate these integrals for $\Delta \neq 0$ we parameterise the path of integration in the complex plane through

$$p = t + i\tan^{-1} t, \qquad dp = dt\left(1 + \frac{i}{1+t^2}\right) \tag{A3.198}$$

so that the integral representation of the Hankel function becomes

$$H_0^{(1)}(ky') = \frac{2}{i\pi} \int_0^\infty dt \left(1 + \frac{i}{1+t^2}\right) \exp\left(iky'\left(\frac{\cosh t + it\sinh t}{\sqrt{1+t^2}}\right)\right) \tag{A3.199}$$

We see that the exponential now decays for positive y. If we substitute this into (A3.194) we find that

$$\begin{aligned}&\Xi_{0,\pm}(0, \Delta) \\ &= \frac{2}{\pi k} \int_0^\infty dt \left(1 + \frac{i}{1+t^2}\right) \frac{\sqrt{1+t^2}}{\cosh t + it\sinh t \pm \sin\theta_i \sqrt{1+t^2}} \\ &\quad \times \exp\left(ik\Delta \frac{\cosh t + it\sinh t \pm \sin\theta_i\sqrt{1+t^2}}{\sqrt{1+t^2}}\right)\end{aligned} \tag{A3.200}$$

It is straightforward to code up and numerically integrate expressions of this type. This procedure allows us to evaluate Holliday *et al.*'s Z and T type integrals in the same way, quickly and accurately.

Thus far we have considered only the driving integrals evaluated on a flat plane surface. When the driving term is to be evaluated at a point $\{y, \eta(y)\}$ on the scattering surface we can proceed in much the same way. We now consider the integral representation

$$H_0^{(1)}\left(k\sqrt{y^2+z^2}\right) = \frac{1}{i\pi}\int_{-\infty-i\pi/2}^{\infty+i\pi/2} \exp(ik(y\cosh p + iz\sinh p))dp \quad \text{(A3.201)}$$

Because the angle subtended by the highest point on the scattering surface at the adjunct plane is small, the integrand does not change qualitatively, and is still confined to a small portion of the integration path. The general driving integral can be evaluated as

$$\begin{aligned}&\Xi_{0,\pm}(\eta(y), \Delta)\\ &= \frac{1}{\pi k}\int_{-\infty-i\pi/2}^{\infty+i\pi/2} dp\frac{\exp(ik(\Delta(\cosh p \pm \sin\theta_i) + i\eta(y)\sinh p))}{\cosh p \pm \sin\theta_i}\end{aligned} \quad \text{(A3.202)}$$

with the path integration in the complex plane parameterised through

$$\begin{aligned}&\frac{1}{\pi k}\int_{-\infty}^{\infty} dt\left(1+\frac{i}{1+t^2}\right)\frac{\sqrt{1+t^2}}{\cosh t + it\sinh t \pm \sqrt{1+t^2}\sin\theta_i}\exp\left[\frac{ik}{\sqrt{1+t^2}}\right.\\ &\quad\left.\times\left\{\Delta(\cosh t + it\sinh t \pm \sqrt{1+t^2}\sin\theta_i) + i\eta(y)(\sinh t + it\cosh t)\right\}\right]\end{aligned} \quad \text{(A3.203)}$$

This method can also be applied with very little modification to the evaluation of the V and H integrals considered originally by Holliday *et al.* Rather than evaluate the gradient in (A3.193) explicitly in terms of first order Hankel functions, as was done in References 28 and 29 we introduce the integral representation

$$\begin{aligned}&H_0^{(1)}\left(k\sqrt{(y-y')^2+(z-z')^2}\right)\\ &= \frac{1}{\pi i}\int_{-\infty-i\pi/2}^{\infty+i\pi/2} \exp(ik((y'-y)\cosh p + i(z'-z)\sinh p))dp \qquad y' > y\end{aligned} \quad \text{(A3.204)}$$

and differentiate it directly. This leads us to

$$\begin{aligned}
&\Xi_{V,\pm}(\eta(y),\Delta)\\
&= -\frac{k}{\pi}\int\limits_{-\infty-i\pi/2}^{\infty+i\pi/2} dp\cosh p\int\limits_{\Delta}^{\infty}\exp(ik(y'\cosh p - i\eta(y)\sinh p))\\
&\quad\times\exp(\pm ik\sin\theta_i y')dy'\\
&\Xi_{H,\pm}(\eta(y),\Delta)\\
&= \frac{k}{i\pi}\int\limits_{-\infty-i\pi/2}^{\infty+i\pi/2} dp\sinh p\int\limits_{\Delta}^{\infty}\exp(ik(y'\cosh p - i\eta(y)\sinh p))\\
&\quad\times\exp(\pm ik\sin\theta_i y')dy'
\end{aligned}\tag{A3.205}$$

The integral over y' is once again straightforward so that we get

$$\begin{aligned}
&\Xi_{V,\pm}(\eta(y),\Delta)\\
&= \frac{1}{\pi i}\int\limits_{-\infty-i\pi/2}^{\infty+i\pi/2} dp\frac{\cosh p\exp(ik(\Delta\cosh p - i\eta(y)\sinh p)\pm ik\Delta\sin\theta_i)}{\cosh p\pm\sin\theta_i}\\
&\Xi_{H,\pm}(\eta(y),\Delta)\\
&= \frac{1}{\pi}\int\limits_{-\infty-i\pi/2}^{\infty+i\pi/2} dp\frac{\sinh p\exp(ik(\Delta\cosh p - i\eta(y)\sinh p)\pm ik\Delta\sin\theta_i)}{\cosh p\pm\sin\theta_i}
\end{aligned}\tag{A3.206}$$

In this way Holliday's T and Z driving integrals, appropriate to both V and H polarisations, can be evaluated numerically by parameterising the contour of integration as in (A3.198). This method for evaluating the adjunct plane contributions has proved to be both quicker and more stable and controlled than those devised by Holliday *et al.*

A3.10 Summary

In this appendix we have set up the integral equation formulation of the scalar scattering problem (Sections A3.1–A3.2); complete solutions, such as those describing the reflection and transmission of radiation by a planar interface, can be extracted from this only in the most idealised of circumstances (Section A3.3). Nonetheless, the integral equation formulation provides us with a framework within which many aspects of scattering by a rough surface can be developed systematically. Approximate descriptions of scattering by a perfectly conducting surface can be derived (Section A3.5), both through iterative solution of the integral equation (physical optics and small height perturbation theory) and progressive refinement of the Green's function (half-space and Lorentz reciprocity based calculations). The latter Lorentz reciprocity based approach also allows us to derive the small height perturbation theory result for an imperfectly conducting rough surface with relatively little trouble; this in turn can be modified to take account of the large scale swell structure of the sea surface through its incorporation into the composite model (Section A3.6). Methods for the numerical solution of the scattering integral equations have been developed in Sections A3.7 and 3.8; sufficient detail has been given for explicit expressions to be identified and coded up for both perfectly and imperfectly conducting surfaces. The modelling of low grazing angle scattering invariably encounters difficulties that arise from the finite size of scattering surface that can be accommodated within the computer; these we have ameliorated by introducing semi-infinite adjunct planes, much as was done by Holliday and his co-workers. Here, however, we have described a different and improved method for the calculation of the terms in the integral equations that arise from these planes (Section A3.9), which also allows us to treat the imperfectly conducting surface through the use of the impedance boundary condition derived in Section A3.4. All this should provide the background material required to support a careful reading of Chapter 3, where 'real-life' application of scattering calculations to the study of sea clutter, and motivation to work through appropriate parts of this appendix, can be found.

References

1 W.C. Meecham, 'On the use of the Kirchhoff approximation for the solution of reflection problems', *J. Rational Mech. Anal.*, **5**, 332–334, 1956.
2 E.T. Copson, '*An introduction to the theory of functions of a complex variable*', Chaps. 5 and 6, Clarendon Press, Oxford, 1970.
3 D. Holliday, 'Resolution of a controversy surrounding the Kirchhoff approach and the small perturbation method in rough surface scattering theory', *IEEE Trans. Antennas Propagat.*, **AP-35** , 120–122, 1987.
4 J.A. Stratton, 'Electromagnetic theory', Section 8.14, McGraw Hill, New York, 1945.
5 J.T. Johnson, 'A numerical study of low grazing angle backscatter from ocean like impedance surfaces with the canonical grid method', *IEEE Trans. Antennas Propagat.*, **46**, 114–120, 1998.
6 T.B.A. Senior and J.L. Volakis, '*Approximate boundary conditions in electromagnetics*', IEE Electromagnetic Waves Series 41, IEE Press, Stevenage, UK, 1995.
7 W.T. Shaw and A.J. Dougan, 'Curvature corrected impedance boundary conditions in an arbitrary basis', *IEEE Trans. Antennas Propagat.*, **53**, 1699–1705, 2005.

8 H.D. Ngo and C.L. Rino, 'Application of beam simulation to scattering at low grazing angles: methodology and verification', *Radio Sci.*, **29**, 1365–1379, 1994.

9 D.H. Berman and J.S. Perkins, 'The Kirchhoff approximation and first-order perturbation theory for rough surface scattering', *J. Acoust. Soc. Am.*, **78**, 1045, 1985.

10 W.T. Shaw and A.J. Dougan, 'Half-space Green's function and applications to scattering of electromagnetic waves from ocean like surfaces', *Waves Random Media*, **5**, 341–359, 1995.

11 W.T. Shaw and A.J. Dougan, 'Green's function refinement as an approach to radar back-scatter: general theory and applications to LGA scattering', *IEEE Trans. Antenna and Propagat.*, **46**, 57, 1998.

12 D.E. Kerr, '*Propagation of short radio waves*', McGraw-Hill, New York, 1951.

13 H. Levine and J.S. Schwinger, 'On the theory of electromagnetic wave diffraction by an aperture in an infinite plane conducting screen', *Commun. Pure Appl. Math.*, **3**, 355–391, 1950.

14 D. Lovelock and H. Rund, '*Tensors, differential forms and variational principles*', John Wiley and Sons, New York, 1975.

15 W.H. Peake, 'Theory of radar return from terrain', 1959 IRE Convention Record, Vol. 7, pp. 27–41.

16 H. and B.S. Jeffreys, '*Mathematical physics*', Section 9.08, University Press, Cambridge, 1966.

17 R.F. Harrington, '*Field computation by moments methods*', Macmillan, New York, 1968.

18 E.F. Knot, J.F. Shaeffer and M.T. Tuley, '*Radar cross section*', Artech House, Boston, 1993.

19 C.L. Rino and H.D. Ngo, 'Application of beam simulation to scattering at low grazing angles: oceanic surfaces', *Radio Sci.*, **29**, 1381–1391, 1994.

20 R. Petit (ed.), '*Electromagnetic theory of gratings*', Springer, Berlin, 1980.

21 E.T. Goodwin *et al.*, '*Modern computing methods.*', Notes on Applied Science, No. **16**, Chap. 4, HMSO, 1961.

22 A. Messiah, '*Quantum mechanics*', Vol. 2, Section 16.16, North Holland, Amsterdam, 1961.

23 W.H Press, S.A. Teukolsky, W.T. Vetterling and B.P. Flannery, '*Numerical recipes in FORTRAN; The art of scientific computing.*', 2nd. edn., University Press, Cambridge, 1992.

24 D. Holliday, L.L. DeRaad, Jr. and G.J. St.-Cyr, 'Forward-backward: a new method for computing low grazing angle scattering', *IEEE Trans. Antennas Propagat.*, **44**, 722–729, 1996.

25 M. Spivack, 'A numerical approach to rough surface scattering by the parabolic wave equation method', *J. Acoust. Soc. Am.*, **87**, 1999, 1990.

26 D. Holliday, L.L. DeRaad, Jr. and G.J. St.-Cyr, 'Forward-backward method for scattering from imperfect conductors', *IEEE Trans. Antennas Propagat.*, **46**, 101–107, 1998.

27 D. Holliday, L.L. DeRaad, Jr. and G.J. St.-Cyr, 'Wedge scattering by the method of iteration', *IEEE Trans. Geosci. Remote Sens.*, **33**, 449–456, 1995.

28 D. Holliday, L.L. DeRaad, Jr. and G.J. St.-Cyr, 'Volterra approximation for low grazing angle shadowing on smooth ocean-like surfaces, *IEEE Trans. Antennas Propagat.*, **43**, 1199–1206, 1995.

Index

Note: Page numbers in **bold** show pages with tables or figures separated from text.
Page numbers followed by 'A' indicates the entry is an appendix entry.

www.ingramcontent.com/pod-product-compliance
Lightning Source LLC
LaVergne TN
LVHW010630300926
849162LV00001B/113

* 9 7 8 0 8 6 3 4 1 5 0 3 6 *